Estuaries of the World

Series Editor

Jean-Paul Ducrotoy

For further volumes:
http://www.springer.com/series/11705

Eric Wolanski
Editor

Estuaries of Australia in 2050 and Beyond

This book is endorsed by the International EMECS Center (Environmental Management of Enclosed Seas; http://www.emecs.or.jp/englishver2/aboutemecs_e.html), the Estuarine and Coastal Sciences Association, and LOICZ (Land-Ocean Interactions in the Coastal Zone global Earth system research project and network (http://www.loicz.org/) in the context of their worldwide scientific and educational mandate.

Springer

Editor
Eric Wolanski
TropWATER and School of Marine
and Tropical Biology
James Cook University
Townsville, QLD 4811
Australia

ISSN 2214-1553 ISSN 2214-1561 (electronic)
ISBN 978-94-007-7018-8 ISBN 978-94-007-7019-5 (eBook)
DOI 10.1007/978-94-007-7019-5
Springer Dordrecht Heidelberg New York London

Library of Congress Control Number: 2013946905

Printed on acid-free paper

Springer is part of Springer Science+Business Media (www.springer.com)

I dedicate this book to my grandson Oliver, born and being brought up on the shores of Australian estuaries; I hope that some will remain healthy for him to entrust to his children.

Eric Wolanski

Foreword

Why This Book Series? Why This Book?

Over the last decade, there have been numerous advances in both understanding and managing estuaries, with an increasing focus on multidisciplinary studies, through numerous case studies and projects at local and national levels. In addition, regional and global programmes have been developed; some are being implemented and some are in evolution. However, despite the rapidly increasing knowledge about estuarine ecosystems, crucial questions on the causes of variability and the effects of global change versus local anthropogenic pressures are still poorly understood. At the same time, courses at university increasingly focus on environmental science and management but with comparatively very little emphasis on estuaries. There are excellent textbooks in this field, mostly process-based and about synthesising science, but by and large they do not reflect the great variety of estuaries around the world; so the practical application of this knowledge to one estuary or one coast with several estuaries is very complex and not straightforward. As a result, most of the time, students studying a particular estuary or coast use ill-assorted websites, sometimes of doubtful quality. The situation is comparable to that of decision-makers and managers. They are submerged by all sorts of publications, very few concentrating on one estuary or on specific problems.

Because the perception of politicians and managers of coasts is slowly shifting from a mainly short-term economic approach towards a long-term socio-ecological perspective, Springer Publishers recognised the need to make existing scientific information much more manageable to non-specialists, without compromising the quality of the information. The series *Estuaries of the World* was established in such a context, giving the scientific community the opportunity to assemble and put in order (sometimes disorganised) existing knowledge. Overall, the series will encompass all scientific aspects of estuaries through a multidisciplinary approach.

This book (the first in the collection) deals with a selection of estuaries which are characteristic of a whole continent: Australia. The country is so large that it spreads from the tropics (10th parallel) to the temperate zone in Tasmania. Estuaries themselves differ by an order of magnitude in terms of size; yet, they all have common properties and processes. In Australia, as anywhere else in the world, the coastal zone and its estuaries, large or small, are amongst the most endangered areas. Pollution, eutrophication, urbanisation, land reclamation, dams, irrigation, over-fishing and exploitation continuously threaten the future of some estuaries, which bear the full pressure of these developments. However, unaffected systems still exist in Australia and, if not strictly pristine, enjoy an exceptional ecological quality. In between these two categories, unfortunately some high-quality environments are currently being degraded because of loose management. The major challenge that humans face today is protecting estuaries, which benefit from a good ecological status, by managing their use. Preventing other systems to further degrade and restoring them require immediate action so that future generations can also enjoy the fantastic visual, cultural and edible products that they provide. Such an approach assumes that all users of the environment share views and are able to communicate wisely on the basis of robust science. Current changes in climate (e.g. temperature rise, sea-level rise, increased risks of floods and droughts and ocean acidification) may increase the risk of abrupt and non-linear changes in many estuarine ecosystems, which would affect their composition, function, biodiversity and productivity. In order to provide a solid scientific background to future debates, this book does not just attempt compiling case studies but puts into light best practice both in scientific research and coastal management.

Institute of Estuarine and Coastal Studies, The University of Hull, Hull, UK — Jean-Paul Ducrotoy
James Cook University, Townsville, QLD, Australia — Eric Wolanski

Contents

About the Editor

Professor Eric Wolanski
TropWATER and School of Marine and Tropical Biology, James Cook University

Dr. Eric Wolanski is a coastal oceanographer and ecohydrologist. Eric has 360 publications; he is a fellow of the Australian Academy of Technological Sciences and Engineering, the Institution of Engineers Australia (ret.) and l'Académie Royale des Sciences d'Outre-Mer. He was awarded an Australian Centenary Medal for services in estuarine and coastal oceanography, a Doctorate Honoris Causa from the Catholic University of Louvain, a Queensland Information Technology and Telecommunication award for excellence and the Estuarine and Coastal Sciences Association (ECSA) Lifetime Achievement Award. Eric is a member of the IGBP-IHDP Scientific Steering Committee of Land-Ocean Interactions in the Coastal Zone (LOICZ) and a member of the Scientific Planning Committee of Japan's Environmental Management of Enclosed Coastal Seas (EMECS). He is chief editor of *Estuarine, Coastal and Shelf Science*, *Wetlands Ecology and Management*, and the *Treatise on Estuarine and Coastal Science*.

Prologue

The majority of the Australian population lives near estuaries and the coast. Many Australian estuaries were historically degraded and others are at risk of degrading as the population and the economy are increasing rapidly. Is Australia's development ecologically sustainable for estuaries and coasts? This book addresses this question by detailed studies of a number of iconic Australian estuaries and bays. This book demonstrates, through the writings of eminent Australian scientists, how these estuaries function by merging the physical oceanography, the ecosystem processes and the socio-economic science. The chapters describe most types of Australian estuaries from pristine in the tropics to those impacted by irrigation, urbanisation and industrialisation. The key message is that the basic science has been done, and this makes it possible to understand how these brackish water ecosystems function. This enables the scientists to forecast with some confidence what these estuaries will look like by 2050 based on political and socio-economic decisions that are now made.

This book offers science-based solutions to achieve ecologically sustainable development. It is a wake-up call that every Australian estuary faces present and future socio-economic and environmental problems with various scales. This book shows that we have much to learn by understanding the lessons from the past and from each other as they apply to the wide variety of Australian estuaries in order to ensure that future developments do not occur at the cost of the environment. To help achieve this outcome, this book demonstrates how to use science to balance the socio-economic imperatives with the ecological needs of the estuaries so that they can deliver the full range of ecosystem services – such as a high quality of life – that the population expects.

I commend this book for its comprehensive coverage of the variety of estuaries in Australia and for using the best science available. I hope that it will create constructive discussions and awareness of the opportunities and risks for Australian estuaries and the human population living on its shores and the need for integrating our efforts to deal with these development issues.

This book is especially important because both major political parties have virtually adopted a policy of a "little Australia", probably as a means of avoiding environmental and developmental difficulties and the investment that would be needed for both.

There are some people in Australia who believe that this country is already fully populated. Nobody in any other country in the world, having regard to the world's population pressures, would hold a similar view.

In other words, Australia's current policies are not sustainable.

At the end of the Second World War, Australian political leaders of all persuasions knew that Australia was indefensible with its resources and population. They set about expanding both in a vigorous and farsighted manner which has done much to benefit Australia and to diversify our society.

We need to embrace the future with a similar commitment. We need to aim for a much larger population, not only to justify our holding a large and wealthy continent in the eyes of the world but also to give us greater weight to advance Australian values, to make a greater contribution to security and to peace throughout the Western Pacific.

In short, an Australia of 40 or 50 million would find it much easier to be independent of major powers. We have too often followed major powers into their wars of no direct interest to Australia. That policy which has dogged Australia since Federation needs now, quite desperately, to be ended.

This book is valuable because the research gives us intimate knowledge of how to handle the consequent population pressures, how to protect and enhance the environment and what investments will be necessary to do so. It can therefore be most helpful as a guide for the future of Australia.

The Right Honourable Malcolm Fraser, AC CH
Former Prime Minister of Australia, 1975–1983

Contributors

Kátya G. Abrantes Centre for Tropical Water and Aquatic Ecosystem Research, School of Marine and Tropical Biology, James Cook University, Townsville, QLD, Australia

F.P. Andutta School of Physical, Environmental and Mathematical Sciences, University of New South Wales at Australian Defence Force Academy UNSW-ADFA, ACT, Canberra, Australia

Zoë T. Bainbridge Catchment to Reef Research Group, Centre for Tropical Water and Aquatic Ecosystem Research (TropWATER), Australian Tropical Science and Innovation Precinct, James Cook University, Townsville, Australia

Diane Bell Department of Anthropology, George Washington University, Washington, DC, USA

School of Social Sciences, University of Adelaide, Adelaide, Australia

School of Humanities, Flinders University, Adelaide, Australia

Nathan P. Benfer Asia-Pacific ASA Pty Ltd, Surfers Paradise, QLD, Australia

Christie Bentley National Research Centre for Environmental Toxicology (Entox), University of Queensland, Coopers Plains, QLD, Australia

Gavin F. Birch Environmental Geology Group, School of Geoscience, University of Sydney, Sydney, Australia

Jon E. Brodie Catchment to Reef Research Group, Centre for Tropical Water and Aquatic Ecosystem Research (TropWATER), Australian Tropical Science and Innovation Precinct, James Cook University, Townsville, Australia

Richard J. Brown Faculty of Science and Engineering, Queensland University of Technology, Brisbane, QLD, Australia

Michele A. Burford Australian Rivers Institute, Griffith University, Nathan, Australia

Hubert Chanson School Civil Engineering, The University of Queensland, Brisbane, QLD, Australia

Peter Collis School of Biological Science, Sydney University. Now at Upper Caboolture, QLD, Australia

Rod Connolly Australian Rivers Institute – Coast and Estuaries and School of Environment, Griffith University, Gold Coast, Australia

Aaron M. Davis Catchment to Reef Research Group, Centre for Tropical Water and Aquatic Ecosystem Research (TropWATER), Australian Tropical Science and Innovation Precinct, James Cook University, Townsville, Australia

Jean-Paul Ducrotoy Institute of Estuarine and Coastal Studies, The University of Hull, Hull, UK

Ryan J.K. Dunn Asia-Pacific ASA Pty Ltd, Surfers Paradise, QLD, Australia

Alan Easton Academic and Student Affairs, University of Papua New Guinea, Port Moresby, Papua New Guinea

Joanna C. Ellison School of Geography and Environmental Studies, University of Tasmania, Launceston, TAS, Australia

Peter C. Gehrke SMEC Australia Pty Ltd, South Brisbane, Australia

Badin Gibbes School Civil Engineering, The University of Queensland, Brisbane, QLD, Australia

Alistair Grinham School of Civil Engineering, The University of Queensland, Brisbane, Australia

Ross Johnston Centre for Tropical Water and Aquatic Ecosystem Research, School of Marine and Tropical Biology, James Cook University, Townsville, QLD, Australia

Jochen Kämpf School of the Environment, Flinders University, Adelaide, SA, Australia

Brian A. King Asia-Pacific ASA Pty Ltd, Surfers Paradise, QLD, Australia

Serena B. Lee Environmental Geology Group, School of Geoscience, University of Sydney, Sydney, Australia

Charles J. Lemckert Griffith School of Engineering, Griffith University, Nathan, QLD, Australia

Stephen E. Lewis Catchment to Reef Research Group, Centre for Tropical Water and Aquatic Ecosystem Research (TropWATER), Australian Tropical Science and Innovation Precinct, James Cook University, Townsville, Australia

Li Li School of Physical, Environmental and Mathematical Sciences, University of New South Wales at Australian Defence Force Academy UNSW-ADFA, ACT, Canberra, Australia

Paul Maxwell Australian Rivers Institute – Coast and Estuaries and School of Environment, Griffith University, Gold Coast, Australia

Jochen F. Mueller National Research Centre for Environmental Toxicology (Entox), University of Queensland, Coopers Plains, QLD, Australia

David Neil School of Geography Planning and Environmental Management, The University of Queensland, Brisbane, Australia

Dominique S. O'Brien Catchment to Reef Research Group, Centre for Tropical Water and Aquatic Ecosystem Research (TropWATER), Australian Tropical Science and Innovation Precinct, James Cook University, Townsville, Australia

Andrew Olds Australian Rivers Institute – Coast and Estuaries and School of Environment, Griffith University, Gold Coast, Australia

Duncan W. Palmer Department of Water, Government of Western Australia, Kununurra, WA, Australia

Andy T. Revill CSIRO Marine and Atmospheric Research, Hobart, Australia

Joachim Ribbe Department of Biological and Physical Sciences, University of Southern Queensland, Toowoomba, QLD, Australia

Barbara J. Robson CSIRO Land and Water, Canberra, ACT, Australia

Joe Sampson Mathematics Discipline, Faculty of Engineering and Industrial Sciences, Swinburne University of Technology, Hawthorn, VIC, Australia

Marcus Sheaves Centre for Tropical Water and Aquatic Ecosystem Research, School of Marine and Tropical Biology, James Cook University, Townsville, QLD, Australia

Matthew R. Sheehan School of Geography and Environmental Studies, University of Tasmania, Launceston, TAS, Australia

Manmohan Singh Mathematics Discipline, Faculty of Engineering and Industrial Sciences, Swinburne University of Technology, Hawthorn, VIC, Australia

James Udy Science and Innovation Program, Healthy Waterways Pty Ltd, Brisbane, QLD, Australia

Nicola Udy Marine Resource Management, Queensland Department of National Parks, Recreation, Sport and Racing, Brisbane, Australia

Nathan J. Waltham Centre for Tropical Water and Aquatic Ecosystem Research (TropWATER), James Cook University, Townsville, QLD, Australia

X.H. Wang School of Physical, Environmental and Mathematical Sciences, University of New South Wales at Australian Defence Force Academy UNSW-ADFA, ACT, Canberra, Australia

Tony Weber Catchments and Receiving Environments, BMT WBM, Brisbane, Australia

Ian T. Webster CSIRO Land and Water, Canberra, ACT, Australia

David Williams Australian Institute of Marine Science (AIMS), Brinkin, NT, Australia

Eric Wolanski TropWATER and School of Marine and Tropical Biology, James Cook University, Townsville, QLD, Australia

Sasha Zigic Asia-Pacific ASA Pty Ltd, Surfers Paradise, QLD, Australia

Estuaries of Australia in 2050 and Beyond – A Synthesis

Eric Wolanski and Jean-Paul Ducrotoy

Abstract

This book "*Estuaries of Australia in 2050 and Beyond*" in the series "*Estuaries of the World*" addresses the question: Is Australia's growing human population and economy environmentally sustainable for its estuaries and coasts by 2050? To answer this question, this chapter summarises detailed studies of a number of iconic Australian estuaries and bays. They can be divided in three types based on the human impact, namely (1) estuaries that bore the full pressure of the historical developments, (2) estuaries being degraded, and (3) estuaries that are still relatively pristine. For type (1) the case studies focus on Sydney Estuary, the Coorong/Murray-Darling Estuary, Port Philip Bay, and the Tamar Estuary. For type (2) the case studies focus on the Gold Coast Broadwater, the Hawkesbury Estuary, the Burdekin flood plains, Moreton Bay, the Ord River estuary, Brisbane peri-urban estuaries, South Australia gulfs, Hervey Bay, and Darwin Harbour. For type (3) the case studies focus on the Mary River estuary and floodplains in the Northern Territory and Deluge Inlet in Queensland. In addition, summaries are also provided of the state of the environment and the management strategy for a number of other estuaries and coastal waters. Overall, this chapter synthesises multidisciplinary scientific knowledge in time and space across Australia to suggest what Australian estuaries may look like in 2050 based on socio-economic decisions that are made now, and the changes that are needed to ensure sustainability.

Keywords

Rainfall • Evaporation • Population • Ecosystem services • Estuaries • Quality of life • Development • Environmental impact • Sustainability • Australia

E. Wolanski (✉)
TropWATER and School of Marine and Tropical Biology, James Cook University, James Cook Drive, Townsville, QLD 4811, Australia
e-mail: eric.wolanski@jcu.edu.au

J.-P. Ducrotoy
Institute of Estuarine and Coastal Studies, The University of Hull, Hull HU6 7RX, UK
e-mail: jean-paul.ducrotoy@hull.ac.uk

E. Wolanski (ed.), *Estuaries of Australia in 2050 and Beyond*, Estuaries of the World,
DOI 10.1007/978-94-007-7019-5_1,

Introduction

Is Australia's growing human population and economy environmentally sustainable for its estuaries and coasts? To answer this question, this book reports detailed studies of a number of iconic Australian estuaries and bays. This chapter is a synthesis that integrates this knowledge in time and space across Australia to suggest what the Australian estuaries may look like in 2050 and beyond based on socio-economic decisions that are made now, and the changes that are needed to ensure sustainability.

For this synthesis, it is necessary to remember that Australia is a vast continent with a wide variety of natural settings that shaped the estuaries before the human impact. The human impact is an additional recent impact. As a result no two estuaries are alike. It is necessary to take a step back and understand the importance for the estuaries of the geography and the recent history.

The Geography and the Recent History of Australian Estuaries

Australia is a continent with a land surface area of 7.7×10^6 km^2. It has a long history of human inhabitation lasting several tens of thousands of years and a very short modern history that started with British settlement in 1788 (Flannery 1994). The human population at that time was very small, estimated at about 0.35×10^6. It is likely that most rivers and estuaries were then healthy ecosystems. That is not to say that they were in a virgin state, because the human population, even if small compared to present values, had substantially modified the vegetation cover over the watersheds through the use of man-made bushfires.

The large watersheds of Australia are shown in Fig. 1, the largest watershed being that of the Murray-Darling River, which comprises about 14 % of the total area of Australia. In addition there are numerous small watersheds, many of them are mangrove-fringed in the tropics and sub-tropics, and their number has not been determined.

The human population of Australia in modern history has increased rapidly, and is still increasing rapidly by one person every 1 min 21 s (Fig. 2). By 2100, the population curve is expected to resemble a S-curve characterised by a rapid, non-linear growth followed by a flattening out to levels between 30 and 100×10^6 that depend on the rates of immigration (Fig. 2). The population is now nearing 23×10^6.

While this population S-curve somewhat resembles that in North America following mass immigration (Daniels 1990), the spread of the population differs greatly between the North American and the Australian continents. While there are numerous large inland cities in North America, such is not the case in Australia. The main reason for this difference is the rainfall distribution that in North America enables human settlement over much of the continent while in Australia rainfall is largely restricted to a narrow coastal strip (Fig. 3a). The country is divided between a tropical North and a temperate South (Fig. 3b). As a result much of the human population is concentrated along the temperate coast where rainfall allows it (Fig. 3c), but not along the tropical coast where rainfall is abundant but historically people have been reluctant to settle. There are several reasons for that reluctance, including: firstly the harsh tropical climate, secondly the remoteness of the area, thirdly the devastating monsoonal floods, fourthly in Australia's tropics evaporation commonly exceeds rainfall (Fig. 3d) and as result many tropical rivers are dry in the dry season. All these facts combine to make Australia have scarcely more than two persons per km^2 of total land area. However the population density is large along the temperate coast where reliable rainfall occurs. Australia is one of the world's most urbanised countries as 89 % of its population live in urban areas mainly near estuaries and coasts.

It is apparent from Fig. 1 that the political map of Australia was drawn with no regard to the hydrology. This leads to economic, administrative and political constraints that result in the Australian State governments cooperating little with each other and with the Australian government in water resources management. It is still as if the land, the river, the estuary and the sea were not part of the same system. Large-scale, intensive irrigation farming has been developed in the most suitable catchments in the temperate south. Agriculture requires water, but the availability of water in Australia varies markedly from year to year, with rainfall characterised by a succession of 'good years' (i.e. years with well above average rainfall) and 'poor years' (i.e. years with well below average rainfall; Fig. 4). 'Poor years' commonly occur in succession of several years. Using short term rainfall data from 'good years', water resources managers have commonly over-estimated the availability of water during 'poor years' and thus commonly they have over-allocated water that can be used for irrigation with insufficient water available in 'poor years' for both irrigation and the environment. During 'poor years' the usual management policy in Australia has been to satisfy the needs for irrigation first and neglect the environmental needs of the rivers and estuaries.

The resulting impact of such water usage policies on the estuary is most evident in the Murray-Darling River (Fig. 1). The river catchment extends to several States and in practice there is no catchment water authority; the water resources managers in the various riparian States upstream of South Australia commonly have over-allocated irrigation water licences in their own States and ignored the cumulative impacts on the river from similar allocation in the other riparian States. During 2005–06, which were 'poor years', water extractions were over 9,000 GL from 6,530 GL of

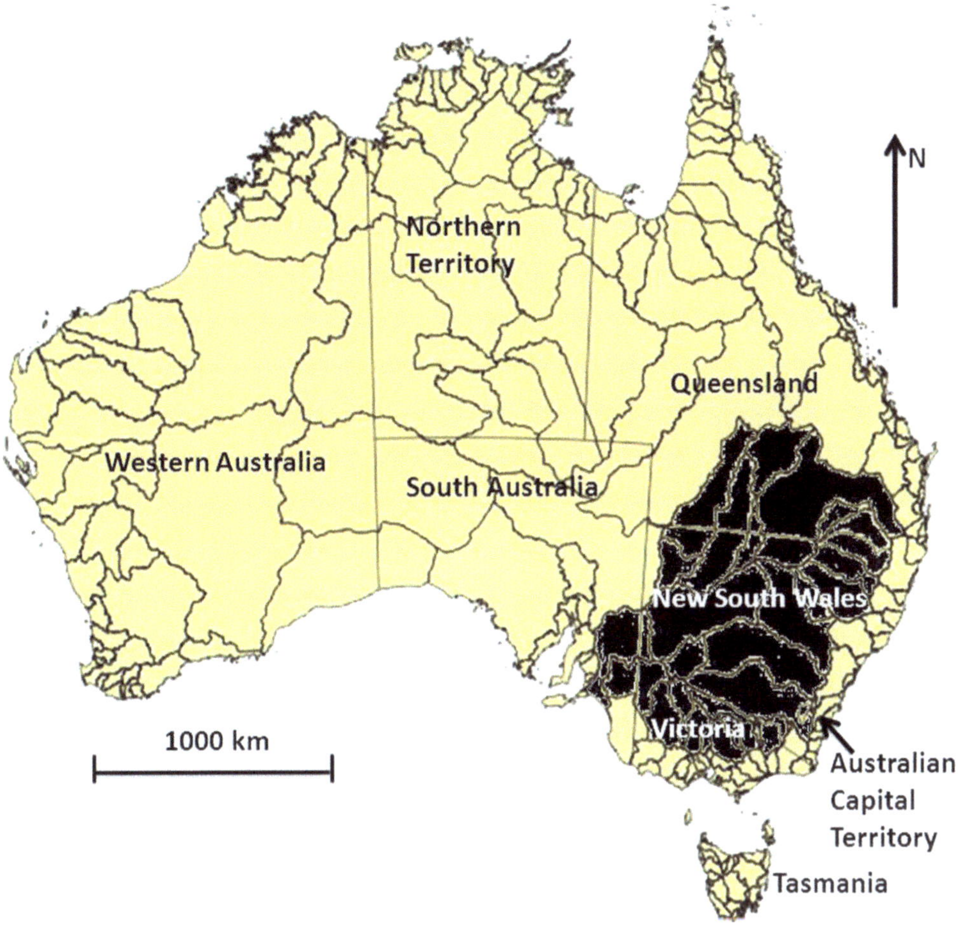

Fig. 1 The political map and the watershed map of Australia. The *darkened area* is the watershed of the Murray-Darling river basin

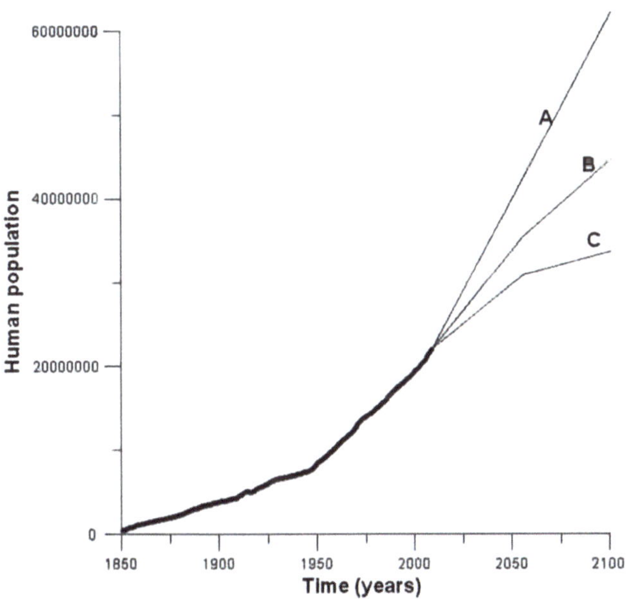

Fig. 2 Time series-plot of the human population in Australia (*thick line*) since 1850 until present, and in the future as scenarios *A*, *B* and *C* that depend on likely rates of immigration (The data were provided by the Australian Bureau of Statistics)

inflow, the balance being due to using of water stored in impoundments in previous 'good years' (Murray–Darling Basin Ministerial Council 2007). As a result during such 'poor years' zero or negligible river flows reached the sea for months at a time. This is totally a man-made environmental crisis and the results for the estuary are devastating, and this is described in chapter "The Murray/Coorong Estuary Meeting of the Waters?" by Jochen Kaempf about the Coorong/Murray Estuary in this book.

Rivers are impacted by changes to river flows and pollution by nutrients, pesticides and herbicides from large-scale irrigation farming that was encouraged by State governments policy of expanding the water supply while environmental management was a secondary consideration (Hussey and Dovers 2006; Petheram et al. 2008). As a result, although European settlement in Australia is only slightly over 200 years old, its environmental impact on Australian rivers is dramatic. From an Australia-wide survey of the river environment index, approximately 85 % of the rivers' length is affected by catchment disturbance (NLWRA 2000). Of the regulated and unregulated rivers for which data are available, over 80 % are modified to some extent

a

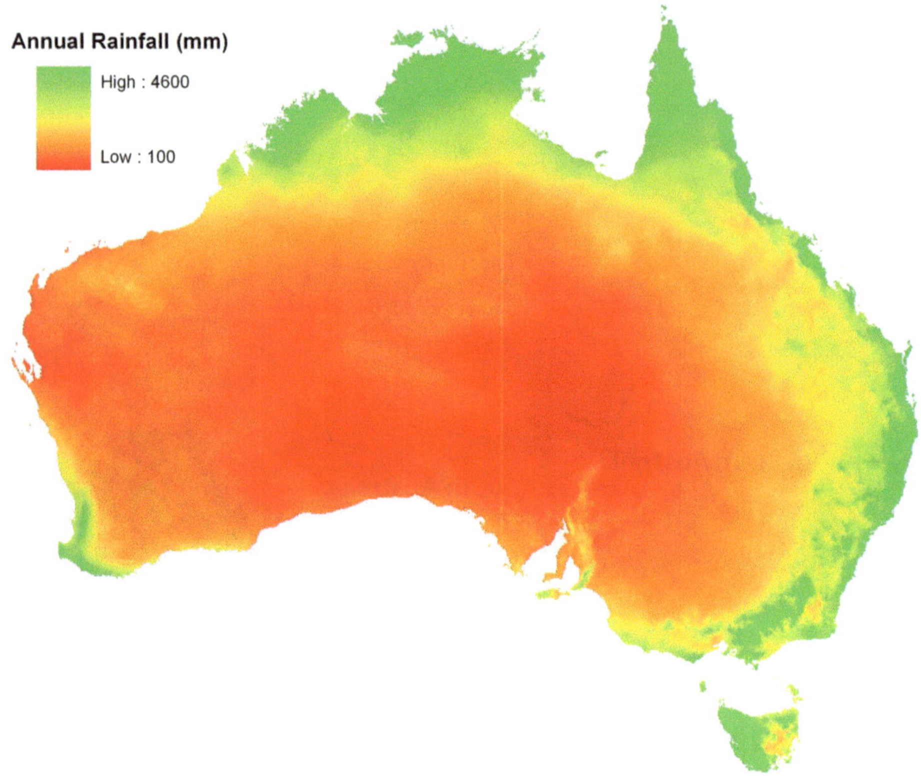

b

Fig. 3 Distribution map of (**a**) mean annual rainfall (mm/year), (**b**) mean temperature (°C), (**c**) human population density and location of capital cities, (**d**) mean annual evaporation (mm/year) ((**a**) is modified from the National Land and Water resources Audit, (**b**) and (**d**) are modified from the Bureau of Meteorology, and (**c**) is modified from Regional Population Growth, Australia)

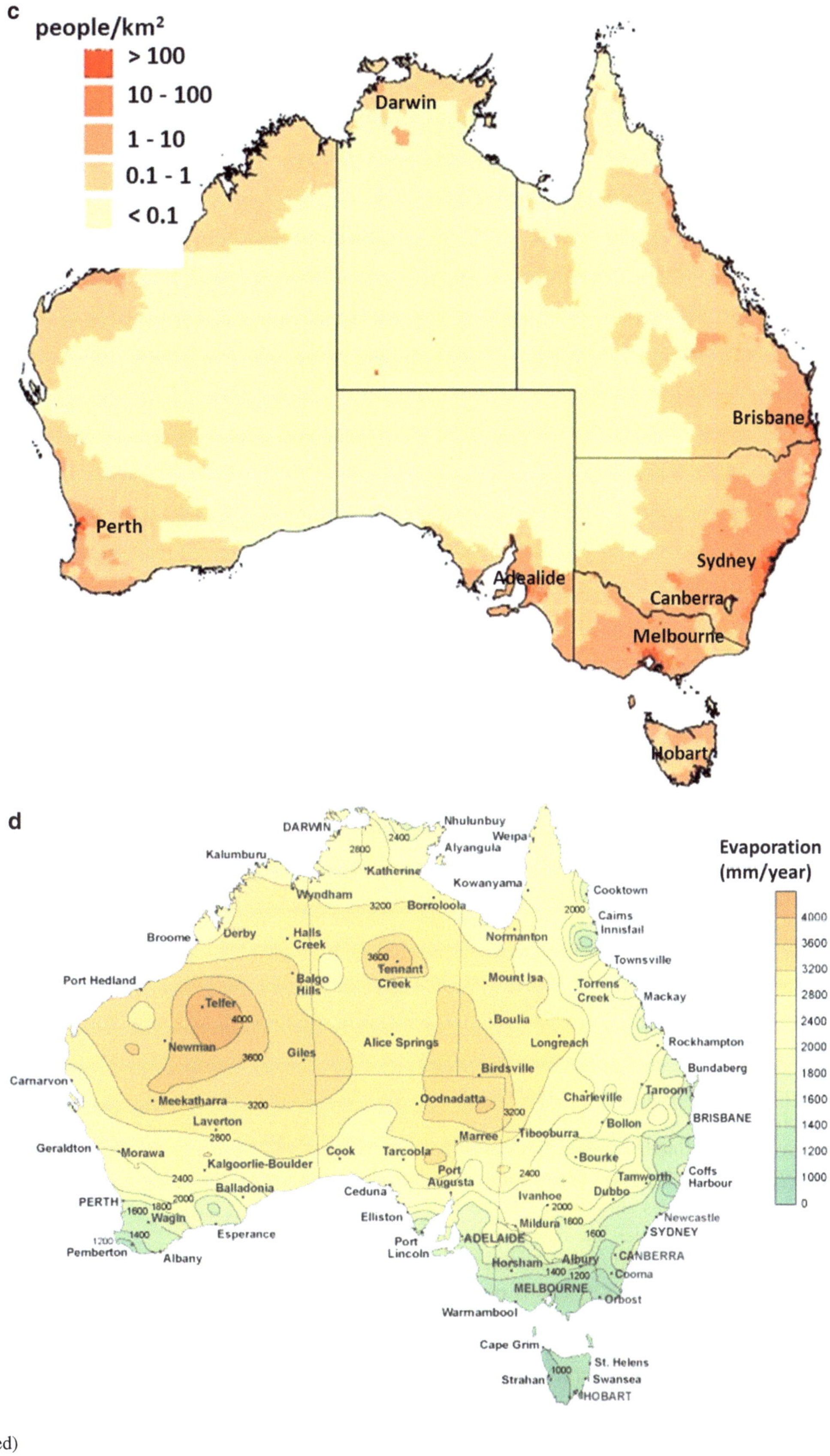

Fig. 3 (continued)

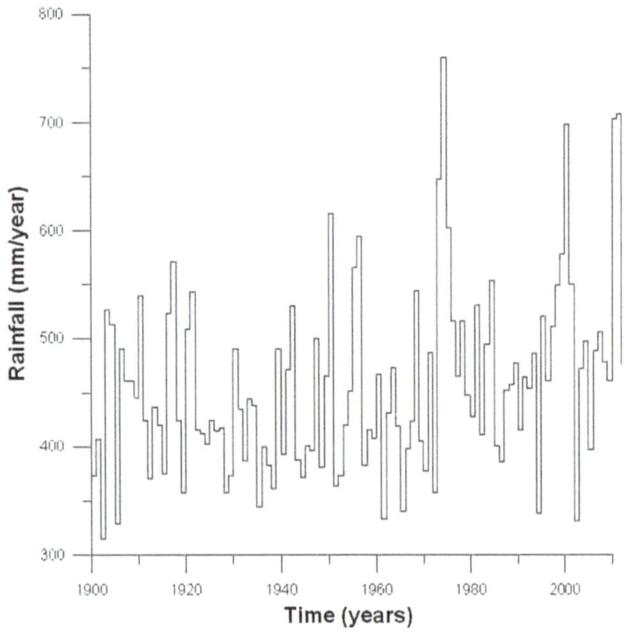

Fig. 4 Time-series plot of Australia-wide annual rainfall from 1900 to present (The data were provided by the Commonwealth Bureau of Meteorology)

and nearly 30 % are substantially modified. Changes to riverine habitats are severe in the Murray-Darling Basin, South Australia, and parts of the Western Australian wheat belt. Nutrient (mainly phosphorus) and suspended sediment loads are greater than natural levels for over 90 % of the length of Australian rivers and are severely modified in almost 10 % of total river length (NLWRA 2000).

These farm-derived nutrients, sediment and also pesticides and herbicides reach the estuaries and coastal waters, which they degrade. In most cases the environmental degradation is local, e.g. a particular estuary or coast is degraded and this affects mainly the local community. In some cases this degradation is significant for Australia; such is the case for the Great Barrier Reef of Australia. Seagrasses in key areas such as Cairns are now at their smallest ever recorded distribution (Rasheed et al. 2013). Coral calcification rates have declined by 15 % since 1990 (De'ath et al. 2009), and coral cover has declined by nearly 50 % over the last 27 years (Dea'th et al. 2012). Farm-derived nutrients, sediment, pesticides and herbicides are responsible for this degradation of the Great Barrier Reef, the management of which does not include managing land-use in the adjoining catchments (Brodie and Waterhouse 2012). This is probably the most spectacular case of failed coastal management in Australia from ignoring to manage the entire ecosystem including the watershed as one system (Wolanski 2007; Mee 2012). This degradation will be further exacerbated by planned, massive coal export ports and industrial developments in or adjacent to the Great Barrier Reef Marine Park, an issue about which in 2012 the UNESCO World Heritage Committee expressed particular concern and (at the time of writing this synthesis) is evaluating if the Great Barrier Reef should be officially listed as a World Heritage *in danger*.

Can the Swan River Estuary Health Be Restored?

The microtidal Swan (also called Swan-Canning) Estuary flows through the city of Perth. The estuary is highly eutrophicated (Peters and Donohue 1999; Hamilton et al. 2001; Brearley 2005) because of slow flushing (Stephens and Imberger 1996; Etemad and Imberger 2005) and runoff from rural and urban catchments (John 2013). The State government recognises the iconic value of the estuary and created the Swan River Trust, which provides funding for a number of mitigation measures that include supporting community groups undertake on ground restoration projects, pilot projects to reduce sedimentation from building sites, improve local government processes to reduce the impact of light industry on the rivers, facilitate community involvements and the review of new development proposals, and facilitate the creation of urban wetlands that help filter urban stormwater (Swan River Trust 2012). It is also advocating progressively phasing out highly water-soluble phosphate fertilisers and improving fertiliser practices in rural and urban catchments. Despite these restoration efforts, the frequency of algal bloom occurrences has remained relatively stationary and the annual average Index of Sustainable Functionality (an index of the sustainable health of the estuary) did not improve from 1995 to 2009 (Kristiana et al. 2012). Thus the mitigation measures to counter eutrophication at best have countered the increased pressures on the health of the estuary from new land development and climate change, but they did not improve the health of the estuary.

Case Studies: The State of Knowledge

Only a limited number of Australian estuaries have been studied in great detail – indeed there are not enough estuarine scientists in Australia to study all the estuaries, neither is there enough research funding. Nevertheless this book brings together the majority of the detailed studies of

Fig. 5 A location map of the estuaries and coastal waters featured in this book. The Swan, Keep, and Peel-Harvey estuaries, and the estuary and coastal waters of Townsville are discussed in this synthesis; the other sites are described in a chapter each in this book

Australian estuaries and coastal waters. The study sites are shown in Fig. 5.

The case studies of Australian estuaries in this book can be divided in three types based on the human impact, namely (1) estuaries that bore the full pressure of the historical developments, (2) estuaries being degraded, and (3) estuaries that are still relatively pristine. For type (1) the case studies focus on Sydney Estuary, the Coorong/Murray-Darling Estuary, Port Philip Bay, and the Tamar Estuary. For type (2) the case studies focus on the Gold Coast Broadwater, the Hawkesbury Estuary, the Burdekin flood plains, Moreton Bay, the Ord River estuary, Brisbane peri-urban estuaries, South Australia gulfs, Hervey Bay, and Darwin Harbour. For type (3) the case studies focus on the Mary River estuary and floodplains in the Northern Territory and Deluge Inlet.

The response of these estuaries to the human impact is described below, it depends strongly on the geomorphology and hydrology, and thus on the geographic location of each estuary.

A Socio-economic Classification of Australian Estuaries

These case studies demonstrate, through the writing of eminent Australian estuarine scientists, how these estuaries function and this knowledge requires merging the physical, chemical and biological oceanography, the ecosystem processes, and the human impact. The studies describe most types of Australian estuaries from pristine estuaries to estuaries heavily impacted by urbanisation, harbour operations, industrialisation, and intensive irrigation and water management schemes in the river catchments. The basic science has been done in some estuaries particularly those in the capital cities, and is being done in other estuaries. This makes it possible to understand, at least as a first order approximation, how these brackish water ecosystems function. This enables the scientists to forecast with some confidence and some uncertainty what these estuaries may look like by 2050 based on political and socio-economic decisions that are made now, just like the decisions made a few decades ago dictate what these estuaries look like now. This book shows that we have much to learn by understanding the lessons from the past and from each estuary. It is hoped that these lessons can then be applied to all Australian estuaries in order to ensure an environmentally sustainable Australia where the estuaries will keep delivering the full range of ecosystem services that the population expects in order to maintain a high quality of life.

The Baggage of History

Historically Australian estuaries were seen as little more than navigation channels and convenient waste dumping sites, a viewpoint similar to that held during the development of ports and harbours in the Asia Pacific region (Wolanski 2006). The impact on Australian estuarine environments from developments along their shores and from land-use in the river catchment was seen as inconsequential. That was the old practice of development at all costs.

Shared Water Resources

Where an estuary depends on the good will of various State governments and the Commonwealth governments, no political compromise has yet been fully agreed by all governments to enable ecological restoration and the system is allowed to further degrade, if not collapse (e.g. see the chapter "The Murray/Coorong Estuary Meeting of the Waters?" by Jochen Kaempf about the Coorong/Murray estuary).

Urbanisation

All State capital cities are harbours built along estuaries or sheltered bays. These are important cities (see Fig. 3c) with large industrial and urban centres, whose respective footprints on the estuaries have now merged. However Australia has become a developed modern, savvy society and in its State capital cities the population is now demanding a high quality of life that includes enjoying the benefits of a healthy estuarine ecosystem. Responding to this public demand, governments do actually attempt to restore some of the ecological functions of degraded urban estuaries (e.g. Port Philip Bay, Sydney Estuary, and Gulf St. Vincent) and they implement policies to try to prevent further degradation of estuaries from creeping urbanisation (e.g. the Gold Coast Broadbeach Estuary, Hervey Bay, and Brisbane small peri-urban estuaries).

In the State capital cities, with the possible exception of Darwin, large efforts are made to restore, or at least preserve, some sort of viable ecosystem in the estuaries and coastal waters. The results of such efforts are exemplified in the case of Perth, described in the box above.

In some capital cities with a population demanding healthy waterways, the State government push for developments can conflict with the population demanding healthy estuaries. A compromise is then sought. For instance in Brisbane – see the chapters "Turbulent Mixing and Sediment Processes in Peri-Urban Estuaries in South-East Queensland (Australia)" and "Moreton Bay and Its Estuaries: A Subtropical System Under Pressure From Rapid Population Growth" by Hubert Chanson and colleagues and Badin Gibbes and colleagues in this book – the government policy seems to attempt to preserve the health of coastal waters and small peri-urban estuaries. However no remediation measures are implemented for the heavily degraded Brisbane River estuary where the Port of Brisbane is located and whose dredged mud is still allowed to be dumped in high-value, urbanised, coastal waters.

Chemical pollution from industrialisation can be remediated by technology. However water pollution by stormwater runoff from urbanised areas is an extremely difficult problem to solve (Beach 2002). The Gold Coast Local Government and those surrounding Port Philip Bay may be the pioneer Local Governments in Australia actually addressing that problem by requiring new developments to treat wet weather runoff (e.g. see the chapters "Gold Coast Broadwater: Southern Moreton Bay, Southeast Queensland (Australia)" and "Port Phillip Bay" by Ryan Dunn et al. and Joe Sampson et al. in this book) to oblige new urban developments to treat urban stormwater at the source, while discussions have started to retrofit stormwater runoff treatment at the Gold Coast Broadwater and in Sydney (see the chapters "Gold Coast Broadwater: Southern Moreton Bay, Southeast Queensland (Australia)" and "Sydney Estuary, Australia: Geology, Anthropogenic Development And Hydrodynamic Processes /Attributes" by Ryan Dunn et al. and Lee and Birch in this book). Several other Local Governments in Australia have developed, or are developing, strategies to improve stormwater quality, such as harvesting and use options for the local park irrigation or urban wetlands. So far however there are no policies to retrofit existing urban areas with such features. Such features will be expensive but are necessary to maintain the quality of life of the urbanised population; for instance it is estimated that retrofitting such features along the 15 km long Saltwater Creek on the Gold Coast to achieve the desired water quality load reductions would cost about A$ 65 million. An additional funding challenge of such features will be the high on-going maintenance costs of urban wetlands and gross pollutant traps (N. Waltham, personal communication 2013).

Limits to Restoration

Complete restoration to the original pre-European conditions of the urban estuary is however generally impossible on ecological grounds (Duarte et al. 2009) and there is thus a limit to what can be realistically achieved. For Australian urban estuaries, preventing further degradation is a substantial achievement by itself. The urban population seems satisfied that the ecosystem is relatively healthy even if it does not resemble the pre-European system (e.g. see the chapter "Past, Present and Futures of the Tamar Estuary, Tasmania" in this book by Joanna Ellison and Matthew Sheehan about the Tamar River estuary).

Social Inequity

In Darwin Harbour the government can safely encourage rapid port developments and industrialisation on the east side of the harbour with minimal remediation measures. This is because about half of the harbour's watershed and mangroves are protected by Aboriginal traditional owners on the west side of the harbour and this, together with swift tidal flushing, keeps Darwin Harbour ecologically healthy (Wolanski 2006). This may lead to social inequity as the Aborigines may not benefit from the ecosystem services (such as the rich fishing grounds in the harbour) that they provide to Darwin. This healthy state of Darwin Harbour could also change if there is ever an accident leading to a major pollution event, because the pollutants may remain trapped in the mangrove wetlands for decades or even centuries, as is shown in the chapter "Hydrodynamics and Sediment Transport in a Macro-tidal Estuary: Darwin Harbour, Australia" by Fernando Andutta and colleagues in this book.

Suburbs and Regional Towns

State governments respond to public demands for healthy waterways. Such pressure is lacking in some suburbs of capital cities and in regional towns. Regional towns generally have single industries that drive the local economy and/or have a lower socio-economic status than capital cities. The population is generally unwilling to hurt their economy to improve the health of rivers and estuaries and it sees environmental health as a lower priority. Consequently, with no votes to gain in an election, State governments are unwilling to commit money to prevent the degradation of estuaries. This is the case of the Hawkesbury Estuary described by Peter Collis in this book. The present sewage treatment plants are already unable to prevent the on-going eutrophication of the Hawkesbury Estuary. Even with the upgrades of all the sewage treatment plans there will still be an increase of about 20–50 % of the discharge of nutrients to the estuary when the population doubles by 2050; some of this nutrient loading will come from stormwater runoff, a problem which the Local Governments in the Hawkesbury River catchment have chosen to ignore. All this will further increase the eutrophication of the estuary by 2050. No solution is proposed by the State government for the Hawkesbury, it is to be increasingly eutrophicated.

A similar situation of a lack of effective action to prevent further eutrophication also exists in the microtidal Peel-Harvey Estuary (also known as the Peel Inlet-Harvey Estuary) located south of Perth. A new channel was constructed to improve flushing, but land use remediation measures were not implemented. As a result, eutrophication still persists as evidenced by the sharp decline of the density of crustaceans and molluscs and the occurrence of algal blooms and occasional fish kills (Brearley 2005; Davis and Kloop 2006; Wildsmith et al. 2009).

A similar situation also occurs with the industrialisation of Spencer Gulf, even though science suggests that this poses the biggest threat to the gulfs' ecosystem health, as is shown by Jochen Kaempf in his chapter "The Murray/Coorong Estuary, Meeting of the Waters?" in this book.

Irrigation in the Tropics

In the tropics, the government push for developments is the least constrained by environmental considerations, and this is facilitated by these areas being the furthest away from capital cities and by the local economy being dominated by one single driver, such as an irrigation project. The two major tropical irrigation projects in Australia are in the Ord river and Burdekin river floodplains and they are discussed in this book in the chapters "The Ord River Estuary: A Regulated Wet-Dry Tropical River System" and "Water Resource Development and High Value Coastal Wetlands on the Lower Burdekin Floodplain, Australia" by Barbara Robson et al. and Aaron Davis et al. Both projects are young but already suffer from increased salinization problems that may impact their long-term sustainability. Nevertheless both schemes are expanding. In the case of the Burdekin, the tail waters from irrigated areas are affecting with nutrients, pesticides and herbicides, the whole length of the stream and, further they flow into, and pollute, a RAMSAR wetlands site. Remediation measures are discussed but at present they are not implemented (see the chapter "Water Resource Development and High Value Coastal Wetlands on the Lower Burdekin Floodplain, Australia" by Aaron Davis and colleagues in this book). In the case of the Ord, detailed studies have been undertaken of the fate of the nutrients in the tailwaters from irrigated areas, as described in the chapter "The Ord River Estuary: A Regulated Wet-Dry Tropical River System" by Barbara Robson and colleagues in this book. These studies were used to forecast as small the likely impact of the on-going increase by 236 % of the irrigated areas. This sounds reassuring and is indeed backed by science, until one realises that most of the tailwaters from these new irrigated areas will flow not to the Ord River in Western Australia but to the Keep River in the Northern Territory, which is not governed by Western Australian legislation. The Northern Territory government chose to ignore the finding from two independent studies, by Wolanski (unpublished data) and by its own hydrologist, that the Keep River estuary is susceptible to eutrophication from nutrient-enriched tailwaters from irrigation because at neap tides it forms poorly-flushed pools of water separated by rock bars (D. Williams, personal communication 2011). Instead it simply required the developer to undertake a comprehensive water quality monitoring programme for the Keep River but it stipulated no water quality criteria to be met (Northern Territory Government 2012). Further The Northern Territory Environmental Assessment Act does not cater for appeals. All this gives the green light to development at all costs.

Conflicts of Interest

The Queensland government has a conflict of interest because it owns several ports (and thus it wants to maximise economic returns) and at the same time it has a duty of care to the population and the environment. The port of Townsville is government owned. The port seawalls have created a stagnant zone in coastal waters where mud accumulates (Lambrechts et al. 2010) and the mass resuspension of this mud during occasional storms and its later mass deposition on seagrass beds may be responsible for the severe recent decline in seagrass and resulting mass mortality sea turtles in coastal waters off Townsville (Elmore 2011). The Queensland government proposes no remediation measures for the port operations. Further, the port is used to export mining ore, including Lead (Pb). This generates pollution and Pb-enriched marine sediment is evident downwind of the Port (Doherty et al. 2000a, b). No remediation measures are proposed for the port operations. Finally, airborne black dust deposits in the suburbs downwind of the Port of Townsville (Fig. 6a). Some members of the local community feel threatened by this dust both from a quality of life perspective and from a health/toxicology perspective, particularly with regards to Pb. With the Queensland government dismissing these concerns outright with no studies, concerned people collected black dust around their houses on clean plates kept outside the houses or scooping the dust from window sills and outside tables and benches or from floors. The samples do not seem to have been contaminated by lead paint or pipes. These dust samples were analysed for Pb at certified laboratories in Australia. Figure 6b shows the spatial distribution of the particulate Pb concentration in Townsville during the tradewind season of 2008. The suggested pollution plume was consistent with the pattern expected from a Pb source in the port area. Figure 6c shows the time-series plot during from 2005 to 2009 of particulate Pb in the black dust collected at a house in Townsville downwind of the port; this figure suggests that there is a Pb pollution threat and that the 'safe' limit may have been exceeded all the time during 5 years.

In reply to a letter by concerned citizens, in August 2011 the Queensland Minister for the Environment stated that Pb pollution was safe because Pb does not build up in the human blood after a few months of exposure. This statement however is contradicted by medical evidence (EPA 1998; Balch and Balch 2000; Grandjean 2010) that there is probably no safe exposure to Pb and further that:

> Lead remains in the blood stream for weeks, then is absorbed in the bones, where it can collect for a lifetime. Lead that is not excreted through the digestive system accumulates and is absorbed directly from the blood into other tissues. When lead leaves the blood stream, it is stored in the bones where it continues to build up over a lifetime. Lead from the bones may then reenter the blood stream at any time as a result of severe biologic stress, such as renal failure, pregnancy, menopause, or prolonged immobilization or illness.

The Queensland government has refused to comment ever since. Also no remediation measures are taken in the port since the Pb-rich black dust continues to settle (S. van Grinsven, personal communication 2012).

The Queensland government also has a conflict of interest in the coal export port of Gladstone because it owns both the port and the railway transporting coal to the port from mines inland. The wagons are uncovered and thus Queensland Rail loses 4.8 million tons of coal dust a year to the wind (MacDonald 2010). The port stores the coal in open areas and the conveyor belts to the ships are uncovered, thus the wind blows away coal dust that deposits over the city of Gladstone (K. Burns, personal communication 2012). The Queensland Government (2010) reported mean air pollution levels within acceptable levels for an industrial area but, by not showing coal dust deposition data, it simply brushed over spikes in the air quality data, the excess of self-reported symptoms of asthma in both adults and children, and the 100 % increase in the incidence of the cancer Chronic Lymphoid Leukaemia (Queensland Government 2007). No remediation measures are apparently implemented in the rush to develop the coal export industry.

With such industry-friendly State governments with conflicts of interest, there is no independent agency examining the links between environmental pollutants and human health, suggesting Australia tropical States have ignored the overseas lessons of the need to battle pollution for the sake of public health and quality of life (Davis 2002).

Water pollution issues are treated with the same cavalier attitude in Australia tropics in the policy of development at all costs. For instance Ross Creek, the estuary of Townsville, is the most polluted by heavy metals and hydrocarbons of all the North Queensland estuaries that were sampled (Inglis and Kross 2000). This pollution was known for two decades and is particularly high near the mouth of the estuary downstream of the port (Doherty et al. 2000a, b; da Silva et al. 2004). Yet no remediation measures are seemingly implemented.

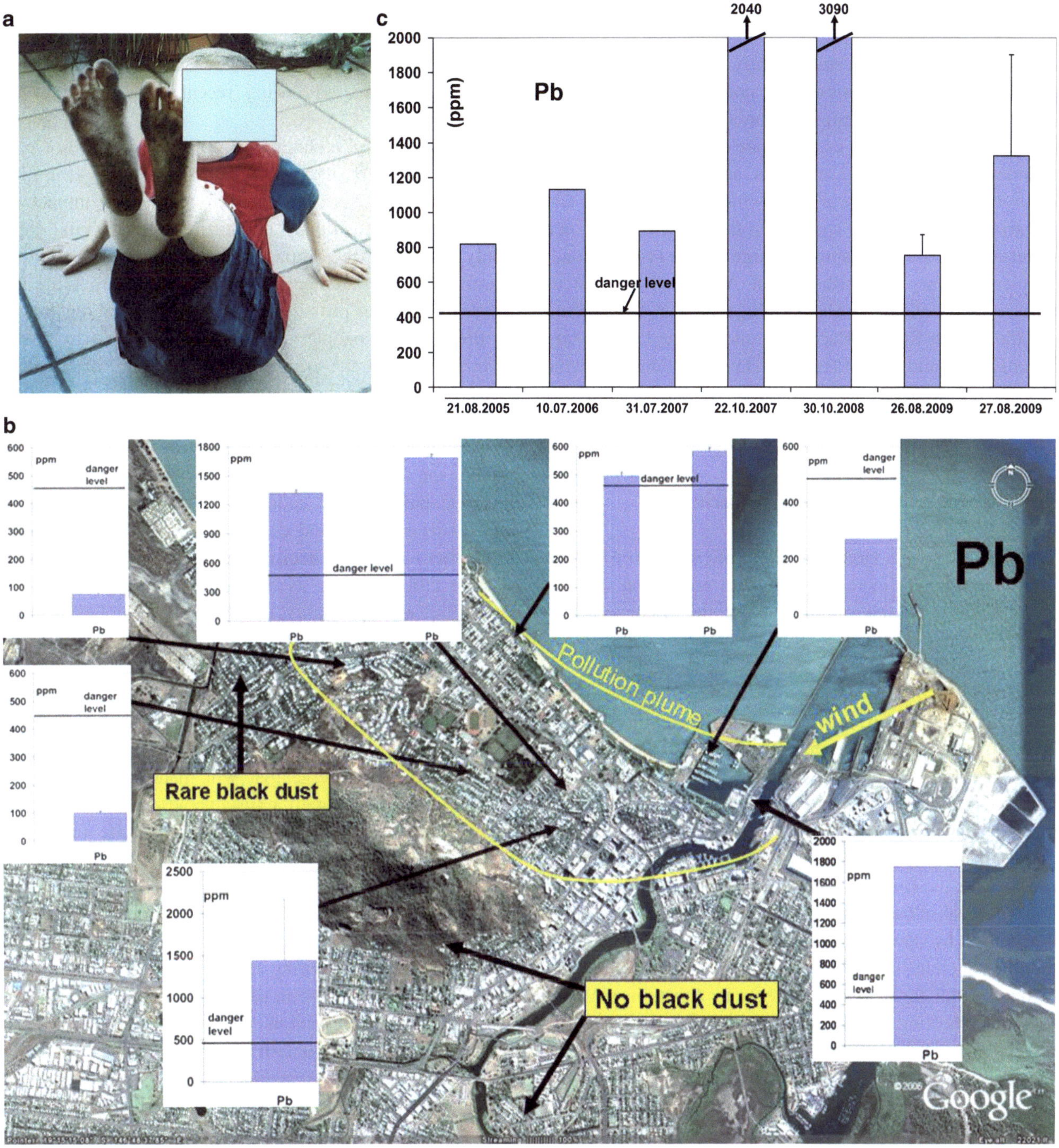

Fig. 6 (**a**) *Black dust* deposited on a patio and sticking on a child foot. This is the result of only one night of deposition at a house near the beach downwind of the port of Townsville in 2008. (**b**) The spatial distribution of the particulate Pb concentration in the black dust deposits in Townsville during the tradewind season of 2008. Also drawn is the local wind direction and the suggested, approximate limit of the pollution plume based on these data. (**c**) A time-series plot of particulate Pb concentration in the *black dust* collected at a house in Townsville downwind of the port from 2005 to 2009. To these data are added the State of Washington EPA legal 'danger level' for polluted soils (Data from Wolanski 2011)

The same philosophy of development at any cost for estuarine water quality is evidenced by the Queensland government allowing in 2013 the discharge of highly chemically polluted mine wastewater in rivers (Queensland Government 2013), in spite of the scientific knowledge that most of these chemical pollutants will be trapped in the estuaries (Wolanski 2007).

Pristine Estuaries

Australia still has several slightly affected and even pristine rivers and estuaries in far north Western Australia and Queensland, western Tasmania, and the Northern Territory –such pristine estuaries are a rarity in the world outside Australia. These estuaries are by themselves an Australian gift to the world of what pristine rivers and estuaries look like. Two chapters in this book address such systems; these are the chapter "Deluge Inlet, a Pristine Small Tropical Estuary in North-Eastern Australia" by Marcus Sheaves and colleagues about Deluge Inlet in Queensland and the chapter Recent, Rapid Evolution of the Lower Mary River Estuary and Flood Plains" by David Williams about the Mary River estuary in the Northern Territory. These pristine estuaries are precious and should not be used and abused.

The Future of Australia's Estuaries

There are calls for a large increase in migration to Australia in order to rapidly increase the human population so as to create a 'big Australia', as explained in the prologue to this book by Mr. Malcolm Fraser, former Prime Minister of Australia.

At the same time, to develop their economy the States and Territories with water and vast tracts of potentially arable land – i.e. mostly in the tropical regions of Queensland, Western Australia and the Northern Territory where the human population is presently small (Fig. 3c) – are pushing for large-scale irrigation projects to be made possible by the proposed construction of up to 100 dams, so that Australia can become the food bowl of Asia if not the world.

As a result most estuaries throughout Australia are now threatened by new projects of urbanisation, irrigation, mining, or industrialisation. Australian States generally recognise that threat. Developments in Australia are now subject to meeting environmental criteria set by legislation and this requires environmental impact studies, although as shown in case studies in this book State governments can and do routinely bypass this process when they want.

Encouragingly however more recently the concept of sustainable development is starting to be discussed in Australia.

Additional changes may also come from climate changes that may impact the water yield of river catchments as a result of changes of the temperature, evaporation and rainfall. Climate change sceptics thrive in the atmosphere of industry-friendly development-focused State governments and their outcome of seeding doubt (Michaels 2005) has been used by several State and Local Governments to dismiss planning for climate change when considering some development projects. Nevertheless climate change is seemingly happening (BOM 2012; CSIRO 2012). For instance rainfall in the southwest corner of Western Australia, that includes the State capital city Perth, has significantly decreased during recent decades (Li et al. 2005).

Engineering and technology by themselves do not provide a solution to sustainable development of Australian estuaries. When assessing the environmental impact of developments, Australian engineering consultants commonly state that the impact will be minimal – without quantifying this statement – and rarely do they value the ecosystem services provided by the estuaries and the quality of life to the people. Common statements by State governments and the engineering community of 'world's best practices' when discussing environmental sustainability of estuaries are nonsense and are not backed by facts as the case studies in this book illustrate.

Thus the answer to the question "Is Australia's growing human population and economy environmentally sustainable for its estuaries and coasts by 2050?" may be, based on the socio-economic decisions made now, (1) ***possibly yes*** in large cities as long as the population is pro-active in demanding a high quality of life, which implies healthy waterways, and (2) ***probably not*** in rural and remote areas and especially so in the tropics.

Hopefully this pessimistic prediction may turn out to be incorrect. Australia is privileged to have a number of eminent estuarine scientists. Hopefully future enlightened governments will emerge that will made socio-economic decisions compatible with a sustainable Australia and will call on these scientists to help provide a safe future for Australian estuaries in 2050 and beyond. This will require adopting an ecosystem-ecohydrology-based management approach at the watershed scale, to avoid repeating all over Australia the mistakes done over the last 200 years. In partnership with engineering and technology, as opposed to mutual exclusion at present, science has key role to play to help develop this approach to ensure a sustainable Australia.

One of us (EW) is a father and grandfather of young Australians and views Australian estuaries as a bank account that we hold in trust for future generations. If Australia does not learn from past mistakes that degraded our estuaries, Australia fails a critical test as a society and we will leave our children and grandchildren a serious debt to pay.

References

Balch PA, Balch JF (2000) Prescription for nutritional healing, 3rd edn. Avery, New York, 784 pp

Beach D (2002) Coastal sprawl: the effects of urban design on aquatic ecosystems in the U.S. Pew Ocean Commission, Arlington

BOM (Bureau of Meteorology) (2012) http://bom.gov.au

Brearley A (2005) Ernest Hodgkin's Swanland: estuaries and coastal lagoons of south-western Australia. University of Western Australia Press, Perth, 550 pp

Brodie J, Waterhouse J (2012) A critical review of environmental management of the 'not so Great' Barrier Reef. Estuarine Coastal Shelf Sci 104–105:1–22

CSIRO (2012) http://www.csiro.au/Outcomes/Climate/Understanding/State-of-the-Climate-2012.aspx

da Silva T, Ridd M, Klumpp D, Ridd P (2004) Relative contribution of food and water to the Cd burden in *Balanus amphitrite* in an urban tidal creek discharging into the Great Barrier Reef lagoon. Estuarine Coastal Shelf Sci 60:313–324

Daniels R (1990) Coming to America: a history of immigration and ethnicity in American life. Harper Collins, Princeton, 576 pp

Davis D (2002) When smoke ran like water: tales of environmental deception and the battle against pollution. Basic Books, New York, 316 pp

Davis JR, Kloop K (2006) Eutrophication in Australian rivers, reservoirs and estuaries – a southern hemisphere perspective on the science and its implications. Hydrobiologia 559:23–76

De'ath G, Lough JM, Fabricius KE (2009) Declining coral calcification on the Great Barrier Reef. Science 323:116–119

De'ath G, Fabricius KE, Sweatman H, Puotinen M (2012) The 27–year decline of coral cover on the Great Barrier Reef and its causes. PNAS 109:17995–17999

Doherty GB, Brunskill GJ, Ridd MJ (2000a) Natural and enhanced concentrations of trace metals in sediments of Cleveland Bay, Great Barrier Reef Lagoon, Australia. Mar Pollut Bull 41:337–344

Doherty GB, Coomans D, Brunskill GJ (2000b) Modelling natural and enhanced trace metal concentrations in sediments of Cleveland Bay. Aust Mar Freshw Res 51:739–747

Duarte CM, Conley DJ, Carstensen J, Sánchez-Camacho M (2009) Return to *Neverland*: shifting baselines affect eutrophication restoration targets. Estuar Coast 32:29–36

Elmore L (2011) http://www.crikey.com.au/2011/09/22/bureaucrats-calm-in-face-of-possible-6000-turtle-1500-dugong-deaths/

EPA (1998) Lead management in NSW – an application of multicriteria analysis. NSW Environment Protection Authority, Sydney

Etemad S, Imberger J (2005) Vertical eddy diffusivity estimations in Swan River Estuary. Dyn Atmos Ocean 39:175–187

Flannery T (1994) The future eaters: an ecological history of the Australasian lands and people. Grove Press, Sydney, 432 pp

Grandjean P (2010) Even low-dose lead exposure is hazardous. Lancet 376:855–856

Hamilton DP, Chan T, Robb MS, Pattiaratchi CB, Herzfeld M (2001) The hydrology of the upper Swan River Estuary with focus on an artificial destratification trial. Hydrol Processes 15:2465–2480

Hussey K, Dovers S (2006) Trajectories in Australian water policy. J Contemp Water Res Educ 135:36–50

Inglis GJ, Kross JE (2000) Evidence for systemic changes in the benthic fauna of tropical estuaries as a result of urbanization. Mar Pollut Bull 41:367–376

John J (2013) A diatom prediction model and classification for urban streams from Perth, Western Australia. Koeltz Scientific Publishers, Koenigstein, 166 pp

Kristiana RR, Antenucci JP, Imberger J (2012) Using a multi-component indicator toward reducing phytoplankton bloom occurrences in the Swan River estuary. Environ Manag 50:237–56

Lambrechts J, Humphrey C, McKinna L, Gourge O, Fabricius K, Mehta A, Wolanski E (2010) The importance of wave-induced bed fluidisation in the fine sediment budget of Cleveland Bay, Great Barrier Reef. Estuarine Coastal Shelf Sci 89:154–162

Li Y, Cai W, Campbell EP (2005) Statistical modeling of extreme rainfall in southwest Western Australia. J Clim 18:852–863

MacDonald C (2010) Tackling dust from coal and iron ore trains. Australian Bulk Handling Review. http://www.bulkhandling.com.au/news/print-editions/january-february-2010/tackling-dust-from-coal-and-iron-ore-trains

Mee L (2012) Between the Devil and the Deep Blue Sea: the coastal zone in an era of globalisation. Estuarine Coastal Shelf Sci 96:1–8

Michaels D (2005) Doubt is their product. Sci Am 292:96–101

Murray–Darling Basin Ministerial Council (2007) http://www.mdba.gov.au/

NLWRA (National Land and Water Resources Audit) (2000) http://lwa.gov.au/files/products/national-land-and-water-resources-audit/pn20386/pn20386.pdf

Northern Territory Government (2012) http://www.nretas.nt.gov.au/__data/assets/pdf_file/0008/20420/assessmentreport34b.pdf

Peters NE, Donohue R (1999) Nutrient concentration and fluxes in tributaries to the Swan-Canning estuary, Western Australia. in 'Impact of land-use change on nutrient loads from diffuse sources'. IAHS Publication No. 157, pp 87–94

Petheram C, Bristow KL, Nelson PN (2008) Understanding and managing groundwater and salinity in a tropical conjunctive water use irrigation district. Agric Water Manag 95:1167–1179

Queensland Government (2007) http://www.health.qld.gov.au/ph/documents/caphs/cll_summary_report.pdf

Queensland Government (2010) http://www.ehp.qld.gov.au/air/gladstone/pdf/final-health-risk-assessment-report.pdf

Queensland Government (2013) http://statements.qld.gov.au/Statement/2013/5/28/fitzroy-mine-water-strategy

Rasheed M, McKenna S, Tol S (2013) Seagrass habitat of cairns harbour and trinity inlet. https://research.jcu.edu.au/research/tropwater/resources/13%2017%20Cairns%20Monitoring%20and%20Baseline%20report%202012.pdf

Stephens R, Imberger J (1996) Dynamics of the Swan River Estuary; the seasonal variability. Mar Freshw Res 47:517–529

Swan River Trust (2012) http://www.swanrivertrust.wa.gov.au/home

Wildsmith MD, Rose TH, Potter IC, Warwick RM, Clarke KR, Valesini FJ (2009) Changes in the benthic macroinvertebrate fauna of a large microtidal estuary following extreme modifications aimed at reducing eutrophication. Mar Pollut Bull 58:1250–1262

Wolanski E (2006) The environment in Asia pacific harbours. Springer, Dordrecht, 497 pp

Wolanski E (2007) Estuarine ecohydrology. Elsevier, Amsterdam, 157 pp

Wolanski E (2011) The possible threat of lead pollution from atmospheric deposition in Townsville. EnviroTox Conference, Darwin, 17–20 Apr 2011

Part I

Estuaries that Bore the Full Pressure of the Historical Developments

Sydney Estuary, Australia: Geology, Anthropogenic Development and Hydrodynamic Processes/Attributes

Serena B. Lee and Gavin F. Birch

Abstract

The Sydney Estuary is the focal point of the intensely developed city of Sydney. Since European settlement in 1788 the waterway has undergone many changes, including reclamation, contamination, modified fresh-water flow regimes and altered rates of sedimentation. The various alterations and their impact on the system is the focus of this chapter. Research undertaken over the past thirty years identified the threat of contamination on estuary health. This issue came to a head in 2006 with the closure of the Sydney commercial fin fish and prawn industries due to high concentrations of dioxins detected in fish and prawn tissue. Improved understanding of the impact of different chemicals on estuarine species has led to changes in policy and practices within the waterway and adjacent catchment. Despite better practices contaminants continue to be supplied to the estuary via the complex stormwater network draining the surrounding highly urbanised catchment. Stormwater runoff represents the major contemporary source of estuary contamination. Recent field and numerical investigations show that in order to reduce contaminant concentrations stormwater runoff must treated before being discharged into the waterway. Due to the hydrodynamic behaviour of this geometrically complex waterway rather than rapidly flushing out of the estuary to the open ocean contaminants supplied via stormwater runoff become entrained down the water column and settle on the estuary bed. Whilst many improvements have been made to address processes affecting estuary health, continued monitoring of contaminant concentrations within estuary waters, bed sediments and species are required to determine the success of past management strategies and to better inform decisions about the future management of this highly prized waterway.

Keywords

Estuary • Sediment • Toxicity • Urbanisation • Reclamation • Stratification

S.B. Lee (✉) • G.F. Birch
Environmental Geology Group, School of Geoscience, University of Sydney, Sydney, Australia
e-mail: serena.lee@usyd.edu.au; gavin.birch@sydney.edu.au

E. Wolanski (ed.), *Estuaries of Australia in 2050 and Beyond*, Estuaries of the World,
DOI 10.1007/978-94-007-7019-5_2,

Box 1

Serena Lee and Gavin Birch studied Sydney Estuary, an urban estuary dissecting Sydney. Since European settlement in 1788 the waterway has undergone many changes, including reclamation, contamination, modified fresh-water flow regimes and altered rates of sedimentation, that now threaten the estuary health. In 2006 Sydney commercial fin fish and prawn industries were closed due to high concentrations of dioxins detected in fish and prawn tissue. Improved understanding of the impact of different chemicals on estuarine species has led to changes in policy and practices within the waterway and adjacent catchment. Stormwater runoff represents the major contemporary source of estuary contamination. Recent field and numerical investigations show that stormwater runoff must treated before being discharged into the waterway because contaminants supplied via stormwater runoff are not flushed to the sea, but instead they are entrained down the water column and settle on the estuary bed.

Whilst many improvements have been made to address processes affecting estuary health, continued monitoring of contaminant concentrations within estuary waters, bed sediments and species are required to determine the success of management strategies and to better inform decisions about the future management of this highly prized waterway.

Introduction

Much of the beauty of Sydney can be attributed to four deeply-incised estuaries, which dissect the raised coastal margin of the region. These waterways have provided an extensive shoreline and have brought marine conditions deep into the catchment. These attributes have made Sydney one of the most beautiful cities in the world. This chapter describes one of these waterways – Sydney Estuary.

Physical Description

The Sydney Estuary, classified as a ria, is approximately 30 km in length, ranging in width from approximately 60 m near the headwaters to approximately 3 km approaching the estuary mouth. The total surface area of the water body is approximately 50 km^2 and surrounding catchment is close to 500 km^2. Parramatta River, Duck River, Lane Cove River and Middle Harbour Creek are the four principal estuary tributaries, however additional creeks and canals flow into numerous off-channel embayments, which join the main channel along the length of the waterway (Fig. 1).

The estuary has a convoluted shape owing to the inherited dendritic drainage pattern. Bathymetry is irregular, ranging from less than 3 m within shallow off-channel embayments to over 40 m within deep water holes along the main channel of the middle to lower estuary. The irregular bathymetry, presence of islands along the main channel and geometrically intricate shoreline all contribute to the complex three dimensional current patterns within the water body.

Sydney Estuary has a microtidal regime with a maximum tidal range of 2.1 m. Tides are semidiurnal with a diurnal inequity, and maximum tidal lag from the mouth to the headwaters is approximately 10 min. During quiescent weather conditions, tides are the dominant process affecting hydrodynamic behaviour, with the influence of ocean swell restricted to the lower estuary. While residence times within the lower estuary are in the order of days, it is estimated that complete tidal flushing for embayments in the upper estuary may take up to 7 months (Das 2000).

Generally, the water body is well-mixed, with salinity ranging from ~27 PSU at the headwaters to ~35 PSU at the mouth, due to low fluvial supply from the catchment and tidal forcing (Irvine 1980). During quiescent weather conditions when rainfall is low (<5 mm day^{-1}), fresh water flow exerts little influence on hydrodynamic behaviour. Following high precipitation intermittent stratification develops due to the rapid influx of stormwater runoff from the urbanised catchment. The resulting high freshwater runoff volumes act to increase surface water current speeds in the down-estuary direction and increase near-bed current speeds in the up-estuary direction due to gravitational circulation (Lee and Birch 2012).

Winds within Sydney Estuary are complex, influenced by sea breezes, synoptic flow and the larger Sydney basin drainage flow superimposed onto intricate terrain (Spark and Connor 2004). The surrounding topography generates local wind patterns within the estuary, which are capable of

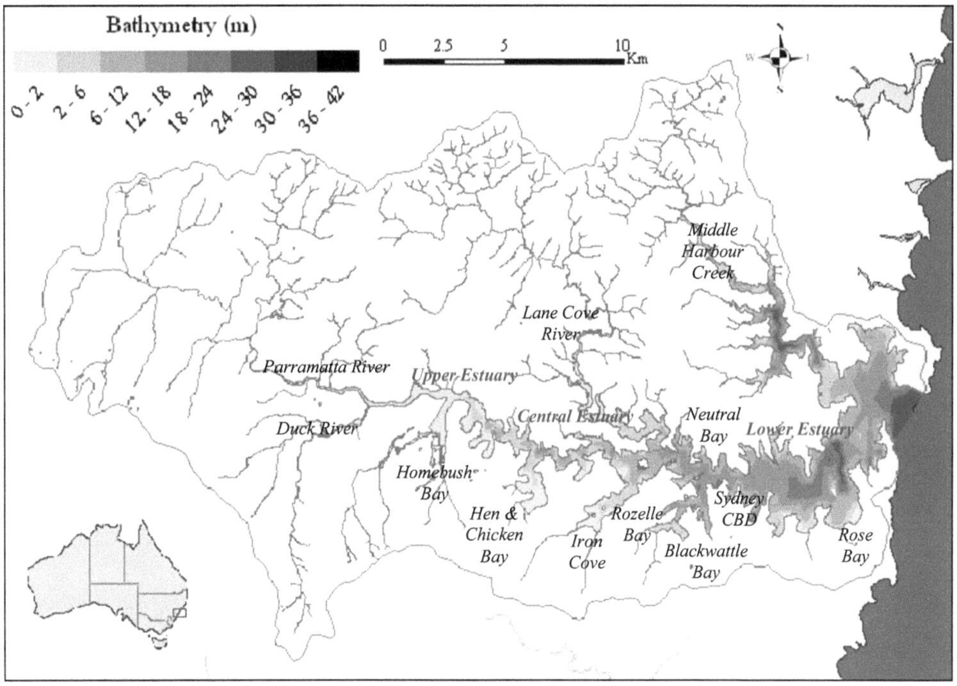

Fig. 1 Site map displaying the extent of the Sydney Estuary, tributaries catchment and bathymetry

generating surface wind waves of up to 0.5 m during periods of prolonged high wind. Wind also plays a role in estuarine circulation patterns (Lee and Birch 2012) and wind waves are largely responsible for resuspension events within shallow embayments along the southern shoreline (Taylor 2000; Birch and O'Hea 2007).

Geological History

Sydney estuary is located in the Permian to Triassic age (300–220 Ma) Sydney Basin. Most of the high land in the catchment is comprised of Ashfield Shale, which overlies the Hawkesbury Sandstone. Sydney Estuary is a large drowned river valley, which has cut up to 85 m into Hawkesbury Sandstone.

The ancient river, which is now Sydney estuary, was eroded down into an elevated coastal plain forming steep-sided banks and the old, mature river that meandered across a flat plain 80 Ma ago. This river was once considerably larger than it is today, but was 'captured' by the Hawkesbury River leaving it considerably smaller. During interglacial periods, sea level rose and the 'river' was flooded to form an estuary.

During the Quaternary period, sea level oscillated from 5 m above to 120 m below the present day position every approximately 100,000–150,000 years due to global climate change. However, for the majority of the last 135,000 years sea level was 20–70 m below the present and erosion was more pronounced than deposition during this period. The last glacial period ended about 17,000 years ago and sea level started to rise from a position about 30 km east of its present location. By 8,000 years ago, sea level stood at 5 m below present and the sea reached its present position about 6,000 years ago. Sand swept ahead of the advancing sea was pushed into embayments to form spits behind which lagoons and estuaries formed. Some of this sand was transported into the mouth of Sydney estuary forming a tidal delta, while a fluvial delta was deposited in the upper parts of the estuary. The central basin became mantled in fine-grained sediment carried in suspension after floods.

Anthropogenic Modification of the Estuary

Reclamation

Sydney estuary has been extensively modified by reclamation since settlement, especially in the upper regions. Reclamation and infilling of intertidal areas has reduced the shoreline by approximately 77 km of the 322 km of original length (Pitblado 1978) and 13.4 km^2 (23 %) of the total 50 km^2 area of the estuary has been lost (Murray 2003; Birch et al. 2009). Approximately 9×10^6 m^3 of water has been lost on each tidal cycle through reclamation resulting in poor water quality, sedimentation and loss of habitat, e.g mud flats, and mangroves and salt marsh.

The head of Sydney Cove was the first to be reclaimed and remodelled into a semi-circular sandstone quay with the Tank Stream channelled in the period 1835–1854 and Mort and Walsh Bays were infilled (Fig. 1). Settlement began to spread out from the city with the introduction

Fig. 2 (*Top left*) Photograph displaying the urbanised nature of the Sydney Estuary. (*Top right*) Aerial image of the central/lower Sydney Estuary. Examples of estuarine areas reclaimed for parks or other land uses are indicated at the bayends of Rozelle, Blackwattle Bay and Darling Harbour. This image exemplifies extensively modified estuary shorelines approaching the city centre. (*Bottom left*) Gross pollutant stormwater trap device installed along the Duck River. (*Bottom right*) Remediation works, Homebush Bay. Following remediation formerly industrial sites were converted to residential land uses

of trams and trains during the period 1854–1889 resulting in reclamation of Blackwattle Bay, Pyrmont Bay, Darling Harbour, Woolloomooloo Bay and Rushcutters Bay (Fig. 2) (Stephensen 1966; Shore 1981). Reclamation was undertaken using domestic waste, sewerage, offal and dead animals in the second half of the eighteenth century which lead to foul odours and the fear of disease (Solling and Reynolds 1977).

A major contributing factor to rat infestation and outbreak of bubonic plague in 1898/99 was considered to be due to the continued widespread disposal of garbage in intertidal swamps (Coward 1988). During the period 1889–1922 dilapidated and rat infested foreshores in Walsh Bay and Darling Harbour were replaced by Sydney Harbour Trust (SHT) under an Act of Parliament in 1900. At the same time extensive reclamation took place further afield, i.e. in Canada Bay, Kings Bay, Hen and Chicken Bay, Iron Cove, White Bay, Rozelle Bay, and Rose Bay. Most reclamation took place in Sydney estuary when the foreshore was extended at Silverwater, Homebush Bay, Garden Island, Exile Bay, Kings Bay, Iron Cove, Glebe Island and Darling Harbour between 1922 and 1955 to create 5.7 km^2 of new land.

Environmental Impacts of Reclamation

Approximately 100×10^6 tonnes of garbage, industrial waste and contaminated estuarine sediments were used to undertake 11.35 km^2 of reclamation in Sydney Estuary. Not much is known about the composition of this material, except at Homebush Bay, which was remediated in association with the Sydney 2000 Olympic Games (Suh et al. 2003a, b, 2004a, b). A total of $137 million was allocated for clean-up of the site in one of the largest remediation projects carried out in Australia. Here waste comprised putrescible, building, chemical and garbage municipal waste, construction debris, household garbage, demolition waste, ash fill and dredged sediment containing heavy metals, asbestos, a range of hydrocarbons, including dioxins benzene, toluene, ethylbenzene and xylene (BTEX) compounds and polycyclic aromatic hydrocarbons, as well as organochlorine pesticides. A total clean-up of 400 tonnes of hazardous waste classified as Scheduled Chemical Waste, which had to be destroyed by a thermal/catalytic treatment under NSW EPA license. Estuarine sediment from the adjacent bay used as infill material contained elevated concentrations of metals, which polluted groundwater. Leachate produced in reclaimed lands due to rainwater filtration and tidal action was studied by Suh et al. (2003a, b) at Bicentennial Park adjacent to Rozelle Bay. Results showed that during dry periods when water tables recede, oxygen ingress led to decreasing acidity (pH) and increasing metal (copper, lead, zinc, arsenic and chrome) concentrations and that metals enter the estuary by tidal action and during periods of rainfall. High metal concentrations in sediments at the heads of most estuary embayments are juxtaposed adjacent reclaimed lands. The total mass of metals associated with reclaimed land is unknown, however 1Mt of this estuarine material was used for this purpose in Iron Cove; 4.6 Mt in Homebush Bay and 2.8 Mt on the banks of the Parramatta River (McLoughlin 2000), which gives an idea of the potential of this source. However, stormwater canals also discharge to the estuary at these locations (Barry et al. 1999, 2000; Birch et al. 1999) and differentiating the relative magnitude of each source has not been attempted.

Ecological Effects of Reclamation

Approximately 50 % of the shore of Sydney Estuary is composed of retaining seawalls or other built structures (Chapman and Bulleri 2003). Construction of seawalls alters intertidal habitat by reducing the intertidal area and producing fewer crevices and overhangs compared to natural rocky shores and some rock pools and other habitats are absent. The majority of seawalls in Sydney estuary have been constructed to support reclamation activities at the heads of embayments (Fig. 2). The change from muddy, mangrove and saltmarsh wetlands with gentle slopes to vertical seawalls has resulted in major alterations to ecological function and biological productivity, as well as changes in hydrology and physio-chemical attributes of the estuary.

Effect of Urbanization and Industrialization on Sydney Estuary

Studies of historic changes in land use in the Sydney Estuary catchment and potential adverse effects on estuary condition have been undertaken by the School of Geosciences at Sydney University over a considerable period (Birch 2000; Murray 2003; Jolley 2005; Taylor et al. 2004; Birch et al. 2007, 2013; Townsend 2011; Lee et al. 2011; Lee and Birch 2012; Lee 2012). Sediment cores were used to determine effects on the estuary and historical maps and charts provided information on changes to land use. Hydrological modelling of catchment attributes (landuse, rainfall, runoff coefficients) provided metal loading to the estuary via stormwater discharge for various time slices. Twelve sedimentary cores taken in nine highly contaminated embayments were analysed for metals and dated using radioisotopes lead 210 and caesium 137 (Taylor et al. 2004).

Sediment in Blackwattle Bay, Iron Cove and Homebush Bay are highly contaminated by metals and these bays are located 2, 5 and 12 km, respectively from central Sydney. Metal concentrations (copper, lead and zinc) at the bottom of cores from these bays show low values and at some depth concentrations increase markedly. The depth at which metal concentrations begin to increase is the point or time where contaminants start to be discharged from the catchment, e.g. the onset of contamination. Contamination commenced in about 1860 in Blackwattle Bay, in approximately 1910 in Iron Cove and at about 1925 in Homebush. This spatio-temporal distribution reflects the spread of urbanization and development of industry outwards from central Sydney. Similar down-hole trends are evident for organic contaminants (organochlorine compounds) which are entirely man-made and introduced into Australia after the Second World War (Taylor et al. 2004). The onset of contamination by these chemicals was dated at 1945 in these cores, i.e. the same time as they were introduced into the catchment.

Data from cores and surficial sediments indicate that metal concentrations in surficial sediment have been declining in the upper estuary and increased in the lower harbour, in Lane Cove and in Middle Harbour over the last 25 years (Birch and Taylor 2004). Introduction of the Clean Waters Act in 1978, which required industry to discharge waste into the sewerage system and to reduce waste, contributed to reduction of metal supply to the harbour. Relocating industry away from the water front has also reduced contaminants discharged directly to the estuary. Increased concentration of metals in sediments of the lower estuary is due to rapid expansion of residential and commercial property and increased transport services (Birch and Taylor 2004).

Sediment Quality and Toxicity

Sediments mantling the Sydney Estuary contain high concentrations of a wide range of contaminants, including metals (Irvine and Birch 1998; Birch and Taylor 1999) organochlorine pesticides (Birch and Taylor 2000) polycyclic aromatic hydrocarbons (PAHs), polychlorinated biphenyls

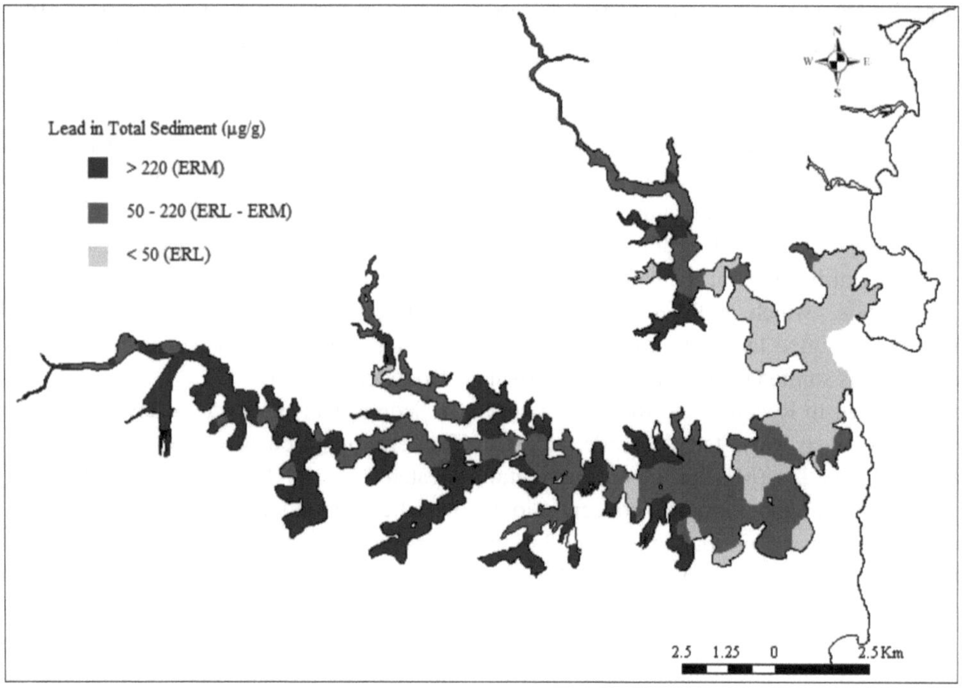

Fig. 3 Distribution of Pb in total surficial sediment (Adapted from Birch et al. 2008)

(PCBs) (McCready et al. 2000), and polychlorinated dibenzo-*p*-dioxins and dibenzofurans (Birch et al. 2006). The major source of these chemicals is related to historic industrial activities, and past and present stormwater discharge with minor leachates from reclaimed areas. High concentrations of contaminants, mainly metals, discharged to the receiving basin indicate continued supply of stormwater-related anthropogenic materials.

The quality of sediment mantling Sydney Estuary was determined in a three-tiered approach (Birch et al. 2008). The chemistry of surficial sediment was initially established through detailed sampling and analysis of >1,000 samples for a suite of metals (Cd, Cr, Cu, Co, Fe, Mn, Ni, Pb and Zn), and a smaller number of samples (n = 140) analysed for organochlorine pesticides (OCs) (DDT, DDD, DDE, chlordane, aldrin, heptachlor, dieldrin, heptachlor epoxide, lindane), hexachlorobenzene (HCB) and total polychlorinated biphenyls (PCBs, reported as Aroclors), whereas 16 priority pollutant polycyclic aromatic hydrocarbons (PAHs) (acenaphthene, acenaphthylene, anthracene, benz(a)anthracene, benzo[*a*]pyrene, benzo[*b*+*k*]fluoranthene, benzo[*ghi*] perylene, chrysene, dibenz[*ah*]anthracene, fluoranthene, fluorene, indeno[*1,2,3-cd*]pyrene, napthalene, phenanthrene and pyrene), as well as 2-methylnaphthalene were determined on 124 samples (Birch and Taylor 1999, 2000; McCready et al. 2000).

Studies in the first tier of assessment showed sediments of the upper estuary, landward of the Sydney Harbour Bridge, contain some of the highest reported concentrations of a wide range of contaminants. Contaminant concentrations increased markedly in the upper parts of most embayments and in the western tributaries of Middle Harbour close to major stormwater inputs (Fig. 3 for lead). Contaminant concentrations were elevated due to proximity to source, a mainly muddy substrate and poor flushing by tides and currents in these areas. Discharge from industries located on the shores of the estuary resulted in contaminant 'hot spots'.

Sediment quality guidelines (Long et al. 1995; ANZECC/ ARMCANZ 2000; Simpson et al. 2005) were used to assess the probability of toxicity in the second tier of investigation. These guidelines consist of two concentrations, namely the lower level (Effects Range Low, or ERL), which denotes the concentration below which adverse biological effects are seldom observed and the Effects Range Median (ERM), which distinguishes concentrations above which adverse biological effects are expected to occur frequently. Concentrations between ERL and ERM guidelines indicate intermediate irregular biological response. Chemicals were assessed for possible adverse biological effects by comparing concentrations at each site to the respective ERL and ERM values. Australia has adopted similar guideline values named Interim Sediment Quality Guidelines–Low and –High (ISQG-L and -H), respectively.

Areas exceeding SQGs were determined for single chemicals and contaminant mixtures in the second tier of assessment (Birch and Taylor 2002a, b, c). Sediment in approximately 2, 50, and 36 % of the estuary exceeded the high SQG value (ERM or ISQG-H) for Cu, Pb and Zn,

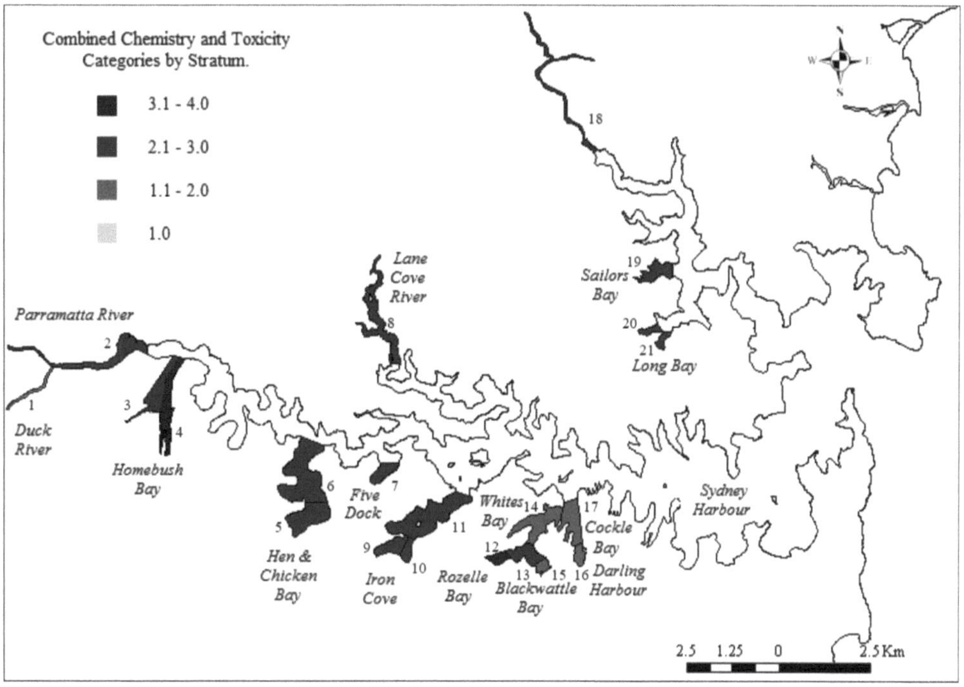

Fig. 4 Combined chemistry and toxicity categories by stratum based on mean chemistry/toxicity scores. Stratums are numbered from 1 to 21 (Adapted from Birch et al. 2008)

respectively. Estuarine sediment exceeded ERL (or ISQG-L) concentration for at least one metal in all but one small area near the entrance. Organochlorine compounds exceeded ERM concentrations over most of Sydney Estuary, however PCB concentrations were above the ERM value in only a small part of the waterway (Birch and Taylor 2002a, b, c). Sediments in almost all upper and middle parts of Sydney Estuary and Middle Harbour had at least one OC or PAH concentration which exceeded ERM values.

Results of the sediment quality study were used to compartmentalize the estuary for detailed chemical/ecotoxicological studies in the third and final tier of investigation. The five bays exhibiting high contaminant concentrations were divided into 12 strata and 7 strata were selected from areas with intermediate contaminant concentrations. An additional two strata were chosen from areas with low contaminant concentrations to determine sediment toxicity in the least impacted locations. Samples were randomly collected within each stratum.

Surficial sediments ($n = 65$) were analysed for 12 metals, 21 OCs, 24 PAHs, 7 Aroclor mixtures of PCBs and total organic carbon (TOC). Samples were tested in a 10-day, whole-sediment amphipod survival test (*Corophium colo*) and a pore water sea urchin (*Heliocidaris tuberculata*) fertilisation test to assess both contaminant uptake routes, i.e. via the gills for the dissolved phase and by ingestion for the solid phase. Pore water were also tested in sea urchin larval development ($n = 61$) and microbial bioluminescence Microtox© ($n = 57$) tests (McCready et al. 2004, 2005, 2006a, b, c; Spyrakis 2002).

Amphipod survival, Microtox© pore water, sea urchin larval development and sea urchin fertilisation tests resulted in 17 %, 98 %, 59 %, and 98 % of the samples in Sydney estuary being toxic ($p < 0.05$) relative to negative controls, respectively. All samples were toxic in at least one test. In the sea urchin fertilisation test, 2 %, 31 %, 18 % and 49 % of samples were non-toxic, slightly toxic, moderately toxic and highly toxic, respectively. Sediments in Parramatta River (Stratum 2), the southern embayments of the central estuary (Strata 3, 4, 5, 6 and 7) and Long Bay (Strata 20 and 21) were highly toxic in the sea urchin fertilisation test, whereas sediments in Blackwattle Bay (Stratum 15), Iron Cove east (Stratum 10) and Upper Middle Harbour (Stratum 18) were moderately toxic (Fig. 3). Sediment in the remaining strata (Strata 3, 9, 11, 14, 16 and 17) were toxic. Only two sediment samples in Homebush Bay east and Hen and Chicken Bay were highly toxic in the amphipod survival test. Sediments in Five Dock Bay (Stratum 7), Homebush Bay east (Stratum 4) and Rozelle Bay (Stratum 12) were moderately toxic in this test. Microtox© test of pore water was less discriminative than the other three tests used in the investigation. Sediment containing high metal concentrations (Iron Cove, Rozelle Bay and Five Dock Bay) showed high toxicity.

The areas assigned highly toxic (score 3.1–4.0, Fig. 4), moderately toxic (score 2.1–3.0, Fig. 4) and slightly toxic (score 1.1–2.0, Fig. 4) comprised 17 %, 52 % and 31 % of the estuary investigated, respectively. These results were similar to the sediment quality assessment in phase 2, which subdivided the area into high, medium-high and medium-low priority classes for 15 %, 54 % and 31 % of the same area, respectively. Although only 16 % of the estuary was investigated, it was estimated that 2.7 %, 8.3 % and 5 % of the total area of the estuary was highly toxic, moderately toxic and slightly toxic, respectively. Of 25 estuaries surveyed in a similar way in North America, 7 % of the total area of the estuaries were found to be toxic (Long 2000), i.e. similar to the 11 % proportion of Sydney Estuary, which was found to be toxic plus moderately toxic by Birch et al. (2008).

Stormwater

Annualised Loading

The average annual discharge of stormwater from the Sydney Estuary catchment was predicted using the Model for Urban Stormwater Improvement Conceptualisation (MUSIC) to be 215,300 ML and average annual loadings was As, Cd, Cr, Cu, Ni, Pb, and Zn were 0.8, 0.5, 1.7, 3.2, 1.1, 3.6 and 17.7 tonnes, respectively (Birch and Rochford 2010). The proportion of metals discharged under low- (<5 mm day^{-1} rainfall), medium- (>5 <50 mm day^{-1} rainfall), and high-flow conditions (>50 mm rainfall/day) was predicted to be approximately 10 %, 60 % and 30 %, respectively. Metal loading characteristics were determined to assist in the development of future, second-generation remediation technologies and science-based strategies. High metal concentrations in fluvial particulates and estuarine sediments adjacent to stormwater discharge points suggest creeks entering the upper (Duck and Parramatta Rivers) and central Sydney Estuary (Homebush and Hen and Chicken and Neutral Bays and Iron Cove) and rivers discharging to the western shores of Middle Harbour (Long and Sugarloaf Bays) be prioritised for remediation. Metals discharged under low-flow conditions (~10 % of total load) are trapped in adjacent embayments, but may be effectively remediated, however, metals associated with medium-flow events present a major future challenge for remediation.

High Precipitation Events

Typically, Sydney Estuary waters are predominantly saline since fresh-water supply is extremely low, reflecting the erratic rainfall regime consisting of long dry periods punctuated by short duration high-precipitation events. Based upon rainfall records from Sydney Observatory from 1858 to 2012, average daily rainfall in the Sydney region is less than 4 mm, with zero rainfall 65 % of the time and rainfall in excess of 50 mm day^{-1} 1 % of the time. The typically low rainfall is reflected by low fresh-water runoff from the Sydney Estuary catchment. This runoff regime alters when the extended dry spells are broken by intense short period high-rainfall events. When high-precipitation events occur, rainfall is rapidly transported to the estuary due to the small, highly-urbanised catchment and efficient stormwater drainage system. When rainfall ceases it takes less than a day for fresh-water runoff rates to return to baseflow conditions. These short-lived high-runoff events cause intermittent stratification, which is most defined in the upper estuary (Fig. 5) (Wolanski 1977; Lee et al. 2011; Lee and Birch 2012; Lee 2012). The largest subcatchments of this system drain to the narrowest sections of the estuary, while the smaller subcatchments drain to the wider embayments. As a consequence, following high rainfall, freshwater dominates the water column within the Lane Cove, Duck and Parramatta River upper reaches with only a small component near the bed remaining brackish/saline. Within embayments, saline waters dominate the water column with a thin layer of freshwater forming on top of marine waters. A fresh-water plume develops along the main estuary channel from the confluence of the Parramatta and Duck Rivers (Figs. 5 and 6). A second fresh-water plume also develops along the Lane Cove River reaching the main channel in the central estuary (Figs. 5 and 6). While stratification may be well-defined in the upper estuary where surface water salinities can be lower than 5 PSU, typically the surface-water plume becomes brackish (~10–25 PSU) in the central estuary with salinities only slightly reduced (>28 PSU) in the lower estuary (Fig. 6). Rather than rapidly escaping the estuary via the mouth in a distinct lower-density surface plume, stormwater mixes with saline waters and is largely retained within the estuary until flushed due to tidal forces (Lee and Birch 2012). Since fresh-water runoff volumes are not sustained for more than one or two days, stratification begins to break down within days of rainfall ceasing. The time taken for stratification to break down depends upon the volume of freshwater discharged, the tidal regime and to a lesser extent wind conditions. Stratification is most defined when high catchment runoff coincides with neap tides and down-estuary winds. Spring tides facilitate mixing between saline and fresh waters as do high winds directed across or up the main channel (Lee and Birch 2012).

The time required for the system to return to quiescent salinity conditions varies between events with the shortest recovery times predicted when high runoff coincide with spring tides. High fluvial supply contributes to increasing gravitational circulation increasing down-estuary currents at the surface and increasing up-estuary currents near the bed. Elevated current velocities are experienced during spring

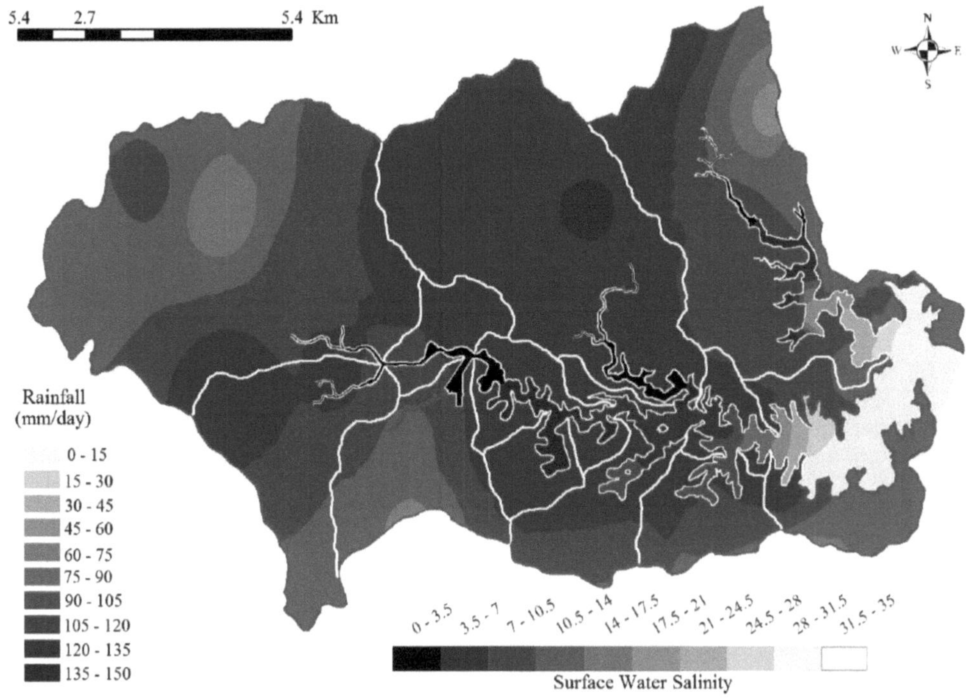

Fig. 5 Example of catchment rainfall and estuary response following a high-precipitation event in June 2007. Surface-water salinities reflect values 0.5 m below the water surface one day after the day of highest precipitation. The major subcatchments of the Sydney estuary are *highlighted* in *white*

tides enhancing mixing and flushing. Numerical modelling predictions show recovery times for a June 2007 high-precipitation event (displayed in Figs. 5 and 6) was 19 days, while recovery time for an August 2007 event (Fig. 6) was 43 days. Runoff volumes for these events were similar, however the June event coincided with a spring tidal cycle and the August event coincided with a neap tidal cycle. While quiescent salinities are attained between 3 and 8 weeks, Das (2000) estimated complete tidal flushing of the waterway may take up to 225 days.

During quiescent conditions, suspended sediment concentrations throughout the estuary are relatively low (0.5–40.5 mg/l) and the low salinity zone where most flocculation processes occur is not found in the main channel (Hatje et al. 2001). Following high rainfall, surface water suspended sediment concentrations (SSC) up to 130 mg/l have been observed in upper estuary sections of the main channel. The amount of sediment generated by the catchment varies between high-precipitation events and between subcatchments, consequently stormwater SSC are temporally and spatially variable (Lee and Birch 2012). The length of the antecedent dry period and rainfall intensity influences stormwater runoff SSC, with longer dry periods and higher rainfall acting to increase SSC. Unlike salinity, SSC return to quiescent conditions within a few days of rainfall ceasing. This is due to particle/floc settling since stormwater does not rapidly migrate beyond the estuary therefore the majority of sediments discharged into the waterway following high rainfall are retained within the system, settling on the estuary bed.

Sydney Estuary embayments are depositional environments. Terrigenous sediment discharged into embayments during low-flow conditions deposit close to stormwater outlets (Taylor 2000; Birch 2011). Sediments from adjacent subcatchments incrementally fill off-channel embayments, increasing bed thickness by between 0.6 and 2.7 cm/year (Taylor et al. 2004). Under high-flow conditions sand, silt and clay aggregates continue to settle to the bed within embayments, close to discharge locations, while individual clay particles remain suspended within the lower-density surface water plume, migrating with the plume beyond the low-flow regions of deposition. While the plume remains well-defined, clay particles are able to migrate beyond the embayments into the main estuary channel. In this way, sediment from subcatchments may be transported far from source, settling to the bed in adjacent embayments, in low-energy regions of the main channel or remaining suspended in the water column eventually flushing from the system.

Effects on Estuary Condition

Heavy metals, organochlorine compounds and polyaromatic hydrocarbons preferentially adhere to fine clay particles.

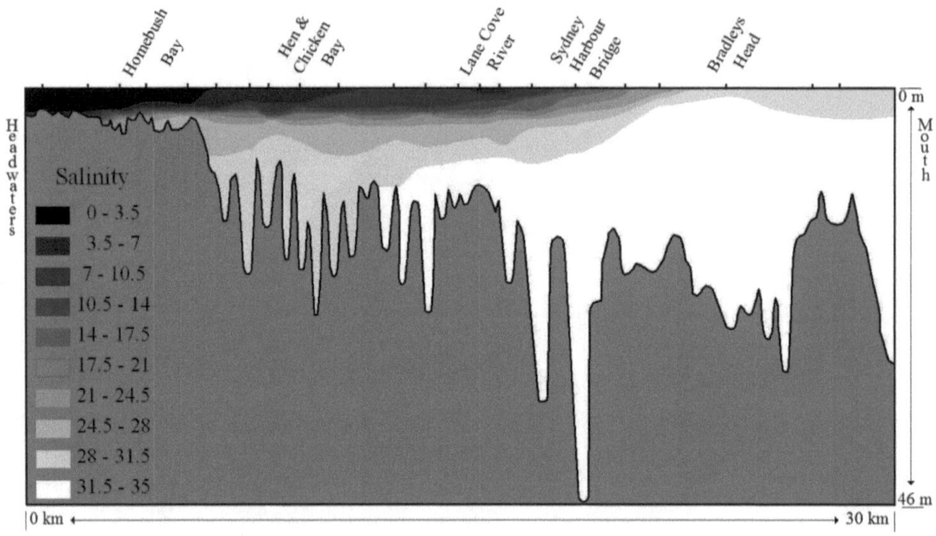

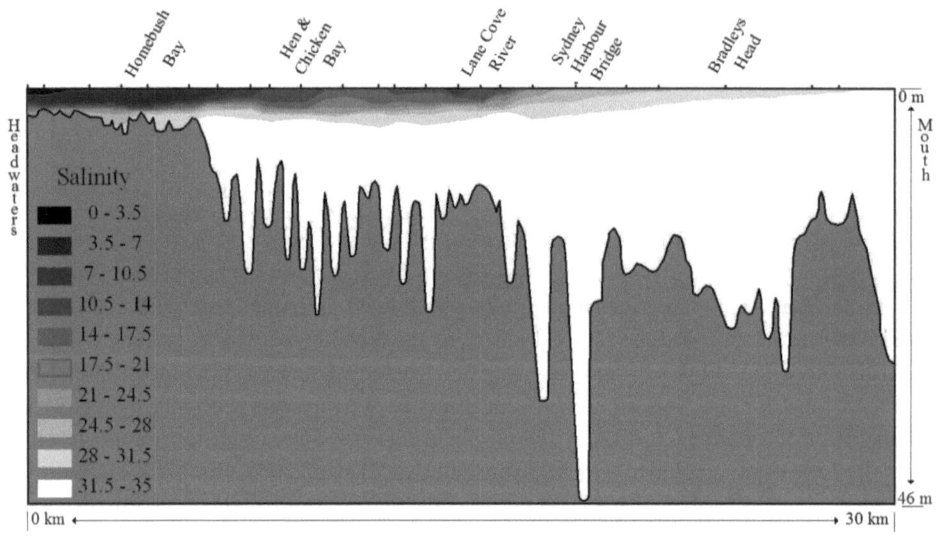

Fig. 6 Axial salinity distributions due to two separate high-precipitation events. The top image displays salinity distributions measured a day after high precipitation in June 2007, during a spring tidal cycle. The bottom image displays salinity distributions measured a day after high precipitation in August 2007, during a neap tidal cycle

As a consequence, high-precipitation events and the resulting fresh-water plumes provide a mechanism via which particulate contaminants may be transported beyond the embayments into which they typically deposit. Of the material escaping an embayment via the fresh-water plume, the majority deposit on the bed further down estuary, while a smaller component deposit in the up-estuary direction (Lee 2012). A minor component is flushed from the estuary altogether. This mechanism contributes to the further degradation of the estuary bed by moving contaminated sediment from urbanised catchments well-beyond ranges of deposition experienced under baseflow runoff conditions.

Management Issues

Management of Dioxins

Sydney Estuary acts as a nursery for numerous fish species and is home to a wide array of marine organisms. Formerly these waterways were commercially fished, however in 2007 these activities were halted due to high concentrations of dioxins measured in fish and prawn tissue. Congeners of dioxins and furans are known to be highly toxic to animals. Animal feeding studies show that 2,3,7,8-Tetrachlorodibenzo-p-dioxins are amongst the most toxic chemicals ever tested (Bopp et al. 1991), and both 2,3,7,8-TCDD and 2,3,7,8-TCDF are included in the list of persistent organic pollutants (POP) included in the Stockholm Convention 2001, of which Australia is a signatory. Signatories of this agreement are required to reduce and eventually eliminate all anthropogenic sourced POP (Ritter et al. 1995). The presence of 2,3,7,8-TCDD within estuary bed sediments poses a significant risk to the health of all species which inhabit the estuary or rely upon it as a food source, including humans.

The National Dioxin Study conducted by the Australian Government Department of Environment and Heritage in 2001 identified high concentrations of toxic dioxin and furan congeners within Sydney Estuary bed sediments (Birch et al. 2007). Sediments within Homebush Bay, an off-channel embayment in the upper estuary have the highest dioxin concentrations recorded in Australia, and amongst the highest reported in the world, second only to the Frierfjorden Estuary in Norway (Mueller et al. 2004; Bellucci et al. 2000). Homebush Bay was identified as the source of dioxin contamination of bed sediments, with contaminants sourced from this embayment detected a further 5 km up-estuary and 12 km down-estuary (Birch et al. 2007).

Contamination of Homebush Bay foreshore sediments occurred between 1928 and 1986 and subsequently led to contamination of bed sediment within the embayment. The region within Homebush Bay most affected by dioxin contamination is approximately 800 m long and 100 m wide, beginning near the eastern side of the embayment mouth encompassing the area directly adjacent to the reclaimed eastern foreshores (Fig. 7). Three processes are likely responsible for the migration of dioxins throughout the rest of the estuary:

1. Aerial deposition during the period when dioxins were produced as a by-product of combustion during pesticide production at Rhodes Peninsular.
2. Diffuse and point source stormwater runoff generated by periods of heavy rain may transport dioxin-laden sediment from the contaminated Homebush Bay foreshore areas into the embayment where they travel as part of the fresh-water plume on top of the marine water, escaping the embayment and moving down/up-estuary.
3. Spring tidal currents near the mouth of the Homebush Bay are sufficient to resuspend and mobilise newly deposited sediments, which may then escape into the main estuary channel during ebb tides. Once in the main channel successive flood and ebb tide currents may transport these contaminated sediments to other sections of the estuary.

In order to address dioxin contamination, Homebush Bay foreshore sediments and bed sediments within the section of the embayment containing the highest dioxin concentrations have been remediated (Fig. 2). This involved removal and

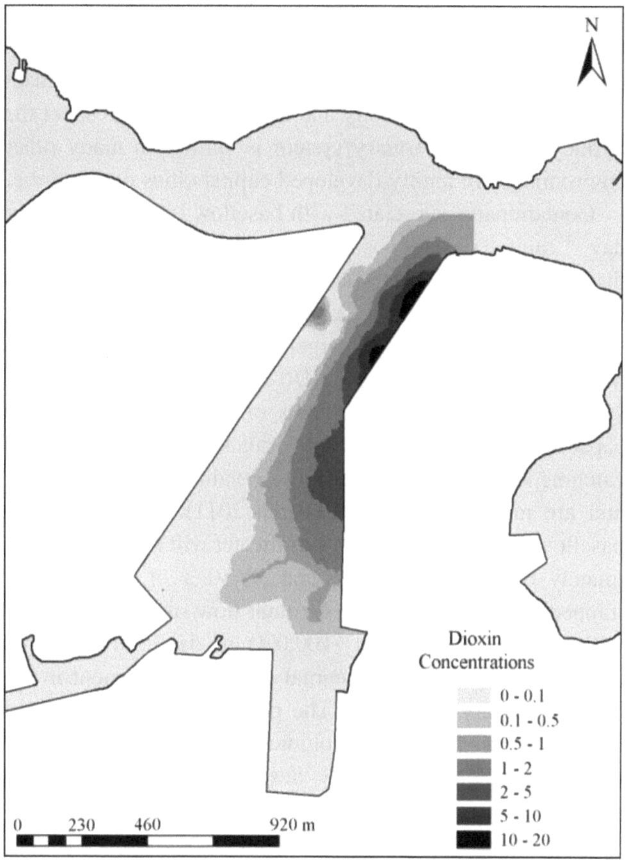

Fig. 7 Homebush Bay bed sediment dioxin concentrations before remediation commenced. 2,3,7,8 TCDD contours displayed are based on 1997 Surface Sediment Composites (μg/kg dry weight) (Adapted from EVS 1998)

treatment of contaminated foreshore and estuary bed sediment using thermal/catalytic treatment, a high temperature heating process which converts the organic contaminants to carbon dioxide and water. The area from which bed sediment was removed was refilled with rubble-sized Virgin Excavated Natural Material. Presently the effectiveness of the Homebush Bay remediation program is unknown. Long-term monitoring of dioxin concentrations within suspended sediments and fish and prawn tissue is required to determine if the strategy implemented has been able to reduce the uptake of dioxins in species inhabiting the estuary.

Stormwater Management

Increasing population, higher housing density and more transport infrastructure have enhanced the generation of contaminants in the Sydney Estuary catchment, as well as the supply of contaminants to the adjacent receiving basin. Increasing contaminant loading via stormwater to the adjoining estuary has reduced water and sediment quality over the past 100 years and stormwater remains a major source of concern. Retrofitting remedial devices in old, highly-urbanised catchments dominated by diffuse contaminant sources is long-term, costly and complex. In this respect the Sydney catchment-estuary system is similar to many other environments in highly-developed capital cities of the globe.

Contaminants associated with baseflow (none, or <5 mm day^{-1} rainfall) are deposited close to the mouths of canals discharging to the estuary and this sediment is often toxic to benthic animals, whereas chemicals carried by high flow (>50 mm day^{-1} rainfall) largely bypass embayments in a buoyant, freshwater plume. Off-channel sand infiltration systems located near the mouths of stormwater canals are capable of remediating very large quantities of contaminated catchment water during baseflow conditions inexpensively and are rapidly constructed (Birch 2011). Remediation of baseflow and some first-flush stormwater will remove approximately 10 % of total metals and 30–50 % of TN presently trapped in embayments. A continual flow of baseflow water will provide approximately 100,000 m^3 day^{-1} of harvested urban water runoff with minimal storage requirement in the Sydney Estuary catchment. The potential for recycling and harvesting of enormous volumes of non-environmental, treated stormwater is highly desirable in such intensely-urbanised environments to supply large recreational and sports facilities in the vicinity, especially with increased dry periods predicted by current climate models. The concept of converting contaminated waste water into a valuable resource, while cleaning up the environment, is highly desirable. The potential for recycling and harvesting enormous volumes of treated stormwater in intensely-urbanised environments is in keeping with the concept of 'cities as water supply catchments'.

Conclusions and Future Directions

Since implementing the Clean Waters Act, water quality within Sydney Estuary has markedly improved. Removal of industry from the estuary foreshore has further contributed to estuary health. Improved understanding of the impact of different anthropogenic chemicals on estuarine species has led to policy changes, including the restricted use of tributyl-tin, enabling species such as oysters to begin repopulating the intertidal zones of the estuary. Despite these improvements historical contamination and the continual supply of new contaminants via stormwater runoff present management challenges which must be addressed. Intense research is currently underway investigating the impact of different chemical contaminants on organisms within the estuary. The results of these studies may provide the impetus required to justify implementing an effective stormwater remediation strategy. Presently water sensitive urban design is a favoured management option particularly for new developments. This strategy is an effective management tool for small sites, however when considering stormwater management of old, intensively-urbanised subcatchments, this method it is too slow and too costly to implement on the scale required to achieve a marked improvement in stormwater quality. End-of-pipe remediation regimes are able to treat runoff from entire subcatchments and may be readily implemented and are relatively inexpensive. The foreshore areas adjoining most Sydney Estuary embayments are occupied by reclaimed parklands, consequently retrofitting stormwater remediation devices located in these areas may be easily achieved with little disturbance to the surrounding community. In order to manage stormwater in a way which best meets the needs of the community, while improving the health of the estuary, water sensitive urban design measures should continue to be implemented at new or redeveloped sites and end-of-pipe remediation devices should be put in place in old, densely urbanised catchments at the sites where stormwater discharges to the Sydney Estuary. Presently, sewerage overflows account for about one third of nutrient input to Sydney Estuary. The sewerage infrastructure is in need of reparations to replace derelict pipes and upgrading to increase capacity due to increased demand to reduce pathogen supply.

Due to the publicity surrounding dioxin contamination in fish caught in Sydney Estuary waters much has been done to attempt to remediate Homebush Bay bed sediments, the embayment from which the dioxins principally derived. With respect to the impact of dioxins on human health, long-term monitoring of fish tissue is required to determine the success of the dredging and capping program undertaken in the embayment. Long-term monitoring of bed sediment dioxin concentrations is required throughout the estuary to ensure the source of dioxin contamination has been effectively capped. Modelling investigations encompassing chemical, sediment and hydrodynamic transport would facilitate improved understanding of the long-term impact of dioxins and the potential recovery time for the estuary with respect to these chemicals. Uptake studies would further advance our understanding of the impact of these chemicals on estuarine species and enable appropriate guidelines for dioxin concentrations within estuary waters and bed sediment to be put in place.

As an urban water body, Sydney Estuary is subject to many pressures impacting upon the health of the system and the organisms living therein. On the whole, water quality has improved since the implementation of the Clean Waters Act, however illegal discharges, historical contamination and continued stormwater contamination present management challenges. Recent research investigating current patterns, stratification, mixing and residence times has improved our understanding of this complex system. Continued research addressing sediment transport, particularly resuspension and subsequent dispersal of bed sediments, chemical partitioning, and the uptake of contaminants by estuarine species is required to facilitate improved understanding of long-term contaminant transport processes and the impact of contaminants on the health of estuarine species.

Acknowledgements This study was supported by an Australian Postgraduate Award (Industry) from the Australian Research Council (No. LP0455486). Thanks to As Prof Charles Lemckert, Dr Craig Jones, Dr Andy Manning and Dr Paul Craig for their assistance with field equipment, laboratory analysis and modelling expertise.

References

ANZECC/ARMCANZ (2000) Australian and New Zealand guidelines for fresh and marine water quality. Volume 1, section 3.5 – sediment quality guidelines. Australian and New Zealand Environment and Conservation Council, and Agriculture and Resource Management Council of Australia and New Zealand. Canberra, Australia

Barry SC, Taylor SE, Birch GF (1999) Heavy metals in urban stormwater canals entering Port Jackson, Australia and their impact on the estuarine environment. In: Joliffe, Ball JE (eds) Proceedings of the Eight International Conference on Urban Storm Drainage, Sydney. August 1999, vol 4. pp 1825–1832

Barry SC, Taylor SE, Birch GF (2000) Heavy metal supply to the Port Jackson Estuary, Australia by stormwater. Aust J Water Res 4 (2):147–154

Bellucci LG, Frignani M, Raccanelli S, Carraro C (2000) Polychlorinated dibenzo-p-dioxins and dibenzofurans in surficial sediments of the Venice Lagoon (Italy). Mar Pollut Bull 40(1):65–76

Birch GF (2000) Marine pollution in Australia, with special emphasis on central New South Wales estuaries and adjacent continental margin. Int J Environ Pollut 13(1–6):411–423

Birch GF (2011) Contaminated soil and sediments in a highly-developed catchment-estuary system (Sydney Estuary, Australia) – an innovative stormwater remediation strategy. J Soil Sediment 11(1):194–208

Birch GF, O'Hea L (2007) The chemistry of suspended particulate material in a highly contaminated embayment of Port Jackson (Australia) under various meteorological conditions. Environ Geol 53:501–516

Birch GF, Rochford L (2010) Stormwater metal loads to the Sydney estuary (Australia). Environ Model Assess 169:531–551

Birch GF, Taylor SE (1999) Source of heavy metals in sediments of Port Jackson estuary, Australia. Sci Total Environ 227:123–138

Birch GF, Taylor SE (2000) The distribution and possible sources of organochlorine residues in sediments of a large urban estuary, Port Jackson, Sydney. Aust J Earth Sci 47:749–756

Birch GF, Taylor SE (2002a) Possible biological significance of contaminated sediments in Port Jackson, Sydney, Australia. Environ Monit Assess 77:179–190

Birch GF, Taylor SE (2002b) Application of sediment quality guidelines in the assessment of contaminated surficial sediments in Port Jackson (Sydney Harbour), Australia. Environ Manag 29(6):860–870

Birch GF, Taylor SE (2002c) Assessment of possible sediment toxicity of contaminated sediments in Port Jackson estuary, Sydney, Australia. Hydrobiologia 472:19–27

Birch GF, Taylor SE (2004) Sydney Harbour and catchment: contaminant status of Sydney Harbour sediments: a handbook for the public and professionals. Published by the Geological Society of Australia, Environmental, Engineering and Hydrogeology Specialist Group, p 101

Birch GF, Scollen A, Snowden R, Suh J (1999) Sources of heavy metals in stormwater draining into Port Jackson, Australia. In: Joliffe, Ball JE (eds) Proceedings of the eighth international conference on urban storm drainage, Sydney, August 1999, vol 4, pp 2202–2209

Birch GF, Matthai C, Fazeli MS (2006) Efficiency of a retention/detention basin to remove contaminants from urban stormwater. Urban Water J 3(2):69–77

Birch GF, Harrington C, Symons RK, Hunt JW (2007) The source and distribution of polychlorinated dibenzo-p-dioxin and polychlorinated dibenzofurans in sediments of Port Jackson, Australia. Mar Pollut Bull 54:295–308

Birch GF, McCready S, Long ER, Taylor SE, Spyrakis G (2008) Contaminant chemistry and toxicity of sediments in Sydney Harbour, Australia: spatial extent and chemistry-toxicity relationships. Mar Ecol Prog Ser 363:71–87

Birch GF, Murray OA, Johnson I, Wilson A (2009) Reclamation in Sydney Harbour, 1788–2002. Aust Geogr 40(3):347–368

Birch GF, Chang C-H, Lee J-H, Churchill LJ (2013) The use of vintage surficial sediment data and sedimentary cores to determine past and future trends in estuarine metal contamination (Sydney estuary, Australia). Sci Total Environ 454–455:542–561

Bopp RF, Gross ML, Tong H, Simpson HJ, Monson SJ, Deck BL, Moser FC (1991) A major incident of dioxin contamination: sediments of New Jersey estuaries. Environ Sci Technol 25 (5):951–956

Chapman MG, Bulleri F (2003) Intertidal seawalls: new features of landscape in intertidal environments. Landsc Urban Plan 62:159–172

Coward D (1988) Out of sight: Sydney's environmental history 1851–1981. Department of Economic History, Research School of Social Sciences, Australian National University, ACT, Canberra

Das P, Marchesiello P, Middleton JH (2000) Numerical modelling of tide-induced residual circulation in Sydney Harbour. Mar Freshw Res 51:97–112

Hatje V, Birch GF, Hill DM (2001) Spatial and temporal variability of particulate trace metals in Port Jackson estuary, Australia. Estuarine Coastal Shelf Sci 53:63–77

Irvine I (1980) Sydney Harbour: sediments and heavy metal pollution. PhD thesis, University of Sydney, Sydney

Irvine I, Birch GF (1998) Distribution of heavy metals in surficial sediments of Port Jackson, Sydney, Australia. Aust J Earth Sci 45 (1):169–174

Jolley R (2005) Onset and flux of Cu, Pb and Zn loads in runoff in relation to historical changes in land-use. Iron Cove, Port Jackson, Australia 1885–1996. Unpubl Hons thesis, School of Geosciences. Sydney University, Sydney

Lee SB (2012) Catchment and estuarine response to high precipitation in a well-mixed/stratified estuary (Sydney Estuary, Australia). PhD thesis, the University of Sydney, Sydney

Lee SB, Birch GF (2012) Utilising monitoring and modelling of estuarine environments to investigate catchment conditions responsible for stratification events in a typically well-mixed urbanised estuary. Estuarine Coastal Shelf Sci 111:1–16

Lee SB, Birch GF, Lemckert CJ (2011) Field and modeling investigations of fresh-water plume behaviour in response to infrequent high-precipitation events, Sydney Estuary, Australia. Estuarine Coastal Shelf Sci 92(3):389–402

Long ER (2000) Spatial extent of sediment toxicity in U.S. estuaries and marine bays. Environ Monit Assess 64:391–407

Long ER, MacDonald DD, Smith SL, Calder ED (1995) Incidence of adverse biological effects within ranges of chemical concentrations in marine and estuarine sediments. Environ Manag 19:81–97

McCready S, Slee D, Birch GF, Taylor SE (2000) The distribution of polycyclic aromatic hydrocarbons in surficial sediments of Sydney Harbour, *Australia*. Mar Pollut Bull 40(11):999–1006
McCready S, Spyrakis G, Greely CR, Birch GF, Long EL (2004) Toxicity of surficial sediments from Sydney Harbour and vicinity, Australia. Environ Monit Assess 96:53–83
McCready S, Greely CR, Hyne RV, Birch GF, Long ER (2005) Sensitivity of an indigenous amphipod, *Corophium* sp. to chemical contaminants in laboratory toxicity tests conducted with field collected sediment from Sydney Harbour, Australia and vicinity. Environ Toxicol Chem 24(10):2545–2552
McCready S, Birch GF, Long ER (2006a) Metallic and organic contaminants in sediments of Sydney Harbour and vicinity – a chemical dataset for evaluating sediment quality guidelines. Environ Int 32:455–465
McCready S, Birch GF, Long EL, Spyrakis G, Greely CR (2006b) Predictive abilities of numerical sediment quality guidelines for Sydney Harbour, Australia and vicinity. Environ Int 32:638–649
McCready S, Birch GF, Long ER, Spyrakis G, Greely CR (2006c) Relationships between toxicity and concentrations of chemical contaminants in sediments from Sydney Harbour, Australia, and vicinity. Environ Monit Assess 120:187–220
McLoughlin LC (2000) Shaping Sydney Harbour: sedimentation, dredging and reclamation 1788–1990s. Aust Geogr 31(2):183–208
Mueller J, Muller R, Goudkamp K, Shaw M, Mortimer M, Haynes D, Paxman C, Hyne R, McTaggart A, Burniston D, Symons R, Moore M (2004) Dioxins in aquatic environments in Australia. National Dioxins Program Technical Report 6, Australian Government Department of the Environment and Heritage, Canberra, 105 pp
Murray OA (2003) A history of reclamation in Sydney Harbour 1788–2002. MApplSci thesis, School of Geosciences, University of Sydney, Australia
Pitbaldo RM (1978) Estuarine modelling: the development and validation of a two layer, real-time model for the Parramatta river estuary. PhD thesis, University of Sydney
Ritter L, Solomon KR, Forget J (1995) Persistent organic pollutants. Report for the International Programme on Chemical Safety (IPCS) within the framework of the Inter-Organisational Programme for the Sound Management of Chemicals (IOMC)
Shore H (1981) From the quays. Sydney metals in soils of a highly urbanised sub-catchment (Iron Cove) of Port Jackson (Sydney). Aust J Soil Res 42:329–338
Simpson SL, Batley GE, Chariton AA, Stauber JL, King CK, Chapman JC, Hyne RV, Gale SA, Roach AC, Maher WA (2005) Handbook for sediment quality assessment. Publication of Centre for Environmental Contaminant Research, CSIRO, Sydney
Solling M, Reynolds P (1977) Leichhardt: on the margins of the city: a social history of Leichhardt and the former municipalities of Annandale, Balmain and Glebe. Allen and Unwin, Sydney
Spark E, Connor GJ (2004) Wind forecasting for the sailing events at the Sydney 2000 Olympic and Paralympic games. Weather Forecast 19:181–199
Spyrakis G (2002) Assessment of the predictive ability of sediment quality guidelines using sea urchin toxicity tests. MSc thesis, University of Sydney, Australia
Stephensen PR (1966) History and description of Sydney Harbour. Rigby Limited, Sydney
Suh J, Brown P, Birch GF (2003a) Geochemical factors affecting leachate composition derived from soils in reclaimed land using laboratory freshwater and saline water column experiments. Mar Freshw Res 54:885–893
Suh J, Brown P, Birch GF (2003b) Hydrogeochemical characteristics and importance of natural and anthropogenic influences on soil and groundwater in a reclaimed land adjacent to Port Jackson, Sydney, Australia. Mar Freshw Res 54:767–779
Suh J, Birch GF, Matthai C, Hughes K (2004a) Spatial distribution and sources of heavy metals in reclaimed lands of Homebush Bay; the venue of the 2000 Olympic games, Sydney, New South Wales. Aust J Earth Sci 51:53–66
Suh J, Birch GF, Hughes K (2004b) Hydrochemistry in reclaimed lands of the 2000 Olympic games site, Sydney, Australia. J Coast Res 20 (3):709–721
Taylor SE (2000) The source and remobilisation of contaminated sediment in Port Jackson, Australia. PhD thesis, the University of Sydney, Sydney
Taylor SE, Birch GF, Links E (2004) Historical catchment changes and temporal impact on sediment of the receiving basin, Port Jackson, New South Wales. Aust J Earth Sci 51(2):233–246
Townsend J (2011) Industrial and urban development in hen and chicken Bay, Sydney Estuary, NSW, Australia and impact of contamination on the estuary. Unpubl. Master of Applied Science thesis, School of Geosciences, Sydney University, Sydney, Australia, 2011
Wolanski E (1977) The fate of storm water and stormwater pollution in the Parramatta Estuary, Sydney. Aust J Mar Freshwat Res 28:67–78

The Murray/Coorong Estuary: Meeting of the Waters?

Jochen Kämpf and Diane Bell

The land and waters is a living body.
We the Ngarrindjeri people are a part of its existence.
The land and waters must be healthy for the Ngarrindjeri people to be healthy.
We are hurting for our Country.
The Land is dying, the River is dying, the Kurangk (Coorong) is dying
and the Murray Mouth is closing.
What does the future hold for us?

(Tom Trevorrow, Ngarrindjeri Elder, Camp Coorong, 2002)

Abstract

This chapter gives an overview of natural and anthropogenic factors that have shaped the lower reaches of the *River Murray* – the *Murray Estuary* (nowadays called the *Coorong Estuary*) including the *Lower Lakes* – *Lake Alexandri*na and *Lake Albert*. The reader will learn of the traditional owners, the *Ngarrindjeri* peoples, whose enduring stories recall the transformation of the landscape by their pioneering culture heroes whose deeds widened the River Murray and created the distinctive ecology of the Lower Lakes and the Coorong some 10–12,000 years ago. The reader will also learn that the natural environment of the river has been severely degraded over the last 150 years through extensive water extraction used for irrigation and the construction of barrages. It becomes obvious that modifications to the system have been so detrimental and far reaching that a return to natural conditions is an almost impossible task. It is uncertain whether current water management plans will prevent an irreversible collapse of the system. This chapter celebrates the ecological richness of the watershed that the Murray once was and pays respect to the wise stewardship of its traditional owners.

Keywords

Coorong estuary • River Murray • Murray mouth • Lower lakes • Natural conditions • Human impacts • Ngarrindjeri • Water management • Australia • Drought • Meeting of the Waters

J. Kämpf (✉)
School of the Environment, Flinders University, PO Box 2100,
Adelaide, SA 5001, Australia
e-mail: jochen.kaempf@flinders.edu.au

D. Bell
Department of Anthropology, George Washington University,
Washington, DC, USA

School of Social Sciences, University of Adelaide, Adelaide, Australia

School of Humanities, Flinders University, Adelaide, Australia

E. Wolanski (ed.), *Estuaries of Australia in 2050 and Beyond*, Estuaries of the World,
DOI 10.1007/978-94-007-7019-5_3,

Box 1

Jochen Kämpf studied the Coorong/Murray estuary. Diane Bell works with the Ngarrindjeri traditional owners of the region. The enduring stories of the traditional owners recall the transformation of the landscape by their pioneering culture heroes whose deeds widened the River Murray and created the distinctive ecology of the Lower Lakes and the Coorong some 10-12,000 years ago. The natural environment of the river has been severely degraded over the last 150 years through extensive water extraction used for irrigation and the construction of barrages. It becomes obvious that modifications to the system have been so detrimental and far reaching that a return to natural conditions is an almost impossible task. It is uncertain whether current water management plans will prevent an irreversible collapse of the system. This chapter celebrates the ecological richness of the watershed that the Murray once was and pays respect to the wise stewardship of its traditional owners.

Overview

The *Murray Estuary* (also called the *Coorong Estuary*) is the lower reach of Australia's longest river – the *River Murray*. The ecosystem health of the estuary and adjacent internationally significant wetlands of the *Lower Lakes* has deteriorated over the last 70 years mainly due to excessive water extractions used for irrigated agriculture. The situation has dramatically worsened during the last severe drought from 1995 to 2009. Currently the Murray-Darling Basin Authority, the Basin states and the Commonwealth are in the process of crafting a *Basin Plan* that will set diversion limits and establish watering plans needed to maintain and protect important environmental assets (see MDBA 2012). Details of this policy are subject to ongoing heated debates.

This chapter is composed of two parts. The first part is an overview of the natural and modified states of the Murray Estuary system together with a projection for the future. The second part explores ways in which the 'Meeting of the Waters' has shaped and is shaped by the culture of the traditional owners, the *Ngarrindjeri* Aboriginal people who still inhibit the lower reaches of the River Murray. Together these two parts address a central tenet of ecology and core principle of Ngarrindjeri culture that 'All things are connected'. 'Meeting of the Waters' refers to the *zone* where riverine fresh water meets salty seawater and the *time* when peak river flows flush the Lower Lakes with freshwater. Naturally, this event, being subject to pronounced seasonal and longer-term variations, is strongly coupled with the aquatic ecosystem such as the breeding cycle of the iconic Murray Cod (*Maccullochella peeli*, called *Pondi* by the Ngarrindjeri) – a long-lived (>45 years) freshwater fish that plays a central role in the Ngarrindjeri culture. Read together, our perspectives facilitate an appreciation of how severe and devastating the impacts of past and still ongoing modifications of natural environments can be in the Australian context and for humankind.

The Murray/Coorong Estuary

Introduction

The *Murray Estuary* (often called *Coorong Estuary*) is located in the lower reaches of the *River Murray* of the *Murray-Darling Basin* (Fig. 1) that originates in the *Australian Alps*. The Australian Alps, being part of the *Great Dividing Range*, is the highest mountain range of Australia. This range is located in southeastern Australia, and it straddles southeastern New South Wales, eastern Victoria, and the Australian Capital Territory. The alpine environment and general downslope seepage makes the Australian Alps an important water storage for the Murray-Darling Basin and the River Murray. Precipitation occurs all year round but is greatest in winter and spring. During winter, much water is held as snow and ice and held back from streams until it thaws in warmer weather. Hence, natural river flows of the Murray usually peak during late austral spring (July–November) (CSIRO 2008).

The River Murray is Australia's longest river. It originates from the catchment of the Australian Alps and flows over a total distance of 2,375 km into the *Lower Lakes* – Lake Alexandria and Lake Albert (Fig. 2). In wetter years,

Fig. 1 Map of the Murray–Darling river basin (Courtesy of Shane Strudwick, Discover Murray http://www.murrayriver.com.au. Slightly modified)

Lake Alexandria occupies an area of approximately 600 km^2 and has average and maximum depths of 2.4 and 6 m, respectively. Lake Albert has an area of 140 km^2 and an average depth of 1.7 m. Since the 1940s, a number of barrages – the *Murray barrages*–have semi-permanently isolated the Lower Lakes from seawater and waters from the hypersaline *Coorong Lagoon* that forms an *inverse estuary* (Wolanski 1987).

The Coorong is a coastal lagoon that is separated from the sea by Younghusband Peninsula to the southeast of the Murray Mouth and Sir Richard Peninsula to the northwest of the Mouth. Both peninsulas are coastal dune barriers covered by sand dunes and scrubby vegetation. The lagoon is about 150 km long, with a width that varies from 5 km to just 100 m. The water depth is between 1 and 2.5 m. Parnka Point separates the lagoon into the *North Lagoon* and the *South Lagoon* (see Fig. 2).

In earlier European settlement, the *Murray Mouth* was referred to as the region where the River Murray entered the waters of Lake Alexandrina. Today it is referred to as the section where water from Lake Alexandrina passes through to the sea (Phillips and Muller 2006) (Fig. 3). Dependant on flow dynamics, climate conditions and sediment movement, the width of the Murray Mouth is highly variable. For instance, the position of the Murray Mouth has migrated over 1.6 km since the 1830s, with migrations up to 6 km over the past 3,000 years. Movements of 14 m over 12 h have been observed. Thus, the Murray Estuary is naturally a geomorphologically highly dynamic area (Bourman 2000). During times of dramatically reduced river flows, the mouth closed in 1981 for the first time in recorded history.

The water sources into the Coorong include freshwater flows from the River Murray (controlled by barrages), marine water via tidal inflows through the Murray Mouth, and more localized rainfall and groundwater inputs (Webster 2006). In addition, it receives fresh to brackish inputs from the Upper South East (USE) drainage scheme of the south-eastern region of South Australia via Salt Creek near the southeastern end of the system. These water sources, the effect of evaporation and tidal stirring determine spatial and temporal salinity variations in the lagoon. When the

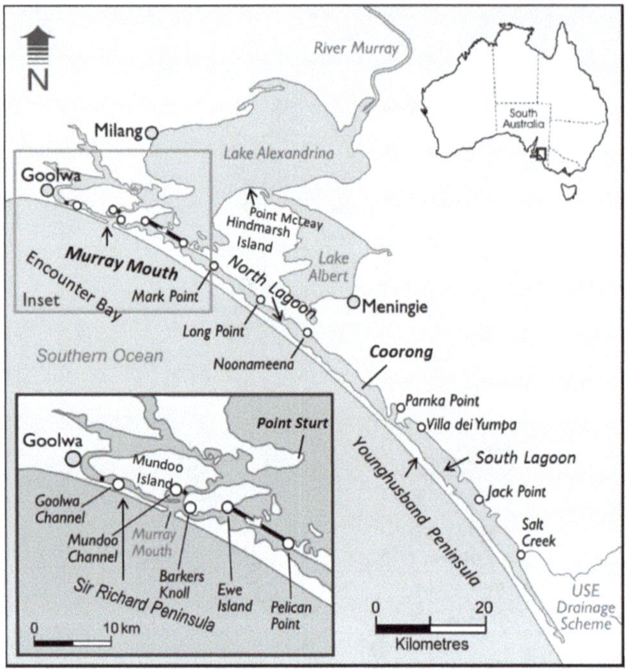

Fig. 2 Map of the Murray/Coorong estuary, South Australia (Modified from Brookes et al. 1999)

Fig. 3 Murray Mouth, South Australia (Photo courtesy of CSIRO)

barrages are flowing, water can be near fresh at its mouth end and when they are not seawater salinity prevails in this region. Towards the southeastern end of the Coorong, salinity increases steadily due to evaporation. The salinity variation representing estuarine, marine and hypersaline conditions supports different ecological communities.

The state of the Murray Mouth can be regarded as an integral measure of rainfall variability in the Australian Alps and upstream water losses by evaporation, interaction with groundwater aquifers, and water diversions, noting that about 30 % of the total evaporative loss in the Murray-Darling Basin occurs in the Lower Lakes (McJannet et al. 2008).

Natural Conditions

The natural River Murray Estuary is a Holocene (last 10,000 years) feature occupying Quaternary inter-dune areas, and formed as a result of a sea level rise of about 150 m that accompanied deglaciation from 17,000 to 7,000 years BP (Bourman 2000). The coastal barrier system developed from about 7,000 years ago and subsequently the barriers have migrated landward (Bourman 2000). The simple model of Holocene sea level change in the Southern Hemisphere proposes that the rise in sea level after the last glaciation ceased about 6,000 years ago, when the sea reached its current level, and it has been relatively steady since that time. However, Baker and Haworth (2000a, b) recently proposed that sea level in Australia has declined significantly over the past 6,000 years, and that this change has been oscillatory. In agreement with this hypothesis, Bourman (2000) described the formation of extensive marine sand flats in the Murray Estuary approximately 5,000–6,000 years ago when sea level was about 1 m higher than at present. Aeolian (wind-driven) processes have been, and still are, active in this environment, creating migrating dune systems (Bourman 2000).

It was during the last period of significant sea level rise, about 7000 years ago, that the Southern Ocean flooded in to the area we now refer to as Lake Alexandrina, spreading out across an area of natural subsidence (Bourman et al. 2000). A new estuary began to form with the Southern Ocean regularly washing into the entire area through a wide opening (Barnett 1994). From the beginning, localized wave action would have deposited sand at the margins of the seaward opening of the young estuary slowly building sand flats, then beaches. Beaches build as sand from wave action gradually accumulates. Sand dunes start to form when dry sand from the beach is blown beyond the reach of the waves. The shoreline across the Murray Estuary has accumulated so much sand over the last 6000 years that 5,000 tonnes of sand may be in motion at any one time along just 10 km of beach (Bourman 2000). The sand peninsula then formed from water currents running along the early beach face forming shoals that consolidated into a sand-spit, developing into a beach and then a sand dune extending across the entire estuary. Studies of the deposition of sand and sediments within Lake Alexandrina suggest that sand pit formation was complete by 2,300 years ago (Barnett 1994). Other studies indicate that the sand barrier was already in place about 3,600 years ago and that at this time Lake Alexandrina was predominately a freshwater lake (Cann et al. 2000). This period is thought to have corresponded with a period of climatic aridity within the Murray-Darling Basin when there were no large floods to break open the sand barrier and so it consolidated.

Provided there is adequate freshwater inflow, lakes and lagoons protected by sand barriers, even without a sea inlet, can remain healthy functioning freshwater ecosystems. Paradoxically it is when freshwater inflows are too high that 'overtopping' of the sand barrier will typically occur, followed by 'lagoon breakout', and then a rise in salinity levels as the lagoon is reconnected with the ocean (Pollard 1994). Hence, coastal processes of beach and dune formation across the Murray Estuary would have occasionally been interrupted when there was significant flooding in the Murray-Darling Basin. Thus in the early stages of development of the modern Murray Estuary, such flooding events have created more than one outlet to the Southern Ocean and many discrete sand islands (Bourman et al. 2000).

The formation of a transition zone (or *mixing zone*) between riverine freshwater and seawater is a characteristic feature of positive estuaries. Diluted water in this transition zone is of referred to as *brackish water*. A typical feature of this mixing zone is the formation of a *turbidity region*– a localized area where sediment accumulates over time. Turbidity regions usually coincide with the brackish-water zone of low salinities in a range between 2 and 10 ppt (Bowden 1984). When river flows are strong and extend into the adjacent sea, turbidity regions can be 'flushed' into the sea (Grabemann and Krause 1994).

The history of the Lower Lakes records these seasonal variations. When explorer Charles Sturt noted Lake Alexandrina's waters south of Point Sturt as salty, those in the central lake as brackish and those in the northern part as fresh (Sturt 1833) in February 1830, he was describing a period outside the river's flooding season; that is, he visited during a period when intermittent wind-driven saltwater intrusions into the central lake were likely to occur.

Local knowledge recorded in the poetry of Lower Lake fishermen and accounts of early visitors and settlers indicate that flushing of the Lower Lakes and the Coorong lagoon occurred from September until maybe Christmas, when peak river flows entered these regions and exited through the mouth (Sim and Muller 2004; Wood 2007). Owing to this natural flushing, waters of the Coorong lagoon were likely far less salty than they are today.

During periods of reduced natural river flows (January to July), on the other hand, seawater entered through the mouth and the brackish water zone shifted towards the Lower Lakes. According to reports by early visitors and settlers (Sim and Muller 2004), the seawater intrusion rarely extended into the central portion of Lake Alexandrina past Point Sturt (see Fig. 2), except when strong southeasterly winds drove seawater transiently into this region (Wood 2007). Ngarrindjeri stories of the 'Meeting of the Waters' record a similar limit to seawater intrusions. The oral accounts of the traditional owners and the written records of early non-Indigenous observers are consistent with a recent paleolimnological analysis indicating that salinity in Lake Alexandrina was only moderately influenced by seawater inflow, particularly over the past ca. 2000 years (Fluin et al. 2007).

The Era of Urbanization

The lower reaches of the River Murray channel in South Australia represent the most modified river sections of the entire river system. When the explorer Sturt first saw the lower reaches of the River Murray in South Australia in 1830, it was a period of low flows and he was unable to imagine a navigable estuarine system. In later periods as European settlement developed, the options for navigation and irrigation saw the emergence of plans to build a system of barrages to keep the seawater out of the estuarine system (MDBA 2011b). Increased salinities in the Lower Lakes became an increasing problem during the Federation Drought (1895–1903). At the peak of this drought an editorial in the *Southern Argus* – summarized the situation as follows (Elliot 1903).

> Through the joint influences of long continued drought and an increasing diversion of its waters in its upper course, the River Murray has so steadily lowered its levels that its lower reaches and the lakes which for centuries it had supplied with a constant flow of fresh water, have fallen to sea level, with the result that instead of the river 'rushing out to sea' the tides of the ocean have flowed in, changing the freshwater lakes to salt ones, and carrying the ocean waters so far upstream that as far as Murray Bridge it is 'salt as the sea'. The effects of this change have been disastrous to residents on the shores of the lakes and on the banks of the stream, the cutting off of supplies of freshwater very materially affecting stockbreeders, and thus deteriorating land values.

Wetlands around the Wellington area were 'reclaimed' for agriculture from about 1880s onward. Between Mannum and Wellington, large scale levee banks were constructed either side of the main bed of the River Murray and by 1929, South Australian Government planning had reclaimed most of the wetland and low lying swamp areas for agriculture (Marsden 1985). This area today is represented by a 'reclaimed' perched river system, where agricultural land lies below the main river channel.

The River Murray Waters Agreement (1915) established the first River Murray Commission and water sharing arrangements between New South Wales, South Australia and Victoria. The Mundoo channel was closed in 1915 by a barrage with a timber sluiceway to restrict the ingress of saline water to Lake Alexandrina (Bourman 2000). In 1933/34, the River Murray Waters Agreement was amended to provide for the construction of five barrages. Construction of the Murray barrages began in 1935 and was completed by 1940. The names of these barrages are Goolwa, Mundoo, Boundary Creek, Ewe Island and Tauwitchere (Fig. 4).

The barrages comprise five low head weirs and earthen causeways linking the islands that once formed a previous shoreline. The barrages, blocking 7.6 km of previously navigable channels, are operated for several purposes; that is, to reduce salinity levels in the lower reaches of the River Murray and associated lakes; to stabilise the river level, and normally maintain it above the level of reclaimed river flats between Wellington and Mannum, so as to provide irrigation by gravitation rather than pumping; during low flows, to concentrate releases to the ocean to a small area, and so scour a channel for navigation; and to maintain pool water that can be pumped to Adelaide and other regions of South Australia (MDBA 2011b).

In addition there are a total ten weirs (and associated locks) along the River Murray, constructed between 1920 and 1929. The purpose of these weirs (and the associated locks) is to provide permanent navigation between the Lower Lakes and Wentworth (in New South Wales) and relatively constant pool level to facilitate pumping for irrigation and water supply. The weir at Kulninie (Weir 9) also raises the water level high enough to allow gravity diversion to Lake Victoria.

A number of pipelines, constructed from the 1940s onward, divert river water for use in Adelaide, the Iron Triangle (Port Augusta, Port Pirie and Whyalla), Yorke Peninsula and southeastern South Australia. The water in this reach is also directly drawn for towns and agriculture around the Lower Lakes and the River Murray. On average, Adelaide receives 55 % of its water supply from the Murray.

In the Lower Lakes and Coorong there is now an abrupt interface between the fluvial and tidal reaches, reducing the size of the estuarine component to 11 % of its pre-barrage scale (Bourman 2000). The remnant estuary can change abruptly from saline/estuarine to fresh conditions and back again in an unseasonal and unnatural pattern (Newman 2000). The Mundoo barrage resulted in the development, growth and consolidation of a permanent sand island – Bird Island. This sand island had 4 m high sand dunes before 1988 (Harvey 1988) and was a kilometre in diameter by 2000 (Bourman 2000). Bird Island has changed the geomorphology of the region, interrupted the estuaries' evolution to a fully tidal system, and blocks water releases through the gates of the Goolwa sea dyke.

Changes to the flow regime of the River Murray have resulted in significant changes to the ecological character of the region since its listing as a Ramsar wetland in 1985 (Phillips and Muller 2006). Underlying changes in the ecological character of the Coorong probably started earlier, especially with the construction of the Lower Lakes barrages (Jensen et al. 2000). It is now recognized that the environmental assets of the region are degraded and require management intervention to meet international and national obligations in the protection of migratory waterbirds and other species.

Fig. 4 Map of barrages in the Lower Lakes and pipelines along the Murray river, South Australia (Image source: MDBA 2011b)

Ecological Significance of the Lower Lakes

The Coorong and the Lower Lakes are recognized both nationally and internationally as significant in their role in supporting critical aquatic ecosystems within the Murray-Darling Basin, and for providing habitat for migratory avifauna listed under various international agreements. This recognition includes their designation as a wetland of international importance under the Ramsar Convention in 1985. The region consists of a unique mosaic of 23 Ramsar wetland types which include intertidal mud, sand or salt flats, coastal brackish/saline lagoons, permanent freshwater lakes,

Fig. 5 Australian Pelicans in the Coorong, South Australia (Photo courtesy of Australian Broadcasting Corporation (ABC))

permanent freshwater marshes/pools, shrub-dominated wetlands, and water storage areas (Phillips and Muller 2006). It supports a large number of fish and bird species, such as the Australian Pelican (*Pelecanus conspicillatus*) (Fig. 5), during critical stages of their life cycles.

Of the 49 species of native fish recorded in the Lower Lakes, 20 species utilise the site at critical stages of their life cycle. This site is considered significant because of the diversity of its fish species and the diversity of their form, structure and breeding styles, including their migration habits between fresh, estuarine and marine waters (Phillips and Muller 2006). This includes seven diadromous fish species such as common galaxias and estuary perch that move between fresh, estuarine and marine waters at various stages of their life to breed.

Changes in estuarine and wetland habitats, as well as in the benthos as their major food, have been shown to effect shorebird numbers and distributions (Raffaelli 1999; Stillman et al. 2005). Shorebirds can thus be used to indicate the health of a system (Martinez Fernandez et al. 2005; West et al. 2005). Coastal wetlands in the southern hemisphere constitute non-breeding habitats for many species of migratory shorebirds (Piersma et al. 1993). These migratory shorebirds rely on sufficient food supply in the mudflats along their flyways. Coastal wetlands containing high (>100,000) numbers of shorebirds are characterized by extensive sand- and mudflats in the vicinity of regions with high coastal zone productivity (Butler et al. 2001).

A total of 77 bird species have been recorded in the Lower Lakes and Coorong wetlands, most being waterbirds (Phillips and Muller 2006; Paton et al. 2009). The site is important as waterbird habitat at a global, national and state scale. Of these, 49 species of birds including 25 species listed under international migratory conservation agreements, rely on the wetland at critical life stages, such as migration stop-over, for breeding habitat or as refuge during times of drought. A significant number of waterbirds use this Ramsar site, at times reaching 200,000–400,000 individuals—far in excess of 20,000 or more waterbirds required to meet the Ramsar criteria. A number of species that frequent this site regularly occur in abundances greater than 1,000 individuals. Sixteen species of waterbirds have been recorded in numbers greater than 1 % of the global population, including the Cape Barren goose (*Cereopsis novaehollandiae*), curlew sandpiper (*Calidris ferruginea*), red-necked avocet (*Recurvirostra novaehollandiae*) and fairy tern (*Sterna nereis*) (Phillips and Muller 2006).

Murray Mouth Closures: Early Warning Signals

The Murray Mouth closed in 1981–1983 for the first time in recorded history (MDBA 2011a). Alternating periods of droughts and floods are common and natural features of the Murray-Darling Basin (Fig. 6), noting that the first severe drought after European settlement – the *Federation Drought*– lasted from 1895 to 1903. Years of pronounced droughts and flooding events in the Murray-Darling Basin are natural features of climate variability. On the other hand, the amount of river water diverted for agriculture and other purposes has gradually increased over the years (Fig. 7). For instance, this amount has quadrupled from the 1920s to 1980s. As a consequence of this water extraction, prolonged drought conditions in the 1970s have led to a 'collapse' of the system, leading to reduced barrage-controlled flows and a full closure of Murray Mouth. During this period, Adelaide's reliance on River Murray water resources to over 90 %, noting that South Australia is minor user of water resources from the Murray-Darling Basin. Major users are New South Wales (~50 %) and Victoria (~33 %). South Australia uses about 10 % of the total water diversion, and the remaining 17 % are shared by Queensland and the Australian Capital Territory.

The mouth again approached closure during protracted drought conditions beginning in 1998. In December 2000, a natural flood peak to South Australia was enhanced by an appropriately timed release from Lake Victoria. This resulted in unintentional scouring of sand at the mouth (Gippel 2003). Mundoo Barrage was opened fully (for the first time since 1996) in the belief that its proximity to the mouth would offer the most effective means of removing sediment from the mouth. Nevertheless, this flushing operation did not scour significant volumes of sand from the mouth, partly because the flow rates were not sufficiently high for long enough duration.

Studies show that reduced flows due to upstream abstraction and the alteration of the natural flow regime by the presence of weirs and barrages have reduced flows to such an extent that, in a typical year, flow to the sea through the

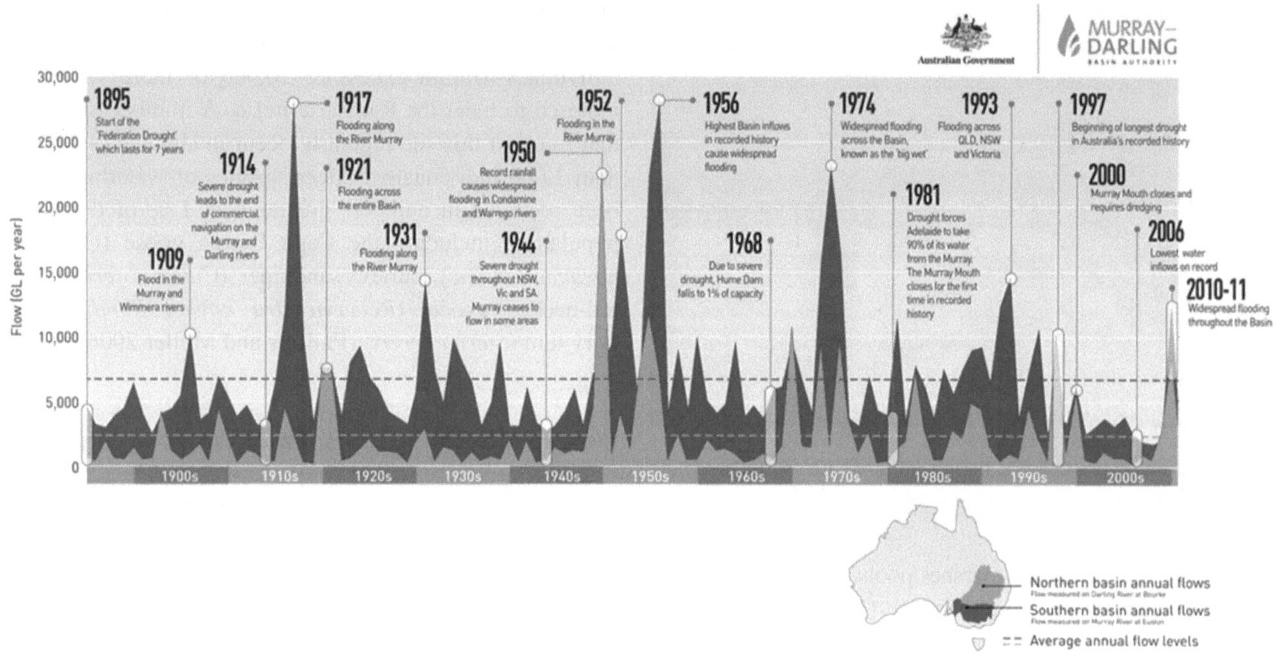

Fig. 6 Historical river flows within the Murray-Darling Basin (Image source: Jason Alexandra, MDBA 2012)

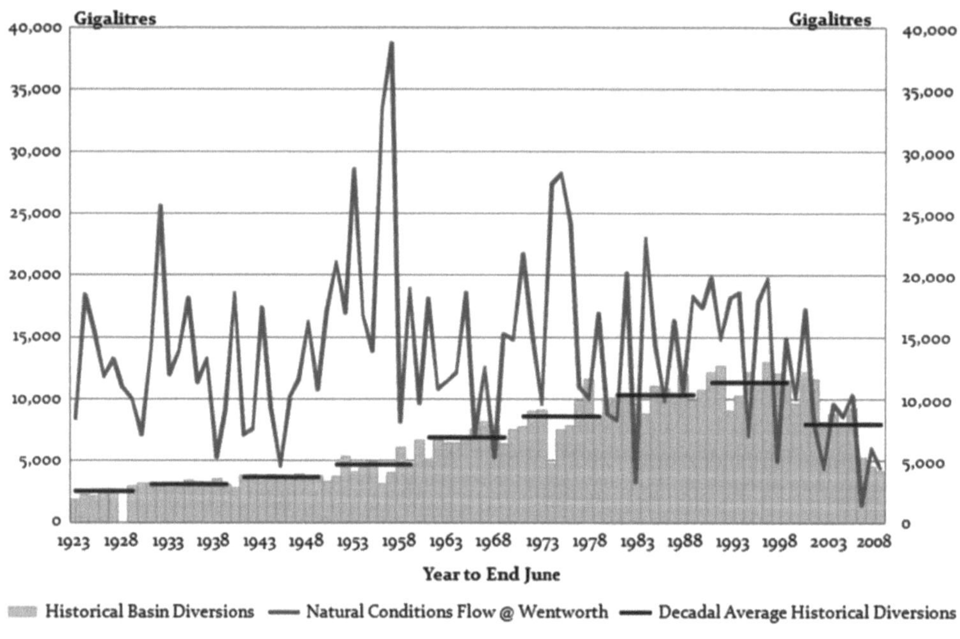

Fig. 7 Time series of historical basin diversions and natural conditions flow at Wentworth, lying at the junction of the Murray and the Darling rivers (Image source: Jason Alexandra, MDBA 2012)

Murray Mouth is only 27 % of the natural median flow (Lamontagne et al. 2004). This flow reduction has caused marine sand to build up inside the Murray Mouth so that it is now at a much greater risk of completely closing resulting in the restriction of fish migration. An analysis by Close (2002) found that under natural conditions the risk of mouth closure was negligible, while under current conditions 31.5 % of years have a risk of mouth closure; that is, mouth closures are expected to occur on an average of every 3 years.

After years of debate on environmental problems facing the River Murray, South Australia passed the *River Murray Act 2003* Section "The Millennium Drought (1995–2009): the kiss of death?" of which establishes 15 Objectives for *a Healthy River Murray*. In addition, extensive dredging operations commenced in October 2002 to keep the Murray Mouth open to maintain marine inflows to the Coorong (Shuttleworth et al. 2005). Dredging operations continued until December 2010.

The Millennium Drought (1995–2009): The Kiss of Death?

Australia's *Millennium Drought* commenced in 1997 and lasted Australia-wide until late 2009. The end of the drought was officially declared in April 2012. Drought in Australia is defined as rainfall over a 3 month period being in the lowest decile (1/10th) of what has been recorded for that region in the past. Beginning in the second half of 1991, a very severe drought occurred throughout Queensland which intensified in 1994 and 1995 to become the worst on record. By October 1994, part of the upper Darling River system had collapsed and 40 % of Queensland was drought declared. By 1995 the drought had spread across many parts of Australia and by 2003 was recognized as the worst on record. Most Australian mainland capital cities faced a major water crisis with less than 50 % of water storages remaining.

In combination with a regional drought in the Basin, inflows to the Coorong from the Lower Lakes barrages have been especially low since 2002, with no inflow recorded during the period 2006–2010. This has resulted in the siltation of the Murray Mouth channel and hypersalinisation of the South Lagoon with salinities exceeding four times that of seawater.

Years of drought and poor water management further upstream in the Murray-Darling river system (Goss 2003), reduced water levels, saltwater incursion in the southeast, and poor environmental conditions in the Coorong Lagoon and Murray Estuary have raised concerns about the health of this wetland (Kingsford et al. 2011) and its function as an overwintering site for shorebirds (e.g. Gosbell and Christie 2005).

Salinity is the most important environmental factor in estuaries, and together with temperature and accompanying physico-chemical properties, it determines species distribution and abundances (Kinne 1966, Mackay and Cyrus 2001; Ysebaert et al. 2003). In 2004, the salinity of the North Lagoon of 26–32 ppt was slightly below that of adjacent shelf water (~35 ppt) indicating marine influences, whereas the salinity in the South Lagoon reached extremely high levels of 100–130 ppt (3–4 times seawater salinity) (Dittman et al. 2006).

The Coorong, Lower Lakes and Murray Mouth (CLLAMMecology) Research Cluster found that increased salinities along the Coorong (particularly South Lagoon) have led to a decline in key food resources such as the aquatic plant *Ruppia tuberosa* and the fish species Smallmouth Hardyhead *Atherinosoma microstoma*. In the South Lagoon, these declines showed a significant correlation with declines in the abundance of many bird species (relative to census data from 1985) including Black Swan (59 % decline) and piscivorous species such as Fairy Tern (82 % decline) and Australian Pelican (77 % decline) (Rogers and Paton2009). Interestingly, Rogers and Paton (2009) found that declines in the abundance of migratory shorebirds such as Curlew Sandpiper (94 %), Sharp-tailed Sandpiper (63 %) and Red-necked Stint (68 %) in South Lagoon were not correlated with changes in the density of key food resources for these species. While it is clear that the majority of bird species that utilise the Coorong have declined in abundance over the last 24 years, high salinities have allowed the Australian Brine Shrimp *Parartemia zietziana* to flourish and become a major food source for Banded Stilt. In 2006, Banded Stilts nested at the Coorong for the first time.

Fig. 8 State of the Darling river at Wilcannia, New South Wales, in 2006 (running completely dry in 2008) (Photo credit: Simon Cotter)

The 2004 macrobenthic survey by Dittmann et al. (2006) indicated that the species richness was highest near the Murray Mouth between the Goolwa Barrage and the northern end of the North Lagoon (29 species), where abundances were also high (>27,000 individuals per m^2). Nevertheless, although abundances of macrofauna were high in the Murray Mouth region, the biomass available for bird consumption was about 10 times lower compared to other estuaries along the flyway of the shorebirds. Hence, the benthic food available for migratory shorebirds in mudflats of the Coorong may not sustain their food requirements during drought years and the absence of freshwater inflows.

In 2007–2008 and at the peak of the drought, flows in the River Murray were so low (see Fig. 8) that the water levels of the Lower Lakes had dropped by more than a meter and became so low that areas of dried-up lake beds started to expose potential acid sulphate soils (Fitzpatrick et al. 2010). Release of this sulfuric acid from the soil can in turn release iron, aluminium and other heavy metals (particularly arsenic) within the soil. Once mobilized in this way, the acid and metals can create a variety of adverse impacts including severe habitat degradation and pollution and acidification of groundwater and aquatic water bodies. Indeed, artificial water diversions from the natural flows in the Murray-Darling Basin

Fig. 9 Collapse of Long Island river bank (near Murray Bridge) in March 2009 (Photo courtesy of News Limited, Australia)

played a dominant role in dramatically reduced water levels in the Lower Lakes. In addition, reduced river flows have led to substantial riverbank collapses (Fig. 9) and destruction of native river red gums (*Eucalyptus camaldelensis*) and other riverine vegetation. Overuse and poor irrigation practices have also led to increased salt content in the soil, reducing the productivity of the land.

Management Responses

Apart from dredging operations in the Murray Mouth, a number of policies and water management and engineering options were implemented in the last decades with the overall goal to secure water resources in Australia and to improve the health of the River Murray and Ramsar-listed wetlands.

From 1979 to 2011, a total number of 18 salt interception schemes have been constructed in the Murray-Darling Basin to reduce river and soil salinities via different engineering options such as flushing with deep aquifer groundwater and the use of evaporation ponds for salt harvesting.

In 1999, the Murray-Darling Basin Ministerial Council introduced the *Murray-Darling Cap* which is a policy limiting the water diversions in the Murray-Darling Basin 1993 levels. This cap did not prevent the environmental disaster occurring during the Millennium Drought. In addition, the decision was made to build large seawater desalination plants in all major Australian cities. Three plants are currently operating in Perth, Sydney and the Gold Coast at Tugun. The plant in Adelaide has been completed but its use has been delayed until the next severe drought conditions. Future use of the Melbourne plant at Wonthaggi is uncertain. Environmentalists and scientists have strongly criticized the use of seawater desalination technology in Australia given the associated increases in carbon dioxide emissions and social implications associated with the investment recovery (e.g. Barnett and O'Neill 2010) and pollution risks for marine ecosystems (e.g. Kämpf 2011).

In 2007 construction of a temporary weir was proposed near Pomanda Island (Wellington) to allow for marine flooding of the Lower Lakes. Environmentalists (and local industries that relied on the Lakes) were strongly opposed to this weir arguing that the sudden flooding of the lakes with seawater would irreversibly alter and damage the ecology of the Lower Lakes and not solve the problem of increasing river and soil salinities.

In 2009, environmental flow regulators – the *Goolwa Regulators* – were constructed in the Goolwa Channel and the Currency Creek. The stated aim of these flow regulators was to capture winter inflows, which would raise water levels in the Goolwa Channel, cover exposed acid sulphate soils and prevent the flow of acidic water into the Lower Lakes. As drought conditions ceased by the end of 2010, the Wellington Weir project was cancelled and the Goolwa Regulators were partially removed.

In 2008, the Murray-Darling Basin Authority assumed the functions of the former Murray–Darling Basin Commission and, as required by the *Water Act 2007*, began the task of drafting a Basin Plan that addresses a number of criteria such as setting sustainable diversion limits, planning for the protection and restoration of wetlands, establishing environmental watering plans, rules for trading of water rights, and arrangements for meeting 'critical' human water needs.

The currently proposed Basin Plan (see MDBA 2012) is based on a fixed 'long-term sustainable level of take' for all consumptive use, so-called 'long-term average sustainable diversion limits' (SDLs), certain allocations to 'Basin states', water resource plans, and water trading options. A key aspect of the Plan is 'water recovery' through market-based, infrastructure and regulatory measures. The Basin Plan is currently subject to heated debates between state and federal politicians, scientists, irrigation-based industries and local communities.

Projections for the Future

The CLLAMMecology Research Cluster undertook model-based projections of river flows for the next 100 years (Lester et al. 2009). Scenarios investigated included a mixture of climate change, sea-level rise and various management options. They investigated the effect of current extraction levels, dredging at the Murray Mouth and a proposed increase in the flow volumes at Salt Creek via the Upper South East Drainage (USED) scheme. The clear messages from this study are that past and current extraction levels are the main reason for the currently poor environmental condition of the Coorong and that, without actions, climate change has the potential to be devastating for the ecology of the Coorong. For instance, the number of days without flow over the barrages would peak in the thousands (i.e. up to 8 years) with sites being predicted to be in a degraded ecosystem state for almost half of all years. Hence, the future of the River Murray, Lower Lakes and the Coorong estuary is in the hands of governmental regulators, and only the future will show whether the currently proposed 'balance' between environmental minimum flows and extraction permits was sufficient for the protection of this iconic and important Australian environment.

The 'Meeting of the Waters'

Introduction

Aboriginal occupation of the Australian continent is subject to debate but a time depth of 40,000–60,000 years is generally accepted (Hiscock 2008). Prior to the establishment of a penal colony at Sydney Cove by the British in 1788, some 250–300 languages, most of which had several dialects, were spoken by the various Indigenous nations (Walsh and Yallop 2005) whose beliefs and practices were attuned to local conditions that ranged from the tropical north, through the desert reaches of central Australia to cooler climes of the southeast. The traditional lands of the Ngarrindjeri nation reached from about 30 km north of Wellington on the River Murray to Cape Jervis in the west and Kingston in the southeast (Fig. 10). With reliable all-year round access to fresh water, these lands sustained relatively high population densities and the Ngarrindjeri pursued a lifestyle of 'seasonal sedentariness' (Berndt et al. 1993: 19). According to Ngarrindjeri belief, in the *Kaldowinyeri* (Creative Era), a larger than life pioneering hero called *Ngurunderi* gave form and shape to their lands and ordered their society (Bell 2008). Through care of their *ngatji*, (totems; also translated as 'friend' and 'countryman' by Meyer 1843: 86) which include key local species such as *Ngori*, the pelican (see Fig. 5), and *Pondi*, Murray Cod, the Ngarrindjeri continue to demonstrate their responsibility for their lands and waters.

The Ngarrindjeri conceptualization of the Murray Estuary as the 'Meeting of the Waters', as an area where Ngarrindjeri *ngatji* (totems) breed, where the myriad mixings and minglings of sea, river, tributaries, lakes and Coorong energize the waters, is of an awe-inspiring, spiritually-charged zone, rich in *Kaldowinyeri* stories. Key Ngarrindjeri stories of the 'Meeting of the Waters' celebrate fertility, new life and rites of passage for departing spirits. They establish seasonal connections between Earth and Sky, differentiate important species, languages and cultural zones, and are touchstones for the underlying unity of Ngarrindjeri peoples and their *ruwi* (country). In the Ngarrindjeri language *ruwi* is the singular form of the word for body, *ruwar*. Thus land and body are one: damage to one registers on the other.

Ngarrindjeri narratives generate an interfunctional, interconnected, interdependent world for the living and those who have passed away. Ngarrindjeri knowledge of their *yarluwar-ruwi* (sea-land country) is made manifest in the cultural teachings of this generation of elders. It echoes the wisdom of their forebears and reaches back to the *Kaldowinyeri* (Ngarrindjeri Nation 2007). Respect for one's elders, the old people who have passed away, stories and *ruwi* are key Ngarrindjeri values.

Cultural knowledge is embedded in social relations and hence it is important to know not only the content but also the context of the telling. Stories are gender-inflected, age-appropriate, family and place specific. In an oral culture, one earns the right to know and to retell certain stories, the circulation of knowledge is circumscribed, and not everyone knows everything. The various written records of explorers, settlers, missionaries, anthropologists, museums, bureaucrats, journalists and a host of other commentators and adjudicators must likewise be interrogated. Who asked what question, where, when and of whom? Language skills, sensitivity to cultural difference and so on also inflect the written record (Bell 1998: 419ff).

So what can we learn of the history and ecology of Murray Mouth, Goolwa Channel, Lakes Alexandrina and Albert and Coorong area from the cultural knowledge and practices of the Ngarrindjeri from existing sources? What guide might Ngarrindjeri cultural knowledge and practices pose for our understanding of the past, present and future of the 'Meeting of the Waters'?

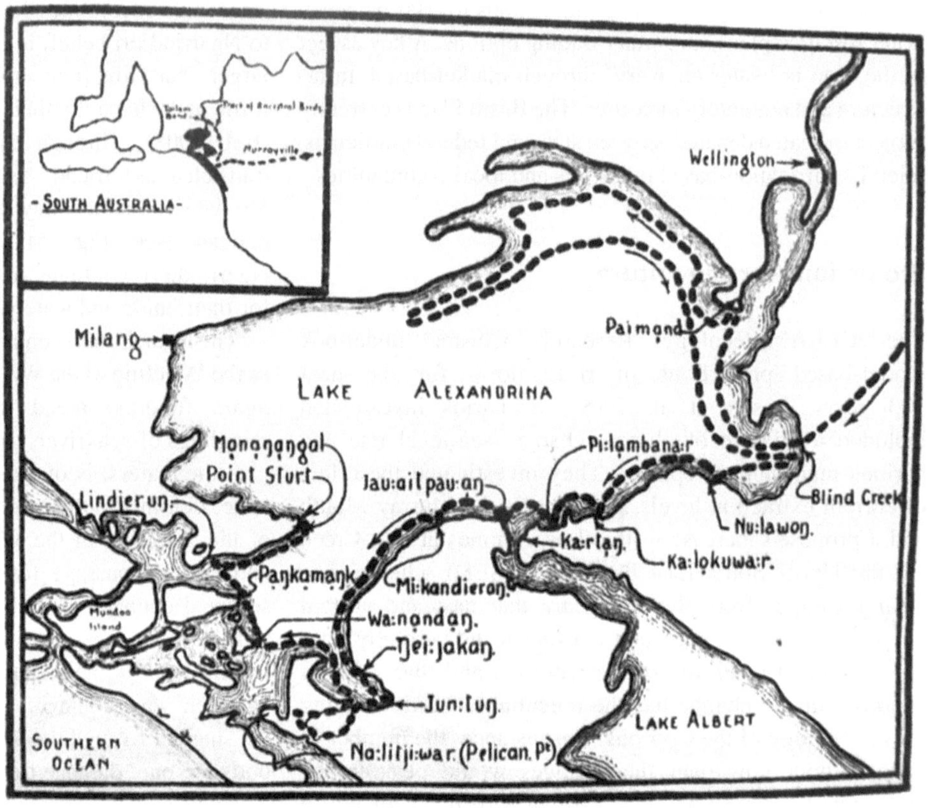

Fig. 10 Map of places named in the Ngarrindjeri fishing legend recorded by Harvey (1943)

Mapping the Fresh Water: A Legend of a Lakes People

In a time when birds were men, *Tenetjar*, two red-billed gulls (*Larus novae-hollandiae*) began a migration from their home at Murrayville in Mallee country in search for new fishing grounds. At Tailem Bend and in Lake Alexandrina, the gull brothers noted 'the water was calm and fresh, an ideal fishing ground' and this was confirmed by their catch of *Pondi*, the Murray Cod (Harvey 1943: 109). The brothers joined with other birds, black gulls, shags, small white gulls, divers, weets-weets, Blue coots, coots and pelicans and, as a flock, migrated south-westwards to Lake Alexandrina, thereby demonstrating the social, co-operative nature of getting a living. This story records the 'intense concern' of the travellers with access to freshwater fish and is, as South Australian Museum-based ethnologist Alison Harvey (1939, 1943: 112) notes, a 'legend of a lakes people'. They did not dally once they reached the salty environs of the Coorong. The magpie would no longer light a cooking fire and the travellers knew they needed a different kind of net to catch salt-water mullet (*ibid.:* 111). Continuing on their way northwards, they crossed to *Lindjeran*, a rocky limestone cliff 3 mile south of Point Sturt ... Here they found fresh water and continued their fishing and reached the tall cliffs of Point Sturt, where, after an altercation with Magpie over fire and distribution of the catch, the birds including became men.

Thus, as can be seen on Harvey's map (1943: 108) (Fig. 10), the activities of these Ngarrindjeri *ngatji* trace an eastern boundary of the 'Meeting of the Waters' site, delineate the extent of seawater incursions into the lake, map the resources of a freshwater lake, explain the collaborative behaviours of fish and birds and present injunctions to their fellow lake people to avoid the salty water which they first encountered at 'Cold and Wet'(near Hack's Point/Parnka Point on the Coorong) and later in the islands near the Murray Mouth (Harvey 1943: 112).

Naming the Species: From the Body of the Murray Cod

Ngurunderi's story is of an epoch-defining pioneering creative hero who gives form, shape and meaning to the land; transforms a meandering stream into the mighty River Murray; calls upon the seas to rise and separate Kangaroo Island from the mainland; establishes major ceremonies; and enunciates societal rules. Not surprisingly the story had been told from a number of perspectives and his travels intersect with localized and other travelling ancestral beings (Bell 1998: 91ff; Berndt 1940; Hemming 1994; Hemming et al. 1989; Meyer 1846). *Ngurunderi*'s power infuses the 'Meeting of the Waters' with significance. It is from the limestone caves at Goolwa that he farewells the spirits on their journey to the

Land of the Dead, Kaarta, Kangaroo Island, and on the Goolwa beach that *Ngurunderi* praises the handicraft of a localized ancestor *Jekejere* and extols the good fishing (Bell 1998: 570ff; Tindale 1937: 115–6).

The importance of kinship ties in Ngarrindjeri narratives points to the embedded and relational nature of cultural knowledge. It is *Ngurunderi* who, while looking for his sons, sees them at the end of the Coorong and wishing them to return 'beckoned' to the water which rushes in and 'that is how the Coorong originated' (Tindale 1930–1952: 102). It is through *Ngurunderi*'s interactions with *Nepele*, his brother-in-law, that the various species of fresh and seawater fish are named. Having chased *Pondi*, the giant Murray Cod, down the river, each swish of his tail making one reach after another of the river, *Ngurunderi* finally spears *Pondi* in Lake Alexandrina and, with the assistance of his brother-in-law, who was already camped at Raukkan (Point McLeay), cuts the fish into pieces and names the species. As he names each fish, he casts it into its home. *Ngurunderi* made all the fresh fish and saltwater fish of the Ngarrindjeri. To the last piece he said, '*Nung Pondi* – you keep being Murray cod'. (Bell 1998: 564–5).

As brothers-in-law these two would be from different clans and have different *ngatji* but they are brought into a collaborative creative endeavour in the differentiation of fresh and salt water species from the body of *Pondi*. The kinship reciprocities point to social bases of the story, the injunction to the living is to work together and to know and nurture one's territory and its resources.

The Seasonal Cycles: Fresh Water Flows

Kaldowinyeri stories recall multiple interests in the region of the 'Meeting of the Waters' as a resource for food and fresh water but they also evoke feelings of belonging, an aesthetic appreciation of the beauty of country, and of the spiritual power of the *ruwi* where the waters meet. Through the story of the pregnant turtle, *Krawi Thukubi* (a giant short necked turtle; Murray turtle, *Emydura macquarii*), who travels across the country before the time of *Ngurunderi*, new life reaches the Murray Mouth. The drag of her tail made the river and her flippers carved out the lagoons and backwaters. She smells the salt, pushes on and, in giving birth, flushes the Murray Mouth. In her life cycle the seasonal round is made manifest and tied to cultural knowledge about water quality and quantity (Bell 1998: 99).

The reflections of the Pleiades, *Mantjiga*, in the fresh waters of the spring melt in the Goolwa channel herald the return to the Skyworld of the Seven Sisters, and recall the proud history of the ordeals undergone by these young women, *Yartuka*, on behalf of all women (Bell 1998: 573ff; Unaipon 1925). When the Pleiades reappear from behind the Milky Way in September/October (Berndt et al. 1993: 164), their rising along with the end of the blooming of Billy Buttons signals it is now safe for people to enter the waters (Rankin 1969). Swimming before this seasonal prompt would bring disease and Ngarrindjeri children still observe this prohibition on entering the waters.

Waiyungari, seen at his brightest in the spring as the red planet Mars, was once camped near Raukkan on the shores of Lake Alexandrina when he stole the wives of his older brother *Nepele* (Berndt et al. 1993: 227–31; Tindale 1935: 265–73). For both the women and *Waiyungari*, their relationship violated a key Ngarrindjeri taboo because *Waiyungari* who appears in the story as a red-orchred youth, denoting his *narambi* status as an initiation novice, should have remained separate and secluded. Thus, while *Waiyungari* energizes spring, as a punishment for his transgression, the fishing is poor when he disappears in October-November but improves as the *Yartuka* rise in the sky. In summer, the freshwater fish are attracted to the insects among the reed beds of the River and the Lakes and the fishing is good (Berndt et al. 1993: 367).

Lake Stories: Nurtured by Fresh Water

Freshwater mussels (*Velesunio ambiguous*) of the shores of Lake Alexandrina are noted as a staple food in the *Kaldowinyeri*. Having made camp at *Lalanganggel* (Mt Misery) *Ngurunderi* 'obtained water from the lake in his canoe and filled it with fresh water mussels, on which he lived' (Berndt et al.1993: 224). *Nepele*'s wives who had been diving for mussels near Raukkan, having filled their net bags returned to the shore to cook where they caught sight of ground reddened by *Waiyungari* presence (*ibid.*: 191, 228).

The freshwaters of Lake Alexandrina nourish the sedges (Spiny flat sedge, *Cyperus gymnocaulos*) used by cultural weavers who extol the virtues of this pliable durable species and lament the loss of reed beds with rising salinity levels and clearing of the land for agriculture. Rights to collect the freshwater rushes must be negotiated with traditional owners and there are strict rules about how and when to collect this precious resource. The earliest contacts make note of Ngarrindjeri woven artefacts (Sturt 1833: 155; Angas 1847). Ngarrindjeri weaving is noteworthy for its quality and range: mats, eel traps, scoops, baskets, ritual items. In early twentieth century photographs of Louisa Karpany with her baskets, we see the same items as cultural weavers are making today (Bell 2008: 9). But weaving is also when stories are told, when the Ngarrindjeri tighten the circle of family and knowledge (Bell 1998: 591–4).

Multiple Meanings and Meetings

In the region of the Murray Mouth, Goolwa, Hindmarsh and Mundoo Island, the intersection of clan boundaries and dialects suggests a resource rich area for locals and the need for mechanisms by which neighbouring groups would be able to gain legitimate access. The exact nature of the clan boundaries as recorded in Norman Tindale (1974) and Berndt et al. (1993) are not identical but their mapping shows that four dialects of the Ngarrindjeri Nation – Ramindjeri, Yaraldi (Yaraldekald), Tangani (Tangenkald) and Warki (Warkend) – have clans in the region (Bell 1998: 551). Each of these clans will have *ngatji* for which they are responsible and thus each will be able to make claim to the place as their *ruwi*.

Broadly speaking, these dialect groups reflect major ecological divisions: Ramindjeri are the people of Encounter Bay; Tangani from Coorong; Yaraldi from the eastern side of the Lakes and lower River Murray; and Warki from the western side of Lake Alexandrina. The underlying unity of these dialect groups is articulated in the story of *Wuruwi*, the Huntsman spider *ngatji* who was a bad-tempered woman living near Goolwa. When she died people gathered to celebrate and feast. As each group arrived and ate a different part of her body, they began to speak a distinctive language (Bell 1998: 137–8; Berndt et al.1993: 237–8; Meyer 1846: 14).

According to Taplin (1873: 130), *Gutungald* (Goolwa) denoted the 'place of cockles' (*Donax deltoids*) and Ngarrindjeri cockling rituals recorded by ethnomusicologist Catherine Ellis in 1964 are being taught to children today (Bell 2008: 15). Ngarrindjeri elders who recall the Goolwa Channel as a prized cockling site lament that the freshwater flows that once brought down the nutrients needed to trigger spawning in the mature cockles have diminished. Recent research explores the impact of the closure of the Murray Mouth on the cockle life cycle (Murray-Jones et al. 2002). Ngarrindjeri cite the building of the barrages as a major interruption in the life cycle of the once vibrant cockle population at Goolwa on the ocean beach.

Goolwa was a meeting place for ceremonies, arranging marriages and trade (Bell 1998: 549ff). The middens and burials on the Goolwa foreshore (Draper 2000) and the trade-routes map in Tindale's papers (see Bell 1998: 556) indicate groups gathered at Goolwa. To sustain large gatherings there must have been reliable sources of freshwater at the time of the gatherings. The limestone cups in the Goolwa Channel would have been one source and the freshwater flows another.

In diverse sources recorded by a range of interested parties, strands of Ngarrindjeri cultural knowledge of the 'Meeting of the Waters' have been recorded. Missionary George Taplin (Aborigines' Friends' Association) at Raukkan from 1859 to 1879, offers a lake-centric view and James Ngunaitponi, father of David Unaipon, was one of his sources (Taplin 1859–1879, 1873, 1879), while Lutheran Missionary H.A.E. Meyer at Encounter Bay from 1840 to 1846 provides a Ramindjeri-centric view (Meyer 1846). Both worked on the local languages (Taplin 1879; Meyer 1843) both reflected the sensibilities of their age and socialisation. Researchers based at the South Australian Museum including Alison Harvey (1939, 1943), Philip Clarke (1999), and Steve Hemming et al. (1989) have complied a formidable record. Through published and unpublished papers of Norman Tindale (1930–1952, 1931–1934, 1935, 1937), who spent decades in the region and worked closely with Milerum, Clarence Long, a senior man of the Coorong, from the early 1930s till Milerum's death in 1941 (Tindale and Long n.d.), we gain a Tangani-centric perspective (Tindale 1986). Anthropologists Ronald and Catherine Berndt (Berndt 1940; Berndt et al.1993) worked primarily with Yaraldi in the early 1940s and their interlinear translations of stories told by Margaret 'Pinkie' Mack (Berndt 1994a) and Albert Karloan (Berndt 1994b) highlight their wealth of knowledge of the natural world. More recently anthropologist Diane Bell (1998, 2008) worked with Ngarrindjeri (Tangani, Yaraldi and Ramindjeri) with a primary focus on women's cultural knowledge and practices. David Unaipon (1924–1925, 1925), the first published Australian Aboriginal author, compiled a rich corpus of stories, while researcher Doreen Kartinyeri (2006) traced Ngarrindjeri history in a series of family histories. The tradition of Ngarrindjeri telling their own stories is well established (Ngarrindjeri Nation 2007; Bell 1998: 448ff). Ngarrindjeri cultural knowledge regarding the past, present and future of the region offers a grounded and enduring basis for future research. The 'Meeting of the Waters' emerges as central to Ngarrindjeri identity.

On 29 April 1994, Neale Draper, Senior Archaeologist, Culture and Site Services Branch, Department of State Aboriginal Affairs made a preliminary report to the State Minister of regarding the cultural significance of the 'Meeting of the Waters' (Draper 1994). The final version was presented November 2000. On 4 May 2009 the site was finally registered (662–4727) under section "Naming the Species: From the Body of the Murray Cod" of the *Aboriginal Heritage Act1988 (SA)*.

Conclusion

The verisimilitude of the narratives of the time when the creative heroes strode the land; when characters were transmogrified into physical features; when the stars and planets were human; and when breaches of the law were punished is evident in the continued order of the Ngarrindjeri world. The seasons repeat, the celestial order is secure, birds and fish play out their foretold behavioural scripts, Ngarrindjeri *ngatji* (totems) continue to reproduce. If this order is disturbed people sicken and die. Today Ngarrindjeri mourn their imperilled *ngatji*. In the Ngarrindjeri worldview, loss of species, degraded habitats, and regulated and diminished fresh water flows contribute to the ill-health and pervasive social malaise of local communities. Damage to land registers on the bodies of the living.

Ecological indicators and the wisdom of the traditional owners of the Coorong, Lakes and Murray Mouth offer complementary perspectives on the deep history of the region, the present degradation of Australia's River and point towards the need for an integrated approach if the health of the estuary and the communities that rely on the Murray-Darling system is to be achieved and sustained.

Acknowledgements This work did not receive financial support.

References

Angas GF (1847) Savage life and scenes in Australia and New Zealand: being an artist's impression of countries and people at the Antipodes. Smith Elder and Co, London

Baker RGV, Haworth RJ (2000a) Smooth or oscillating late Holocene sea-level curve? Evidence from cross-regional statistical regressions of fixed biological indicators. Mar Geol 163:353–365

Baker RGV, Haworth RJ (2000b) Smooth or oscillating late Holocene sea-level curve? Evidence from the palaeozoology of fixed biological indicators in south east Australia and beyond. Mar Geol 163:367–386

Barnett EJ (1994) A Holocene paleoenvironmental history of Lake Alexandrina, South Australia. J Paleoclimnol 12:259–268

Barnett J, O'Neill S (2010) Maladaptation (Editorial). Glob Environ Chang 20:211–213

Bell D (1998) Ngarrindjerri Wurruwarrin: a world that is, was, and will be. Spinifex Press, Melbourne

Bell D (ed) (2008) Kungun Ngarrindjeri Miminar Yunnan: listen to Ngarrindjeri women speaking. Spinifex Press, Melbourne

Berndt RM (1940) Some aspects of Jaraldi culture, South Australia. Oceania 11(2):164–185

Berndt CH (1994a) Pinkie Mack. In: Horton D (ed) The encyclopaedia of Aboriginal Australia: Aboriginal and Torres Strait Islander history, society and culture. Aboriginal Studies Press, Canberra, pp 639–640

Berndt CH (1994b) Albert Karloan. In: Horton D (ed) The encyclopaedia of Aboriginal Australia: Aboriginal and Torres Strait Islander history, society and culture. Aboriginal Studies Press, Canberra, pp 536–537

Berndt RM, Berndt CH, Stanton J (1993) A world that was: the Yaraldi of the Murray River and the Lakes, South Australia. Melbourne University Press at the Miegunyah Press, Melbourne

Bourman RP (2000) Geomorphology of the lower Murray lakes and Coorong. In: Jensen A, Good M, Tucker P, Long M (eds) River Murray Barrages, pp 23–29.

Bourman RP, Murray-WallaceCV BAP, Harvey N (2000) Rapid coastal geomorphic change in the River Murray estuary of Australia. Mar Geol 170:141–168

Bowden KF (1984) Turbulence and mixing in estuaries. In: Kennnedy VS (ed) The estuary as a filter. Academic, Orlando, pp 15–26

Brookes JD et al (1999) An ecosystem assessment framework to guide management of the Coorong. Final report of the CLLAMMecology Research Cluster. CSIRO: Water for a Healthy Country National Research Flagship, Canberra, 47 pp

Butler RW, Davidson NC, Morrison RIG (2001) Global-scale shorebird distribution in relation to productivity of near-shore ocean waters. Waterbirds 24:224–232

Cann JH, BourmanRP BEJ (2000) Holocene foraminifera as indicators of relative estuarine-lagoonal and oceanic influences in estuarine sediments of the River Murray, South Australia. Quat Res 53:378–391

Clarke PA (1999) Waiyungari and his relationship to Aboriginal mythology of the Lower Murray, South Australia. Rec S Aust Mus 32(1):51–67

Close AF (2002) Options for reducing the risk of closure of the River Murray mouth. MDBC technical report 2002/2, options, version 3, Murray-Darling Basin Commission, Canberra, ACT, 24 April 2002, 18 pp

CSIRO (2008) Water availability in the Murray. A report to the Australian Government from the CSIRO Murray-Darling Basin Sustainable yields project. CSIRO, Melbourne, p 217

Dittmann S, Cantin A, Noble W, Pocklington J (2006) Macrobenthic survey 2004 in the Murray Mouth, Coorong and Lower Lakes Ramsar site, with an evaluation of food availability for shorebirds and possible indicator functions of benthic species. Technical report. Department for Environment and Heritage, Adelaide, p 120

Draper N (1994) Aboriginal heritage sites directly impacted by the Kumarangk (Hindmarsh Island) bridge project. A report to the South Australian Minister for aboriginal affairs, April 1994, Department of State Aboriginal Affairs, Adelaide

Draper N (2000) Kumarangk: (Hindmarsh Island, South Australia): Aboriginal heritage assessment. Report prepared for the South Australian Department of State Aboriginal Affairs, and the Ngarrindjeri Traditional Owners of the Lower Murray Region

Elliot JW (1903) The influx of the sea. Editorial. *Southern Argus*. 26 February

Ellis C (1964) Archive tape. LA 120a. Australian Institute of Aboriginal and Torres Strait Islanders, Canberra

Fitzpatrick RW, Grealish G, Chappell A, Marvanek S, Shand P (2010) Spatial variability of subaqueous and terrestrial acid sulfate soils and their properties, for the lower lakes South Australia. Technical report, prepared by the CSIRO for the South Australian Department of Environment and Natural Resources, Adelaide, 122 pp

Fluin J, Gell P, Haynes D, Tibby J, Hancock G (2007) Paleolimnological evidence for the independent evolution of neighbouring terminal lakes, the Murray Darling Basin, Australia. Hydrobiologia 591(1):117–134

Gippel CJ (2003) Review of achievements and outcomes of environmental flow initiatives undertaken on the extended River Murray System to August 2002. Report by Fluvial Systems Pty Ltd, Stockton to The Living Murray initiative of the Murray-Darling Basin Commission, Canberra, ACT, 85 pp

Gosbell K, Christie M (2005) Wader surveys in the Coorong and S.E. Coastal Lakes, February 2005, Australasian Wader Studies Group

Goss KF (2003) Environmental flows, river salinity and biodiversity conservation: managing trade-offs in the Murray–Darling Basin. Aust J Bot 51:619–625

Grabemann I, Krause G (1994) Suspended matter fluxes in the turbidity maximum of the Weser Estuary. In: Dyer KR, Orth RJ (eds) Changes in fluxes in estuaries. Olsen & Olsen, Fredensborg, pp 23–28

Harvey A (1939) Field note book. AA 105 H.K. Fry collection. South Australian Museum, Anthropology Archives, Adelaide

Harvey A (1943) A fishing legend of the Jaralde tribe of Lake Alexandrina, South Australia. Mankind 3(4):108–112

Harvey N (1988) Coastal management issues for the mouth of the River Murray, South Australia. Coast Manag 16:139–149

Hemming SJ (1994) In the tracks of Ngurunderi: the South Australian Museum's Ngurunderi exhibition and cultural tourism. Aust Aborig Stud 2:38–46

Hemming SJ, Jones PG, Clarke PA (1989) Ngurunderi: an Aboriginal dreaming. South Australian Museum, Adelaide

Hiscock P (2008) Archaeology of ancient Australia. Routledge, London, 368 pp

Jensen A, Good M, Harvey P, Tucker P, Long M (2000) River Murray Barrages environmental flows. Report to Murray-Darling Basin Commission. Canberra, ACT, Wetlands Management Program, Department of Environment and Natural Resources, Adelaide, p 118

Kämpf J (2011) The case against seawater desalination in Australia (debate). Aust Water Assoc Water J 38(3):39–40

Kartinyeri D (2006) Ngarrindjeri nation: genealogies of Ngarrindjeri families. Wakefield Press, Adelaide

Kingsford RT, Walker KF, Lester RE, Young WJ, Fairweather PG, Sammut J, Geddes MC (2011) A Ramsar wetland in crisis – the Coorong, Lower Lakes and Murray Mouth, Australia. Mar Freshw Res 62:255–265

Kinne O (1966) Pysiological aspects of animal life in estuaries, with special reference to salinity. Neth J Sea Res 3:222–244

Lamontagne S, McEwan K, Webster I, Ford P, Leaney F, Walker G (2004) Coorong, Lower Lakes and Murray Mouth. Knowledge gaps and knowledge needs for delivering better ecological outcomes. CSIRO: Water for a Healthy Country National Research Flagship, Canberra, 27 pp

Lester RE, Webster IT, Fairweather PG, Langley RA (2009) Predicting the future ecological condition of the Coorong. Effects of management and climate change scenarios. CSIRO: Water for a Healthy Country National Research Flagship, Canberra, 108 pp

Mackay CF, Cyrus DP (2001) Is freshwater quality adequately defined by physico-chemical components? Results from two drought-affected estuaries on the east coast of South Africa. Mar Freshw Res 52:267–281

Marsden S (1985) The River Murray region of South Australia – a short history. In: Dallwitz J, Marsden S (eds) Heritage of the River Murray, South Australian State Historic Preservation Plan Regional Heritage Survey Series: Region 5, SA Department of Environment and Planning, Adelaide 1985. http://www.sahistorians.org.au/175/bm.tags/river-murray/

Martinez Fernandez J, Selma MAE, Aymerich FR, Saez MTP, Fructuoso MFC (2005) Aquatic birds as bioindicators of trophic changes and ecosystem deterioration in the Mar Menor Lagoon (SE Spain). Hydrobiologia 550:221–235

McJannet DL, Webster IT, Stenson MP, Sherman BS (2008) Estimating open water evaporation for the Murray-Darling Basin: A report to the Australian Government from the CSIRO Murray-Darling Basin Sustainable Yields Project, CSIRO, 58 pp

MDBA (2011a) Guide to the proposed basin plan. Murray Darling Basin Authority. http://www.mdba.gov.au/bpkid/guide/

MDBA (2011b) Fact sheet: all about the barrages. Murray–Darling Basin Authority. MDBA publication number: 24/11. http://download.mdba.gov.au/proposed/FS_barrages.pdf

MDBA (2012) Proposed basin plan – a revised draft. Murray–Darling Basin Authority.MDBA publication no: 57/12. http://download.mdba.gov.au/revised-BP/PBP_reviseddraft.pdf

Meyer HAE (1843) Vocabulary of the language spoken by the Aborigines of South Australia. James Allen, Adelaide

Meyer HAE (1846) Manners and customs of the aborigines of the Encounter Bay Tribe: South Australia. South Australian Government Printer, Adelaide. Republished 1963, South Australian Facsimile Editions, No 20, Libraries Board of South Australia

Murray-Jones S, Johnson JE, Ye Q (2002) Goolwa Cockle stock assessment. South Australian Research and Development Institute, Adelaide

Newman R (2000) Hydrology of the Lower Lakes and Coorong. In: Jensen A, Good M, Tucker P, Long M (eds) River Murray barrages environmental flows. Murray-Darling Basin Commission and Department for Environment and Heritage South Australia, Canberra, pp 14–22

Ngarrindjeri Nation (2007) Ngarrindjeri Nation Yarluwar-Ruwe plan: caring for Ngarrindjeri sea country and culture. NLPA, Meningie. http://www.environment.gov.au/indigenous/publications/ngarrindjeri-plan.html

Paton DC, Rogers DJ, Hill BM, Bailey CP, Ziembicki M (2009) Temporal changes to spatially-stratified waterbird communities of the Coorong, South Australia: implications for the management of heterogeneous wetlands. Anim Conserv 12:408–417. doi:10.1111/J.1469-1795.2009.00264.X

Phillips W, Muller K (2006) Ecological character of the Coorong, Lakes Alexandrina and Albert wetland of international importance. Department for Environment and Heritage, Adelaide, p 323

Piersma T, de Goeij P, Tulp I (1993) An evaluation of intertidal feeding habitats from a shorebird perspective: towards relevant comparisons between temperate and tropical mudflats. Neth J Sea Res 31:503–512

Pollard DA (1994) Opening regimes and salinity characteristics of intermittently opening and permanently open coastal lagoons on the south coast of New South Wales. Wetl (Aust) 13:16–35

Raffaelli D (1999) Nutrient enrichment and trophic organisation in an estuarine food web. Acta Oecol 20:449–461

Rankine A (1969) Old ways and new. Ms No. 1439, recorder unknown, 11 March 1969. Australian Institute of Aboriginal and Torres Strait Islander Studies, Canberra

Rogers DJ, Paton DC (2009) Spatiotemporal variation in the waterbird communities of the Coorong. CSIRO: Water for a Healthy Country National Research Flagship, Canberra, 48 pp

Shuttleworth B, Woidt A, Paparella T, Herbig S, Walker D (2005) The dynamic behaviour of a river-dominated tidal inlet, River Murray, Australia. Estuarine Coastal Shelf Sci 64:645–657

Sim T, Muller K (2004) A fresh history of the lakes: Wellington to the Murray Mouth 1800s to 1935. River Murray Catchment Water Management Board, Strathalbyn, 69 pp. http://www.gwlap.org.au/docs/A%20Fresh%20History%20of%20the%20Lakes%202004.pdf

Stillman RA, West AD, Goss-Custard JD, McGrorty S, Frost NJ, Morrisey DJ, Kenny AJ, Drewitt AL (2005) Predicting site quality for shorebird communities: a case study on the Humber estuary, UK. Mar Ecol Prog Ser 305:203–217

Sturt C (1833) Two expeditions into the interior of South Australia, vol 2. Smith, Elder and Co, London

Taplin G (1859–1879) Journal: five volumes as typed from the original by Mrs. Beaumount, Mortlock Library, Adelaide

Taplin G (1873) The Narrinyeri. Reprinted in J.D. Woods (ed) The native tribes of South Australia. E.S. Wigg & Son, Adelaide, pp 1–156

Taplin G (1879) Folklore, manners and customs of the South Australian Aborigines: gathered from inquiries made by authority of the South Australian government. E. Spiller, Acting Government Printer, Adelaide

Tindale NB (1930–1952) Murray River notes. Anthropology Archive, South Australian Museum, Adelaide

Tindale NB (1931–1934) Journal of researches in the south east of South Australia. Anthropology Archive, South Australian Museum, Adelaide

Tindale NB (1935) The legend of Waijungri, Jaralde tribe, Lake Alexandrina. South Australia, and the phonetic system employed in its transcription. Records of the South Australian Museum, 5(3)

Tindale NB (1937) Native songs of the south east of South Australia. R Soc S Aust Trans 61:107–120

Tindale NB (1974) Aboriginal tribes of Australia: their terrain, environmental controls, distribution, limits, and proper names. Australian National University Press, Canberra
Tindale NB (1986) Milerum. In: Nairn B, Serle G (eds), Australian dictionary of biography, vol 11 (1891–1939). Melbourne University Press, Melbourne, pp 498–499
Tindale NB, Long C (n.d.) The world of Milerum. Stage A, vols 1–10. Anthropology Archive, South Australian Museum, Adelaide
Unaipon D (1924–1925) Legendary tales of the Australian Aborigines. Ms copy (MLA 1929 Cyreel 1134). Mitchell Library, Sydney
Unaipon D (1925) The story of the Mungingee. The Home, February, 42–43
Walsh M, Yallop C (eds) (2005) Language and culture in Aboriginal Australia. Aboriginal Studies Press, Canberra, p 226
Webster IT (2006) Hydrodynamic modelling of the Coorong. CSIRO: Water for a Healthy Country National Research Flagship, Canberra, 53 pp
West AD, Goss-Custard JD, Durell S, Stillman RA (2005) Maintaining estuary quality for shorebirds: towards simple guidelines. Biol Conserv 123:211–224
Wolanski E (1987) An evaporation-driven salinity maximum zone in Australian tropical estuaries. Estuarine Coastal Shelf Sci 22:415–424
Wood A (2007) Poor man river: memoirs from the River Murray Estuary. Digital Print, Adelaide, p 163
Ysebaert T, Herman PMJ, Meire PM, Craeymeersch JA, Verbeek H, Heip CHR (2003) Large-scale spatial patterns in estuaries: estuarine macrobenthic communities in the Schelde estuary, NW Europe. Estuarine Coastal Shelf Sci 57:335–355

Port Phillip Bay

Joe Sampson, Alan Easton, and Manmohan Singh

Abstract

Port Phillip Bay (also commonly referred to as Port Phillip) is a large bay in Victoria, Australia. It is an urban waterway as it drains Melbourne. The bay has a narrow entrance, where tidal velocities are high. Port Phillip was formed about 6,000 years ago. Aborigines had lived in the area for tens of thousands of years. Europeans first arrived in the area in 1802. The bay has a temperate climate.

The tides, due to the Moon and Sun, fluctuate vertically by about 2 m at the entrance and less than 1 m for the rest of the bay. Waves in the bay are mainly generated by the wind. The bay has 132 beaches, which are quite popular for swimming. The bay and its environs are home to a variety of flora and fauna. A substantial amount of shipping comes from and goes to the Port of Melbourne. Over four million people live around the bay. The population quite possibly could double in the next 40 years, putting pressures on the environment. The Federal, State and Local governments involved with the bay will have to cope with the effects of global warming, which will likely intensify with time. It could cause flooding of bayside areas regularly. Substantial population growth in the future could cause pollution and put increased demands on the water supply to Melbourne. The Victorian Government has plans to deal with these matters.

Keywords

Port Phillip • History • Tides • Global warming • Storm surges • Waves • Beaches • Ecology • Shipping • Urban areas

J. Sampson (✉) • M. Singh
Mathematics Discipline, Faculty of Engineering and Industrial Sciences, Swinburne University of Technology, Hawthorn, VIC 3122, Australia
e-mail: jsampson@swin.edu.au; msingh@swin.edu.au

A. Easton
Academic and Student Affairs, University of Papua New Guinea, Port Moresby, Papua New Guinea
e-mail: aeaston@upng.ac.pg

E. Wolanski (ed.), *Estuaries of Australia in 2050 and Beyond*, Estuaries of the World,
DOI 10.1007/978-94-007-7019-5_4, © Springer Science+Business Media Dordrecht 2014

Box 1

Joe Sampson and colleagues studied Melbourne-fringed Port Phillip Bay in temperate Australia. Europeans first arrived in the area in 1802; over four million people now live around the bay. A substantial amount of shipping comes and goes to the Port of Melbourne. As a result the bay was substantially degraded but in recent decades the Victorian Governments initiated major water quality remediation measures. In the future The Bay will have to cope with the effects of global warming, which could cause flooding of bayside areas regularly, as well use a likely doubling of the population by 2030 that would exacerbate pollution and put increased demands on the water supply to Melbourne. The Victorian Government has plans to deal with these matters and Melbourne's future looks promising provided relevant Federal, State and Local governments take necessary actions.

Introduction

Port Phillip Bay is a large bay in Victoria, Australia. Melbourne, the capital of Victoria, is at its northern edge. The Bay has a narrow entrance, where tidal velocities are high. The entrance is known as Port Phillip Heads or The Rip. The southern section of the bay, near the Heads, contains extensive sand banks, that are uncovered at low tide. The bay area is about 1,900 km^2. It is about 58 km from North to South and 65 km from East to West at its widest points. Unlike most bays it has a narrow entrance. There are high tidal velocities at the entrance. It is deepest at the entrance, reaching about 100 m. However, most of the bay is fairly shallow, with depths usually below 20 m. The coastline is about 250 km long from Point Lonsdale to Point Nepean, which are on opposite sides of the entrance, which is about 3 km wide. Much of the coastline is urbanised. Figure 1 is a detailed locality map of Port Phillip Bay. Figure 2 is a location map of The Bay. Figure 3 is a detailed map of Port Phillip. Figure 4 shows some locations and depth contours in The Bay.

Rivers and Creeks

A number of rivers and creeks flow into Port Phillip Bay (see Fig. 1). The most significant is the Yarra River, which enters the bay at Melbourne and begins in the Yarra Ranges. The Yarra River is 242 km long. The Yarra has about 50 tributaries feeding into it. There are several species of fish in the Yarra. The Upper Yarra Reservoir was built in 1957, mainly to reduce flooding downstream. This reduced the river's average flow by half.

The following is a summary of information from Melbourne Water (2012). The Yarra is in good condition in the upper sections, where it flows through forested, mountain areas that have been used for water supply for 100 years. But in the lower reaches it is in poor condition due to erosion, pollution, weeds and changes to land use and river flows. As land in the middle and lower sections was cleared for agriculture and urban development clay soils were eroded giving the river a muddy colour. Over the past 20 years, however, there has been a general improvement in the river's water quality resulting from the increased sewering of catchments and the diversion of industrial discharges into the sewerage system. In recent years, loss of habitat has been slowed through revegetation, erosion control and removal of barriers to fish migration. This has helped some animals, such as platypus, which have been found again in areas where they had disappeared.

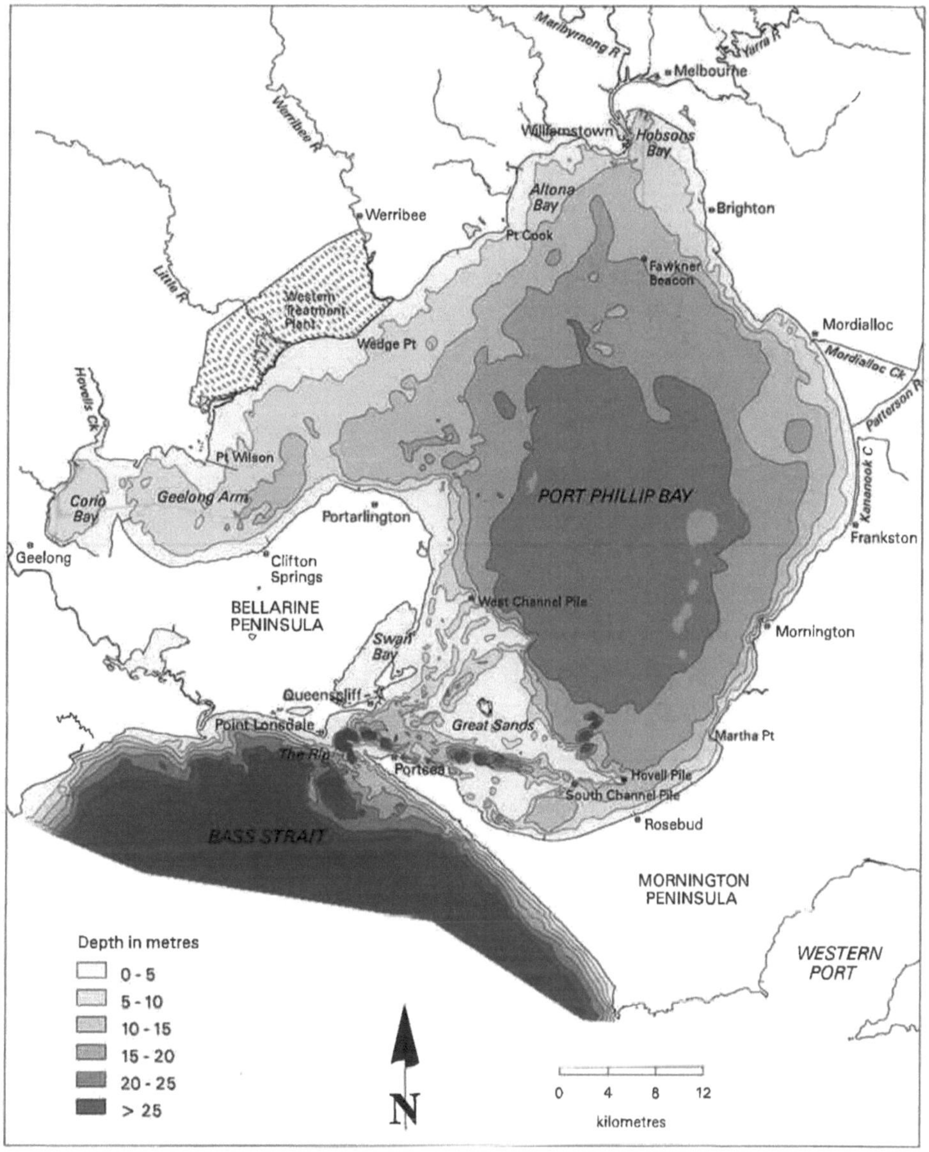

Fig. 1 Port Phillip Bay, showing, bathymetry, locations and rivers (Melbourne Water and CSIRO)

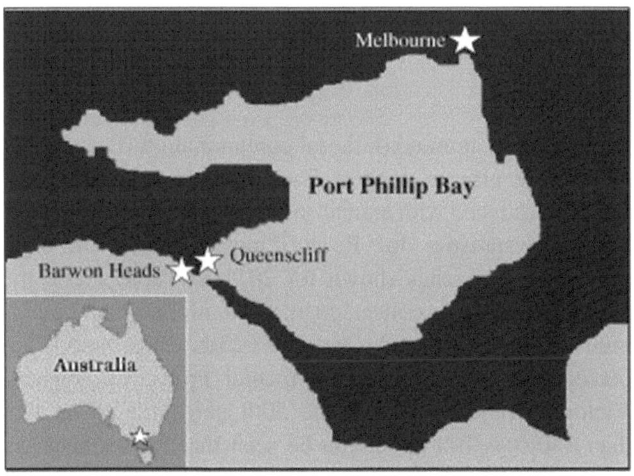

Fig. 2 Location map of Port Phillip Bay (From Ryder et al. 2004)

Other confluents include the Werribee River, Maribyrnong River, Little River, Patterson River and Mordialloc Creek (see Fig. 1).

Rivers' and Creeks' Water Quality

Melbourne Water, a Victorian Government authority, manages Melbourne's water supply catchments, sewage, rivers and major drainage systems throughout the Port Phillip region. It monitors water quality at 136 sites along rivers and creeks in the Port Phillip and Westernport region. The water monitoring plan is designed to assess long-term trends (typically over 10 years). Sites are sampled monthly and tested for the following water quality indicators: water temperature, dissolved oxygen, salinity, pH level, nutrients

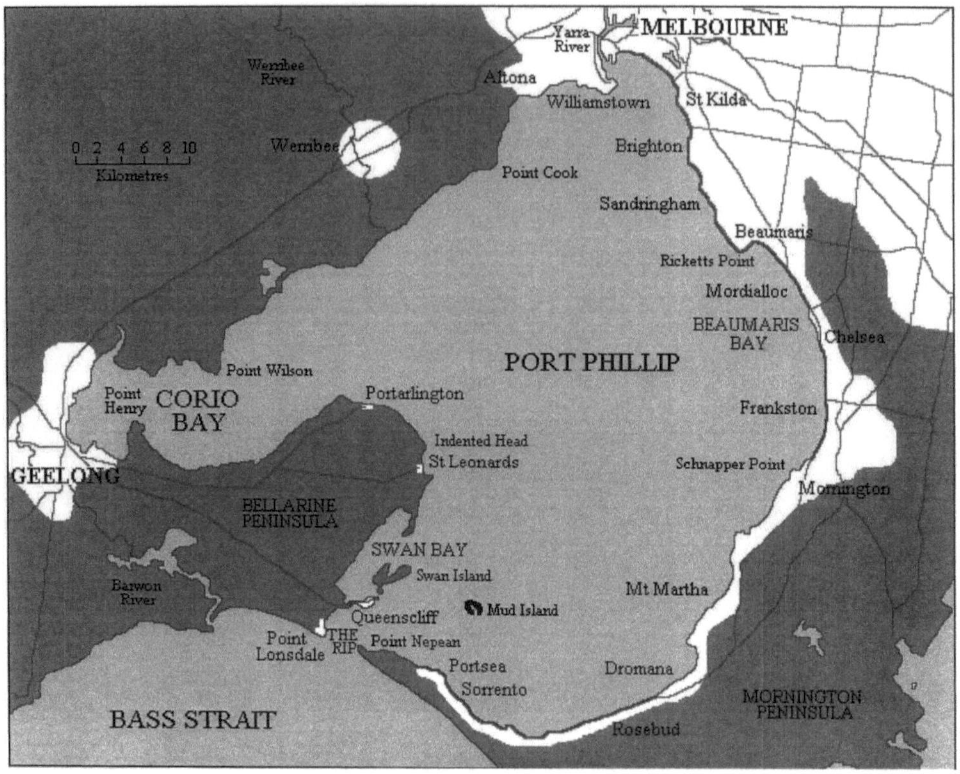

Fig. 3 Detailed map of Port Phillip (Wikipedia). *White* indicates urban areas

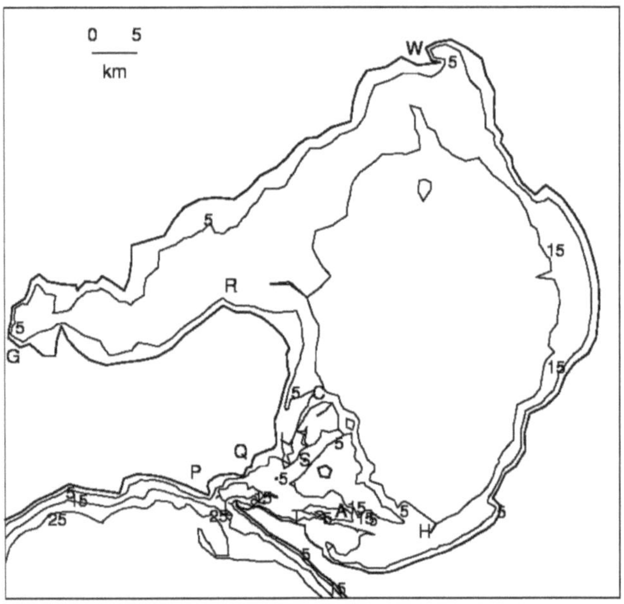

Fig. 4 Depth contours (m) and locations in Port Phillip Bay. Locations are indicated by letters; *P* is for Pt. Lonsdale, *R* for Pt. Richards Channel No. 1, *Q* for Queenscliff, *H* for Hovell Pile, *C* for West Channel Pile, *G* for Geelong, *W* for Williamstown, *A* is in the South Channel, *S* is in the Symonds Channel and *T* is in the Portsea Channel

Table 1 E. coli (orgs/100 mL) at six sites for 2011

E. coli (orgs/100 mL)	n	Min	25th %	50th %	75th %	Max
Kananook Creek	11	200	600	780	5,300	19,000
Little River	11	0	20	80	130	190
Maribyrnong River	11	20	70	100	280	6,000
Mordialloc Creek	10	130	260	370	745	6,500
Werribee River	11	20	160	200	420	760
Yarra River	11	80	255	320	710	5,500

Melbourne Water (2011)

(e.g. a nitrite, ammonia), faecal contamination (E. coli) and metals (e.g. arsenic, copper). Each year a water quality fact sheet is produced with annual summary data for all the sites. Summary statistics for E. coli measurements (in orgs/100 mL) at six sites shown for 2011 in Table 1 are: the number of samples collected ('n'), the minimum and maximum levels measured, and the 25th, 50th and 75th percentiles. The EPA (environmental Protection Agency) Victoria's objective is for the 50th percentile to be less than 150 orgs/100 ml. It can be seen that for the six sites the results overall are not satisfactory, which is true overall of all of the sites.

Geology

According to Short (2005) the uplifted Mornington Peninsula (bordering the south and south-east of the bay) is composed of 200–500 million year old sedimentary rocks (i.e. rock fragments moved by wind, rain and other agents and which in time solidify) and metasedimentary rocks (i.e. sedimentary rocks changed due to changes in physical and chemical conditions) while the bay and the western shore are a depression covered in the north by Miocene (10–25 million year old) coarse sedimentary rocks and in the west by Pliocene (2–10 million year old) coarse sedimentary rocks and to the south the Bellarine Peninsula borders the depression and is composed of Tertiary (3–65 million year old) sediments and volcanics.

History

The bottom of Port Phillip Bay and Bass Strait had been part of a land surface joining Victoria and Tasmania (Keble 1946). Port Phillip Bay was formed about 6,000 years ago during the last Ice Age when the land surface area was flooded (Bird 2006).

The Aboriginal people lived in the area now the base of the bay from 20,000 to 40,000 years ago; in oral history there are stories of the flooding of the bay (Broome 2005). Prior to the arrival of Europeans in the Port Phillip region there were already five aboriginal language groups (the Wathaurong, the Woiwurrung, the Daungwurrung, the Djadjawurung and the Boonwurrung), together known as the Kulin nation, totalling about 10,000 at the time of the Europeans' arrival. The aborigines were semi-nomadic, hunting animals (including kangaroos, emus, wallabies, echidnas, possums, wombats, turkeys, lizards, quolls and grubs) and fish (including crayfish, shellfish, mussels and scallops), making use of boomerangs, spears, fish hooks and nets; they also collected plant foods (Presland 2010). Presland provides an account of life in a Boonwurrung camp close to the shore of Port Phillip Bay, in the warmer months of the year. A camp might comprise 30 people. Campfires were lit. Gum from trees was used not only as a source of sugar but also for plugging leaks in wooden water containers.

The first British to enter the bay were the crew of HMS Lady Nelson, captained by Lieutenant Murray (Lee 1915). Murray sent his first mate, Mr. Bowen, with a small crew of five to examine Port Phillip Bay. The launch returned and the first mate reported that he had found an entrance to the harbor. He saw no aborigines but did see their huts. Murray arrived there on February 15th 1802. He named a high mountain in the south east Arthur's Seat; and a group of islands where swans were plentiful, which he named Swan Isles (they comprise Swan Island and neighbouring islands). He named the bay Port King after Governor King of New South Wales. King renamed it Port Phillip, after the first Governor of New South Wales. Murray noticed a number of aboriginal fires and huts. At one stage onshore Murray and some others met a group of male aboriginals, each in possum skins and each with a bundle of spears. At first relations were friendly but later there was an altercation in which the aboriginals threw spears and the British fired guns. Murray reported favourably on the soil and vegetation. Murray took possession of the Port on the 8th of March. Murray reported back to Governor King on March 24th.

On April 27th 1802 Captain Matthew Flinders, of Royal Navy, entered Port Phillip Bay, thinking that he was the first British person to discover it (Flinders 1814), unaware that Murray had been there 10 weeks earlier. Flinders named Indented Head (on the west coast of the bay) where he met some aborigines with whom he exchanged presents. Flinders stated that the country surrounding the bay was fertile and suitable for agriculture. Flinders said that "were a settlement to be made at Port Phillip, as doubtless there will be some time hereafter, the entrance could be easily defended; and it would not be difficult to establish a friendly intercourse with the natives." When he arrived in Sydney on May 8th he recommended to the Governor that a penal colony be built on Port Phillip Bay.

The British government decided to establish a convict settlement on Port Phillip Bay (Bonwick 1857). In early 1803 Governor King sent Charles Grimes, surveyor-general of New South Wales, to walk round and survey the bay; on February 2 Grimes discovered the Yarra River, which enters the bay at the northern end (Flemming 1803). In late 1803 Colonel Collins was sent out from England to make a convict settlement at Sullivan Bay, near Sorrento. The convicts were 367 males, 17 with their wives, plus 7 children, one of whom, John Fawkner, was a co-founder of Melbourne. Some convicts, escaped including a William Buckley, who lived with the aborigines for several years. According to Broome (2005) the aborigines "believed him to be Murrangurk, a deceased relative, transformed into ghost-like whiteness and strangely bereft of his former language and customs. They took him in, tolerated his oddness and gave him a wife." Presland (2010) says that Buckley lived with them for 32 years and that "the arrival of Batman's party at Indented Head in June 1835 attracted Buckley's interest and awakened in him a desire to re-enter European society. His published reminiscences of his 32 years with the Wataurong today provide invaluable information about the pre-European way of life of Aboriginal people in the Port Phillip area." Collins was not satisfied with the settlement and moved to Tasmania (Bonwick 1857).

John Batman, who had been a settler in Tasmania in 1835, bought from eight aborigine chiefs 600,000 acres of Melbourne for some blankets, knives, looking glasses, tomahawks, beads, scissors, flour and other objects (Bonwick 1857). The British government rejected the deal (Broome 2005). On the 15th of August, 1835, a group of people, including John Fawkner, arrived at Port Phillip Bay. They shortly afterwards settled in the Melbourne area. The Melbourne settlement gradually expanded, with suburbs gradually extending along the west and east coasts and well into the hinterland. In 1838 the town of Geelong, on the southwestern coast of the bay, was proclaimed.

In 1906, G. H. Rogers, then head teacher of Dromana State School, wrote to Gordon McCrae and asked him as one who had lived at Arthur's Seat (an inland suburb adjacent to Dromana), at the south east of the bay, from the 1840s, to give him an account of those days (Daley 1940). McCrae replied in two letters, including a detailed description of the animals. He said: "My father took up Arthur's Seat. . .about 1844. I arrived there. . .with my brothers".

Of aborigines he said "There was a tolerably large tribe of natives on the run, and we have had as many as 200 at a time camped in our paddocks. They were a mild, inoffensive people, largely a fishing tribe, and seemed to enjoy a sense of security when within the posts and rails far larger than that in the open. We could trust them with guns or arms of any kind and found them honest and most useful about the place in aid of other people. The young fellows made excellent stock riders. Some of the women washed well, and the men in several instances shot for the pot and hunted and killed kangaroos for us with our dogs."

He said "Our house we built entirely (save fittings) from hardwood timber, mainly stringy-bark from trees that we felled at the back of the mountain towards the north, where there was any amount of box, stringy-bark, peppermint, and messmate."

Of fish he said "We did a lot of fishing and found a splendid schnapper-ground off Mt. Martha Point. I remember once our filling the boat right up to the sights one afternoon with schnapper weighting from 13 to 16 lb. each. We baited our lines with mutton-fish or Venus' ear, the shells of which we prised off the rocks with our knives. Along with the schnapper we had coat-fish, parrot-fish and leather-jackets, but out on the sandy bottom opposite the house we always got flathead, dog-fish very often, and along the shore we used to have some sport with sting-rays and shark-tailed rays, as also the pig-fish (Castracion) which is said to be as old as the world".

Of the animals he said "when I was in my 'teens, we had kangaroos in immense droves, also brush kangaroos or wallaby, and paddy-melon, bandicoots of two varieties, the great opossum of two sorts, also the ring-tail, the flying squirrel and flying mouse. In the gullies we had the wild dog or dingo, which we caught occasionally in box-traps with a sliding up-and-down door; at the back of the mountain the porcupine ant-eater or echidna in numbers". "Of reptiles we had the great iguana or tree-lizard, running to 5 ft in length; a considerable variety of snakes, including a python; many lizards of different sizes and figures, and with these the rock or sleeping lizard."

On the beaches he said "we had the pelican, the penguin, the grey and grey-white gull, which the blacks called 'bungan', the small white and lavender gull, the pied oyster-catcher, the tern, the cormorant and the little sandpiper and musk duck."

In the swamps were "the Nankeen bird with one long white drooping feather behind the ear; the rail, the bittern, the snipe and jack-snipe, ducks of several sorts – wood duck, black duck, teal, spoonbill, black swan, and geese, cranes blue and white, coots, water-hens, kingfishers here and there, also the swamp or ground-parrot with the barred tail feathers."

On the flats were "spur-wing plovers, minas, and leatherheads, besides in the timber outside many varieties of parrot – lorry, rosella, blue mountain or honeysuckle parrot, sulphur-crested white cockatoo, black cockatoo of two kinds, grey cockatoo with scarlet crest, corella or cockatoo parrot. Among the cherry trees specially the bronzewing pigeon and satin birds."

In the scrubs by the waterholes were "the various honey-eaters, warblers, and red-coat robins, also the emu-wren with the two long emu feathers in its tail; the laughing jackass everywhere, and frequently the butcher-bird (a shrike, but known to some as the whistling jackass). In the bottoms where there was good cover there was any amount of quail. The turkey was not within our limit; none nearer than Boneo or the big swamp. Of birds of prey we had very large eagle-hawks, falcons, also owls, some of the white and of great size. Beyond the honey-eating parakeets and love birds – though there must be more – I find myself at the end of my bird-list."

The timber included coast Banksia (or honeysuckle) and grass-tree.

Climate

Port Phillip Bay has a maritime temperate climate (Sturman and Tapper 1996). The weather is very changeable due to the receiving of air movements both from the warm inland areas and the cold Antarctic Ocean. According to Black and Mourtikas (1992) winds are variable although westerlies prevail in winter and southerlies or south-westerlies in summer. The summers are warm, having a mean maximum temperature of about 25 °C and a mean minimum of 14 °C, while the winter figures are 14 °C and 7 °C. Annual rainfall fluctuates greatly, ranging from about 400 mm on the northwestern shore to about 800 mm on the southeastern shore. Nicholls and Larsen (2011) provided evidence that Melbourne has warmer weather when there are northerly winds from the interior, particularly after a dry period. The hottest temperature ever recorded for Melbourne was 46.4 °C on the 7th of February 2009. The coldest temperature ever for Melbourne was −2.8 °C on the 21st of July 1869. The wettest year in Melbourne was 1916 when 968 mm fell. The wettest month in Melbourne was February 1972 when 238 mm fell. The wettest day in Melbourne was the 3rd of February 2005 when 113 mm fell.

Tides

The tides, the vertical and horizontal motion of the water due to the gravitational pull of the Moon and Sun, have a maximum variation of height of about 2 m at the entrance and generally about 1 m for the rest of the bay. The tide enters the bay at Port Phillip Heads from Bass Strait. Experimental measurements show that the current is fast in the vicinity of the entrance, of the order of 4 m per second, but very small for most of the bay and that the maximum tidal height diminishes sharply from Point Lonsdale at the entrance to Queenscliff, which is about 4 km away inside of the bay. For regions just beyond the Sands region in the South of the bay, high tides occur approximately about 3 h after they occur at Point Lonsdale, at the Port Phillip Heads.

There is an approximate 12 hourly cycle to the tides. The times of greatest current at the entrance are times of greatest water slope; this is either at high water or low water at Port Phillip Heads. Slack water at any point in the bay is followed by high water about 3 h later, then slack water 3 h later then low water 3 h later then slack water 3 h later and so on. At slack water the water slope is close to horizontal and there is little motion throughout the bay. Tidal flow changes direction at Port Phillip Heads slightly after slack water. At high water at Port Phillip Heads the water flows into the bay while at low water it flows out.

The tides are modified by the wind. The Port of Melbourne Corporation issues tide tables for Point Lonsdale, Melbourne and Geelong (Port of Melbourne Corporation 2013) which are useful for boating and shipping. These tables are based on calculations from the National Tidal Centre, which is run by the Bureau of Meteorology. These tables give the times of high tides and low tides (occurring successively approximately) every 6 h, plus at Port Phillip Heads only the flood times (when the tide starts to flow into the bay) and the ebb tides (when the tide starts to flow out of the bay). The tide tables assume a calm bay and hence ignore the effect of winds on heights. The Bureau of Meteorology website gives similar results plus maximum and minimum tidal current speeds.

Tide stream signals are shown by night at the Point Lonsdale lighthouse below the main light. For the first half of the flood stream there is one green light and two green lights for the second half of the flood. For the first half of the ebb stream there is one red light and two red lights for the second half.

The highest and lowest of tides occur in the wake of South Westerly and Northerly winds respectively (Easton 1970). In spite of such effects predictions at Williamstown usually are quite accurate. A highest recorded tide of 2.044 m above chart datum was recorded on 30th November 1934 after a severe storm South Westerly storm. The lowest recorded tide of −0.356 m occurred on 19th September 1926. These extreme levels compare with the mean high water level of 0.79 m and the mean water level of 0.23 m. Under normal weather conditions negative tides are not recorded. In the 22 years from 1934 to 1956 the tide exceeded 1.5 m on four occasions and, each time, it was accompanied by a deep depression of 980 millibars or less. The centre of a depression must be South of Melbourne to cause such extreme tides. The highest recorded tide at Port Phillip Heads was 2.160 m recorded on 4th July 1981. The lowest recorded tide was 0.400 m recorded on 27 October 1972. In contrast the mean highest tide is 1.35 m and the mean lowest tide is 0.45 m. The highest recorded tide at Geelong was 1.619 m, recorded on 29th June 1980. The lowest recorded tide at Geelong was 0.231 m, recorded on 7th March 1972. In contrast the mean highest tide was 0.85 m and the mean lowest tide is 0.35 m.

On 28th November 1954, following a change of wind from North to South, the tide level at Williamstown rose 0.43 m in 30 min and then fell 0.34 m in 55 min (Bradley 1957). Similar short period changes occur frequently. Seiches with ranges of up to 0.46 m are also common in Port Phillip Bay. A seiche is a standing wave. Seiches usually follow the passage of a front across the bay.

An analysis of the Williamstown tide record at the time of the Krakatoa eruption showed oscillations with a range of 0.10 m and a period averaging 86 min; they were not considered to result from that eruption (Wharton 1888). This record contained a wave with a period of 8.3 h (Honda et al. 1908) calculated as the period a seiche would have in Port Phillip Bay.

At Point Lonsdale winds and seiches produce more irregular tides. Tidal currents, which may reach 10–15 km/h, are greatest when the differences between the levels in Port Phillip Bay and Bass Strait are largest.

The sea level at various locations in Port Phillip Bay is measured by tide gauges. The Port of Melbourne Corporation operates six tide gauges in Port Phillip Bay at Point Lonsdale, Queenscliff, West Channel Pile, Hovell Pile, Fawkner Beacon and Williamstown. The Victorian Regional Channel Authority operates tidal gauges at Geelong and Point Richards (see Fig. 4). The data from the gauges is sent daily to the National Tidal centre, operated by the Bureau of Meteorology. The tide record for at least 1 year,

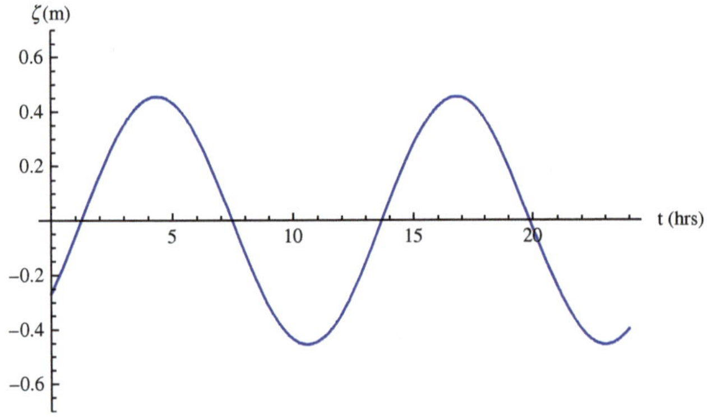

Fig. 5 The M2 partial tide at Point Lonsdale

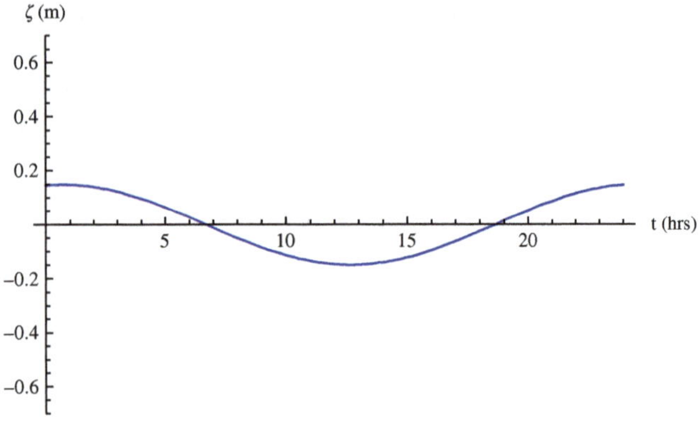

Fig. 6 The K1 partial tide at Point Lonsdale

preferably 3 years, is mathematically analysed into a number of components called tidal constituents (or partial tides). Each constituent is what is called a sine wave, defined by three terms; the amplitude, period and phase. The amplitude is a measure of the strength of the term, the period the time between successive repetitions of the term and the phase measures when it starts repeating itself. The period of each term is known beforehand by gravitational theory, being determined by the motion of the moon or sun or both. On the other hand, amplitudes and phases are obtained by mathematical analysis of tidal gauge data. Once the values of the terms for each constituent are determined at a given site the height of the tides can be predicted at that location for the next calendar year by adding the partial tides. Similar analysis occurs for predicting the tidal current at Point Lonsdale.

Figure 5 shows an example of one partial tide at Point Lonsdale: the M2 component. The vertical coordinate is height and the horizontal coordinate time. The amplitude (i.e. the height of each peak above the horizontal axis) is 0.455 m., the period (i.e. the horizontal distance between successive peaks) is 12.42 h and the phase 36.5° (this is related to when the curve first crosses the time axis at 1.3 h). M2 is caused by the moon's gravitational pull. Figure 6 shows the K1 constituent over 1 day. K1 is caused by the moon's and the sun's gravitational pull. K1 has amplitude 0.149 m, period 23.93 h and phase 280.5°.

The addition of partial tides is illustrated in Fig. 7 by the adding of M2 and K1, the two the largest constituents at Point Lonsdale. The resultant sum is two unequal high tides and two unequal low tides each day.

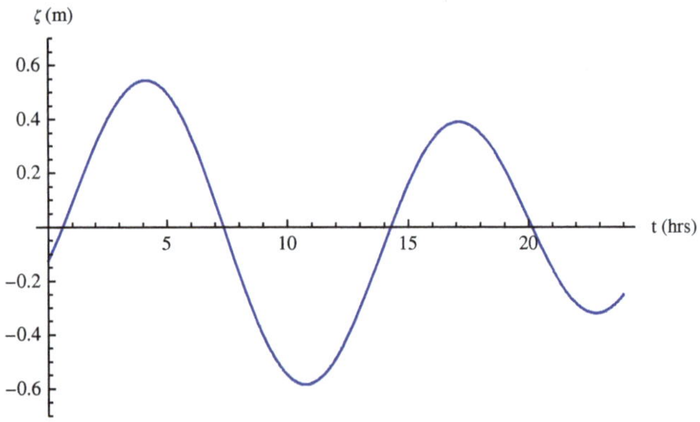

Fig. 7 The sum of the M2 and K1 partial tides at Point Lonsdale

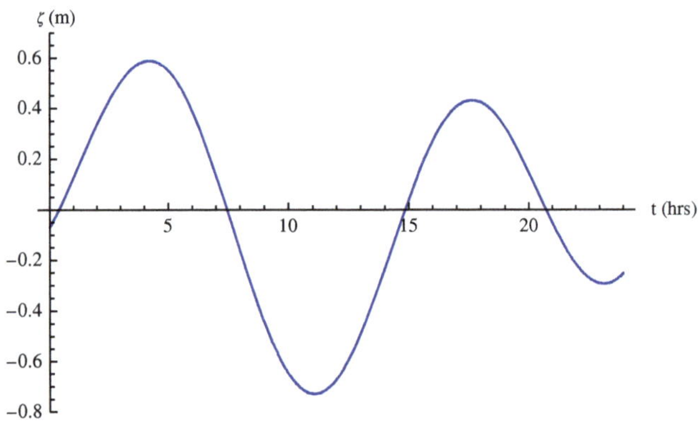

Fig. 8 The sum of the M2, K1, S2, O1 and N2 partial tides at Point Lonsdale over 1 day

Figure 8 shows the sum of the M2, K1, S2, O1 and N2 partial tides at Point Lonsdale over 1 day. These five partial tides give an accurate estimate of the total tide, although the National Tidal Centre uses 112 partial tides to give a more accurate picture. Figure 9 shows the sum of the M2, K1, S2, O1 and N2 partial tides at Point Lonsdale over 31 days. The tidal range around day 4 is almost twice what it is around day 12. This alternation of high tidal and low tidal range is called spring tides and neap tides. This is a common feature throughout Port Phillip Bay.

As only a limited number of tidal gauges exist in Port Phillip Bay to better understand the variation of tidal heights (and tidal currents) throughout the bay the tides can be modelled mathematically using equations derived from physics e.g. the shallow water equations (Vreugdenhil 1998). Because Port Phillip Bay has a mathematically

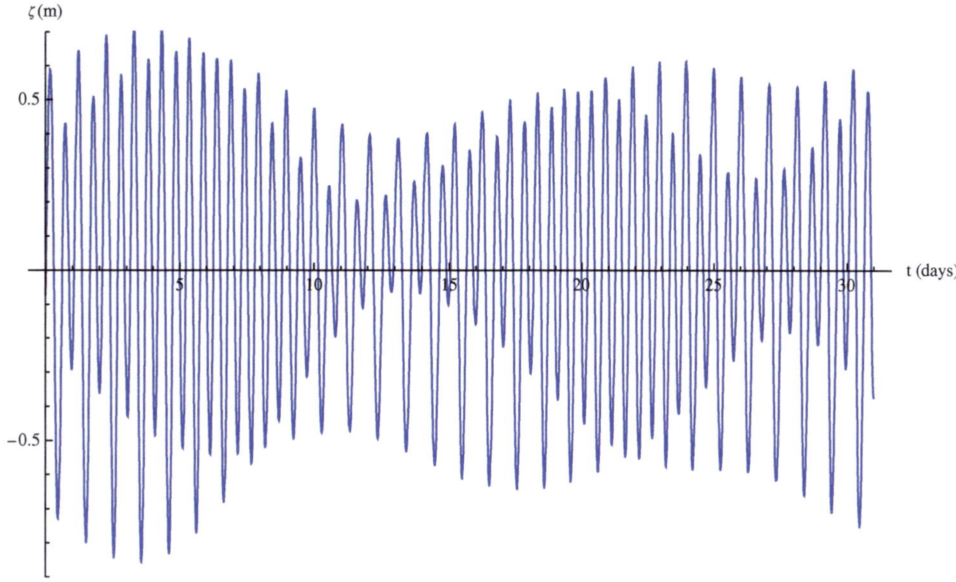

Fig. 9 The sum of the M2, K1, S2, O1 and N2 partial tides at Point Lonsdale over 31 days

complicated shape these equations cannot be solved exactly; they must be solved approximately, although still with great accuracy, using computers to do the large number of calculations required. Such calculations are called numerical modelling. To test the accuracy of any model used the measured tidal heights at places where tidal gauges exist and measured tidal currents at certain locations are compared with modelled results. Input to the model are the values of the main partial tides (e.g. the five most significant) near the entrance to the bay. A number of papers have been published on modelling the tides in Port Phillip Bay. These include papers by Easton (1978), Black et al. (1993), Walker (1997), Lawson and Treloar (2004a, b) and Sampson (2008).

Storm Surges

Storm surges, the flooding of land from sea during a storm, do occur regularly in Port Phillip Bay.

For example, a storm surge occurred in May 1994, which caused inundation of low-lying areas in the suburb of South Melbourne. According to McInnes and Hubbert (1996) storm surges are most frequently caused by periods of sustained westerly winds which occur when cold fronts travel along the southern Australian coast; they mainly occur in Autumn and Spring.

Bird (2006) said “there have been many storms in Port Phillip Bay, and at the end of November 1934 there was a major storm surge. A combination of heavy rainfall and river flooding, low barometric pressure and southerly gales raised high tide water level in the bay by as much as a metre. This caused extensive flooding, rapid erosion of cliffs cut in soft clay and sandstone on the east coast of the bay, and erosion of beaches.”

Hubbert and McInnes (1999) modelled two storm surges that occurred on two occasions in Port Phillip Bay in 1994 using shallow water equations similar to those used in modelling tides; the equations include meteorological effects. The graph of observed and modelled sea levels at one location over a 60 h period shows the two curves to be fairly close to each other.

Waves

The waves in Port Phillip are mainly caused by wind (they can be generated by vessels). According to Lawson and Treloar (2004a) the ocean waves (caused by ocean winds) do not penetrate Port Phillip Bay north of the Sands, the shallow region in the south of the bay. For the rest of the bay the waves are generated by winds blowing over the bay. The period of a wave is the time for one wave to pass a fixed point, typically about 10 s. A wave recording system has been deployed off the coast of Point Lonsdale since 1993, being currently operated by the Port of Melbourne Corporation. Wave data is available online from www.portofmelbourne.com/shipping/weather/waves/wavedadata.aspx The data shown is: the maximum wave height over a half-hour length of time (typically about 4 m); the significant height (i.e. the average height of the highest one third of waves in the half-hour, typically about 3 m), the peak wave period (the maximum wave period over the half-hour, typically 15 s) and the wave direction (generally in an approximate north or south direction depending on whether it is high tide or low tide).

The wind-generated water waves can be mathematically modelled using the shallow water equations, which are used to model tides; the effect of wind is included in the equations. The modelling shows the effect of the wind superimposed on the tide. Black et al. (1993) used the shallow water equations to model currents and sea level variations in Port Phillip Bay. As data showed that wind directions and speed are fairly similar at any given time, the 1-hourly measurements from Point Cook, a low-lying peninsula (in the north-west of the bay), exposed to all wind directions, were used in the model. Comparison of model results for sea level and current with experimental results showed a generally good match. They concluded that better results would have been obtained if the model had been refined. Wilson (1982) tested three models for the prediction of wave height and period in Port Phillip Bay due to local winds. The model compared favourably with experimental data. According to Lawson and Treloar (2004b) the wind driven currents in Port Phillip Bay are not very strong, reaching about 0.2 m s^{-1} during storm events and typical values in the centre of the bay are less than 0.4 m s^{-1} during per second. Bird (2011) wrote that the waves in Port Phillip Bay generate on the east coast a longshore drift, i.e. movement parallel to the shore, of sand from north to south in summer and vice versa in winter. Belski et al. (2012) mathematically modelled waves generated by a storm in the bay on October 9 and 10, 2009. Input winds for the model were provided by the Bureau of Meteorology. The winds were predominantly south-easterly, being usually about 10 m/s. The average wave height was about 0.7 m and usually of similar direction to the wind.

Tsunami

A tsunami is a series of water waves generated by such disturbances as underground earthquakes, landslides or volcanic eruptions (Bryant 2001). Unlike normal water waves tsunamis are very long, with hundreds of kilometres between crests. Although small in height on the ocean (about 1 m) a tsunami grows dramatically in height as it reaches the shore. According to Bird (2006) a tsunami could be generated by an earthquake in the Southern ocean reaching Port Phillip Bay with a large tsunami causing inundation similar to that of the 1934 storm surge in the bay that was mentioned previously in this chapter. Bryant, the author of "Tsunami: The underrated hazard" (Bryant 2001) also said (The Age 2005a) that a tsunami could be generated by an earthquake in the Southern ocean and "Port Phillip Bay would be susceptible to the full force of a tsunami if the coastline was struck. It would get in the bay and start wreaking havoc around the shoreline". Appelbe (The Age 2005b) disagreed with Bryant, saying that the energy delivered through the narrow entrance to the bay would be small and would dissipate quickly aided by the network of narrow channels, islands and sandbars right inside the heads. Greenslade (2012, Centre for Australian Weather and Climate Research, personal communication), who has done numerical modelling on the effects of tsunamis on Australia but has not modelled tsunamis inside Port Phillip Bay, is in agreement with Appelbe. She said "Port Phillip Bay should be relatively low risk due to: (1) distance from subduction zones; (2) shallowness of Bass Strait; (3) narrowness of the Heads at least for earthquake generated tsunamis, which is where our priorities are". Greenslade has modelled a tsunami generated by a large earthquake occurring south of New Zealand. The maximum amplitude dropped from 2 m in the Tasman Sea to less than 0.25 m outside Port Phillip Bay. Greenslade says "the model resolution here is 4 arc minutes (1.4 km), so not enough to resolve the Heads (3.5 km wide), so we can't really say anything specific about the amplitudes inside the Bay without further high resolution modelling, but it is certainly the case that Bass Strait will cause the tsunami to slow down and reduce in amplitude." The disagreement between Bryant and Applebe about a tsunami generated in the Southern Ocean could be resolved by numerical modelling of a tsunami in Port Phillip Bay due to an earthquake in the Southern Ocean. Numerical modelling of actual tsunamis has been found to be accurate, in good agreement with experimental results (Mader 2004).

The Australian Government has a Joint Australian Tsunami Warning Centre that warns the community of tsunamis. The Victorian State Emergency Service has a webpage on what to do in the event of a tsunami.

Beaches

Most of Port Phillip Bay is ringed by beaches, alternating sometimes with cliffs, as is shown in a map in Bird (2011); Swan Bay, in the south-west of the bay, is mainly bordered by salt marshes. According to Black and Mourtikas (1992) "broad sandy beaches occur in two main regions on the eastern coastline; one from Mordialloc to Frankston, the other from Dromana to Portsea. On the west coast narrow sandy beaches extend continuously from Williamstown to Corio Bay". Swan Bay, in the south west, contains mudflats that are intertidal, i.e. submerged for part of the day and above the sea the rest of the day because of the rise and fall of the tides.

The beaches are popular in summer months, with many people congregating along them and swimming alongside them. The beaches can be dangerous for swimmers and for this reason are patrolled by 12 surf lifesaving clubs and 15 royal lifesaving clubs. A detailed list of the 132 beaches in Port Phillip Bay is given in Short (2005); a map is given for each beach, with details about winds, waves, swimming and surfing. A photograph of St. Kilda Beach, on the North of The Bay, is shown in Fig. 10a. A photograph of Sandringham Beach, on the North East of The Bay, is shown in Fig. 10b.

Fig. 10 (**a**) St Kilda Beach (The Age June 22 2012) (**b**) Sandringham Beach (The Herald Sun April 4 2012) (**c**) Burrunan Dolphins (The Age Sep 16 2011) (**d**) Swanson Dock, Port of Melbourne, with container terminals; the Yarra River is in the foreground (Port of Melbourne Corporation)

Beach Water Quality

EPA (Environment Protection Authority) Victoria, a government body, monitors levels of entrecocci (bacteria), measured in organisms/100 mL (org/100 mL) weekly at 36 beaches in Port Phillip Bay (see Fig. 11) (EPA Victoria 2012b). During the 2011–2012 summer season the EPA found bacterial water quality suitable for swimming during fine weather at Port Phillip Bay beaches during Summer 2011–2012, with only a handful of sites exceeding the water quality trigger levels, with each of these exceedences following heavy and persistent rain. The EPA assesses beach water quality against bacterial water triggers. If a sample greater than 400 org/100 mL is collected a resample is collected; if the resample is again greater than 400 org/100 mL a swim advisory is issued. If a sample of 1,000 org/100 mL is obtained a swim advisory is issued.

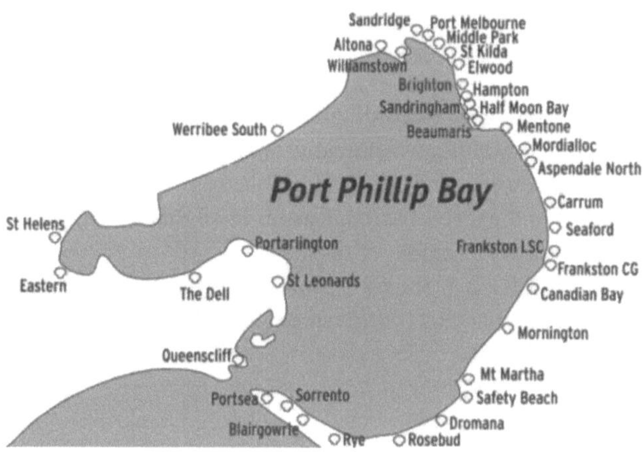

Fig. 11 Beaches monitored by EPA Victoria

Long term trends in water quality at the 36 beaches are assessed by comparisons of annual 75th percentiles against the State Environment Protection Policy (Waters of Victora) (SEPP (WoV)) 75th percentile objective of 150 org/100mlL. Over a period of five summers (from 2007–2008 to 2011–2012) the objective was reached at 16 of the 36 beaches for all five summers, at 11 for four summers, at 11 for three summers and at one for two summers. In 2007–2008 all 36 sites reached the objective; in 2008–2009 35 did; in 2009–2010 34; in 2010–2011 23; in 2011–2012 19. The more rainfall in a summer the fewer the sites that

Table 2 Enterococci (orgs/100 mL) 75th percentiles for a sample of beaches in Port Phillip Bay over five summer periods; SEPP (WoV) objective: 75th percentile of 150 or/100 mL

Beach	2007–2008	2008–2009	2009–2010	2010–2011	2011–2012
Dromana	15	20	26	58	172
Mornington	47	155	20	70	160
Mentone	20	146	195	273	230
Hampton	<10	<10	20	53	74
Williamstown	20	20	26	240	96
St Helens	68	<10	10	43	44
Rainfall (mm) Nov–Mar; average of five sites	224.8	165.0	250.8	463.2	312.7

EPA Victoria (2012b)

Table 3 The results at 6 Port Phillip Bay beaches for 5 of the 16 weekly entrecocci results (orgs/100 mL) over the 2011/2012 summer period

Beach	21 Nov 11	27 Dec 11	10 Jan 12	6 Feb 12	5 Mar 12
Seaford	210	10	52	63	250
Aspendale North	73	41	230	74	63
Beaumaris	96	500	510	460	330
Brighton	86	98	<10	30	41
Port Melbourne	20	16,000	51	330	990
Altona	10	320	1,000	120	52

EPA Victoria (2012b)

Table 4 Annual mean salinity (psu) compared to SEPP (WoV) objectives (2008–2011)

Location	SEPP (WoV) objective	2008	2009	2010	2011
Corio Bay	33.1–36.5	38.1	38.0	37.3	35.0
Hobsons Bay	32.1–35.5	36.8	36.7	35.5	32.4
Werribee	32.6–36.0	37.4	37.3	36.7	34.6
Central Bay	32.1–35.5	37.2	37.1	36.5	34.4
Dromana	33.1–36.5	36.9	36.8	36.0	34.3

met the objective. According to EPA Victoria (2012b) "on days of fine weather water quality was found to be generally good but during and after rain, bacterial levels could be elevated and exceed bacterial water level quality triggers". Table 2 shows a sample of results.

The results at 6 sites for 5 of the 16 weekly entrecocci results over the 2011/2012 summer period are shown in Table 3.

Salinity

According to EPA (2012a) "salinity is an important parameter for the ecology, transport and mixing characteristics of Port Phillip Bay. Salinity has been shown to influence the abundance and composition of biological communities as well as having physiological effects on hatching success and growth of squid and prawns in other systems, even with small (<5 psu) changes". Table 4 shows annual mean salinity at a number of places in The Bay (refer Figs. 5 and 13) and comparisons with State Environment Protection Policy (Waters of Victora) (SEPP (WoV)) objectives.

Islands

Port Phillip Bay has a small number of islands, most of which are in the south of the bay. The largest is Swan Island, of area 1.4 km^2. The island is bordered mainly by marshes. Fortifications were built on the island in the 1870s during the Crimean War to stop a Russian invasion of the bay. The island contains a training area for ASIS (the Australian Secret Intelligence Service).

The Mud Islands in the south of the bay consist of three low-lying islands which enclose a lagoon, which is surrounded by a marsh. A large number of birds are on the island.

Ecology

In the History section of the chapter the flora and fauna in the southeast of the bay in the nineteenth century, according to George McCrae who lived there from the 1840s, was discussed.

Wildlife that exists on land areas around the bay include possums, kookaburras, cockatoos, lorikeets, bats, snakes and foxes.

According to Museum Victoria (2006) there are 350 species of fish in the bay, including sharks, stingrays, perch, mullet, snapper, whiting and perch, plus many mammals,

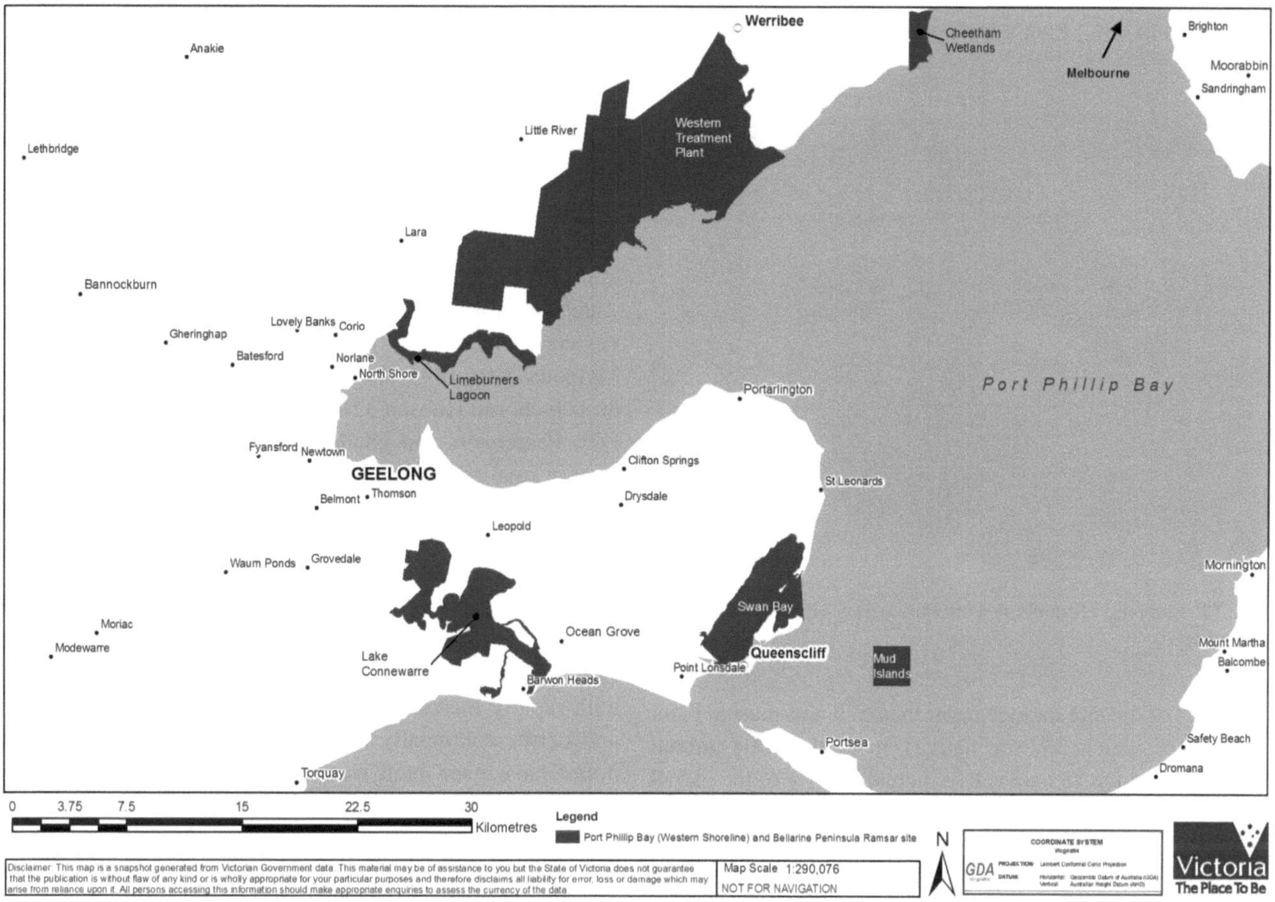

Fig. 12 Port Phillip Bay (Western Shoreline) and Bellarine Peninsula Ramsar site (Department of Sustainability and Environment (Victoria))

including dolphins, sea lions, seals and whales. The Burrunan dolphin species, discovered in 2011, lives only in the bay and Gippsland Lakes, eastern Victoria (see Fig. 10c).

According to Department of Sustainability and Environment (2012a) the bay also supports several hundred species of molluscs, several hundred species of crustaceans, at least 200 species of seaweeds, several hundred species of polychaetes (bristle worms), two species of seagrass, several hundred species of cnidarians (jellyfish, corals, etc.) and several hundred species of sponges.

The Port Phillip Bay Environmental Management Plan (EMP) has a number of objectives including the most important priorities – to reduce nutrient and marine pest risks (The State of Victoria, Department of Natural Resources and Environment 2002). Action on implementation of the EMP is reported annually. Copies of these annual reports can be found at http://www.dse.vic.gov.au/coasts-and-marine/marine/report-library

The Seagrass and Reefs Program for Port Phillip Bay (The Program) (Department of Sustainability and Environment 2012b) is a Victorian Government program. The aim of the program is to better understand the ecology of seagrass and temperate reef habitats to ensure the sustainability of the marine environment and to have Government, industry and community better informed and so better able to manage the bay marine environment. As seagrasses and reefs are important fish habitats, it will be of benefit to have improved management of these ecosystems. A recent study found that the health of the seagrass (measured by seagrass cover, length and stem density) is higher or consistent with past seasonal trends at six of eight monitoring sites (Hirst et al. 2012).

The "Taxonomic Toolkit for Marine Life of Port Phillip Bay" (Museum Victoria 2012) provides information and images for over 1,000 animal species inhabiting The Bay. It is useful for marine scientists, environment managers and the general public.

The Port Phillip Bay (Western Shoreline) and Bellarine Peninsula Ramsar site involves seven separate areas, mainly along the western coast of the bay (see Fig. 12); it includes Swan Bay and the Mud Islands (the Ramsar Convention is an intergovernmental treaty, whose member countries wish to maintain the ecological character of their wetlands and use them sustainably). According to the Victorian Government Department of Sustainability and Environment website

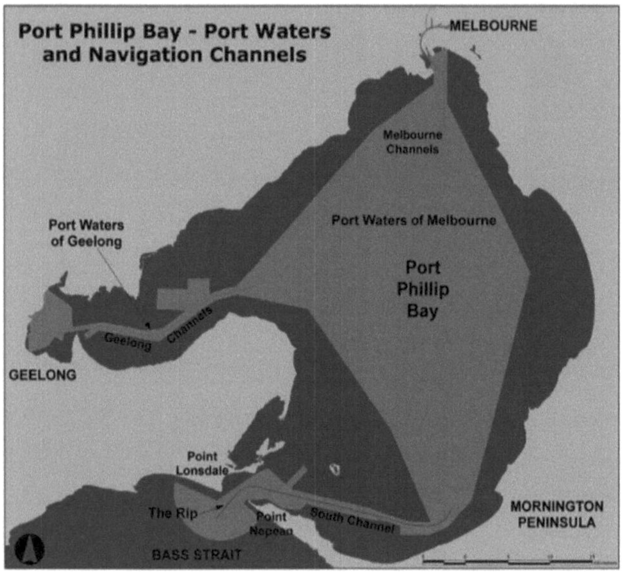

Fig. 13 Shipping channels in Port Phillip Bay (Port of Melbourne Authority)

"features of the site include more than 579 non-marine plant species (of which 332 are native), more than 304 animal species (of which 285 are native including 50 % of the global population of the global population of the orange bellied parrot), a drought refuge for water birds when inland lakes and wetlands dry out and with sites of cultural heritage due to wetlands plentiful in resources that have attracted Aboriginal tribes for thousands of years".

Cohen et al. (2000) found that there is generally a decreasing amount of biomass and species diversity with increasing bay depth.

Costermans (1983) says that trees around Melbourne include gums, stringybark, peppermints, box, coast banksia, tea-trees, wild cherry and wattle.

Shipping

Because sections of Port Phillip Bay are fairly shallow channels have been built, beginning in the late nineteenth century to access berths in Melbourne and Geelong, with a margin of safety beneath the bottoms of their hulls and the sea bed. The South Channel is a shipping channel running in an east–west direction from the bay entrance to near Dromana, on the south-east coast. There are channels running into Melbourne and into Geelong. A map of the navigation channels and port waters is shown in Fig. 13. The Port of Melbourne fronts onto Hobson's Bay The POMC (Port of Melbourne Corporation) is responsible for piers and wharves in Melbourne and the Geelong Port for those in Geelong. The draught for the Melbourne channels is 14 m. A photograph of Swanson Dock, Port of Melbourne, is shown in Fig. 10d.

The Port of Melbourne is Australia's largest container and general cargo port, handling about 37 % of Australia's container trade. More than 40 shipping lines make around 3,100 ship calls a year to Melbourne. In 2011 the Port was handling 2.5 million containers. This is expected to double by 2025.

The history of the Port of Melbourne, which fronts onto Hobson's Bay, as discussed in Port of Melbourne Corporation (2012a, b), is summarised in this paragraph. An underwater bar at the Yarra River entrance ruled out the entry of vessels more than 9 ft of water. After the settlement of Melbourne in 1835 ships arriving from overseas had to drop anchor in Hobson's Bay or Sandridge (Port Melbourne) Pier. Passengers and goods then had to be shipped up the river in smaller vessels. The rapid sixfold growth of Victoria from 1851 to 1861 placed a lot of strain on the port. The Yarra River was widened and deepened in the 1880s. Victoria Dock, a major dock facility, was opened in 1893. By 1908 it was handling 90 % of Victoria's imports. In 1914 its capacity was enlarged by the addition of a central pier and in 1925 the entrance was widened. Between the two world wars with rapidly increasing tonnage and size of ships channels and berths continually required dredging. More and more docks have been built over the years: Appleton in 1956, Webb in 1960 and Swanson in 1969.

Mathematical modelling of tides was useful in predicting beforehand the effect of the proposed Port Phillip Bay Channel Deepening Project, which began on 8 February 2008. Prior to the project the guaranteed minimum depth in the main commercial shipping channels of the bay was 11.6 m at all tides. The government of Victoria, in the late 1990s, announced plans to deepen the channels in the bay so that they could accommodate ships of up to 14 m draft at all times. The plans were to deepen sections of the Great Ships Channel at the bay's entrance, the South Channel and channels going into Port Melbourne and Williamstown. In July 2004 the Port of Melbourne Corporation (which had assumed some of the duties of the former Victorian Channel authority), the Victorian State government authority responsible for the channel deepening project, released the Environmental Effects Statement on the channel deepening project. The section of the Environmental Effects Statement dealing with the effect of the channel deepening on the tides was written by Lawson and Treloar (2004a, b). Lawson and Treloar modelled the astronomical tides in Phillip Bay.

The levels of tides for the existing topography and for the topography including the proposed channel deepening were computed. Their results showed that the effect of the proposed channel deepening on then existing tidal levels in Port Phillip Bay (using the partial tides M2, S2, K1, O1 and N2) would be to increase the maximum tidal height at most locations, with the greatest increase 0.008 m and the greatest reduction 0.002 m. Sampson et al. (2005) also modelled the then existing tides and the post channel deepening tides, giving results very close to those of Lawson and Treloar with the maximum tidal height increasing at most locations, but with a greatest increase of 0.007 m and a greatest reduction of 0.002 m. According to the report published by the Australian Government/Bureau of Meteorology (2011) measured high tides post channel deepening are generally around 0.01 m higher than pre channel deepening, which is in general agreement with the results of Lawson and Treloar and Sampson et al.

There are plans to expand Melbourne's port capacity (Port of Melbourne Corporation 2012a; The Age 2012b). These are discussed in the later The Future section.

Living Communities

Urban Areas

Most of the population surrounding the bay is concentrated in Melbourne and its suburbs (4.17 million people in June 2011, according to the ABS (Australian Bureau of Statistics)). Melbourne is at the north of the bay, while its suburbs extend along the east and west coast and into the hinterland. The rest of the population is in Geelong and its suburbs (0.18 million in 2010, ABS), on the west of the bay.

Impact of Urbanisation/Industrialisation on the Marine Environment and Management of Such

This paragraph summarise information from an Australian Government document (Australian Government 2006). By the end of the nineteenth century, the waste and sewage pollution of the Yarra River was causing serious health problems. This led to the Melbourne Metropolitan Board of Works to initiate a major sewerage program in 1893 with the first Melbourne homes being connected to a waste treatment plant at Werribee. Management of the bay involves a network of government, private sector and community agencies working together. The coordination of actions is through the Association of Bayside Municipalities (ABM), comprising the ten municipalities bordering The Bay. Current management objectives were developed in response to a major environmental study of the bay by CSIRO in 1992–1996 (CSIRO 1996) plus long term trends identified by the EPA (Environment Protection Authority) Victoria from ongoing monitoring data for 1984–1999. Publication of a litter study by Melbourne Water in 1993 led to programs to change behaviour with pet faeces, cigarette butts and litter from take away foods. The CSIRO study compared recent long term monitoring with earlier sampling data and concluded that nutrient levels have been reasonably stable since 1984 and are less than those in the 1970s prior to the major upgrading of sewage treatment. Stormwater pollution is most evident when runoff after summer storms causes beach closures or significant deposit of litter on beaches. Management and monitoring of water quality is by the EPA. The EPA sets targets for water quality through enforceable regulations. Melbourne Water is responsible for water harvesting, supply and treatment. Local government, through the ABM, is responsible for beach cleaning and management of local creeks and drains and promotion of public awareness about water quality. The ABM councils, with support from the federal and state governments, have undertaken capital works programs for installation of gross pollution and sediment traps.

The CSIRO 1992–1999 study concluded that "despite a population of over three million people living around its shores, the Bay is generally cleaner and healthier than other bays around the world near large cities." The study concluded that there has been a long-term decline in the level of most toxicants (e.g. mercury, copper, zinc) in the Bay, probably due to stronger environmental regulations and diversion of liquid waste to the sewerage system.

The Future

Global Warming

The evidence for global warming and its impact on the Bay in the future and what governments need to do to minimise the impact is discussed here in detail.

According to IPCC (2007) global average sea level in the twentieth century rose at a rate of about 1.7 mm per year. Satellite observations since 1993 show that sea level has been rising at a rate of about 3 mm per year. The IPCC says that the sea level is projected to rise at an even greater rate in the twenty-first century, the major causes being global warming due to thermal expansion of the oceans (water expands as it warms) and the increasing melting of ice. One scenario is that by the mid-2090s the global sea level is 0.44 m above 1990 levels. Another scenario (Australian Government/Department of Climate Change 2009), has a sea level rise of up to 1.5 m or more at the end of the century.

Mitchell et al. (2000) give the following results for sea level rises from tide gauges in Port Phillip Bay: Williamstown: 0.26 mm per year (32 years of data), Geelong: 0.97 mm/year (25 years) and Point Lonsdale: −0.63 mm/year (34 years). Lawson and Treloar (2004a) made this comment on the data: "the variability in these numbers indicates the difficulty of determining long-term trends from historical tide-gauge data since there would not be expected to be any variability in the long-term sea-level between these sites. It is more likely that the variability reflects some variation in the relative movements of the tide gauges with respect to sea level." Lawson and Treloar obtained the hourly recorded sea-levels from Williamstown tide gauge for the years 1966–2002 inclusive (37 years). For each year the mean (average) sea level was mathematically calculated making use of all the partial tides. The general trend was for the mean sea level to rise by 2.3 mm per year, i.e. about ten times what Mitchell obtained. Lawson and Treloar says that if one excludes from their data years where there are data gaps and restricts the calculations to the years used by Mitchell et al. a similar result is obtained. Lawson and Treloar conclude that "it is recommended that it be assumed that there is a rise in sea level relative to land in Port Phillip at a rate between 1 and 2 mm per year." However the IPCC has been estimating about 3 mm per year.

According to Schwartz (1967) each centimetre of sea level rise in the Great Lakes in North America results in a metre of beach recession. Using this result a 3 mm sea level rise per year would result in 30 cm recession of beaches per year or 30 m per decade. A 0.44 m sea level rise by the 2090s as estimated by IPCC, would lead to a 44 m beach recession, which, according to Bird (2006), is more than the width of the beaches so that they would have to be artificially nourished to prevent disappearances.

According to Bird (2006) the effects of rising sea level in Port Phillip Bay will be that "the mouths of inflowing creeks and rivers such as the Yarra and the Werribee will become wider and deeper as high tides attain augmented levels. Nearshore water will deepen, allowing larger waves to break on the shore, intensifying erosion of cliffs and beaches. Where the cliffs are in hard rock, such as the granodiorite of Mount Martha, erosion will be slight as the sea rises, but the soft clay and sandstone cliffs of the Bellarine Peninsula and the north-eastern coast between Sandringham and Balcombe Bay are likely to be cut back more rapidly as wave attack reaches higher levels."

Rising sea level and increased wind speed during storms, both caused by global warming, will mean that a storm surge at high tide will cause more coastal flooding. McInnes et al. (2009) modelled a 1 in 100 year storm surge plus tidal height in 2100 with a 19 % increase in the winds forcing the storm surge and 82 cm of sea level rise at five Victorian locations, not including Port Phillip Bay; they showed an increase in overall height of the surge of about 2.4 m. They assumed that tide ranges were not changed with rising sea levels but said that tidal range may actually increase in Port Phillip Bay.

The City of Port Phillip, which has a population of about 90,000, is at the north end of the bay. It produced a study (City of Port Phillip 2007) on planning for climate change. The study showed the St Kilda foreshore and the area around Elwood canal as being prone to storm surge flooding and stormwater flooding.

According to a report by the Australian Government/Department of Climate Change (2009) about 30,000 residential buildings in regions bordering Port Phillip Bay may be at risk of inundation from a sea level rise of 1.1 m and storm tide associated with a 1-in-100 year extreme storm. The current replacement value of such buildings is of the order of $7 billion. A quarter of buildings in Hobson's Bay and Port Phillip local government areas may be at risk of storm tide (storm surge plus tide) by 2100.

The same Australian government report of 2009 says "if it were necessary to protect cities such as Sydney or Melbourne, dykes would potentially need to be constructed around low-lying lands or across estuary entrances, as is done in the Netherlands. A dyke across the entrance to Port Phillip Bay would be challenging to construct. 'The Rip' between the heads would require a dyke of some 3 km long and constructed in some 20 m water depth. Despite this distance, 'The Rip' is relatively narrow compared with the circumference of the rest of the Bay, which is around 220 km. The dyke would need to have locks to allow water and ships to pass. The locks would then be shut if a storm surge or high tide was forecast. However, because of the powerful currents and swells (waves not generated by local winds), constructing a dyke stretching across 'The Rip' would be a difficult engineering challenge and would be very expensive."

Federal, State and Local governments are all aware of the steps that they need to take to minimise the impact of global warming. The Australian Government is a signatory to the Kyoto Protocol, which binds industrial countries to reduce their emissions of greenhouse gases. The carbon tax, introduced in 2012, is designed to reduce greenhouse gas emissions.

New Container Port

An article in The Age newspaper on September 4, 2012, (The Age 2012a), stated that the Victorian government is considering building a massive new 'Bay West' container port, near Point Cook on the north west coast. Industry sources were quoted as being concerned that a hard surface would make any dredging difficult.

Expanding Melbourne's Port Capacity

In 2011 the Port was handling 2.5 million containers. This is expected to double by 2025.

For this reason there are plans to expand Melbourne's port capacity (Port of Melbourne Corporation 2012a; The Age 2012b). There will be a new container terminal at Webb Dock. Two and a half million cubic metres of silt will be dredged from near Webb Dock as part of the $1.2 billion redevelopment plan. In addition 50,000 cubic metres will be

Fig. 14 Proposed new container terminal at Webb Dock (Port of Melbourne Authority)

dumped within the area of the Port of Melbourne Dredged Material Ground, located near the centre of The Bay; there will be no environmental impact assessment because the state government says that it is not necessary. See Fig. 14 for a diagram of the proposed project.

Population Growth

According to the ABS (Australia Bureau of Statistics), Melbourne's population grew by 17.4 % from 2001 to 2010, i.e. at a rate of 1.8 % per year. If this rate continued the population at 2050 would be double that of 2010, i.e. around eight million people. An increased population might increase pollution in Port Phillip and the Yarra River and other rivers and creeks.

As well an increased population will place more demands on the water supply, some of which comes from the Yarra River. The government has what is called "A Water-Supply Demand Strategy for Melbourne 2006–2055" (Melbourne Water 2006). It outlines socially acceptable and cost-effective actions to save and source water for Melbourne till 2055. The actions are:

(i) maintain existing water conservation savings;
(ii) smarter management of existing river supplies;
(iii) save more water at home;
(iv) save more water at work and play;
(v) reduce water leaks and wastage;
(vi) increase the use of local water sources;
(vii) investing in water efficiency opportunities around Melbourne; and
(viii) harnessing alternative strategies.

A Cleaner Yarra River and Port Phillip Bay

In March 2012 the Victorian Government Minister for Environment and Climate Change, Ryan Smith, set up a government taskforce to develop an action plan for the Yarra River and Port Phillip Bay, focussing on four key priorities. The first priority was to enable more effective coordination between government agencies in protecting water quality and providing timely information to communities about water quality events. The second priority was to manage threats to water quality, including pollution, litter and stormwater inputs by identifying new priority actions to address them. The third priority was to develop easier ways for the community to access information about water quality of the Yarra and The Bay. The fourth priority was to support Victorians to take actions that care for and protect the Yarra and the bay.

In October 2012 the Victorian Government Department of Sustainability and Environment released the plan, a five-year plan, called "A Cleaner Yarra River and Port Phillip Bay", to "improve the health of the Yarra River and Port Phillip Bay". The report says that good water quality in the river and bay is important for healthy ecosystems and clean safe swimming.

The report stated that good water quality in the Yarra River and Port Phillip Bay is vital to provide healthy and diverse ecosystems, clean and safe swimming and recreational and commercial fishing and aquaculture industries. The report went on to state that water quality in the Yarra has improved significantly over the past few decades and has been stable in recent years, comparing well to similar rivers in major urban areas overseas. An independent report by the Office of Environment Monitor found that after 4 years of bayside monitoring that overall Port Phillip Bay was in good health, consistent with the previous decade. Water quality in both the Yarra and Port Phillip can be impacted after heavy rain, with pollution washing pollutants into waterways.

The action plan involves 17 priority actions: namely,

1. identify the lead agencies for managing water quality in the Yarra River and Port Phillip;
2. develop a new Environment Management Plan for Port Phillip Bay that will help address key risks to bay health including marine pests and nutrient loads;
3. put in place a Response Plan for water quality events in Port Phillip Bay such as algal blooms;
4. extend the role of the Yarra River and Port Phillip Bay government taskforce to oversee and coordinate implementation of the plan;
5. update best practice environmental guidelines for dredging;
6. work with catchment managers to improve water quality at beaches in Port Phillip Bay;
7. revise the guidelines for urban stormwater management for new urban development, to facilitate and support best practice;
8. identify litter hotspots and develop local partnerships with community, businesses and local government to cut litter in these areas;
9. target pollution hotspots to improve compliance by industry, land developers and other sources of pollution in waterways;
10. investigate and trial improved early warning and community information programs for sewage spills and leaks affecting waterways;
11. develop innovative approaches to reduce wastewater impacts on waterways, including accelerating the sewering of unsewered areas;
12. update guidance and requirements for on-site wastewater management;
13. provide water quality information and advice through a single website so that Victorians can quickly and easily access up-to-date information on beach water quality, including beach closures and algal blooms;
14. prepare a report that summarises Yarra River and Port Phillip Bay water quality and health and actions to improve them;
15. improve key water quality monitoring and reporting programs, including the Port Phillip Bay Beach report;
16. support Victorians to take action to protect water quality and reduce litter; and
17. encourage people to report litter and pollution, learn more about water quality and join community groups to take action to protect the Yarra River and Port Phillip Bay.

Future Melbourne

On the Future Melbourne website, set up by the City of Melbourne, trends and challenges facing Melbourne are discussed, including climate change and adapting to it, significant population change, rapid technological change and an increase in oil prices.

Summary

The population around Port Phillip Bay quite possibly could double in the next 40 years, putting pressures on the environment.

The governments involved with Port Phillip will have to cope with the effects of global warming, which is likely to intensify with time. It could cause flooding of bayside areas regularly. Substantial population growth in the future could cause pollution and put increased demands on the water supply to Melbourne. The Victorian Government has plans to deal with these matters. Port Phillip's future looks promising provided relevant Federal, State and Local governments take necessary actions.

References

Australian Government (2006) Australia's national programme of action for the protection of the marine environment from land-based activities. Case Study 11: Port Phillip Bay. http://www.environment.gov.au/coasts/pollution/npa/pubs/port-phillip.pdf

Australian Government/Bureau of Meteorology (2011) Tide height assessment in Port Phillip Bay, report 9: quarter ending 31 December 2011. National Tidal Centre, Bureau of Meteorology, Commonwealth of Australia. http://www.oem.vic.gov.au/Assets/1076/1/TideHeightAssesment_201112.pdf

Australian Government/Department of Climate Change (2009) Climate change risks to Australia's coast. http://www.climatechange.gov.au/publications/coastline/climate-change-risks-to-australias-coasts.aspx

Belski A, Babanin A, Zieger S, Dobynin M, Plesckachevsky A (2012) Investigation and modelling of suspended particulate matter in Port Phillip Bay. In: Proceedings of the twenty-second (2012) international offshore and polar engineering conference, Rhodes, Greece.

Bird ECF (2006) The effects of a higher sea level on the coast of Port Phillip Bay. Vic Nat 123:49–54
Bird E (2011) Changes on the coastline of Port Phillip bay, Office of the Environmental monitor, State Government Victoria
Black KP, Mourtikas S (1992) Literature review of the physics of Port Phillip Bay, CSIRO Port Phillip Bay Environmental Study, technical report no. 3. C.S.I.R.O, Melbourne
Black K, Hatton D, Rosenberg M (1993) Locally and externally-driven dynamics of a large semi-enclosed bay in southern Australia. J Coast Resour 9(1993):509–538
Bonwick J (1857) Discovery and settlement of Port Phillip, George Robertson. trove.nla.gov.au/work/8484650
Bradley JE (1957) Abnormally high tides in the port of Melbourne. Proc R Soc Vic, new series, 69:37–40, Melbourne
Broome R (2005) Aboriginal Victorians: a history since 1800. Allen and Unwin, Crows Nest, Australia
Bryant E (2001) Tsunami: the underrated hazard. Cambridge University Press, New York, USA
City of Port Phillip (2007) Planning for climate change, a NATCLIM initiative. http://www.environment.gov.au/coasts/pollution/npa/pubs/port-phillip.pdf
Cohen BF, Currie DR, McArthur MA (2000) Epibenthic community structure in Port Phillip Bay, Victoria, Australia. Mar Freshw Res 51(7):689–702
Costermans LF (1983) Trees of Victoria. Costermans Publishing, Frankston
CSIRO (1996) Port Phillip Bay Environmental Study 1992-1996. http://www.melbournewater.com.au/content/library/publications/reports/rivers_and_creeks_reports_pdf/Port_Phillip_Bay_Environmental_Study.pdf
Daley C (1940) Arthur's seat in the forties. Vic Hist Mag 18(3):57–64
Department of Sustainability and Environment (2012a) Port Phillip Bay. http://www.dse.vic.gov.au/coasts-and-marine/marine/bays,-inlets,-estuaries-and-lakes/port-phillip-bay
Department of Sustainability and Environment (2012b) Seagrass and reefs program for Port Phillip Bay. http://www.dse.vic.gov.au/coasts-and-marine/marine/seagrass-and-reefs-program-for-port-phillip-bay
Easton A (1970) The tides of the continent of Australia. PhD thesis, The Flinders University of South Australia
Easton A (1978) A tidal model of Corio Bay, Victoria. In: Noye J (ed) Numerical simulation of fluid motion. North-Holland, Amsterdam, Australia, pp 357–369
EPA Victoria (2012a) Baywide water quality monitoring program. Milestone report no.9. Port Phillip water quality: November 2007–December 2011. March 2012 http://www.oem.vic.gov.au/Assets/1084/1/EPAWQMilestoneRep9_05Apr12_Final.pdf
EPA Victoria (2012b) Beach report 2011–2012, August 2012. http://www.epa.vic.gov.au/our-work/publications/publication/2012/august/1484
Flemming J (1803) A journal of the exploration of Charles Grimes. dalmura.com.au/genealogy/Robbins/Journal.PDF
Flinders M (1814) A voyage to Terra Australis, Book 1 Chapter IX. freeread.com.au/ebooks/e00049.html
Hirst A, Heislers S, Ball D, Blake S, Coots A (2012) Bayside seagrass monitoring program. Milestone report no. 15. Fisheries Victoria technical report series no. 1568. http://www.oem.vic.gov.au/Assets/1080/1/SeagrassMonitoring_MilestoneRep15_27Feb12_Final.pdf
Honda K, Terada T, Yoshida Y, Isitani D (1908) An investigation on the secondary undulations of oceanic tides. J Coll Sci Tokyo 24:1–110
Hubbert GD, McInnes KL (1999) A storm surge inundation model for coastal planning and impact studies, 1999. J Coast Res 15:168–185
IPCC (Intergovernmental Panel on Climate Change) (2007) Fourth assessment report: climate change, 2007. IPCC, Geneva, Switzerland
Keble RA (1946) The sunklands of Port Phillip Bay and Bass Strait. Mem Nat Mus Vic 14:69–122
Lawson and Treloar Pty Ltd, POMC Channel Deepening, Sediment Transport and Water Quality Modelling (2004a) Report prepared for the Port of Melbourne Corporation by Lawson and Treloar Pty Ltd, report Rm2054/J5372 ver 1.0 FINAL June 2004, Notting Hill
Lawson and Treloar Pty Ltd, POMC Channel Deepening, Sediment Transport and Water Quality Modelling Model Calibration (2004b) Report prepared for the Port of Melbourne Corporation by Lawson and Treloar Pty Ltd, report Rm2074/J5372 ver 1.0 FINAL June, Notting Hill
Lee I (1915) The logbooks of The 'Lady Nelson', Grafton and Co, London. http://openlibrary.org/books/OL658013M
Mader CL (2004) Numerical modeling of water waves. CRC Press, New York, USA
McInnes KL, Hubbert GD (1996) Extreme events and the impact of climate change on Victoria's coastline, Environment protection authority, state government of Victoria, publication 488. Environment Protection Authority, Melbourne, Australia, 69 pp
McInnes KL, Macadam I, O'Grady J (2009) The effect of climate change on extreme sea levels along Victoria's coast. CSIRO, Aspendale, Australia
Melbourne Water (2006) A water-supply demand strategy for Melbourne 2006–2055. http://www.melbournewater.com.au/content/library/water_storages/water_supply-demand_strategy.pdf
Melbourne Water (2011) Summary water quality data 2011. http://www.melbournewater.com.au/content/library/publications/reports/rivers_and_creeks_reports_pdf/Summary%20Waterway%20Water%20Quality%20Data%202011.pdf
Melbourne Water (2012) The Yarra River. http://www.melbournewater.com.au/content/rivers_and_creeks/the_rivers_and_creeks_system/the_yarra_river.asp
Mitchell W, Chittleborough J, Ronai B, Lennon GW (2000) Sea level rise in Australia and the pacific. S Pac Sea Lev Clim Change Newsl 5(1):10–19
Museum Victoria (2006) Melbourne's wildlife. CSIRO Publishing, Collingwood
Museum Victoria (2012) Taxonomic toolkit for marine life of Port Phillip Bay. http://portphillipmarinelife.net.au/
Nicholls N, Larsen S (2011) Impact of drought on temperature extremes in Melbourne, Australia. Aust Meteorol J 61:113–116
Port of Melbourne Corporation (2012a) Expanding Melbourne's port capacity. http://portcapacity.portofmelbourne.com/downloads/120502-expanding-port-capacity.pdf
Port of Melbourne Corporation (2012b) History of the port. http://www.portofmelbourne.com/community/maritimeheritage/porthistory.aspx
Port of Melbourne Corporation (2013) Victorian tide tables. Melbourne
Presland G (2010) First people. Museum Victoria, Melbourne
Ryder K, Temara A, Holloway DA (2004) Avoidance of crude-oil contaminated sediment by the Australian seastar, Patiriella exigua (Echinodermata: Asteroidea). Mar Pollut Bull 49(11–12):900–909
Sampson J (2008) Some solutions of the shallow water wave equations. PhD thesis, Swinburne University of Technology, Melbourne
Sampson J, Easton A, Singh M (2005) Modelling the effect of proposed channel deepening on the tides in Port Phillip Bay. Aust N Z Ind Appl Math J 46(E):C888–C901
Schwartz ML (1967) The Brun theory of sea level rise as a cause of shore erosion. J Geol 75:76–92
Short AD (2005) Beaches of the Victorian coast and Port Phillip Bay. Surf Life Saving Australia Ltd, Bondi Beach, Australia
Sturman A, Tapper N (1996) The weather and climate of Australia. Oxford University Press, Melbourne, Australia
The Age (2005a) Are we prepared for the worst? http://www.theage.com.au/news/Asia-tsunami/Are-we-prepared-for-the-worst/2004/12/31/1104344982287.html?from=moreStories. 5 October 2008
The Age (2005b) Port Phillip Bay is in no danger from a tsunami. http://www.theage.com.au/news/Letters/Port-Phillip-Bay-is-in-no-danger-from-a-tsunami/2005/01/04/1104832107215.html

The Age (2012a) Government urged to consider secret 'Bay West' option for port. http://www.theage.com.au/victoria/government-urged-to-consider-secret-bay-west-option-for-port-20120903-25amq.html

The Age (2012b) Bay dredging in $1.2 b dock expansion. http://www.theage.com.au/victoria/bay-dredging-in-12b-dock-expansion-20121208-2b2fx.html

The State of Victoria, Department of Natural Resources and Environment (2002) Port Phillip Bay Environmental Management Plan. http://www.dse.vic.gov.au/__data/assets/pdf_file/0020/97121/Port_Phillip_Bay_Environmental_Management_Plan_-_Background_Document.pdf

Victorian Government Department of Sustainability and Environment (2012) A cleaner Yarra River & Port Phillip Bay. Department of Sustainability and Environment, Melbourne, Australia

Vreugdenhil CB (1998) Numerical methods for shallow-water flow. Kluwer, Dordrecht, Netherlands

Walker SJ (1997) Hydrodynamic models of Port Phillip Bay, Technical report no. 38, Port Phillip Bay environmental study. Port Phillip Bay Environmental Study, CSIRO Environmental Project Office, Yarralumla, Australia

Wharton WLJ (1888) On the seismic sea waves caused by the eruption of Krakatoa, August 26th and 27th, 1883. In: Symons GJ (ed) The eruption of Krakatoa and subsequent phenomena. Trubner and Co, London, pp 89–150

Wilson KJ (1982) Forecasting shallow water wind wave characteristics. Bureau of Meteorology, Melbourne, Australia, 22 pp

Past, Present and Futures of the Tamar Estuary, Tasmania

Joanna C. Ellison and Matthew R. Sheehan

Abstract

The Tamar Estuary is located on the Bass Strait coast of North Tasmania, a drowned river valley of some 71 km in length. The vast majority of fluvial flow enters through two rivers that meet at the estuary head, the South and North Esk Rivers with a catchment area comprising over 20 % of Tasmania (about 11,000 km^2). The Tamar valley is a down-faulted graben structure, giving a bedrock-confined long and narrow shape to the majority of the estuary, causing tidal amplification to give the head of the estuary the largest tidal range in Tasmania. At the time of first European discovery in the early 1800s the upper Tamar Estuary was found to feature extensive mud banks, with a channel that was difficult to navigate in the 1.7 m draft *Lady Nelson*. With establishment of the city of Launceston, the channel was dredged starting in the late 1870s until the 1960s allowing ship passage, until the major port was moved to the lower estuary. During this period contamination of the upper estuary increased, from organic and inorganic wastes from industrial, mining and domestic sources, as well as heavy metals from mining industries in the catchments, combined with high sediment yield. The Tamar has a high conservation significance being the only mesotidal drowned river valley in Tasmania, along with recording a large number of species not found elsewhere. There are significant threats to native species habitats from introduced species in the estuary, including Australia's largest area of introduced Rice grass. This has caused a dramatic change to the physiography of the intertidal zone, with previous beaches or rock shorelines converted to accreting mud banks under *Spartina*. Sedimentation and water quality issues have long been a concern to the community, and over the last 15 years Natural Resource Management of the estuary and its catchments has greatly improved, including introduction of systematic monitoring. Aspirations for the estuary's future are based on community consensus to maintain and improve biophysical values of the estuary, although preference remains for an upper estuary that resembles its early twentieth century dredged state rather than how it was first described 200 years ago.

Keywords

Tamar River • Geomorphology • History • Sediment • Rice grass • Natural resource management

J.C. Ellison (✉) • M.R. Sheehan
School of Geography and Environmental Studies, University of Tasmania, Locked Bag 1376, Launceston, TAS 7250, Australia
e-mail: Joanna.Ellison@utas.edu.au; matt@wildmatters.com.au

E. Wolanski (ed.), *Estuaries of Australia in 2050 and Beyond*, Estuaries of the World,
DOI 10.1007/978-94-007-7019-5_5, © Springer Science+Business Media Dordrecht 2014

Box 1

Joanna Ellison and Matthew Sheehan studied the Tamar Estuary, a drowned river valley 71 km in length. With establishment of the city of Launceston, the estuary was dredged starting in the late 1870s until the 1960s allowing ship passage, until the major port was moved to the lower estuary. During this period contamination of the upper estuary increased, from organic and inorganic wastes from industrial, mining and domestic sources, as well as heavy metals from mining industries in the catchments, combined with high sediment yield. The Tamar Estuary still has high conservation significance. There are significant threats to native species habitats from introduced species in the estuary, with previous beaches or rock shorelines converted to accreting mud banks under *Spartina*. Sedimentation and water quality issues have long been a concern to the community.

Aspirations for the estuary's future are based on community consensus to maintain and improve biophysical values of the estuary, although preference remains for an upper estuary that resembles its early twentieth century dredged state rather than how it was first described 200 years ago.

Introduction

Tasmania is Australia's highest latitude state with the main island 39° 40′–43° 20′ S, and as the only island state, it has the most coastline per unit land area. Its c. 2,200 km coast exhibits strong contrasts in environmental geomorphology in response to wide variation in geological structure and exposure (Ellison 2010). The west coast faces prevailing westerly winds and swell coming onshore across the South Australian Bight, and can be of high energy owing uninterrupted fetch. Higher mountains on the west of Tasmania capture high precipitation and create a rain shadow towards the east. The east coast facing the Tasman Sea has calmer swell conditions and lower rainfall, though is subject to occasional low pressure systems bringing onshore storms. The north coast facing Bass Strait is also sheltered from westerly swell and has the lowest bathymetric gradients offshore, bringing lower wave conditions but higher tidal ranges. Longshore drift is generally from west to east (Bugg 1990; Sharples 2006), however, the central inflection of the north coast greatly reduces long-shore movement in this section (Davies 1978), limited to circulation within headland confined dissipative beaches.

Discharging into Bass Strait at the centre of the north coast is the Tamar (Fig. 1), which is Tasmania's largest estuary with an area of 100 km^2 (Pirzl and Coughanowr 1997). It is a drowned river valley, a classification of estuaries with wide river mouths, rocky headlands and deep channels (Edgar et al. 1999). During the last glacial period peaking 25,000 years ago Bass Strait was exposed as a land bridge between Tasmania and Victoria (Lambeck and Chappell 2002), which facilitated the first human occupation of Tasmania around 40,000 to 36,000 years ago (O'Connell and Allen 1998). Owing to deeper bathymetry, the eastern Bass Strait was inundated later than the west, with a land bridge still existing to the east up to 14,000 years ago (Fig. 2; Lambeck and Chappell 2002). Subsequent sea level rise until 6,500 years ago caused progressive submergence of the Tamar valley.

Having no Tamar "river" the estuary forms at Launceston where the large rivers of the South Esk and North Esk both meet (Fig. 1). The estuary extends 71 km from this to Bass Strait through a wider lower estuary 12 km section known as Port Dalrymple. Located here are the industrial port of Bell Bay and the town of George Town on the east side, and Beauty Point and the town of Beaconsfield on the west side.

Geomorphological Setting

The Tamar upstream of Port Dalrymple is a geologically well confined estuary with a zig-zag path through headlands from both side (Fig. 1), but following a predominant NNE to SSW direction controlled by geological fault lines.

The geological structure of the Tamar valley consists of three major units (Fig. 3), the oldest is the Mathinna series of Silurian sandstone/mudstone and slate horizons. This was succeeded by the Permian deposition of limestone, sandstone and conglomerate beds, and later over lain with Jurassic dolerite, which is the most common out cropping bed rock surrounding the Tamar region. The Tamar Valley was then initiated by a period of extensive north west-trending block-faulting (Walsh 1990), in which the Western Tiers mountains

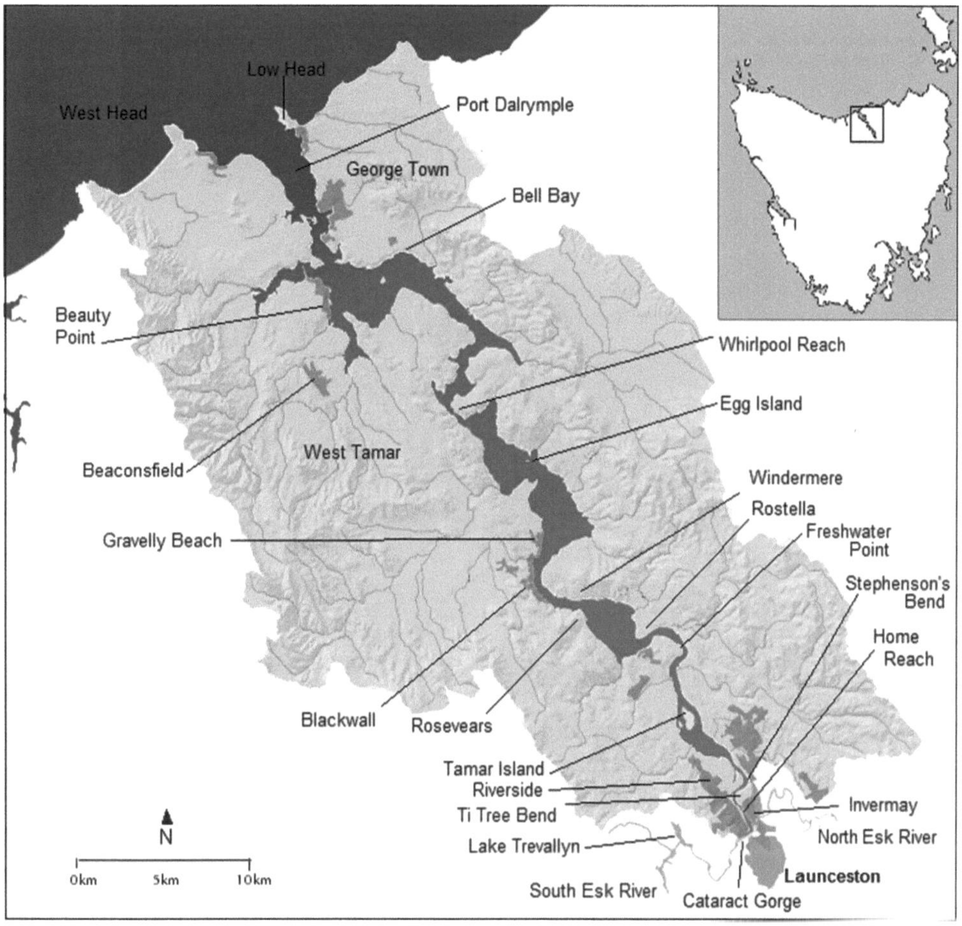

Fig. 1 The Tamar estuary, Tasmania. The stippled area shows minor estuarine catchments, while major fluvial discharge is from the South and North Esk Rivers (Adapted from TEER 2011)

such as Western Bluff, Drys Bluff and Millers Bluff and Ben Lomond to the east (Fig. 4) were uplifted (Fig. 3), the graben containing Palaeocene-Eocene sediments indicating that faulting occurred prior to this in the early Tertiary or late Cretaceous (Baillie1989). Subsequent to faulting was an extensive period of lacustrine and fluvial sedimentation (Fig. 3c) resulting in the deposition of clays, sands and occasional gravels.

Volcanic episodes during the Miocene gave rise to small areas of basaltic dykes and sills intruding fractures associated with earlier faulting. This caused changes to the original course of the Tamar River and its tributaries as well as lacustrine conditions of the upper Tertiary. Basalt flows dammed the Tamar at Whirlpool Reach to form a lake, in which deposited sands, clays and occasional gravels (Walsh 1990). The South Esk flowing into this lake through faulted resistant dolerite deepened its valley just above the Tamar without widening it, to form the beginnings of Cataract Gorge. An erosional period continued during the melting stage of the last Pleistocene glaciation, when the river systems carried large volumes of water which resulted in the stepping down of the valley wall sand erosion of many of the Tertiary clay, sand and gravel deposits. This returned the Tamar to a fluvial system until sea level rose to near present levels during the Holocene, 6,500 ± 250 years ago (Thom and Roy 1985), when the Tamar River became an estuarine system (Fig. 3d).

Shell-rich sands have been found near the head of the estuary in Launceston, 3.6 m below the present surface during excavation for road construction (Goede et al. 1993). Dating of these deposits have indicated that infilling of marine sediment was continuous for approximately 4,000 years following the post-glacial marine transgression (Goede et al. 1993). The fine grain silts and muds deposited on top of the sands suggest an increased input of terrestrial sediments transported from the North and South Esk catchment areas.

Geomorphic criteria were used to analyse 111 moderate or large sized estuaries in Tasmania (Edgar et al. 1999), finding the Tamar to be among the most common type of river dominated estuaries, but among only six of drowned

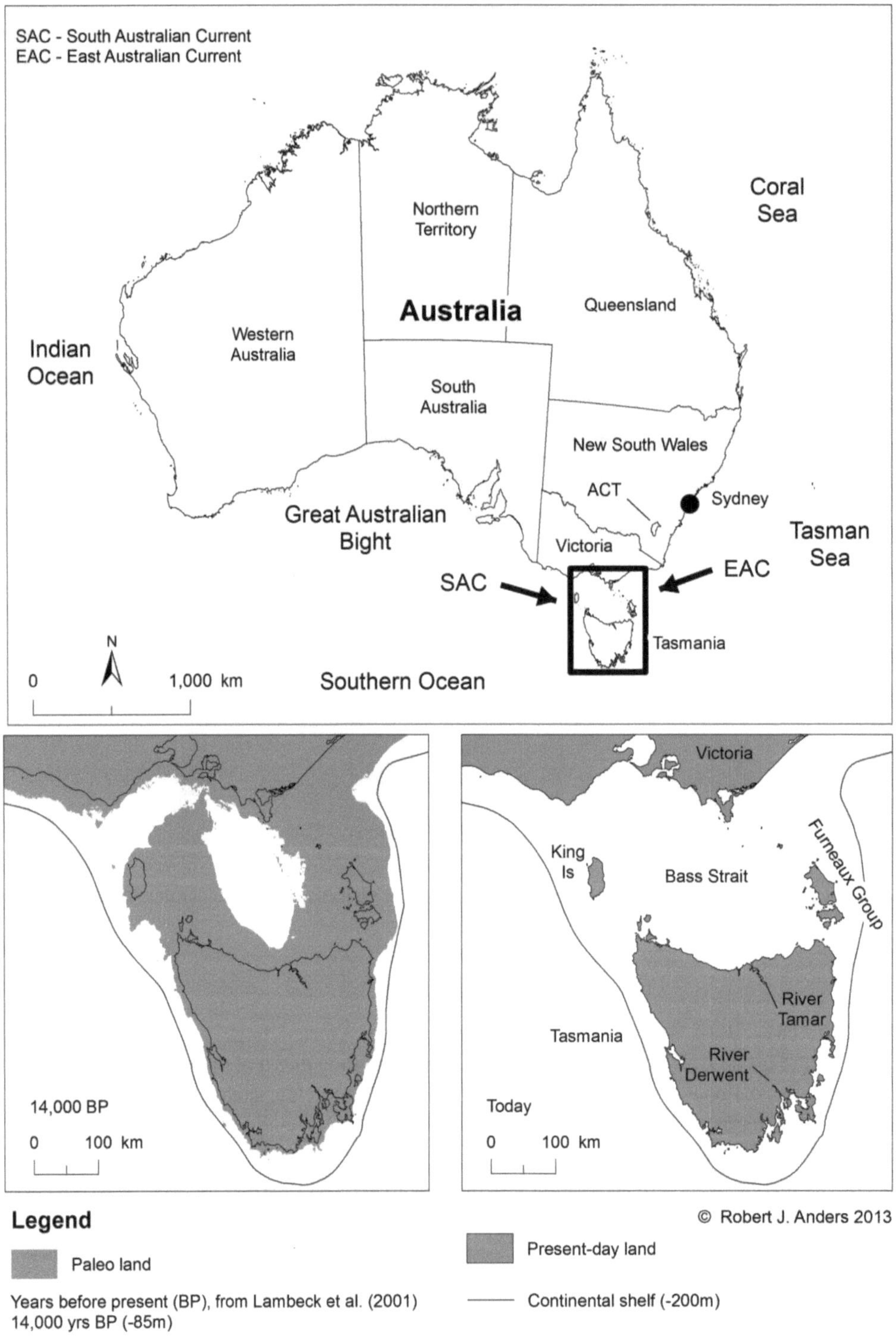

Fig. 2 Location of the Tamar relative to adjacent coastlines, during the early Holocene 14,000 years ago and today (Adapted by R. Anders from Lambeck and Chappell 2002)

river valley types. The Tamar Estuary has a surface area of water of 97.9 km^2, distance around estuary perimeter of 252.7 km, lacking any seaward barrier and has a mid-estuary tidal range of 2.3 m. Cluster analysis of such criteria found that the Tamar Estuary was unique enough to form a single member group of a drowned river valley, differentiated from others by having a significantly higher tidal range compared with others.

Hydrological Setting

Waves, Currents and Tides

Bass Strait located between Tasmania and Victoria has a mean depth of 50–70 m, steeply deepening to over 4 km in depth on each side. Lateral flushing of Bass Strait results

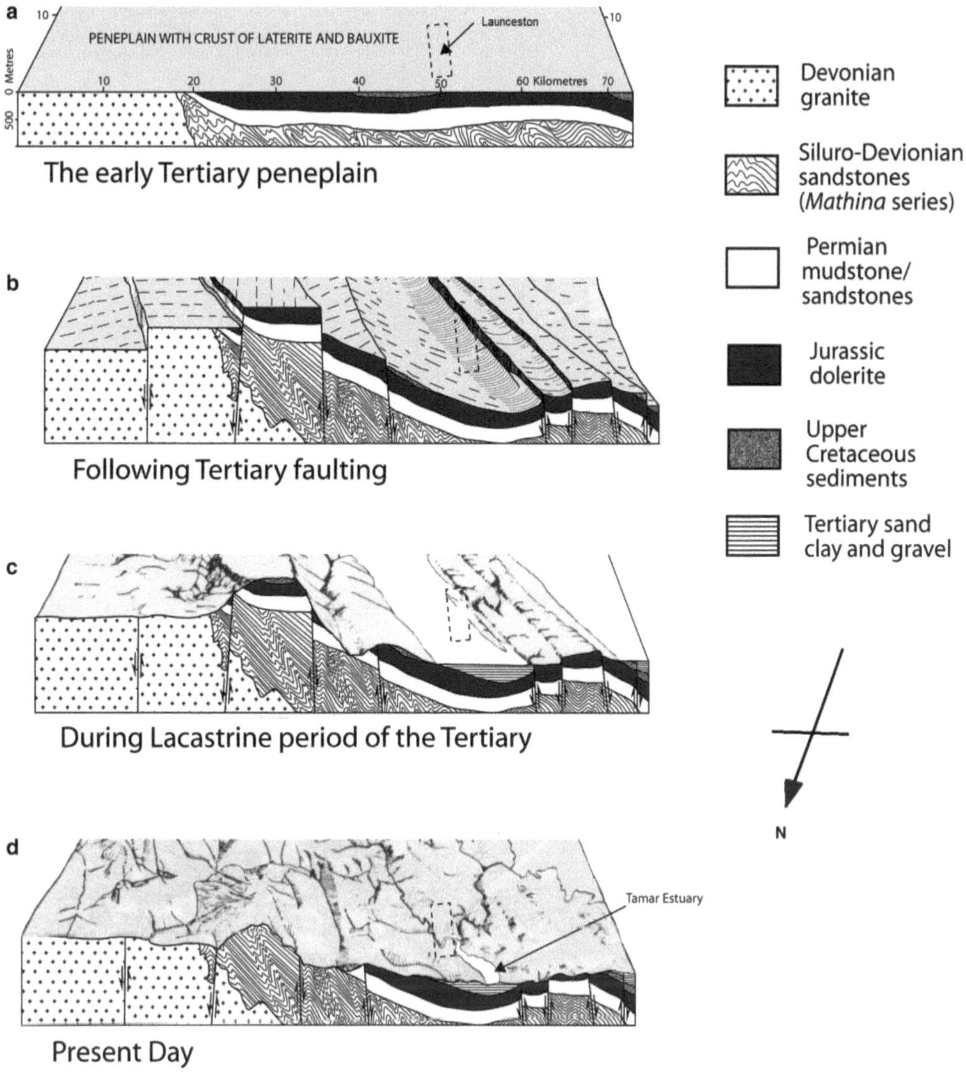

Fig. 3 Geomorphological development of the Tamar valley (Adapted from Walsh 1990). (**a**) Geological units at the end of the Cretaceous priort of aulting. (**b**) Topography immediately after faulting in the early Tertiary. (**c**) Lacustrine periods of the Tertiary, which led to in filling of the graben, followed by an erosional period of the Quaternary. (**d**) The last 6,500 years to present following the postglacial marine transgression

from inflow of three primary water masses, the South Australian Current from the west, the East Australian Current from the north east which has greater influence in summer, and the sub-Antarctic surface water from the south (Sandery and Kampf 2007). Tidal currents also operate in the shallow Bass Strait, which can be an important sand transporting mechanism in mesotidal regions such as Bass Strait (Porter-Smith et al. 2004).

While estuarine definitions can focus on the upper and lower limits of salinity (Cameron and Prtichard 1963), consideration of the influence of tides better suits estuaries such as the Tamar, where tides exert a physical and ecological influence higher in the system than that of the inland influence of salinity. The Tamar is a low mesotidal semidiurnal system with tidal amplification up the estuary, with the mean tidal range increasing from 2.34 m at George Town to 3.25 m at Launceston (Foster et al. 1986) as shown by tide gauge data (Fig. 5). High tide at Launceston occurs about an hour after George Town. Estuaries of this type have been described as hypersynchronous, attributing the increase in tidal range to convergence of the estuary being greater than friction encountered by incoming tides (Nichols and Biggs 1985). Higher tidal ranges allow more widespread tidal flats and salt marsh habitats.

Relative sea level in Tasmania has been within half a metre of present sea level for much of the last 6,000 years (Gehrels et al. 2012). Analysis from a tidal mark struck in Southern Tasmania in 1841 shows that mean sea level has risen by around 14 cm between 1841 and 2002, at an overall rate of 0.8 ± 0.2 mm/year (Hunter et al. 2003).

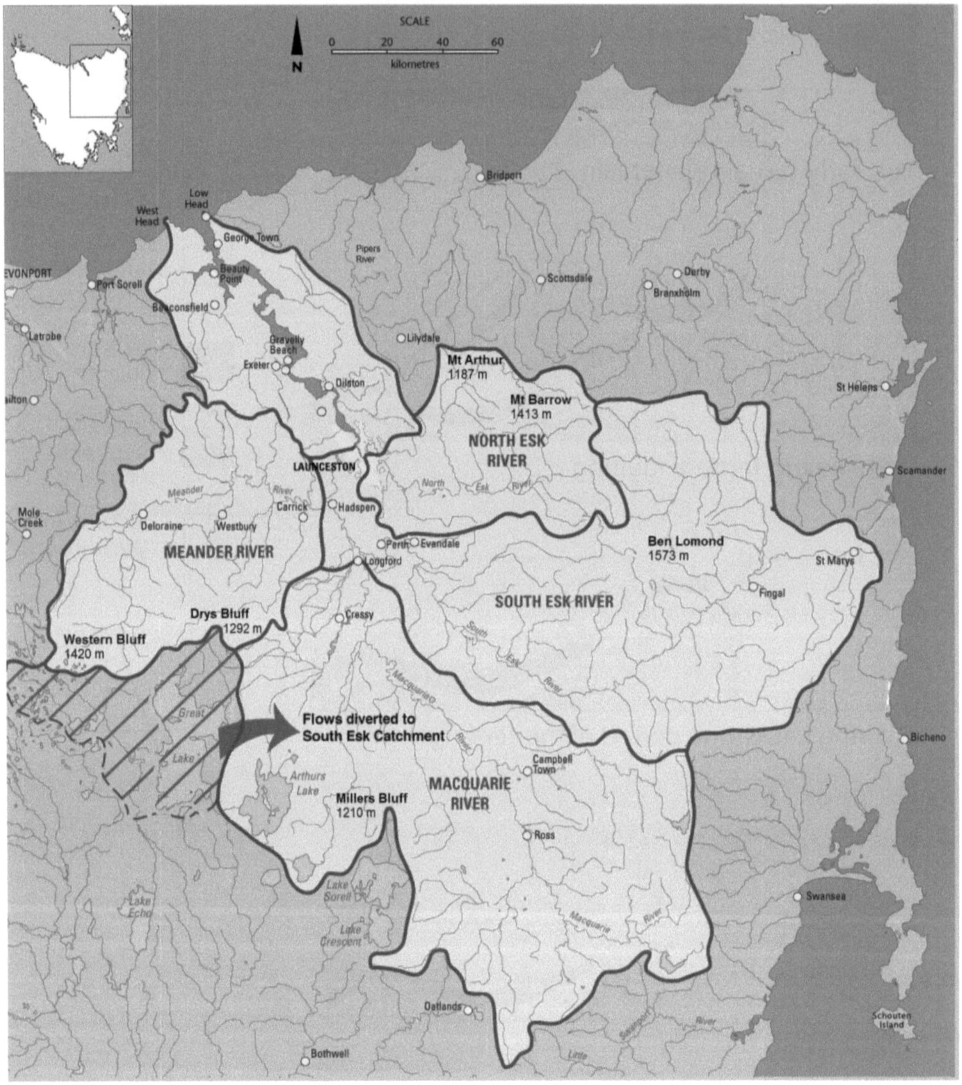

Fig. 4 Map of North east Tasmania showing catchments contributing river flow to the Tamar estuary (Adapted from Pirzl and Coughanowr 1997)

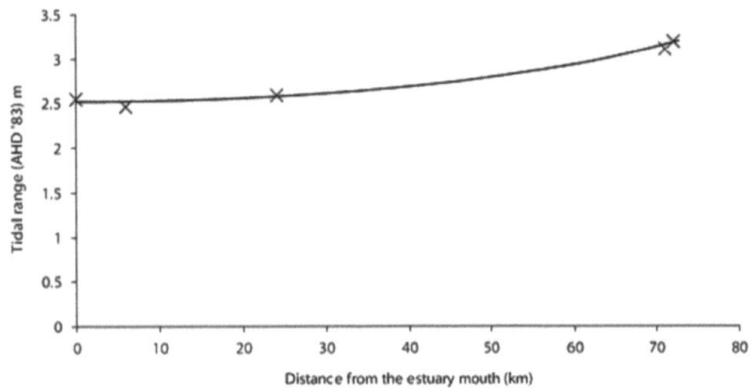

Fig. 5 Tide Gauges of the Tamar from the mouth on the *left* to inland on the *right*: Low Head (Bass Strait), George Town, Bell Bay, Stephenson's Bend and Ti Tree Bend. Tidal range initially decreases relative to Bass Strait on entering the Tamar, and then steadily increases with distance from the mouth

The George Town tide gauge with over 30 years of recordings has demonstrated relative sea-level rise at a rate of 0.3 mm/year (Mitchell et al. 2000).

Climate

Bureau of Meteorology weather stations are located at Low Head and near Launceston at Ti Tree Bend (Fig. 1). The Ti Tree Bend station was established in 1980 and climate records have been taken from this time to the present. The original Low Head station operated between 1877 and 2001, and was relocated in 1998 to where climate data are now collected. Monthly rainfall and temperature data for the old Low Head station (used owing to the longer record) and from Ti Tree Bend (1980–2005) are shown in Figs. 6 and 7.

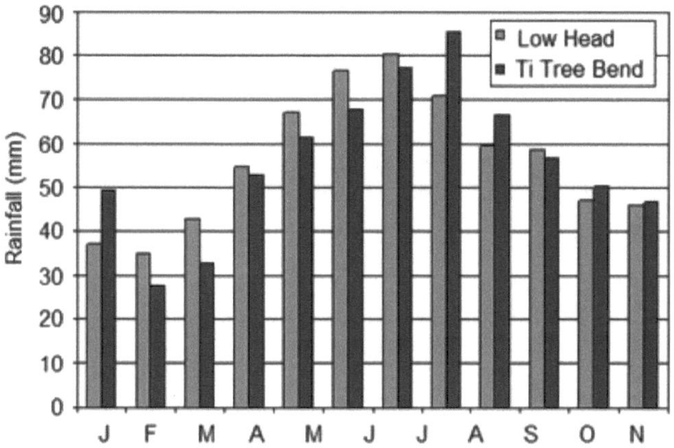

Fig. 6 Mean monthly rainfall data from Low Head and Ti Tree Bend

Rainfall at both Low Head and Ti Tree Bend show respectively annual averages of 676.4 mm/year and 675.7 mm/year, with generally less than 10 mm monthly variation between the stations (Fig. 6). The Tamar therefore has a winter wet season but the summer also receives rainfall, this tending to occur during low pressure storms.

Monthly temperature data is summarised in Fig. 7, showing the in land station (Ti Tree Bend) to experience greater temperature extremes throughout the year than Low Head, however temperature differences between the two sites are generally less than 5 °C. The highest and lowest temperature trends show temperate conditions, however a small amount of climate warming would render the seaward Tamar suitable habitat for mangrove colonisation, frost being the limiting factor.

The graben down faulted Tamar Valley is fringed by ridges of higher land on both sides (Fig. 3), and dominant winds typically align south-east and north-west within the valley. This air movement moderates climate throughout the valley, with most variation between the two weather stations most likely attributable to coastal proximity.

Catchments

The main tributaries to the Tamar Estuary are the North and South Esk Rivers, and Macquarie and Meander sub-catchments join the South Esk between Longford and Hadspen (Fig. 4). The catchment surrounding the Great Lake in the Central Highlands naturally flows south to the Derwent River, and for hydroelectric power generation was diverted to join the Macquarie River. The Tamar overall has an estuarine catchment area of 11,589 km^2 (Edgar et al. 1999), over 20 % of Tasmania. These catchments feature a larger scale horst block fault topography as shown for the Tamar valley in Fig. 3, with peaks of 1,200–1,500 m to the west and east (Fig. 4), below which are steep escarpments then rolling floodplains interrupted by bedrock confinement (Fig. 8).

The South Esk was dammed above Cataract Gorge in 1955 to create Lake Trevallyn (Fig. 1), diverting river flow through a power station to enter the Tamar downstream of Home Reach. Lake Trevallyn is a small storage and can only regulate part of inflows, causing flow over the dam spillway during catchment flood events. At these times spectacular supercritical flow occurs down the Gorge (Fig. 8) which attracts crowds of onlookers and scours silt from the upper estuary.

Agricultural capability of the lands of the Tamar Estuary region (Fig. 1) was classified using soil and slope criteria to show that <6 % had better than low cropping suitability, and most was only suitable for pastoral uses (Noble 1992). The majority of soils are poor for cropping use owing to poor structure, low fertility or stoniness, with fairly extensive podsolic duplex soils. The catchment was recorded in the mid-1990s as principally containing Forest (woodland forest and rainforest) and agricultural land (Pirzl and Coughanowr 1997). Urban areas occupied a small percentage of the total catchment, concentrated around the estuary, and industry is largely restricted to the Launceston region and Bell Bay.

Land use categories for the Tamar catchments (Fig. 4) for 2009 are shown in Table 1, from BMT (2010) using data from Hydro Tasmania Consulting. The total catchment area of 1,123,721 ha (excluding the diverted Great lake catchment) was categorised into 14 major land uses, with concentration of production forestry and conservation areas around the highlands east of Launceston in the headwaters of the North Esk river, concentration of grazing in the drier Midlands to the south of Launceston, patchy irrigated intensive agriculture in the Midlands and increasing towards the Launceston and to its west, and the Tamar valley dominated by urban centres surrounded by rural residential, interdispersed mainly by grazed natural vegetation.

Salinity and Turbidity

Systematic water quality monitoring commenced in the Tamar in 1997 (Dowson and Rushton 2006), though before this a number of programs had included non-systematic surveys (Pirzl and Coughanowr 1997; Aquenal and DEPHA 2008). From 2009 monthly water quality monitoring at 20 stations along the estuary has been continued by the Tamar Estuary and Esk Rivers (TEER) Program (Attard et al. 2011).

The majority of freshwater inflow is from the North and South Esk at the estuary head, with river outflow further north into the Tamar being minor. Geomorphic confinement of the upper estuary causes salinity to be close to fresh despite a 3.25 m tidal range, and is fresh during river floods.

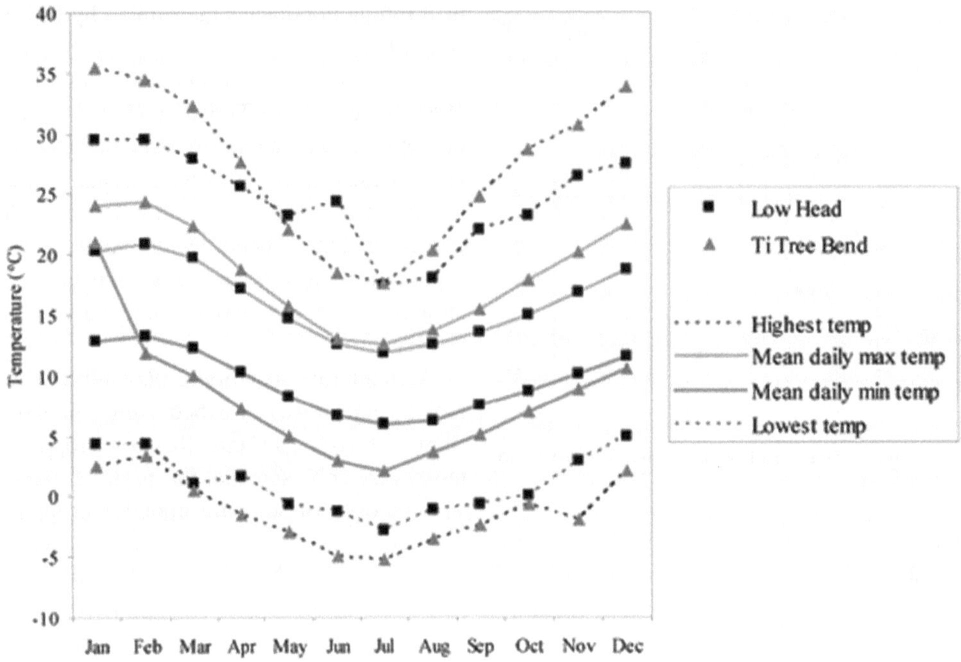

Fig. 7 Summary of monthly temperature data from Low Head and Ti Tree Bend

Fig. 8 Cataract Gorge in flood, April 2011, with Launceston in the background. This section is intertidal, normally so calm that large catamaran ferries can bring tourists to the foreground and rotate (Photograph: J. Ellison)

As the estuary becomes wider downstream salinity increases to about 20 ppt around Windermere (Fig. 1), and is close to seawater salinity downstream of George Town (Attard et al. 2011). This varies seasonally, being fresher in winter and after floods, and more saline in summer. It is a partially mixed estuary (Pritchard 1955; Cameron and Prichard 1963), having a salt wedge where freshwater flow remains above tidal seawater inflow, however entrainment and

Table 1 Landuses of the Tamar catchments

Landuse	Area (ha)	Percentage
Dryland grazing	367,799	33
Production forestry	249,530	22
Grazing natural vegetation	118,483	11
Residual native cover	129,013	11
Rehabilitation		
Stock route		
Nature conservation	87,078	8
Managed resource protection	57,619	5
Irrigated intensive agriculture	47,083	4
Dryland plantation forestry	23,534	2
Water	18,277	2
Dense urban	8,636	1
Rural residential	7,264	1
Urban residential	6,031	1
Irrigated grazing	3,088	0.3
Dryland intensive agriculture	286	0.03

Adapted from BMT WBN (2010: Table 3–2)

turbulence occurs due to a significant tidal influence and resultant fluctuations in water volumes. This results in a greater degree of mixing and a less well defined salt wedge.

The most significant external source of suspended particulate matter to the Tamar is the North and South Esk rivers (Aquenal and DEPHA 2008), with 58,000–80,000 t/year of total suspended solids estimated to be generated from the Tamar Estuary catchments (BMT WBM 2010). This is predominately associated with grazing and forestry related land uses (BMT WBM 2010), and has occurred for decades. Foster et al. (1986) estimated a mean annual sediment supply of 39,000 tonnes per year, equating to approximately 120,000 m^3 of bulk sediment volume (GHD 2009), though this was variable up to more than 100,000 m^3. Turbidity monitoring shows highest levels in the upper estuary with median values ranging between 40 and 60 NTU (Nephelometric Turbidity Units) above Freshwater Point (Attard et al. 2011), tending to be considerably higher in winter than summer downstream of this. Turbidity values are low in the lower estuary, which is the section valued for its biodiversity.

Ecology

The unique geomorphic setting of the Tamar (Edgar et al. 1999) supports a high conservation status owing to its high biodiversity, particularly of fish and invertebrates. Species diversity is facilitated by a range of estuarine habitats of rocky reefs and shores, embayments with sediment, shingle or rocky substrates, tidal creeks, sand flats and beaches, and intertidal salt marsh and mudflats (Smith 1995). Extensive seagrass beds occur at the mouth of the estuary, mostly of *Posidonia australia* and *Amphibolis antarctica* (Rees 1994; DPIPWE 2000) providing habitat for a diversity of fish and invertebrates (Edgar et al. 1999). Sub-tidal reefs are also extensive in the estuary mouth, providing habitat for fish, crustaceans, molluscs, sea stars, urchins and segmented worms (Meynard and Gaston 2010).

The classification of the Tamar as high conservation significance estuary was considered problematic as while the Tamar showed significant conservation values through being the only mesotidal drowned river valley in the State, along with recording a large number of species not found elsewhere, it is also degraded as a result of human pressures (Edgar et al. 1999). This has resulted from the urbanisation of the landward segment of the estuary, along with extensive land clearance for forestry and agriculture in the catchments. There are also significant threats to native species habitats from introduced species in the estuary, including Australia's largest area of introduced rice grass (*Spartina anglica*), and wide occurrence of the Pacific Oyster *Crassostrea gigas* and the east Asian Mussel, *Musculista senhousia*. A further aggressive introduced species that threatens native ecology is the mosquito fish (*Gambusia holbrooki*), which was found at places in the Tamar in the early 2000s.

Native salt marsh vegetation, such as *Sarcocornia quinqueflora, Sclerostegia arbuscula*, and *Suaeda australis* is today limited to a narrow fringe below the high water mark and in sheltered embayments, with extensive salt marshes occurring only near Bell Bay, on relatively sandy substrates in the lower estuary. The ability of *Spartina anglica* to establish at elevations lower than native salt marsh plants has provided it a competitive advantage (Gray et al. 1991; Guenegou et al. 1991). Native salt marshes are restricted to the lower estuary and have been progressively invaded at the seaward margin by *Spartina*, where it establishes in isolated clumps and coalesces to form laterally extensive swards seaward of the native salt marsh (Fig. 9), then progressively moves landward into the native vegetation.

Anthropological Influences: Resources, Pressures, Impacts, and Remediation

The Tamar Valley (Fig. 1) is administrated by three local councils, the largest being Launceston with population of 64,193 (ABS 2012), along with West Tamar with a population of 21,817 and George Town with a population of 6,636. It includes Tasmania's second largest town of Launceston, which today is a regional population and industrial centre, being one of Australia's oldest cities with a rich industrial past dating back to the discovery of tin. The Tamar region has a population of 92,646 of a total Tasmanian population of 511,200 (ABS 2012).

Fig. 9 *Spartina anglica* colonising mud flats in the lower Tamar estuary, with remnants of native salt marsh at the landward edge (Photograph: M. Sheehan)

Human History

Owing to modification of the Tamar post European settlement, and the range of opinions that exist today on what its natural condition is or should be regarding its sedimentary morphology, it is worthwhile to explore its earliest descriptions. This has not been fully done in previous reviews (Foster et al. 1986; Pirzl and Couganowr 1997; Aquenal and DEPHA 2008).

Tasmania was first settled by people around 40,000 years ago (O'Connell and Allen 1998), however Aboriginal settlement sites in the Tamar area from before 7,100 BP are likely no longer accessible owing to the post-glacial sea level rise drowning of the valley, as these sites were likely close to the river's edge (Breen and Summers 2006). At the time of European arrival, the Tamar valley including Launceston was occupied by the *Letteremairrener* people, mostly based near the present location of George Town (Breen and Summers 2006). This clan had a seasonal hunter-gatherer economy, used fire to create pasture that attracted game, and spoke one of at least five aboriginal languages in Tasmania, named *mairremenner* (Taylor 2005). The clan of the Tamar Valley had connections with those further East through a defined walking corridor (Cameron 2011), which had tributary branches into the northern Tamar at Pipers River and also from around Windermere.

The Tamar Estuary was discovered by Europeans when the *Norfolk* entered its mouth late in 1798, under the command of Captain Matthew Flinders (Library Committee of the Commonwealth of Australia [LCCA] 1921a; Macknight 1998; Sprod 2009). They entered a broad inlet and basin, and this discovery was named Port Dalrymple by Governor Hunter. The *Norfolk* explored the estuary for 16 days (Whitfeld 1912; Percy 1993), and the chart by Flinders in 1798 (Flinders 1958) extends as far inland as Windermere today, showing depths along the passage of the ship. Mud banks are noted offshore of Gravelly Beach and Blackwall.

Port Dalrymple was reported to the British colonial base in Sydney as providing good anchorages and potential for a settlement site (McIntyre 1987; Macknight 1998). The *Lady Nelson* visited Port Dalrymple in January 1804, and further reports were sent on its suitability for settlement. By late 1804 with perceived competition from French potential colonisers, Port Dalrymple was chosen as a British settlement site in Bass Strait, and four vessels were sent from Sydney under the command of Lieutenant Governor William Paterson to commence a settlement.

Lieutenant Symons along with William Collins and Thomas Clark explored up the estuary in January 1804 aboard the *Lady Nelson*, reaching the site of Launceston at the estuary head. Collins described the Tamar in some detail

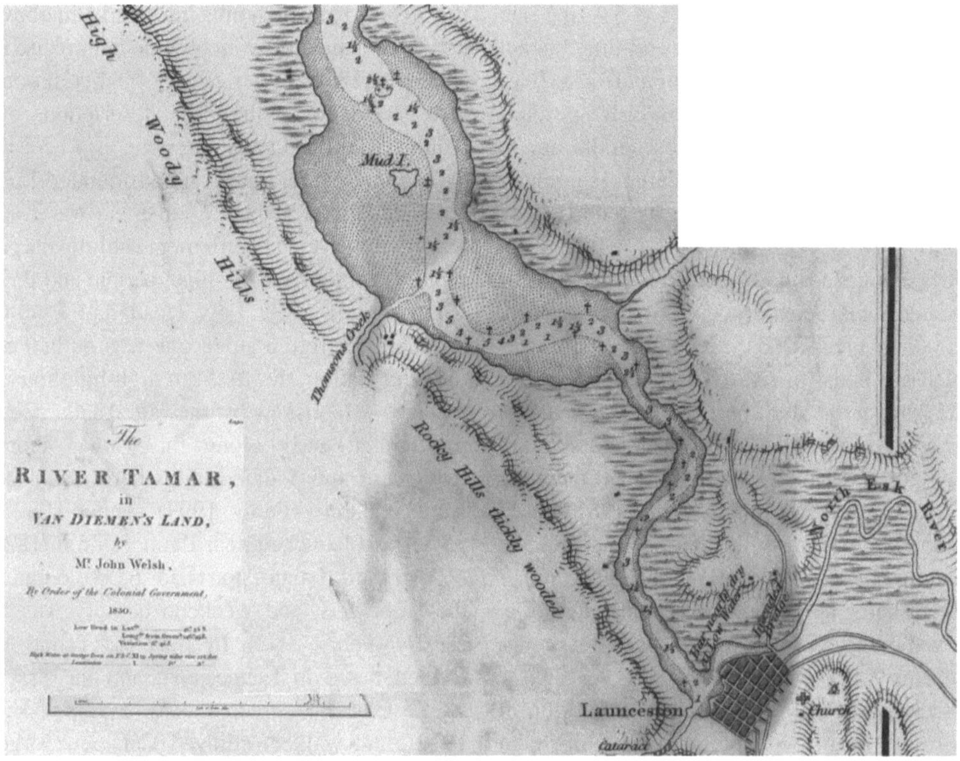

Fig. 10 Partial composite of Welsh's 1830 chart, showing upper Tamar channel depths in fathoms (1 fathom = 1.8 m) (Adapted from a digital image held by the National Library of Australia)

(LCCA 1921b), referring to the channel as "good and navigable for the largest vessels" as far as Egg Island (Fig. 1), upstream the channel was described as "very good although in the reach but narrow; nevertheless any sized vessels may get up and down with safety". Towards Upper Island (known today as Tamar Island) "the channel is still good and navigable for ships, beyond which place only for small craft". Clark described deep water near the high rocks (of Cataract Gorge) but otherwise "The River is very shallow with broad mud banks and reedy swamps beyond them…" (LCCA 1921c).

Lady Nelson was a specially designed vessel of shallow draft for survey work in coastal waters, being equipped with retractable sliding keels, used for exploration of the Australian coast (Grant 1974). The draft without the sliding keels was about 5 1/2 feet; with the keels about 10 1/2 feet (Thomas, 2012, Chair of the Tasmanian Sail Training Association, personal communication), so the draft was 1.7 m with the keels retracted.

Later in 1804 Paterson and others explored up the Tamar (so named by Paterson) on the *Lady Nelson*, as far as where Launceston is today, and further up the North Esk River (LCCA 1921d). Paterson described today's Tamar Island as having low and muddy banks, and ran aground to the south of this. From there they took a tender boat as far as Cataract Gorge, where waterfalls of great force were described. Sections of the estuary coast upstream of where the *Lady Nelson* was left were described as having deep mud and reeds with which the banks were completely covered for some distance inland. The tide lifted the *Lady Nelson* and they proceeded towards the south expecting to reach the Cataract, however it was found that they could not take the *Lady Nelson* any further. It was noted "in the event of it becoming a considerable settlement it will be requisite to have flat-bottomed craft to have any communication with the river below- at least as far as the Cataract." (LCCA1921d). The upper Tamar however remained preferable for settlement, being the longest river in Australia in terms of navigability and the early settlers were under the leadership of seafarers whose main assets and skills were in ships and boats.

The earliest charts of the Tamar record its depths and extent of mud banks. John Welsh's 1830 chart shows depths in fathoms as noted near Hebe reef (Welsh and Walker 1831), and stated under the chart title and date is "High water at George Town on F. & C. XI 15. Spring tides rise 12 ½ feet", indicating that the datum is a high water level. In the upper Tamar extensive floodplains are shown around the North Esk River to the east of Launceston (Fig. 10), and on the inside bends of meanders in the upper Tamar where today the suburbs of Invermay and Riverside are located. Depths of the upper Tamar home reach channel are noted at 1–2 fathoms (1.8–3.6 m) and the channel between mud banks about 150 m wide, and further downstream to below

Tamar Island depths of 1–3 fathoms (1.8–5.4 m) are recorded, with the occasional deeper spot, and the channel between mud banks 100–300 m wide. Allan's (1833) Chart of the Tamar River has many similarities to Welsh's perhaps more of an artist's version, and also notes depths are in fathoms in a note near Hebe Reef but the datum is unclear. Depths of the upper Tamar are the same as recorded by Welsh and Walker (1831).

Stanley's (1881) chart is adapted from Welsh and Walker (1831), and updated, with clearer depth datum control. Soundings are in fathoms related to the level of low ordinary spring tides, which are noted to rise 10 feet. Depths of the upper Tamar are shown the be consistently 0.5 fathoms (0.9 m) shallower than as noted by Welsh and Walker (1831), though with an occasional deeper spot than recorded by Welsh in 1831. However Welsh and Walker (1831) used high water as datum while Stanley (1881) used low water, so this does not mean that the upper channel is shallower, rather slightly deeper there being an upper Tamar tidal range of c. 3.25 m. As reviewed in the next section, this was after dredging of the upper Tamar had commenced.

Two further factors must be taken into consideration in comparing these older charts with present. First, the tidal range of the Tamar increases by c. 1 m from George Town to Launceston (Fig. 5), and if this was not realised by the early chart makers then high tide soundings would have overestimated the channel depth up river relative to the lower estuary where the datum was located. Secondly, sea-level rise has occurred of 14 cm since 1841–2002, at a rate of 0.8 $\pm$ 0.2 mm/year (Hunter et al. 2003). This factor is probably too minor to affect comparison but must be eliminated rather than ignored.

In summary, analysing the early descriptions of Collins, Clark and Paterson, along with the chart of Welsh showing home reach depths of 1.8–3.6 m below a high water mark, it seems that the upper Tamar was at the time of European discovery to feature a narrow and shallow main channel surrounded by extensive mud banks. This was very difficult to navigate through with a 1.7 m draft vessel.

Recent charts show Home Reach depths to be deeper, of 2–3 m below approximately the level of lowest astronomical tide, with channel areas of depth greater than 1 m shown to be up to 100 m wide (Hydrographic Service 1971). The deepest location in the upper estuary is a plunge pool of 50–100 m diameter where the South Esk enters the Tamar at the base of Cataract Gorge, similar to the description of Clark in 1804. From Ti-Tree bend to Tamar Island the channel is 3–4 m depth with occasional deeper spots of 5–6 m, and the deeper channel is mostly 100–150 m wide.

Foster et al. (1986) discarded use Welsh's 1830 chart (Welsh and Walker 1831) as a reference point to initial river condition "because lack of detail in vertical control and some doubts as to its reliability", commenting that the datum has long since disappeared. However, Welsh states a high tide mark (Fig. 10) which puts it around 3 m above low tide mark and today's datums, which while not precise is indicative of a shallower 1830s upper estuary than commonly thought. It is also confirmed by the early descriptions of the Tamar by Collins, Clark and Paterson.

In March 1805 Paterson had founded Launceston, which was later found to lack good fresh water supply (Whitfeld 1912). However its settlement and development proceeded rapidly, with arrival of ships, people and the construction of wharfs. From 1832 Low Head had a temporary wooden lighthouse before a stone one was built (Percy 1993), with the pilot station, the 15.24 m high lighthouse and associated cottages established by the late 1830s. Boat building also proliferated early along the Tamar shores, such as at Rosevears and Launceston where good stands of timber were available (Percy 1993). The first sea going trading vessel was launched near Launceston in 1826 (Percy 1993).

By 1840 Launceston had 6,000 population, 25 % of these convicts, and weekend activity was focussed on the estuary (Percy 1993). In 1837 the Tamar Yacht Club was formed based in Launceston, and the first regatta held in 1839. Gay boat recreational events occurred (Whitfeld 1912), and walkers enjoyed Cataract Gorge but mud flats never tempted bathers, leading to floating baths in early 1860s moored near the Tamar Yacht Club adjacent to the Cataract Gorge outflow.

Floods, Dredging and Sedimentation Concerns

Historical accounts of the Tamar show indications of differences of opinion regarding river management (Whitfeld 1912), and this has remained the case to the present time. In 1834 the Launceston advertiser reported requests to the Lieutenant Governor concerning navigation issues in the Tamar (Whitfeld 1912). Thomas Scott and 90 residents signed a petition that "The River Tamar if properly bouyed and beaconed, the pilot service rendered more effective… which increasing trade of the port absolutely demand, would be of easy and safe navigation". In 1857 the Marine Board was formed to control the port, including ship owners, importers, and exporters using the port. Large sums were spent on improving the waterway, primarily in dredging (Whitfeld 1912). In 1878 a spoon dredge was procured, and work commenced. In 1884 a Priest man dredge was first used, and 1889 two more ordered, followed by a Platypus (bucket) dredge, used with great success. In 1885 the bar at Town Point was removed (see Welsh's 1830 chart in Fig. 10), to create 14 feet of water at low tide. This led to loss of occupation of boatmen who used to transfer passengers and freight from offshore to shore and back (Whitfeld 1912). By the 1880s larger ships were arriving leading to dredging further down the river, in the boat channel, and in 1891 a rock at Whirlpool Reach was partly removed.

Dredging from 1893 created a 61 m wide channel having a minimum depth of 4.57 m (Foster et al. 1986). Dredging operations continued between 1890 and 1965 to maintain sufficient channel depths to accommodate team vessels (Foster et al. 1986), increasing channel depth offshore of Kings Wharf in Home Reach from 2 to 3 m in 1889 to be 5 m below S.L.W. in 1936, 1949 and 1955, 1957. Until the mid-1900s, dredging operations were mainly restricted to Kings Wharf at Launceston and the mouth of the North Esk, after which time dredging was extended to the main channel from Launceston to Rostella to facilitate the passage of larger vessels (Foster et al. 1986).

In 1912 a consulting engineer had recommended removing rocks and other obstructions to the Tamar shipping channel, at a time when the Launceston port was stagnating because it was becoming too difficult to navigate the river (Millington 2003). It was further recommended to straighten the upper Tamar by cutting a canal of over 4 m depth at low tide through the low lying meander bend at Riverside (Fig. 1). In 1919 a steam driver cutter/suction dredge commenced work on what was called "Hunter's Cut" but was later carried away in a flood, and in 1920 a larger machine made further progress (Millington 2003). Completion however proved too expensive and the dredged channels later silted up.

Launceston experienced its first flood in 1809 causing low land to be inundated by up to 1 m, and after that embankments were constructed (Koshin 2007). A greater flood in 1852 found these prevented the escape of water. Material dredged from the removal of Town Point in the late 1880s was used to raise and make more substantial the embankment alongside Invermay. Floods occurred twice in 1889, and again in 1893, and during the 1910s flood protection levees were extended along the North Esk (Koshin 2007).

The 1929 flood is the largest on record, flooding all low lying areas to some meters of depth including Invermay and other suburbs, and after this protection of low lying areas was investigated. It was recognised that clearance of the South Esk catchment had resulted in faster run-off, which during heavy rainfall posed a threat to Launceston (Rechberger 2007). A canal was proposed to divert the North Esk from its natural approach to Launceston, towards the south through Invermay to enter the Tamar at Stephenson's Bend. This was later not carried out as too expensive, rather the cheaper option of raising the levees was undertaken. Again in the 1950s a diversion proposal was turned down, and the government rather engaged UNSW to undertake modelling (Foster et al. 1986) which resulted in the concrete wall that diverts the flood torrent from the South Esk Gorge down the estuary and away from Launceston (Fig. 11).

Between the mid-1920s and 1969, the Launceston Port Authority dredged Home Reach and Kings Wharf, material

Fig. 11 The South Esk River in flood entering Home Reach at Launceston, where an arcuate concrete wall diverts flow towards the *left* (north) and away from Launceston in the background (Photograph: C. Shepherd)

on barges were then floated downstream and dredge material deposited in the main channel where water depths were greater, likely around Rosevears and Gravelly Beach. Between 1958 and the late 1960s, a cutter-suction dredge was used to pipe dredged material from Home Reach to the intertidal zone at Riverside. However by the 1960s primary port facilities had moved from Launceston to Bell Bay, and regular maintenance dredging of the Tamar was discontinued in 1965 when road and rail links virtually eliminated the need for large ships to use the upper reaches of the estuary (Pirzl and Coughanowr 1997). As a result previously dredged areas started to revert to its natural state of low capacity channel and extensive intertidal mudflats. This brought local concerns about loss of channel capacity to contain flood events, along with the popular view that mud is not as pretty as open water. Dredging in Home Reach recommenced in the 1980s to enlarge the waterway, and maintain the navigation channel, access and visual amenity (Pirzl and Coughanowr 1997), dredging approximately 750,000 m^3 of material per year, which was deposited at one of four silt deposit sites located on flood plains adjacent to the estuary at Trevallyn, Ti Tree Bend and Stephenson's Bend.

Siltation of the upper estuary continues to be an intense concern with respect to flood risk, recreational use and visual amenity. For example, the Northern Tasmania newspaper "*The Examiner*" has articles on this topic very frequently along with reader letters to the Editor. The majority of these view the silt as nuisance and needing to be got rid of. The Launceston Flood Authority commenced raking trials in 2012: concern in these trials has been the potential release of historical contaminants trapped in the sediment.

Contaminants

Tin smelters operated within Launceston from the 1870s to the1920s, during which time textile, engineering, and food processing industries were also established, many of which continue today. Until recent decades these industries discharged directly into the North Esk River without regulation, contributing significantly to the heavy metal loads of the Tamar Estuary (Wood 2000). This combined with suspended solids (Foster et al. 1986) and by 1973 the Tamar was recognised to be grossly polluted particularly between Launceston and Rosevears, from organic and inorganic wastes from industrial, mining and domestic sources, as well as heavy metals such as lead and cadmium, shipping ballast and indiscriminate dumping of miscellaneous garbage, car bodies etc. (Parliament of Australia 1973). The provision of adequate sewage treatment facilities appeared beyond the financial capacity of municipal authorities.

Mining pollution has contributed some 10–25 mg $kg^{-1}$of lead to upper Tamar sediments since the 1890s (Seen et al. 2004). However, before systematic monitoring was undertaken (Attard et al. 2011), contaminant measurement in the Tamar was of a localized nature to support specific industrial situations or research programs. Dowson and Rushton (2006) from water quality monitoring 2002–2004 found highest concentrations of heavy metals in the upper estuary and downstream to Freshwater Point, with zinc median levels approaching Australian and New Zealand Environment Conservation Council (ANZECC) guidelines. This was attributed to point source inputs in that section as well as remobilisation of historical industrial deposits in sediments. Continued monitoring up to 2011 has found that several metals have been monitored to be above ANZECC guidelines in the upper estuary (Attard et al. 2011, 2012).

Contamination in the estuary of pathogens derived from sewage and abattoir wastes, hydrocarbons, heavy metals particularly zinc and cadmium and other contaminants such as fluoride, cyanide and phenols was raised as a concern by Pirlz and Coughanowr (1997). Pathogen levels in the upper estuary frequently exceeded guidelines for secondary contact recreation, and it was recommended that oysters from the upper Tamar not be consumed owing to heavy metal contamination (Gawne and Richardson 1992). Upgrades in waste water treatment plants in 1994 resulted in significant reductions in bacterial levels in the upper Tamar (Aquenal and DEPHA 2008), but monitoring results 2009–2011 showed that enterococci levels remained above guidelines in the upper estuary (Attard et al. 2011, 2012), though decreased down the estuary.

Data on nutrients in the Tamar was historically patchy, infrequent and excluding some forms (Prizl and Coughanohr 1997), and data has been improved with the water quality monitoring commenced by TEER. The upper estuary has been shown to have highly elevated levels of all nutrients, while levels in the middle estuary are also typically quite high and exceeding of ANZECC water quality guidelines (Aquenal and DEPHA 2008).

Water quality data has been complied in the last several years to give an annual report card providing advice to the community on recreational swimming and shellfish consumption from different sections of the estuary. In 2010 the upper estuary continued to be degraded with restricted use (TEER 2011), though improved to be fair during higher catchment inflows in 2011–2012 (TEER 2012). Below this the estuary has good ecosystem health but with some water quality issues, while the lower estuary has excellent recreational water quality (TEER 2011, 2012).

Barrage Proposals

Siltation issues in the upper Tamar have attracted proposals of a Tamar barrage. An undated report likely from the mid-1970s (Hoerner et al. n.d.) notes such proposals were first suggested in 1911, and later re-examined in 1939 and 1970. The proposals are for the creation of a freshwater lake upstream of a barrage at some point across the Tamar Estuary.

The Tamar Regional Planning Authority was a statutory authority established in 1969 and produced a Regional Plan (Tamar Regional Master Planning Authority 1975). This focussed on regional growth, industrial and developmental activities, these involving land use, sea and air communications, road and rail, and engineering such as water and sewage. The environment section included investigation of the feasibility of a barrage, lock and freshwater lake proposal, as well as identifying the need for fundamental data on the environmental values, and listed a number of areas that should remain natural areas. The strategic planning diagrams map three additional bridges across the Tamar and extensive areas for industrial development and residential expansion that all have not come to pass, and provide a context for the more precautionary estuary planning processes that later occurred.

Tamar Lock Committee (1975) reported on the 1970 proposal to the Port of Launceston Authority to construct a barrage, lock and spillway across the Tamar at one of several possible sites. Possible benefits listed included water supply, a constant level freshwater lake for recreation, increase in property values, a bridge crossing and improvement to deep water navigation in the upper Tamar. Possible disadvantages listed included siltation of navigation channels and bays in the lower Tamar due to reduction in tidal scour, algae and weed growth in the upper reaches, progressive siltation of the upper reaches, the need for sewage treatment in the upper reaches and the lifting of groundwater levels which could

Fig. 12 The intertidal zone of the central Tamar at Little Swan Point (**a**) in 1956 showing the sand and gravel substrate (Photograph: T. Cole), and (**b**) in 2004 following *Spartina* colonisation, extending some 140 m seaward (Photograph: M. Sheehan)

aggravate landslip movement in critical areas. This was before studies had identified the biodiversity values of the Tamar, and the need for biological and environment studies is noted a following ecology section. The plan had a focus more on visual amenity of views and natural colours and visual significance, and a section on reserves and sanctuaries relates to the prohibition of shooting to protect wildlife therein, though lists submissions from Queen Victoria Museum personnel on waders.

The barrage proposal re-emerges regularly in local media, and the latest has been put forward by Frith (2012) as providing a solution to the upper Tamar silt problem. A low barrage with spillway is proposed, freshwater preventing flocculation so silt is not deposited. Other benefits listed are freshwater supply, protection from sea level rise particularly Invermay, flood mitigation, tourism attraction and lifestyle living.

Introduced Species

The deliberate and accidental introduction of plants and animals into the Tamar Estuary has resulted in dramatic changes to estuarine ecology and geomorphology, particularly by rice grass (*Spartina anglica*). This was introduced to the Tamar by the Department of Agriculture on behalf of the Launceston Port Authority in 1947 and subsequently elsewhere with the intention to stabilise sediments and reduce channel siltation by concentrating tidal flow of the shipping channel (Ranwell 1967; Phillips 1975; Pringle 1993). It was thought that vegetating the mudflats would promote vertical accretion, define the channel and enhance scour, therefore reducing the frequency of expensive dredging operations.

By 1973 rice grass covered approximately 350 acres in the Tamar and its growth and spread was recognised as uncontrolled (Parliament of Tasmania 1973), with detrimental effects to beach and recreation areas. By 1998 the Tamar infestation was Australia's largest at 420 ha (Kriwoken and Hedge 2000), it continued to spread towards the coast (Hedge 1998), but by 2008 was reduced by 46 ha to 374 ha (Sheehan and Ellison 2010). This was due partly to boat wake erosion, and success of the 2006–2008 *S. anglica* management plan for the Tamar Estuary, which attempted to maintain the Port Dalrymple section as a 'Rice Grass Free Zone' (DPIWE 2006).

The introduction of *S. anglica* to the Tamar Estuary caused a dramatic change to the physiography of the intertidal zone, as demonstrated in Fig. 12. Previously sand or gravel beaches or rock shorelines were converted to silty habitats under *Spartina*, which accreted upwards and built out towards the channel. Using transect based topographic surveys and stratigraphic coring we have determined that the volume of material trapped under Spartina is 1,193,441 m^3, comprised of approximately 17 % *Spartina*-derived organic matter and 83 % silts and clays. Using historical profiles, sedimentation rates since the introduction of *S. anglica* were estimated at between 8.7 and 52.4 mm/year (Sheehan and Ellison 2010).

Management and Future Visions for the Estuary

The Tamar Estuary has been the subject of two detailed reviews, both with an objective of providing background to prioritise management issues. Pirzl and Coughanowr (1997) identified environmental degradation in several areas resulting in poor water quality and sediment contamination from industrial heavy metals and faecal bacteria. Further monitoring was found to be necessary owing to limited information available from previous surveys. Other issues of concern were rapid sedimentation of the upper reaches leading to navigation difficulties and flood risk, and introduced species of rice grass and Pacific Oyster that have colonised intertidal areas throughout the estuary. Aquenal and DEPHA (2008) updated the review, finding that those key environmental issues still remain, despite some improvements in both point and diffuse source contaminant emissions.

Table 2 Goals regarding Tamar estuary management in the Tamar NRM Strategy 1999

Goals		
1. To identify, stabilise and reduce adverse sedimentation.		
2. To identify future opportunities for use of the Tamar Estuary and North Esk River		
Actions	Regional Priority	Goal
Define and document		
Undertake pilot study to identify nature of sediments	High	1
Identify nature and sources of sediments entering the Tamar Estuary	High	1
Fill in R and D holes in dynamics of the River system	High	1
Educate and inform		
Educate urban users on how to minimise sediment generation	Medium	1 & 2
Hold regular forums for stakeholders of the estuary to review degradation issues	Medium	1 & 2
Coordinate activities- shared responsibility		
Develop cooperative programs with upstream catchments	Medium	1
Develop and integrated rehabilitation plan for the upper Tamar and lower Esk	Medium	1
On-ground works		
Revegetate areas actively eroding	High	1
Manage stock access to priority riparian and coastal areas already identified	High	1
Instigate efficient dredging program develop infrastructure (e.g. silt traps)	High	1
Use innovation and inspiration for new approaches		
Find uses for dredge spoil	High	1
Establish Tamar estuary trust to bring together stakeholders to develop strategic management plan for the estuary	High	1
Regulate and control		
Regulate where necessary the discharge of sediment into water courses in the Region	High	1
Enforce land clearance regulations to ensure the maintenance of riparian vegetation	Medium	1

Adapted from Rowland (1999b)

Natural resource management (NRM) in the last 15 years provides the best community information on futures for the Tamar, owing to the high level of stakeholder and community consultation involved. Given the degraded state of the Tamar and some differences between local views on its future, it is worthwhile to examine NRM development of aspirational goals for the future.

NRM with emphasis on improved use of catchments and coasts for the benefit of native values such as biodiversity commenced with Landcare and Coastcare in the mid-1980s. Community led local groups developed rehabilitation projects, such as fencing native vegetation from stock access, weed control, and replanting with native species to reduce coastal and riparian erosion. Community capacity building, attitude and awareness change and property management planning were the foci of the first National Land care Program in 1990–1991 (Hajkowitz 2009).

Since 1990, the Australian federal government has funded seven NRM programs, which have tended to increase in both budget and time span (Hajkowitz 2009). During the Natural Heritage Trust period of 1996–2008, a regional coordination for NRM activity was first proposed in 1999 (Banks 1999) and following rigorous search the Tamar Estuary region was selected as the first trial regional NRM case study (Rowland 1999a). A reference group of the three councils in the area of the Tamar, and other stakeholders and interested parties was formed, and following consultation a strategy was developed which included management activities relating to coastal and marine environments, with consideration of the economic and social implication of issues (Rowland 1999b). One issue of ten priorities in this was "Adverse sedimentation impacts on the Tamar Estuary and North Esk River", and future goals and actions were identified and prioritised as shown in Table 2.

Concern for the water quality of the Tamar Estuary and its catchments was ranked overall as the highest priority issue by the Tamar Region NRM Strategy in 1999, from sources such as current and historic industrial sites, waste-water and storm water runoff, and agricultural wastes. The Tamar Region NRM Water Quality Working Group identified that lack of adequate water quality monitoring data was the major impediment in addressing water quality issues in the Tamar Region. The need for baseline data and information of the general health of the Tamar Estuary and information about the known environmental impacts and threats was subsequently improved by Tamar 2020.

Practically concurrent was the Tamar Estuary 2020 project also partly funded by NHT which produced a Tamar Estuary and Foreshore Management Plan (Watchorn 2000). A large steering committee of organisational stakeholders, local and state government developed a detailed aspirational action plan with the vision of "A vibrant Tamar River valley

Table 3 Aspirational goals and principles developed by the Tamar 2020 project

Aspirational goals	Principles for the Tamar Estuary and foreshore
Encourage appropriate use and development	Multiple uses will be encouraged
Raise awareness of values of crown land	The potential for cumulative adverse effects will be considered and recognised in decision making
Promote coordinated approached	Rare, endangered or vulnerable habitats, natural and physical resources and ecosystems will be protected from adverse effects wherever possible
Involve the community	Significant adverse effects of use and development on natural and physical processes, such as sediment transport, shall be avoided, remedied or mitigated
Preserve and promote distinctiveness of the Tamar estuary	Natural, cultural and historic features and important views of such features will be protected where appropriate
Improve understanding of estuarine management	Environments are recognised as a long term public asset which should not be compromised by inappropriate short term decision or developments
	Where there are threats of serious or irreversible environmental damage, lack of full scientific certainty should not be used as a reason for postponing measures to prevent environmental degradation

Adapted from Watchorn (2000)

managed to provide sustainable employment. Marine and land based activities planned in an ever improving environment, based on the superb delivery of produce and the enjoyment of nature and history for the economic benefit of the entire region, its residents and visitors". It details a complex mix of jurisdictional interests that relate to use, development and management of the Tamar Estuary and foreshore. The action plan states aspirational goals and principles which help define a view for where the estuary should be in 2050 as shown in Table 3.

The goals of Tamar NRM (Rowland 1999b) and Tamar 2020 (Watchorn 2000) indicate a precautionary and conservative regional approach to maintenance and improvement of estuarine biophysical values that has since remained prevalent in the community attitudes to major developmental proposals that appear over time. The steering committee of Tamar 2020 was not of conservationists, rather dominated by government developmental services and planners and business CEOs. The goals and principles along with detailed actions of both these projects were later incorporated into the NRM North 2005 strategy.

Following introduction of the Australian NRM regional model in the early 2000s, and division of Australia into 56 regions, Tasmania formed three regions (Tasmanian Government 2001), which in the case of the north was the Northern Natural Resource Management Association (NRM North). Governance was enabled through the 2002 Natural Resource Management Act, at which time state government agencies prepared situation papers to advise the new NRM committees and staff on the status and threats of natural resources. These were combined with stakeholder and community consultation to develop accredited regional NRM strategies, prioritising actions for future activities of the NRM regions. These activities were largely funded by a combination of federal and state government programs.

The Tamar Estuary and its catchments form a large part of the NRM North region, and in 2005 the first accredited strategy encompassed chapters based on principal assets including Coasts. Consultation occurred through circulation of a draft strategy, presentation of this at a large number of public meetings throughout the region and consideration of verbal comment, receipt of written comments of which well over 1,000 were formally considered, incorporated and replied to, and then formal adoption of the strategy by the committee and the NRM Association. Aspirational targets were prioritised as a result of this process and feedback.

Of relevance to the Tamar Estuary prospects 2050 and beyond, this strategy identified Aspirational Targets, defined as the desired condition of Northern Tasmania's natural resources in the long term (50 years), in order to guide regional planning by setting a context for the measurable Resource Condition Targets (RCT). For estuaries, the aspirational target set was "To achieve integrated management of the estuarine, coastal and marine environments for the benefit of the community, while maintaining the natural environment and associated ecological processes". RCTs (Table 4) were defined as the desired condition of the natural resources in the medium term (10–20 years), and Management Action Targets (MAT) were defined as desired short-term outcomes and outputs of actions over 1–5 years. These targets were set from consideration of background resource condition information, previous plans such as Tamar NRM (Rowland 1999b) and Tamar 2020 (Watchorn 2000), stakeholder consultation, and community consultation.

Most challenging for the region was the Tamar Estuary, owing to its greater size than all others, being the most urbanised estuary, also of highest conservation priority owing to high biodiversity identified near its mouth (Edgar et al. 1999), and high degree of modification by river inflow diversion, industrial history, dredging and introduction of invasive species. Issues of sedimentation have always been high on the community agenda and proven difficult to solve, leading to the management action target of "Implement programs for controlling sedimentation in waterways

Table 4 NRM North (2005) strategy resource condition targets

	Resource condition target	Year to be achieved
RCTM1	The areal extent, distribution, abundance and condition of key coastal, estuarine and marine species, communities and their habitats will be maintained at, or improved above 2006 benchmark levels (includes both target and threatened species)	2020
RCTM2	Estuarine and marine water quality at key estuary and inshore monitoring sites will be at or better than 2010 benchmark levels (as measured by nutrient, TSS and TDS levels), unless other levels are required for ecosystem maintenance and health	2020
RCTM3	The net increase in the extent, abundance and impact of current significant vertebrate, invertebrate and vegetative pests and diseases will be below 2006 benchmarks, and the rate of increase of new invasive pests and diseases will be reduced to below the 2006 rate	2020

(e.g. Tamar Estuary)" in the strategy (NRM North 2005: Table A8:1) of contribution to condition: high and significant.

In 2008 NRM North fostered the development of the Tamar Estuary and Esk Rivers program, a cooperative partnership between state and local governments, water authorities and key industry and supporting partners such as Tamar NRM. The program provides a coordinated management approach with a 2030 vision of "The Tamar Estuary and Esk River's systems healthy, productive, valued and enjoyed- our rivers of life". A key goal is to improve scientific understanding of the issues impacting upon the health of the TEER waterways to better identify and target priority areas requiring investment in on-ground works. TEER has produced reports on estuarine monitoring that have fulfilled the lack of data identified by Rowland (1999b), and allowed the RCT's of NRM North (2005) (Table 4) to be evaluated. The top goals set by the community back in 1999 (Table 2) have now all been far improved or completed by NRM North and Tamar NRM through TEER.

The NRM North strategy was reviewed and updated in 2010 (NRM North 2010) in which the Healthy Seas and Coasts objective is "by 2025, marine/estuarine water quality and health at key inshore monitoring sites to be maintained or improved compared to 2010 benchmark levels". The resource outcome stated is "by 2025, marine and estuarine water quality to be maintained and enhanced to ensure the extent and distribution of estuarine species and communities are maintained or improved, compared to 2006 benchmark levels". These improved natural condition goals are also reflected in the Tamar NRM Strategy 2007–2011 with a condition target of "coastal and estuarine processes maintained with no loss in the abundance and extent of assets by 2025" (Tamar NRM 2007).

Conclusion

Estuarine environments worldwide have long been utilised as centres for urban and industrial development, and the Tamar Estuary over the past 200 years has been affected by a number of anthropogenic activities, including mining, agriculture, urban and industrial development, dredging and the introduction of exotic species. Catchment land use change has resulted in high sediment yield into the Tamar, dredging has periodically increased channel depths, and sedimentation has trapped industrial contaminants from early industrial eras. The most significant geomorphic change has resulted from the introduction of *Spartina anglica*, resulting in conversion of the majority of shorelines from rock or beaches to silt/clay marshes, absorbing much of the catchment sediment yield.

In addressing these management issues, an evolution over time has been apparent. The principles and values related in the first Tamar NRM Strategy (Rowland 1999b) and Tamar 2020 Management Plan (Watchorn 2000) were radically different from those of 25 years earlier relating to a vision of the estuary for the future. The switch was from a 1970s vision of an urbanised industrialised estuary with preserved views and some sanctuaries (e.g. Tamar Regional Master Planning Authority 1975) to a late 1990s focus on the management of natural values to maintain biodiversity. This indicates the impact of biodiversity studies and development of realisation that natural systems create species habitats that occurred in the 25 years between these plans. This change was promoted initially by biologists in the museums and local university/college studies, which through community education and the growth of Land care gained support to become a background value at early stages of NRM. It was also perhaps assisted by the continuance of Tasmania as a State not prone to economic development or growth relative to mainland States, which may not have been anticipated in the 1970s.

For the middle to lower Tamar there is community consensus that its future vision follows closely to the NRM goals, visions and aspirational targets as reviewed above and detailed in Tables 2, 3 and 4 and NRM North (2010), of maintenance and improvement of natural biophysical values. Outside of NRM this has been well exemplified by the community opposition to the now failed pulp mill development proposed in the last few years at Bell Bay, showing the precautionary preference against development and supporting the maintenance and improvement of natural biophysical values. The greatest risk in this section is low lying coastal areas known to be at risk from sea-level rise projections (Sharples 2006), which is addressed in the mechanisms of NRM North (2010) strategy, and increasingly by Local Councils (e.g. Inglis and Ellison 2012).

For the Upper Tamar three alternative scenarios are prevalent for future visions:

1. Barrage conversion to freshwater lake: this recurrent proposal as reviewed above is increasingly against contemporary community views that value natural biophysical conditions.
2. Create more open water by clearing the silt, by increasing flows down Cataract Gorge (Davis and Kidd 2012) and/or raking silt deposits to promote natural clearance in floods. This is the prevalent view also supported by the amenity value and reduction of flood risk provided by wider channels and more open water, but is subject to the removal of the silt not causing problems through remobilisation of contaminants.
3. Tolerate the silt better. This review of the early history establishes that when Europeans first visited the upper Tamar it was heavily silted, hence what is viewed as a nuisance is actually the natural state. Community memory has a view of open water normality that is more perhaps from the period of dredging through last century.

Acknowledgments This research was funded by an Australian Research Council Linkage Grant LP0214145. The authors are grateful to Mrs. Maureen Bennett who pointed out to us the detail available in the earliest historical descriptions of the Tamar.

References

Allan D (1833) Chart of the Tamar River, Van Dieman's land: corrected from the best Authorities. http://nla.gov.au/nla.map-rm1210

Aquenal Pty Ltd, Department of Environment, Parks, Heritage and Arts (2008) State of the Tamar Estuary. DEPHA Tasmania, Hobart

Attard M, Thompson M, Kelly R, Locatelli A (2011) Tamar Estuary & Esk rivers ecosystem health assessment program monitoring report 2011. NRM North, Launceston

Attard M, Thompson M, Kelly R, Locatelli A (2012) Tamar Estuary ecosystem health assessment program monitoring report 2012. NRM North, Launceston

Australian Bureau of Statistics (2012) 2011 census data QuickStats-George Town, Launceston and West Tamar local council areas. www.abs.gov.au

Baillie PW (1989) Jurassic-Cainozoic. In: Burrett CF, Martin EI (eds) Geology and mineral resources of Tasmania. Special publication of Geological Society of Australia 15. Geological Society of Australia, Brisbane, pp 339–409

Banks AJ (1999) Regional natural resource management in Tasmania: a framework for developing strategies and setting priorities. Department Primary Industries, Water and Environment, Hobart

BMT WBM (2010) Tamar Estuary and Esk rivers catchment model: final report. TEER Waterways Management Group, Launceston

Breen S, Summers D (2006) Aboriginal connections with Launceston places. Launceston City Council, Launceston

Bugg R (1990) Waves and coastlines. In: Scanlon A, Fish G, Yaxley M (eds) Behind the scenery. Department of Education and the Arts, Hobart, pp 80–98

Cameron P (2011) Grease and ochre: the blending of two cultures at the colonial sea frontier. Fullers Bookshop, Launceston

Cameron WM, Prichard DW (1963) Estuaries. In: Hill MN (ed) The sea. Wiley, New York, pp 306–324

Davies JL (1978) Beach sand and wave energy in Tasmania. In: Davies JL, Williams MAJ (eds) Landform evolution in Australasia. Australian National University Press, Canberra/Norwalk, pp 158–167

Davis J, Kidd IM (2012) Identifying major stressors: the essential precursor in restoring cultural ecosystem services in a degraded estuary. Estuar Coast 35:1007–1017

Dowson G, Rushton M (2006) Draft Tamar Estuary water quality and sediment study 2002–2005. Department of Tourism, Arts and Environment, Hobart

DPIPWE (2000) Tasmanian marine protected areas strategy. Department of Primary Industries Water and Environment, Hobart

DPIWE [Department of Primary Industries, Water and Environment] (2006) Rice grass area-based management plan: Tamar river/beauty point region (2006–2008). Department of Primary Industries and Water, Hobart

Edgar GJ, Barrett NS, Graddon DJ (1999) A classification of Tasmanian estuaries and assessment of their conservation significance: an analysis using ecological and physical attributes, population and land use. Parks and Wildlife Service, Hobart

Ellison JC (2010) Tasmania. In: Bird EFC (ed) Encyclopedia of the world's coastal landforms. Springer, Dordrecht, pp 1395–1402

Flinders M (1958) Port Dalrymple on the north coast of Van Diemen's Land: with the River Tamar, and its tributary streams 1798. Hydrographic Office, Dyeline reproduction of map, Taunton. Original manuscript chart held in the archives of the Hydrographic Department, Ministry of Defence

Foster DN, Nittim R, Walker J (1986) Tamar river siltation study. Technical report no. 85/07. University of New South Wales Water Research Laboratory

Frith R (2012) The Tamar valley: a vision and strategic plan for its development. Report produced by Tamar Lake Inc., Launceston

Gawne KM, Richardson BJ (1992) An investigation into the distribution and accumulation of selected chemical species in the waters, sediments and biota of Deceitful Cove. Report to TEMCO and Comalco, Launceston

Gehrels WR, Callard SL, Moss PT, Marshall WA, Blaauw M, Hunter J, Andrew Milton J, Garnett MH (2012) Nineteenth and twentieth century sea-level changes in Tasmania and New Zealand. Earth Planet Sci Lett 315–316:94–102

GHD (2009) Report for upper Tamar siltation study: preliminary options analysis. Launceston City Council, Launceston, 46 pp

Goede A, Murray-Wallace CV, Turner E (1993) A diverse Holocene molluscan fauna, including *Anadara trapezia*, from Royal Park, Launceston, Tasmania. Pap Proc R Soc Tasman 127:17–22

Grant J (1974) 1772–1833 The narrative of a voyage of discovery performed in his majesty's vessel the Lady Nelson of sixty tons burthen, with sliding keels; in the years 1800, 1801, and 1802, to New South Wales. Heritage Publications, Melbourne

Gray AJ, Marshall DF, Raybould AF (1991) A century of evolution in *Spartina anglica*. Adv Ecol Res 21:1–62

Guenegou M, Levasseur JE, Bonnot-Courtis C, Lafond LR, Le Rhun J (1991) The geomorphological and botanical changes in Kernic Bay (Brittany, France): influence on coastal management. J Coast Res 7 (2):311–339

Hajkowitz S (2009) The evolution of Australia's natural resource management programs: towards improving targeting and evaluation of investments. Land Use Policy 26:471–478

Hedge P (1998) Strategy for the management of Rice Grass (*Spartina anglica*) in Tasmania, Australia. Department Primary Industries, Water and Environment, Hobart

Hoerner MJR, Crawford DJ, McDonough GJ, O'Bryan WE (n.d.) Investigation and preliminary design for a barrage for the Tamar River. Tasmanian College of Advanced Education, Launceston

Hunter J, Coleman R, Pugh D (2003) The sea level at Port Arthur, Tasmania, from 1841 to the present. Geophys Res Lett 30:1–4. doi:10.1029/2002GL016813

Hydrographic Service (1971) AUS 168 chart River Tamar Bell Bay to Launceston. Royal Australian Navy, Commonwealth of Australia, Woollongong

Inglis J, Ellison JC (2012) Shoreline change at Kelso: geomorphic assessment to identify management priorities. Report to West Tamar Council, Tasmania

Koshin J (2007) Flood protection. In: Richardson P (ed) On the tide 3: more stories of the Tamar. Launceston Library, Launceston, pp 50–51

Kriwoken LK, Hedge P (2000) Exotic species and estuaries: managing *Spartina anglica* in Tasmania, Australia. Ocean Coast Manag 43:573–584

Lambeck K, Chappell J (2002) Sea level change through the last glacial cycle. Science 292:679–686

Library Committee of the Commonwealth of Australia (1921a) Introduction. Lieutenant-governor David Collins. In: Commonwealth of Australia: historical records of Australia series III:1, pp ix–xxxiii

Library Committee of the Commonwealth of Australia (1921b) Report of a survey made by Mr. William Collins, of the harbour of Port Dalrymple, January, 1804. In: Commonwealth of Australia: historical records of Australia series III:1, pp 583–585

Library Committee of the Commonwealth of Australia (1921c) Remarks on the land etc. in the harbour of Port Dalrymple by Thos. Clarke, Superintendent January 1804, In: Commonwealth of Australia: historical records of Australia series III:1, pp 585–587

Library Committee of the Commonwealth of Australia (1921d) Journal of Lieutenant- Governor Paterson, Outer Cove, 28th November 1804. In: Commonwealth of Australia: historical records of Australia series III:1, pp 613–623

Macknight CC (1998) Low head to Launceston: earliest reports of port Dalrymple and the Tamar. Historical Survey of Northern Tasmania 3, Launceston

McIntyre L (1987) Rowella- Kayena (West Bay – Richmond Hill) from 1805. Rowella Book Committee, Launceston

Meynard D, Gaston T (2010) Beneath the Tamar: more than silt. NRM North, Launceston

Millington E (2003) Hunter's cut. In: Richardson P (ed) On the tide 2: more stories of the Tamar. Launceston Library, Launceston, pp 74–75

Mitchell W, Chittleborough J, Ronnai B, Lennon GW (2000) Sea level rise in Australia and the pacific. S Pac Sea Lev Clim Change Newsl 5:10–19

Nichols MM, Biggs RB (1985) Estuaries. In: Davis RA (ed) Coastal sedimentary environments. Springer, New York, pp 77–186

Noble KE (1992) Land capability survey of Tasmania. Tamar report. Department of Primary Industries, Water and Environment, Hobart

NRM North (2005) Northern Tasmania natural resource management strategy. NRM North, Launceston

NRM North (2010) Northern Tasmania natural resource management strategy 2010–2015. NRM North, Launceston

O'Connell JF, Allen J (1998) When did humans first arrive in greater Australia and why is it important to know? Evol Anthropol 6:132–146

Parliament of Tasmania (1973) Tamar river pollution. Report of select committee of the legislative council with minutes of proceedings 86:1–11

Percy BJ (1993) The Tamar boats: a complete reference list. Foot and Playsted, Launceston

Phillips AW (1975) The establishment of *Spartina* in the Tamar Estuary, Tasmania. Pap Proc R Soc Tasman 109:65–75

Pirzl H, Coughanowr C (1997) State of the Tamar Estuary: a review of environmental quality data to 1997. Supervising scientist report 128, supervising scientist, Canberra

Porter-Smith R, Harriss PT, Anderson OB, Coleman R, Greenslade D, Jenkins CJ (2004) Classification of the Australian continental shelf based on predicted sediment threshold exceedance from tidal currents and swell waves. Mar Geol 211:1–20

Pringle A (1993) *Spartina anglica* colonisation and physical effects in the Tamar estuary, Tasmania 1971–91. Pap Proc R Soc Tasman 127:1–10

Pritchard DW (1955) Estuarine circulation patterns. Proc Am Soc Civil Eng 81:717

Ranwell S (1967) World resources of *Spartina townsenii* (sensu lato) and economic use of *Spartina* marshland. J Appl Ecol 4:239–256

Rechberger B (2007) River diversion schemes. In: Richardson P (ed) On the tide 3: more stories of the Tamar. Launceston Library, Launceston, pp 46–47

Rees C (1994) Tasmanian seagrass communities. Unpublished Master's thesis, School of Geography and Environmental Studies, University of Tasmania

Rowland CJ (1999a) Regional natural resource management in Tasmania: case study. Natural Heritage Trust, Local Government Association of Tasmania/Department of Primary Industries, Water and Environment, Hobart

Rowland CJ (1999b) Tamar region natural resource management strategy. Launceston City Council, George Town Council, West Tamar Council/Natural Heritage Trust, Hobart

Sandery PA, Kampf J (2007) Transport timescales for identifying seasonal variation in Bass Strait, south-eastern Australia. Estuar Coast Shelf Sci 74:684–696

Seen A, Townsend A, Atkinson B, Ellison J, Harrison J, Heijnis H (2004) Determining the history and sources of contaminants in sediments in the Tamar Estuary, Tasmania, using ^{210}Pb dating and stable Pb isotope analyses. Environ Chem 1:49–54

Sharples C (2006) Indicative mapping of Tasmanian coastal vulnerability to climate change and sea level rise: explanatory report, 2nd edn. Department of Primary Industries Water and Environment, Hobart

Sheehan M, Ellison JC (2010) Assessment of the potential consequences of large-scale eradication of *Spartina anglica* from the Tamar Estuary, Tasmania. In: Ayres DR, Ker DW, Erickson SD, Olofson PR (eds) Proceedings of the third international *Spartina* conference, San Francisco. San Francisco Estuary Invasive *Spartina* Project of the California Coastal Conservancy, Oakland, pp 129–133

Smith B (1995) Tamar invertebrates: an atlas of the common species. Queen Victoria Museum and Art Gallery, Launceston

Sprod D (2009) Van Dieman's land revealed: Flinders and Bass and their circumnavigation of the island in the colonial sloop Norfolk 1798–1799. Blubber Head Press, Hobart

Stanley HJ (1881) North coast of Tasmania, river Tamar from the sea to Launceston. The Admiralty, London

Tamar Lock Committee (1975) River usage and activities Appendix-Tamar Lock Committee Report. In: Tamar Regional Master Planning Authority Tamar Estuary River Management Plan. Tamar Regional Master Planning Authority, Launceston, pp 19–28

Tamar NRM (2007) Tamar NRM strategy 2007–2011. Tamar region natural resource management. Tamar NRM, Launceston

Tamar Regional Master Planning Authority (1975) Tamar Estuary river management plan. Tamar Regional Master Planning Authority, Launceston

Tasmanian Government (2001) Tasmanian natural resource management framework. Department of Primary Industries, Water and Environment, Hobart

Taylor J (2005) Aboriginal languages. Unpublished Master's Thesis, University of Tasmania, Launceston

TEER (2011) Tamar Estuary 2010 report card. NRM North, Launceston

TEER (2012) Tamar Estuary 2012 report card. NRM North, Launceston

Thom BG, Roy PS (1985) Relative sea levels and coastal sedimentation in southeast Australia in the Holocene. J Sediment Petrol 55:257–264

Walsh L (1990) The development of rivers. In: Scanlon AP, Fish GJ, Yaxley ML (eds) Behind the scenery: Tasmania's landforms and geology. Department of Education and the Arts, Hobart, pp 53–80

Watchorn L (2000) The Tamar Estuary and foreshore management plan: Tamar Estuary 2020. Tamar 2020 Steering Committee, West Tamar Council, Beaconsfield

Welsh J, Walker J (1831) The river Tamar in Van Diemen's Land. Office of the Admiralty, London. http://nla.gov.au/nla.map-nk2456-171

Whitfeld E (1912) The river Tamar. The Examiner, Launceston

Wood W (2000) Tamar Estuary fish and sediment study – first year report for the Tasmanian Department of Health and Human Services. William Wood and Associates, Relbia

Part II

Estuaries Being Degraded

Gold Coast Broadwater: Southern Moreton Bay, Southeast Queensland (Australia)

Ryan J.K. Dunn, Nathan J. Waltham, Nathan P. Benfer, Brian A. King, Charles J. Lemckert, and Sasha Zigic

Abstract

The Gold Coast Broadwater, a large shallow estuarine water body, is a central feature of the Gold Coast City in Southeast Queensland (Australia) and forms the southern part of Moreton Bay. The Broadwater has undergone dramatic changes over the past few decades, including the construction of an extensive number and network of artificial waterways that account for up to 90 % of Australia's canal estates. Positioned in one of the fastest growing regions in the developed world, urbanisation surrounding the Broadwater will continue. The region has important biodiversity values that have led to areas of the Broadwater being listed as an international Ramsar site and inclusion to international migratory bird agreements. The Broadwater provides a vital function in the provision of feeding, spawning and nursery sites for recreationally and commercially important finfish species. Key to the protection of the Broadwater is a reduction of pollutant loads from urban and agricultural stormwater run-off, golf courses and industrial infrastructure/areas and replacement of natural habitats with urban development. Collectively, initiatives undertaken by regulatory authorities have been successful to date and demonstrate that future conservation requires the integration of multidisciplinary science and proactive management driven by the high ecological, economical and community values placed on the Broadwater and adjoining waterways.

Keywords

Gold Coast Broadwater • Urban expansion • Coastal waterways • Artificial residential waterways • Urban run-off • Vessel pollution

R.J.K. Dunn (✉) • N.P. Benfer
B.A. King • S. Zigic
Asia-Pacific ASA Pty Ltd, P.O. Box 1679, Surfers Paradise, QLD 4217, Australia
e-mail: rdunn@apasa.com.au

N.J. Waltham
Centre for Tropical Water and Aquatic Ecosystem Research (TropWATER), James Cook University, Townsville, QLD 4811, Australia

C.J. Lemckert
Griffith School of Engineering, Griffith University, Gold Coast Campus, Nathan, QLD 4222, Australia

E. Wolanski (ed.), *Estuaries of Australia in 2050 and Beyond*, Estuaries of the World,
DOI 10.1007/978-94-007-7019-5_6, © Springer Science+Business Media Dordrecht 2014

Box 1

Ryan Dunn et al. studied the Gold Coast Broadwater, a large shallow estuarine water body with an extensive number and network of artificial waterways that account for up to 90 % of Australia's canal estates. The region is one of the fastest growing regions in the developed world with rampant urbanisation. Nevertheless the region has important biodiversity values that have led to areas of the Broadwater being listed as an international Ramsar site and inclusion to international migratory bird agreements. Key to the protection of the Broadwater is a reduction of pollutant loads from urban and agricultural stormwater run-off, golf courses and industrial infrastructure/areas and replacement of natural habitats with urban development. Evidence from previous modelling suggests that under a scenario of current and future urbanisation, sediment and nutrient loads far exceed that which is sustainable for local waterways. As such, the "business as usual" option is not sustainable and major stormwater capital infrastructure works are therefore required.

Additionally, restoration projects such as weed removal, foreshore stabilisation works, revegetation of cleared areas, community education/and capacity building and flood mitigation programs are necessary to achieve the values that the population demands and expects. This is probably the best case in Australia of planning for the future by integrating multidisciplinary science and proactive management driven by the high community values placed on Broadwater and its waterways.

Site Introduction

Promoted as a leading tourism and lifestyle destination, the Gold Coast City in Southeast Queensland, Australia, boasts an image of the 'Green behind the Gold' with its 52 km of white sandy beaches, in front of a backdrop of subtropical rainforest in hinterland areas. The Gold Coast Broadwater, a large shallow estuarine water body, is a central feature of the Gold Coast City and forms the southern part of Moreton Bay, a national and international significant coastal system (27.88 S; 153.41 E, Fig. 1). The Broadwater is positioned in one of the fastest growing regions in the developed world with the Gold Coast population increasing from 110,900 in 1976 (ABS 1986) to 497,848 in 2008 (ABS 2008). The forecast population is projected to reach 900,000 by 2030. As a consequence of continued population increase, the Broadwater will experience further large-scale urban expansion, including residential canal, marina facilities and commercial infrastructure that are already present along the region's intertidal waterways and catchments.

Geomorphological and Hydrological Setting

Climatic Settings and Physical Characteristics

The climate of the Southeast Queensland region is subtropical, with most rainfall occurring during the summer period (December to February). The Broadwater has an average annual rainfall of 1,094 mm (Gold Coast Seaway) (Eyre et al. 2011a) and an average regional mean air temperature range of 13–29 °C. Winds are predominantly from the southeast to northeast from October to March and from the southwest at other times, with a daily pattern of strengthening afternoon sea-breezes superimposed. The region experiences regular severe summer storms, while tropical cyclones are known to affect the region, although infrequently. Depressions of subtropical origin, such as east coast lows, are more frequent but less intense than tropical cyclones. However, both storm systems are capable of generating severe rainfall and elevated coastal sea levels that can lead to coastal flooding (McInnes et al. 2000).

The Broadwater's catchment covers an area of 108,000 ha of which 30 % has undergone urban development with the remainder being undisturbed forest (40 %), cleared grazing land (12 %), crop land (10 %) and road networks (8 %) (Waltham 2002). The Broadwater catchment has experienced moderate to high soil erosion and land degradation over the past few decades, which represents an

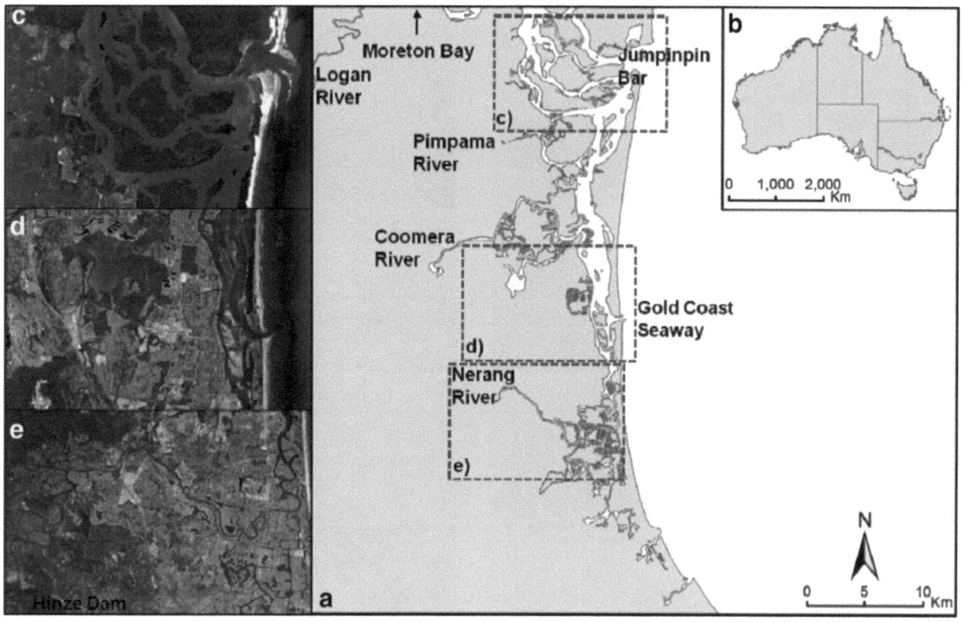

Fig. 1 Location of the Gold Coast Broadwater (**a**); Southern Moreton Bay, Australia (**b**); including the regions of the four principal rivers entering the Broadwater and inset images of; (**c**) Jumpinpin Bar; (**d**) Gold Coast Seaway and (**e**) Nerang River including the Hinze Dam (Satellite images sourced from Google Earth)

approximate 10 % increase in nutrient and sediment loads (BMT WBM Oceanics 2010). Four principal rivers drain the catchment: Nerang River, Coomera River, Pimpama River and the Logan-Albert Rivers, in addition to several smaller creeks (Fig. 1). The volume of water entering the Broadwater from the river systems is of lesser magnitude than tidal inputs, with the exception of periods following heavy rainfall (Moss and Cox 1999).

The Broadwater's main tributary to the south, the Nerang River, is largely urbanised throughout its estuarine reaches and features extensive residential canal developments. Land use upstream from the estuarine zone of the Nerang River is mainly rural residential and grazing, with forested areas upstream of the Hinze Dam (Fig. 1e), approximately 37 km from the river mouth. Upper catchment flow of the Nerang River has been limited since the construction of this dam, which is the major water supply for the Gold Coast. The main catchment to the north is the Coomera. This catchment is currently less urbanised than the Nerang, but rapid urban development in its lower reaches, including extensive residential canal, golf course and periurban development, have occurred on the lower flood plains (Waltham 2002). Land use in the mid Coomera catchment includes acreage housing estates with grazing land and a forested upper catchment. No water storages have been built on the Coomera River. The Pimpama and Logan-Albert Rivers located further to the north drain into the upper portion of the Broadwater.

The Broadwater has undergone dramatic changes, including construction of the Gold Coast Seaway, commercial developments, residential canal and artificial lake estates and dredging operations for navigation purposes. Additionally, increased human activities and intervention within the Broadwater has led to land reclamation to extend portions of the western foreshores to increase recreational and development purpose areas. Indeed a striking feature of the Gold Coast is the extensive number and network of artificial waterways, including canal estates, built for the purpose of increasing useable waterfront property development (Figs. 2 and 3). Since initial canal developments in 1956 (Johnson and Williams 1989) artificial waterways within the Broadwater system have progressively expanded from the initial confines of the Broadwater and lower Nerang River to include a widespread network contributing to approximately 500 km of tidal waterfront (www.ozcoasts.gov.au) with a surface area exceeding 200 km^2, accounting for up to 90 % of Australia's canal estates (Waltham and Connolly 2011). More recently, a shift occurred away from the construction of narrow open canals leading directly off estuarine stretches of coastal rivers, to the construction of artificial tidal lakes with restricted exchange with natural estuarine waters (Zigic et al. 2002) in order to minimise the tidal prism. Presently, the number of artificial lakes connected to the Broadwater system accounts for approximately 95 % of the total Oceania number, where the Oceania number represents 63 % of the global number (Waltham and Connolly 2011). The Gold Coast Broadwater foreshore has also been widely modified during the last decade in response to community aspirations and economic objectives.

Fig. 2 Gold Coast Broadwater and surrounding landscapes; (**a**) skyline of residential and commercial buildings behind the golden beaches; (**b**) expansion of residential suburbs has changed forested catchments to urban residential estates; (**c**) canal estate construction; (**d**) dredging shallow freshwater wetlands for the construction of canal estates; (**e**) construction of canal estate over terrestrial land; (**f**) construction of houses in canal estate including pontoons and jetties; (**g**) artificial lake system separated from estuary via a tidal control device and (**h**) tidal gate controlling exchange of water between estuary and artificial lake system

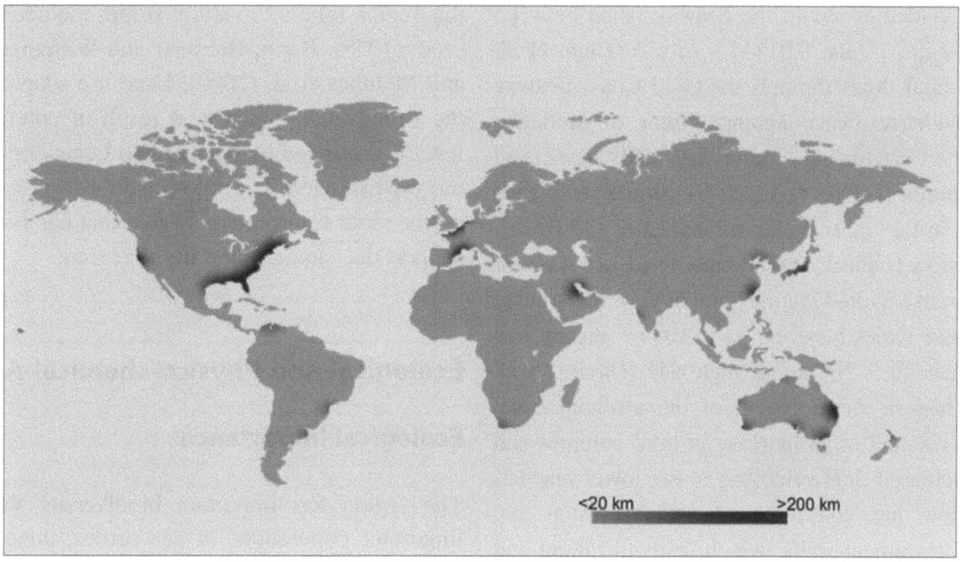

Fig. 3 Global extent of artificial residential estuarine systems (canal and lake estates combined), by country and (where relevant) state or province within country. *Grey scale* gradient represents total linear length (km) (Source: Waltham and Connolly 2011)

The low wave-energy environments of protected beaches, sand flats, mud banks and mangrove habitats of the Broadwater provides economical and recreational benefits to the regional community including: boating, jet skiing, parasailing, fishing, swimming, and diving, supporting a range of vessels and also water float plane activities.

Hydrological Setting and Features

The Broadwater is a micro-tidal, estuarine lagoon, characterised by exposed sandbanks, mangrove systems, islands, and seagrass beds, which are protected from the Pacific Ocean by a barrier island system (South Stradbroke Island, located between Jumpinpin Bar and the Gold Coast seaway) (Fig. 1). Tidal channels within the region are up to 9 m deep, however the system is typically shallow, with a mean mid-tide water depth of 1.74 m (Eyre et al. 2011a). The Broadwater is connected to the adjoining Pacific Ocean through the Gold Coast Seaway in the south, and Jumpinpin Bar to the north (Fig. 1). The tidal exchange of waters through the two oceanic connections is important in the exchange and flushing of the Broadwater (Mirfenderesk and Tomlinson 2008). The hydrodynamics of the Broadwater environment is well known (e.g. Mirfenderesk and Tomlinson 2007, 2008; Mirfenderesk et al. 2007; Sennes et al. 2007; Knight et al. 2008; Ali et al. 2009, 2010; Davies et al. 2009), including the adjoining artificial environments (e.g. Zigic et al. 2002, 2005; Benfer et al. 2010). Collectively, these studies provide important insight into the hydrodynamic characteristics and are important in understanding the hydrodynamic consequences (and related water quality aspects) under past and future land use.

The Gold Coast Seaway (Fig. 1d) is a man-made 250 m wide rock retaining wall entrance constructed in 1985, and serves as the primary navigable connection from the Broadwater to the Pacific Ocean. The mean water depth of the Seaway is 11 m (Mirfenderesk and Tomlinson 2008). In a morphological sense, features of a wave-dominated estuary are visible at the Broadwater, including a barrier system, ebb tide shoals and flood tide shoals in the outer zone of the estuary and mudflats, and mangroves and salt marsh in the central zone of the estuary. The hydrology and geomorphology of the southern region of the Broadwater is influenced by the Seaway entrance. Alternatively, the opening at the Jumpinpin Bar in the north of the Broadwater is not navigable due to the dynamic nature of the opening, with a shallow shifting sand bar dominated entrance.

The central Broadwater is predominantly a marine system (approximately 33 ppt) where the salinity reduces upstream in the adjoining river systems (approximately 10 ppt). At the Gold Coast Seaway and Jumpinin Bar, the tidal range varies between 1 and 2 m (Mirfenderesk and Tomlinson 2008), and this range is the main driving force for horizontal water flow and exchange with the Pacific Ocean (Mirfenderesk and Tomlinson 2008). Tidal characteristics at the Seaway and Jumpinpin Bar have been identified as predominantly semi-diurnal with a diurnal inequality (ebb dominant). Calculation of the total flux through Seaway during a typical flood tide is approximately 66×10^6 m^3, for an approximate cross section area of 3,500 m^2. The same calculation for Jumpinpin Bar is approximately 50×10^6 m^3 (approximate cross section of 3,000 m^2, Mirfenderesk and Tomlinson 2008).

Measured tidal velocities within the Seaway range between 0.001 and 0.909 m s^{-1} and 0.018–1.8 m s^{-1} (Dunn et al. 2012a). Strong tidal flows through the Gold Coast Seaway effectively provide resilience against minor to moderate changes in the catchment conditions and associated stormwater pollution (Davies et al. 2009). Similar resilience is not provided in the upper major tributaries of the Broadwater due to narrow channels and a longer residence time for tidal flushing. From a hydrodynamic point of view the volume of the Broadwater varies between 30×10^6 m^3 during low tide to more than 50×10^6 m^3 at high tide (Davies et al. 2009). Indeed historic development of the artificial canal estates initially resulted in an increase in tidal volumes and subsequently increased tidal velocities in the lower reaches. This altered flow has contributed to bank erosion and undercutting of revetment walls, which leads to failure and damage to residential properties and infrastructure (Zigic et al. 2005). In response, flow structures have been installed to regulate the exchange of water with the adjacent estuary to ensure restricted water exchange, where the tidal range of the adjacent estuary is not desired within the canal system, and in some cases, with careful timing, they can be used upstream to reduce the tidal demands at the mouth of the estuary.

The Broadwater and associated development infrastructure also experience periodic flooding from both intensive and prolonged rainfall events. In fact, Gold Coast region has long been rated as the most vulnerable area subject to flooding in Australia (Smith 2002). Hence flood risk maps are published by the Gold Coast City Council. As a result flood defences of variable style and quality are located throughout the system, including weir structures and restrictions of developments on the floodplains. Current building regulations have been informed by knowledge of past extreme events. The current Disaster Management Plan (GCCC 2010) contains a co-ordinated approach to floods, storms and other potential risks (Cooper and Lemckert 2012). The typical ground elevation in waterfront developments is approximately 2 m above mean high tide level and periodic flooding occurs in the lowest lying areas after extreme rainfall. The floodplains of the Broadwater are developed with a number of dwellings at risk of flooding during major rainfall events. Storm tides within the Broadwater can exacerbate flood events by elevating sea levels at the outflow regions of rivers and streams thereby reducing flow rates. In many situations, the weather conditions that cause storm tide events are also accompanied by severe rainfall. The establishment of design storm tide levels for planning and development purposes is therefore of critical importance to minimise the risk of damage to infrastructure during such events. Previous studies conducted in this region relating to storm surges include those by Harper et al. (1977), Blain, Bremner and Williams Pty Ltd (1985) and McInnes et al. (2000). Land use adaptation options for the Broadwater region as a result of potential rises in sea level is discussed by Cooper and Lemckert (2012). Furthermore, Mirfenderesk (2009) presents discussions of flood risk on the Gold Coast and a system that has been developed to support decision-making in the region.

Ecological and Physico-chemical Aspects

Ecological Importance

The region has important biodiversity values, including important populations of sea turtles, dugongs and annual shorebird migrations that have led to areas of the Broadwater being listed as an international Ramsar site,[1] including Chinese-Australia Migratory Bird (1974) and Japan-Australia Migratory Bird Agreement (1986) status. Additionally, the Broadwater, as part of the *Marine Parks (Moreton Bay) Zoning Plan 2008*, is zoned as a habitat protection zone and contains Marine National Parks including the Coombabah Lake region. Furthermore, the Broadwater also hosts the Southern Moreton Bay Islands National Park, which covers an area over 1,500 ha and supports greater than 50 % of the mangroves of Moreton Bay (DERM 2012). In an effort to protect turtle and dugongs from boat strikes in critical feeding and resting areas, designated vessel "go slow" zones are implemented within the Broadwater. The Broadwater also provides an important nursery for recreationally and commercially important finfish species (*Queensland Fisheries Act 1994*).

Seagrasses, saltmarshes and mangrove communities occupy the Broadwater environment, which are vital in the provision of feeding, spawning and nursery sites for local aquatic fauna (Ross 1999; Thomas and Connolly, 2001; Connolly 2003; Hollingsworth and Connolly 2006; Waltham and Connolly 2006, 2007). The modification or removal of seagrass and mangrove nursery habitats for urban expansion should be viewed with their ecological importance in mind. Large areas of seagrass occur in the northern region of the Broadwater with smaller fragmented areas occurring within the southern sector (Moss and Cox 1999). A 1997 survey of seagrass composition and distribution revealed three species

[1] The Convention on Wetlands, signed in Ramsar, Iran in 1971, is an intergovernmental treaixty dedicated to the conservation and wise use of wetlands. The Convention's mission is the conservation and wise use of wetlands by national action and international cooperation as a means to achieving sustainable development throughout the world (Environment Australia 2003).

of seagrass occurring in the Broadwater; *Zostera capricorni, Halophila ovalis* and *Halophila spinulosa* over an area of 304 ha (McLennan and Sumpton 2005). Mangrove communities include *Avicennia marina* and, to a lesser extent, *Rhizophora stylosa* and *Aegiceras corniculatum,* while salt marsh communities include *Sporobolus virginicus*.

Carbon Flow and Ecology

The determination of organic matter sources and flow within the Broadwater, which provide nutrition for estuarine species remote from carbon sources is important in understanding the functioning and management of the estuarine environment. A number of studies investigating organic matter sources and connectivity within the Broadwater have been undertaken using C/N ratios, stable isotopes (δ^{13}C and δ^{15}N) and fatty acid biomarkers (e.g. Connolly 2003; Thomas and Connolly 2001; Melville and Connolly 2003, 2005; Guest and Connolly 2004; Waltham and Connolly 2006; Werry and Lee 2005; Dunn et al. 2008, Spilmont et al. 2009; Oakes et al. 2010; Werry 2010; Lee et al. 2011). Results of these studies have demonstrated the importance of both autochthonous and allochthonous organic matter sources, including terrestrial and planktonic (i.e. mangroves, seagrasses, zooplankton, diatoms and other algal species) sources.

Resident first order and higher consumers within the Broadwater include recreationally and commercially important fish species, including: *Arrhamphus sclerolepis* (snub-nosed garfish), *Acanthopagrus australis* (yellow fin bream), *Sillago ciliate* (sand whiting), *Platycephalus fuscus* (dusky flathead) and *Mugil cephalus* (mullet). The Broadwater also provides an essential nursery for coastal water fish species (e.g. snapper, mackerel and tailor species). Shark and ray species (e.g. *Carcharhinus leucas* (bull sharks) and *Dasyatis fluviorum* (estuary stingray)), mud crab (*Scylla serrate*) turtles (*Chelonia mydas,* green sea turtle), dolphin (*Tursiops aduncusn*) and dugong (*Dugong dugon*) also inhabit the estuarine environment. The Broadwater is an important region for large communities of resident bird species (e.g. *Phalacrocorax varius* (pied cormorant), *Ardea novaehollandie* (white-faced heron) and *Haematopus longirostris* (pied oyster catcher)) which are routinely found in large numbers on the exposed sand and mud flats. In addition to resident bird species the Broadwater is an important bird staging area along avian migratory flyways (routes) with birds (e.g. *Tringa brevipes* (grey-tailed tattler), *Charadruis leshenaultii* (greater sand plover)) arriving from Europe and Asia during the southern hemisphere summer. Readers are referred to Shorebird Management Strategy Moreton Bay (EPA 2005) for a detailed list of resident and migratory bird species of the Broadwater. Communities of soft sediment benthic infauna which provide food sources for both bird and fish species have been investigated in the Broadwater waterways, as part of baseline and urban development impact assessments (e.g. GHD Pty Ltd. 2003; Dunn 2009) and include amphipod, crab, bivalve, worm and yabby species. Initial information suggests large heterogeneity in assemblages, presumably owing to different grain size and organic matter content in surficial sediments, pollutant accumulation, and also water quality conditions (e.g. Stephenson and Cook 1977; Poiner 1977; Young and Wadley 1979; Stephenson 1980). In a study of the benthic faunal assemblages in canal estates, Cosser (1989) reported 65 taxa present, with 25 taxa comprising approximately 95 % of the total abundance. In that study, two broad community types were identified, one community restricted to dead-end canal locations and characterised by low diversity, low species richness, while the other was distributed in connecting canals and characterised by high species richness and diversity. This spatial arrangement of species followed a progressive transition between community types resulting from deterioration in the concentration of dissolved oxygen; a lower species richness and abundance in dead-end canal sites where dissolved oxygen concentration is low while higher richness and abundance in oxygen rich canal opening areas. Within the Broadwater recreational fishing applies pressures on the benthic communities through the collection of animals for the use of bait. Species particularly targeted and collected by fishing enthusiasts include bloodworms (*Marphysa* sp.), the marine yabby (*Trypaea australiensis*) and soldier crabs (*Mictyris longicarpus*). Additionally, these species are also collected commercially and sold at fishing outlets within the Broadwater region. The removal of benthic species not only alters important trophic links within the Broadwater but potentially influences benthic metabolism, rates of organic matter turnover, efflux rates of regenerated nutrients and also nitrogen cycle pathways (Jordan et al. 2009; Dunn et al. 2009, 2012b; Eyre et al. 2011b).

The construction of extensive artificial residential waterways have replaced natural wetlands and created new estuarine habitats throughout the Broadwater. Comparisons of the fish fauna in the artificial waterways of the Broadwater and in adjacent natural wetlands of mangrove, saltmarsh and seagrass have shown almost complete overlap in the species present (Morton 1989, 1992). Although differences in the relative proportions of species are detectable, all of the economically important species found in adjacent non-disturbed estuarine waters are also present in artificial waterways (Morton 1989, 1992), additionally some critic species to wetlands have also been recorded in canals (e.g. Waltham and Connolly 2007 recorded the Beady Pipefish (*Hippichthys penicillus*)). This same pattern of species overlap is similar to canal developments elsewhere in the world (e.g. Baird et al. 1981; Maxted et al. 1997).

Connolly (2003) provided the first account that some species are able to derive nutrition from local sources in

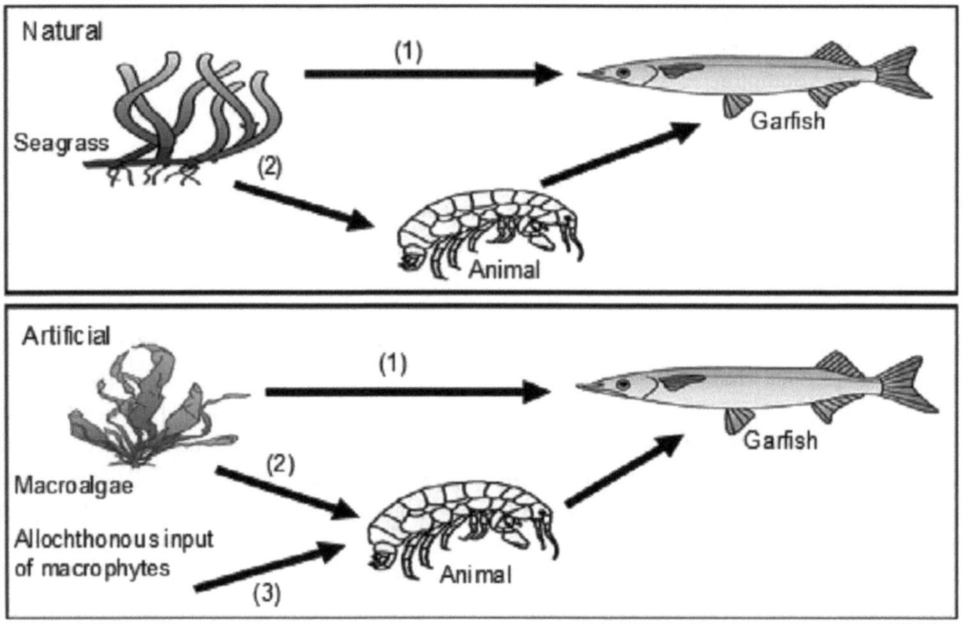

Fig. 4 Trophic models for *Arrhamphus sclerolepis* in natural and artificial urban waterways: (*1*) direct consumption of autotroph; (*2*) direct consumption of an animal intermediary that utilises autotroph; and (*3*) consumption of animal intermediary that utilises detrital macrophytes (having both enriched and depleted $\delta^{13}C$ values) transported from adjacent natural wetlands (Source: Waltham and Connolly 2006)

artificial systems using alternative sources to those available in natural systems within the Broadwater. In an additional study, Waltham and Connolly (2006) provided conclusive evidence of the plasticity of fish to adapt to the created estuary environment within the Broadwater. In that study, the authors combined stomach contents with stable isotope analysis to demonstrate the basal sources of nutrition and uptake pathway in the snub-nosed garfish (*Arrhampuhs sclerolepis*) (Fig. 4).

The Broadwater, including the artificial waterways, provide additional habitat opportunity for Bull Shark (*Carcharhinus leucas*) populations. Sightings of *C. leucas* within the artificial waterways of the Broadwater are often reported (Zeller 1999; Werry 2010), where canal systems, in addition to low salinity river environment, have extended the extent of nursery habitat for newborn and juvenile *C. Leucas* outside the range of conventional natural estuaries (Werry 2010). An extensive investigation by Werry (2010) and Werry et al. (2012) within the Broadwater demonstrated the movement patterns of *C. leucas* within the Broadwater differs, with juveniles remaining within low salinity river reaches and canal systems, while older and larger sharks extend over much wider areas. Recorded movements of the older and larger sharks include extensive coverage of canal and river systems and movements between the Broadwater and adjoining oceanic environment (Werry 2010). Newborn and juvenile individuals remaining resident in single defined areas are susceptible to anthropogenic influences (e.g. fishing pressures, plastics and contaminants). Shark populations and their movement habits pose a safety concern for some local residents of the Broadwater where water activity pursuits are popular. With urban coastal development and recreational use of the Broadwater set to continue to grow, human interactions with *C. leucas* are likely to increase.

Sediment and Water Column Characteristics

Deep sand accumulations derived from long-shore ocean currents characterise the Seaway and eastern margins of the Broadwater. Surface sediments in the eastern shore regions of the Broadwater are predominantly composed of quartz sands, however canal locations typically exhibit relatively finer sediment textures (Burton et al. 2004). Sediment organic carbon content is greater within residential canals and very low within the central regions of the Broadwater. Increased fine sediments and organic carbon within the residential canals is attributable to inputs from urban sources, which receive loads of urban stormwater from surrounding residential areas. These inputs readily remain trapped within the residential canal due to their designs which often have reduced current velocities, and in turn flushing characteristics, particularly in dead-end areas. Northern Broadwater sediments, including Pimpama, Coomera and Merrimac/Carrara floodplains contain soils associated with mangrove and tea-tree wetlands (humic gleys, peaty gleys and meadow podzolics) and areas of pyrite-rich sediments, which when disturbed can produce sulphuric acid and associated elevated aluminium (Al) and

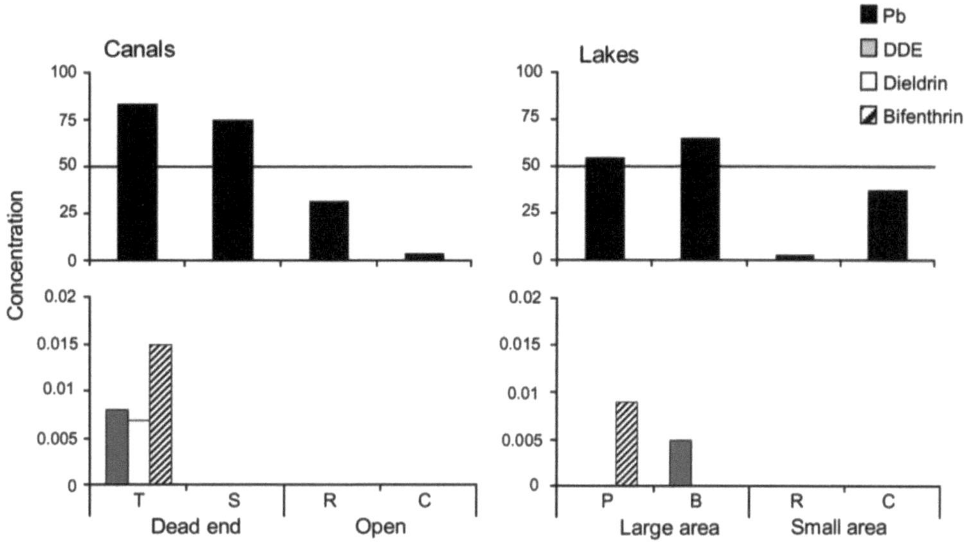

Fig. 5 Example lead (Pb) (mg kg^{-1}) and pesticide (mg kg^{-1}) concentrations in the sediment of single canal and lake systems in Broadwater region. Results shown are for a composite sample from three sediment grabs collected in each system. Interim Sediment Quality Guideline low trigger value for Pb shown, however, DDE, dieldrin and bifenthrin not shown, as concentrations comply with the guidelines (ANZECC/ARMCANZ 2000). For canals, dead end and open labels refer to flow characteristics of each system, while large (~280 ha) and small (~20 ha) area refers to the size of the catchment area draining to each lake system (Source: Waltham et al. 2011)

iron (Fe) concentrations. Discussions of iron-sulfide and trace element concentrations in sediments of the Broadwater region include those by Preda and Cox (1998), Burton et al. (2005, 2008), Robertson et al. (2009) and Pagès et al. (2011).

Salinities and temperature collected throughout the central Broadwater and the oceanic entrances of the Broadwater reflect a well-mixed, well flushed and dynamic system, with no notable stratification (e.g. Mirfenderesk and Tomlinson 2007; Davies et al. 2009). In contrast the prevailing flushing regime within studied canal systems has led to the establishment of an oxycline at a depth of approximately 10 m (depending on season), below which depth hypoxic conditions prevail (Waltham 2002, 2009; Lemckert 2006).

The hydrology, geochemistry and primary productivity of the Broadwater is linked to sediment and nutrient inputs from the catchment, ultimately as a result of freshwater inputs. Within the Broadwater much of the terrestrial loading occurs during episodic high-energy rainfall events. In addition to rainfall events influencing system behaviour, physical parameters, nutrient and trace metal concentrations within the natural and urbanised settings of the Broadwater and Seaway typically demonstrate cyclic variations, with the influence of tidal cycles apparent (e.g. Dunn et al. 2003, 2007a, 2012a).

As expected for a large dynamic estuarine environment, sediment and water column trace metal and nutrient concentrations between and within locations/habitat type in the Broadwater have been shown to vary significantly (for example concentration ranges see Moss and Cox 1999; Dunn et al. 2003, 2007a, b, c, 2012c; Burton et al. 2004, 2005; Warnken et al. 2004; Eyre et al. 2011b; Waltham et al. 2011). In general, undetectable to very low metal concentrations are located in the central Broadwater, whereas elevated concentrations are observed for sites located in residential canals and commercial marinas. This is attributed to the coarse texture of sediments and well-flushed hydrodynamic regime in the central Broadwater, and to the comparatively poorly flushed nature, finer sediment sizes and proximity to traffic and boat maintenance related metal sources in the residential canals and marinas (Burton et al. 2004). Evidence of spikes in sediment pesticide concentrations (some banned over 50 years ago) in some artificial residential waterways of the Broadwater has been reported by Waltham et al. (2011) (see Fig. 5). Nutrient concentrations measured within the Broadwater are typical of concentrations reported in Australian estuarine systems. Sedimentary and water column bacterial concentrations are presented in Pratt et al. (2007) and Dunn et al. (2012a).

The characteristics of sediment transport are important as they play a critical role in the functionality and health of the Broadwater (Webster and Lemckert 2002), as increased suspended sediments also limits light availability through the water column for primary producers. Additionally, when bottom sediments are resuspended trace metals, nutrients and organic contaminants can be released into the water column. The importance of understanding suspended sediment dynamics within the Broadwater has led to studies being completed in an attempt to increase conceptual understanding of sediment dynamics, including turbidity maxima, within the Broadwater (e.g. Hunt and Lemckert 2001; Webster and Lemckert 2002; Hunt et al. 2006; Davies et al. 2009).

Recent studies within the Broadwater have included investigations of baseline nitrogen cycling rates and have

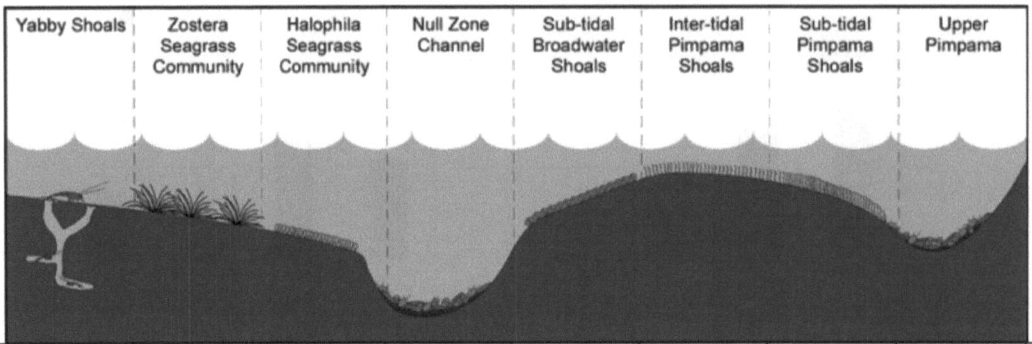

Processes	Yabby Shoals	Zostera Seagrass Community	Halophila Seagrass Community	Null Zone Channel	Sub-tidal Broadwater Shoals	Inter-tidal Pimpama Shoals	Sub-tidal Pimpama Shoals	Upper Pimpama	Total
Denitrification (t N yr^{-1})	103.0 ± 6.8	318.4 ±143.5	176.5 ± 150.3	5.5 ± 0.4	61.2 ± 31.2	4.5 ± 1.5	9.0 ±1.0	10.3 ± 3.0	688.4 ± 337.7
N-fixation (t N yr^{-1})	1.3 ± <0.1	114.6 ± 114.6	147.6 ±147.6	<0.1 ± <0.1	1.0 ± <0.1	<0.1 ± <0.1	0.1 ± <0.1	<0.1 ± <0.1	264.6 ± 264.6
Net N_2 Flux (t N yr^{-1})	101.7 ± 6.8	203.8 ± 28.9	28.9 ± 2.7	5.5 ± 0.4	60.2 ± 31.2	4.5 ± 1.5	9.0 ± 1.0	10.3 ± 3.0	423.8 ±75.6
DIN Flux (t N yr^{-1})	86.7±15.1	14.0±14.3	-5.6± 0.6	0.8 ± 0.5	-7.6±9.4	0.4 ± 1.0	-0.2 ± 0.2	11.6 ± 3.3	100.1 ± 44.7
DON Flux (t N yr^{-1})	-29.2 ±54.9	48.7 ± 49.9	-43.7 ± 15.7	3.6 ±3.0	45.8 ± 73.0	-7.4 ±1.9	1.5 ±7.4	14.4 ±5.6	33.7 ± 211.5
N-Recycling (t N yr^{-1})	135.6 ± 162.1	1013.2 ± 157.3	412.8 ± 89.5	24.0 ± 2.5	584.0 ± 91.7	76.8 ± 10.4	250.7 ± 27.0	21.1 ± 17.2	2909.0 ± 557.9
DIP Flux (t P yr^{-1})	1.8 ± 8.7	1.6 ± 2.4	-0.1 ± 0.3	0.1 ± 0.1	-3.2 ±5.0	0.2 ± 0.2	-0.3 ±0.1	-0.3 ± 0.2	-0.2 ± 27.6
DOP Flux (t P yr^{-1})	10.8 ± 11.2	1.9 ± 9.2	0.7 ± 2.6	0.1 ± 0.1	-3.6 ± 5.0	-0.5 ± 0.5	0.0 ± 0.6	0.2 ± 0.4	9.7 ± 29.5
P-Recycling (t P yr^{-1})	8.5 ± 10.1	62.0 ± 9.8	25.3 ± 5.6	1.5 ± 0.2	36.5 ± 5.7	4.8 ± 0.7	15.7 ± 1.7	1.3 ± 1.1	168.6 ± 34.9
Area (km^2)	5.4 ± 0.3	3.2 ± 0.2	1.7 ± 0.1	0.4 ± <0.1	6.5 ± 0.3	0.6 ± ≤0.1	1.5 ± 0.1	0.9 ± ≤0.1	20.2 ± 1.0

Fig. 6 System wide annual estimates of denitrification, Nitrogen-fixation, net N_2 effluxes, dissolved inorganic nitrogen (*DIN*) fluxes, dissolved organic nitrogen (*DON*) fluxes, nitrogen recycling, dissolved inorganic phosphorus (*DIP*) fluxes, dissolved organic phosphorus (*DOP*) fluxes and phosphorus (*P*) recycling in the eight major open water benthic habitats in the Southern Moreton Bay study area (Source: Eyre et al. 2011b)

shown differences in denitrification rates according to different benthic habitat types and seasonal influences (e.g. Ferguson et al. 2004; Eyre et al. 2011a, b; Teasdale et al. 2009; Dunn et al. 2012c), additionally, manipulative investigations relating to the regulatory influences of burrowing infauna and organic matter have also been performed using Broadwater sediments and inhabiting macrofauna (e.g. Jordan et al. 2009; Dunn et al. 2009, 2012b). Collectively, these studies provide insight into the benthic metabolism and solute fluxes across the sediment-water interface in addition to a recent study by Eyre et al. (2011c) which presents the metabolism of different benthic habitats and their contribution to the carbon budget with regions of the Broadwater. Figure 6 provides annual estimates of biogeochemical processes throughout the Broadwater in varying open water benthic habitats determined by Eyre et al. (2001b). Furthermore, the denitrification efficiencies and ecosystem processes (functional values and mapping) of the Broadwater, is presented by Eyre and Ferguson (2009) and Eyre and Maher (2011), respectively.

Anthropogenic Influences

Key to the protection of the Broadwater is a reduction of pollutant loads from urban and agricultural stormwater runoff, golf courses and industrial infrastructure/areas and replacement of natural wetland habitat with artificial residential canal estates (Waltham et al. 2011). The often restricted flushing nature of the canal and artificial residential lakes developments, relative to nearby natural estuary waters, lends to exacerbation of this issue. This is most evident in dead-end canals where water exchange is low or for waterways that receive stormwater runoff from large urban catchments. Examination of historical data collected within a variety of canals in the Broadwater shows that some estates experience water quality problems mostly due to vertical and horizontal stratification influences (Lemckert 2006). However, canal estates that are shorter and well flushed, better resemble conditions of adjoining estuaries. Increases in nutrient, trace metal, pesticide and bacterial loads entering the Broadwater as a result of

Table 1 Contaminant studies performed within the Gold Coast Broadwater during 2003–2012

Element	Water sampling	Sediment sampling	Biological sampling	Broadwater habitat(s)	Conclusions	Source
Trace metals	X (Water and DGT[a])	–	–	Estuary	Regular pattern of variation in copper and nickel concentrations related to the movement of water past point sources with tidal flows, rather than due to conventional estuarine mixing of end-member waters	Dunn et al. (2003)
Trace metals		X		Estuary, canals and marinas	Sediment metal concentrations undetectable to very low in central Broadwater region, while elevated concentrations were observed in residential canals and marinas	Burton et al. (2004)
Trace metals	X (Water and DGT[a])	–	–	Estuary vessel anchorages	Correlation between recreational boat numbers at anchorage sites and water column copper concentrations for Gold Coast waterways	Warnken et al. (2004)
Nutrients	X	–	–	Creek and Intertidal lake	Nutrient concentrations demonstrated tidal influences, with increased concentrations observed during sampled high tide phases, indicating increased inputs of nutrients originating from external sources other than the study site	Dunn et al. (2007a)
Trace metals	X (Water and DGT[a])	–	–	Estuary, canals, vessel anchorages and marina	Significantly higher concentrations of copper, zinc and nickel concentrations related to the movement of water past point sources with tidal flows	Dunn et al. (2007b)
					DGT-reactive copper concentrations significantly decreased with increased tidal-flushing and vice versa within a marina. DGT measurements also recorded significant increases in copper and zinc after a 24 mm rainfall event	
					DGT-reactive copper increased significantly ($p < 0.001$) during peak boating times, due to increased numbers of Cu-antifouled boats	
Nutrients	–	X	–	Intertidal lake	Lower concentrations recorded for all nutrients in the surface and sub-surface sediments in sample grids dominated by sandy sediment compared to muddy sediments	Dunn et al. (2007c)
					Concentrations compared well with concentrations typically encountered within Australian estuarine systems	
Trace metals	–	–	–	Vessel anchorages	Modelling of copper loading to Broadwater based on boat number observations and literature leaching rates revealed that boat hulls are a major source of copper	Leon and Warnken (2008)
Trace metals	X (DGT[a])	–	X (oysters)	Estuary, canals and marina	Copper concentrations within marina were considerably higher than all other sites	Jordan et al. (2008)
Trace metals	X (DGT[a])	–	X (fish)	Estuary, canals, marina and artificial lakes	With the exception of copper, metal concentrations in water, measured using the DGT technique, complied with relevant Australian guidelines.	Waltham et al. (2011)
					All sediment metal concentrations measured were below the national guidelines, although copper, zinc and lead were found to vary significantly between habitat types.	
					Significantly higher concentrations of copper were found in the gills of fish species from marinas compared to fish caught in other waterways.	
Nutrients, bacteria	X	–	–	Gold Coast Seaway	Nutrient concentrations demonstrated cyclic variations with the influence of tide with near minimum and maximum concentrations generally observed at high and low water, respectively. This demonstrates the influence of oceanic water during flood tides and catchment waters during ebb tides.	Dunn et al. (2012a)

Note: This table is not an exhaustive list, but rather provides examples of previously completed works
[a]Diffusive gradients in thin films is an in-situ time-integrated sampling technique

anthropogenic sources have been reported for receiving sediment, water column and biota (Table 1). A synthesis of data sets reveals important information in understanding key processes and addressing environmental issues. In particular great effort has been made in measuring the natural and anthropogenic processes and associated pollutant concentrations within the Broadwater associated with hydrodynamic conditions (e.g. Dunn et al. 2003, 2007b), developed and undeveloped waterways (Waltham et al. 2011), stormwater runoff (Moss and Cox 1999; Dunn et al. 2007b), marina establishments and antifouled boats numbers (Warnken et al. 2004; Dunn et al. 2007b; Jordan et al. 2008; Leon and Warnken 2008).

The accumulated body of evidence on contaminants within the Broadwater illustrates that there are no major threats of trace metal contamination, with concentrations complying with Australian guidelines (Waltham et al. 2011). The exception is marina facilities which are a source of copper within the Broadwater, additionally anchorages within the Broadwater correspond to elevated trace metal concentrations in the sediment and water column (Warnken et al. 2004). In fact, the potential for recreational and tourism vessels having an impact on ecosystem and public health in Moreton Bay has been one of the key processes identified by the Southeast Queensland Regional Water Quality Management Strategy (a joint whole-of-government and industry initiative) (Warnken and Leon 2006). Due to the large navigable areas of the Broadwater and high usage by recreational and commercial vessels, local marinas and anchorage facilities present notable point and diffuse sources of contaminants, including high loads of contaminants from urban run-off, hard stands and repainting and maintenance outlets (Jordan et al. 2008; Waltham et al. 2011), in addition to contaminant releases sourced directly from the operation of vessels within the Broadwater (Warnken et al. 2004; Leon and Warnken 2008; Dunn et al. 2007b). Warnken et al. (2004) provides data that illustrate relationships between heavy metal concentrations and boating activities at popular Broadwater anchor sites indicating copper emissions from antifouling paints may become an important source with high boat numbers and should be treated with caution by coastal waterway managers. Leon and Warnken (2008) provide an empirical model for the quantification of copper and nitrogen load inputs associated with recreational and tourist vessel numbers in Moreton Bay, including the Broadwater. The key findings of the study reported an estimated input of 141 $\pm$ 46 kg of copper and 1170 $\pm$ 0.38 kg of nitrogen annually. In a related study, Waltham et al. (2011) demonstrated significantly higher concentrations of copper in the gills of three economically important species of fish, with different feeding strategies (partly herbivore *Arrhamphus sclerolepis*, carnivore *Acanthopagrus australis*, detritivore *Mugil cephalus*) from marinas compared to fish caught in other adjacent Broadwater waterways, with fish caught in canals having the second highest copper concentrations and natural waterways the lowest. Concentrations were shown to translate to a low health risk for humans consuming local fish species, with all fish, regardless of feeding strategy (carnivore, herbivore, omnivore), declared safe to eat, complying with Australian Food Standard Code recommended limits for human consumption.

Due to the need for safe navigable waterways for recreational and commercial vessels, dredging regions of the Broadwater is periodically necessary (GCCC 2002). Such practices potentially produce elevated suspended sediment concentrations into the water column as a result of fugitive dredge-related sediment (depending on dredged sediment particle sizes) which in addition to reducing light availability throughout the water column, smother and may ultimately alter benthic communities and trophic linkages, liberate contaminants into the water column and alter nitrogen cycling pathways (Robinson et al. 2005; Erftemeijer and Robin Lewis 2006).

Although the Broadwater is the recipient of numerous point and diffuse sources of contaminants the high degree of mixing and the exchange with the ocean through the Seaway and Jumpinpin Bar aids in the amelioration of water quality (Mirfenderesk and Tomlinson 2008). In addition, this high flushing of the Broadwater is used to expel excess wastewater from the nearby Coombabah waste water treatment plant (WWTP) which services the majority of the Gold Coast City population. Four regional WWTPs have a combined capacity of 1.76 $\times$ 10^5 m^3 a day, with wastewater being treated to a secondary treatment level producing recycled water (Stuart et al. 2009). Presently, the Gold Coast reuses approximately 1.8 $\times$ 10^4 m^3 (10 %) a day of recycled water for irrigation and other commercial uses. The remaining excess treated wastewater (1.1 $\times$ 10^5 m^3 a day) is released to the ocean through a diffuser system (see Stuart et al. 2009) along the northern training wall of the Seaway (Stuart et al. 2009; Dunn et al. 2012a). Under typical operating conditions, treated wastewater is released on the ebb tide, allowing it to be dispersed to the Pacific Ocean, and limiting the recycled water from returning to the Seaway (and ultimately the Broadwater) on the following flood tide (Stuart et al. 2009). Additionally, a second diffuser system exists in the southern sector of the Seaway which also releases treated wastewater from WWTPs located in the south regions of the city. The timing, volume and quality of treated wastewater release are regulated under the *Queensland Environmental Protection Act 1994*. Anticipated population increases have raised concerns about the capacity of the Broadwater to assimilate a greater volume of treated wastewater. As a result, WWTP operators were granted an extension of the existing release licence from 10.5 h per day to 13.3 h per day from the Coombabah wastewater treatment plant (Stuart et al. 2009). The planning load for wastewater treatment on the Gold Coast is expected to grow to 3.51 $\times$ 10^5 m^3 a day over the next 50 years placing significant pressure on the existing release system (Stuart et al. 2009).

The creation of new canals and maintenance dredging of canal sections has resulted in changing the tidal prism of the Broadwater over time. This has resulted in a greater demand on the Seaway to provide greater volumes of water on each tide, increasing tidal currents and creating bank erosion issues. As a preliminary response, all new canal estates require restricted tidal exchange with their adjacent estuary. As a secondary response, in an effort to relieve the stress put on the Seaway, timed gates are installed to connect created brackish lake environments with the upper reaches of the Nerang River. By using carefully timed opening events, the gates are able to reduce the demand on the Seaway by providing the upper reaches with a source of tidal waters during a flood tide and a sink during ebb tide, which has been successful to date (Zigic et al. 2002, 2005). A trade-off though has been reduced flushing potential for the added artificial waterway (Benfer et al. 2010).

Summary

Ongoing urbanisation creates enormous challenges for the management of the Broadwater and thereby threatens the long term sustainability of the system. One example of this challenge is the balance between achieving community expectations and values with continued urban development. In an attempt to achieve this, the Gold Coast City Council has been actively involved with local and regional stakeholders to develop integrated catchment management strategies, which are supported by urban stormwater management plans. These catchment management plans need and indeed follow a total management approach, by combining a range of disciplines (e.g. engineering, environment, social, planning, architecture and management). A major achievement through this process has been the preparation of policies and guidelines relating to stormwater treatment and reuse (Alam et al. 2008). The Gold Coast City Council was one of the first local governments in Queensland (Australia) to implement water sensitive urban design (WSUD) practices (e.g. grass swales/bio-swales, bioretention basins, wetlands and gross pollutants traps) as a statutory requirement under the city's planning framework. In 2007, GCCC adopted WSUD guidelines as part of sustainable development practices for the city for all future urban development. Additionally, attempts have been made to examine opportunities to retrofit WSUD infrastructure to existing urban stormwater network, which involves installing various stormwater quality improvement devices within the existing urban footprint, and works completed have shown improvements to the quality of stormwater runoff, while also extending available habitat for local species (Alam et al. 2008).

As a consequence of the ecological and economic significance of the Broadwater and potential for anthropogenic disturbances, monitoring of the system has been routinely conducted by regulatory authorities (Moss and Cox 1999). Sampling during alternate high and low tides to obtain information on variation in water quality with tidal state, including physiochemical parameters: dissolved oxygen, pH, temperature, conductivity, turbidity, chlorophyll-*a*, nitrogen, phosphorus and faecal coliforms is routinely undertaken at monthly intervals. Routine ecosystem health monitoring within the Broadwater has been ongoing for the past 12 years (see http://www.healthywaterways.org). This program utilises a suite of indicators that provide an understanding of the ecosystem health and response to land use activities and sewage discharge to the Seaway focusing largely on physicochemical parameters along with nutrient and sediment concentrations in the water column, together with seagrass depth/range, nitrogen isotope and coral cover monitoring. Given that evidence suggests contaminants are entering fisheries food webs (e.g. Waltham et al. 2011), there is still a need to understand the ecotoxicological effects of contaminants on local biota, which would provide important data to develop conceptual models. In addition to more conventional water and sediment sampling approaches, time-integrated in-situ sampling (e.g. diffusive gradient in thin films [DGT], diffusive equilibrium in thin films [DET] and semipermeable membrane devices [SPMD]) and routine and experimental biomonitoring approaches including the use of oyster and fish have also been conducted in the Broadwater in an effort to measure and monitor trace metals and metalloids, pesticides, polychlorinated biphenyls (PCBs), polyaromatic hydrocarbons (PAHs) and indicators of bacterial contamination (e.g. Mortimer and Cox 1998; Mortimer 2000; Jordan et al. 2008; Waltham et al. 2011).

Managers are concerned that the region's rapid urban expansion has and will continue to place considerable pressure on the waterway health, which in turn threatens natural processes of the system and the livelihood and lifestyle of residents and tourists. Consequently, the local authority has combined scientific investigations with community consultation to establish a vision and a modified set of water quality objectives (WQOs) for each catchment within the Gold Coast region. An outcome of these assessments is a detailed and targeted management action plan (Waltham 2002). The approach includes predictive computer models to simulate existing and future urban land-use, and to identify opportunities to reduce diffuse pollutant loadings in order to meet each identified WQOs. Under current land use conditions, sediment and nutrient loads may exceed that which is sustainable (Waltham 2002). As such, the "business as usual" option is not sustainable and major stormwater capital infrastructure works are further required. Additionally, restoration projects such as weed removal, foreshore stabilisation works, revegetation of cleared areas, community education/and capacity building and flood mitigation programs have also been implemented in an attempt to improve ecological habitat and aesthetic quality of Broadwater waterways. Such programs are important and necessary in order to protect and achieve the community

Fig. 7 Example restoration projects for the Gold Coast Broadwater and catchment waterways; (**a**) recolonisation of mangrove community at land reclamation beach area; (**b**) restoration of reach along local waterway in the city; (**c**) stabilisation project using soft engineering technology to reduce erosion and encourage natural habitat restoration; (**d**) example of urban sensitive urban design infrastructure to treat stormwater runoff and (**e**) on-going aquatic plant harvesting to improve waterway health

waterway values (Waltham 2002, e.g. Fig. 7). Restoration programs over the past decade have achieved major successes and include community tree planting and weed eradication programs, school group education and capacity programs, landholder financial support and training, and industry support and relations (for example see Griffith Centre for Coastal Management (http://www.griffith.edu.au/environment-planning-architecture/griffith-centre-coastal-management/community-projects/coasted), Gold Coast Catchment Association Inc. (http://www.goldcoast-catchments.org/goals.htm), SEQ Catchments (http://www.seqcatchments.com.au/programs/community-based-ambientwater-water-quality-monitoring-and-rainfall-event-monitoring), Gold Coast Water Watch (http://www.natura-pacific.com/news/article/waterwatch-queensland/) and Gold Coast Management Groups (http://www.goldcoast.qld.gov.au/environment/catchment-management-groups-576.html)). Collectively, initiatives undertaken have been successful to date and demonstrate that future conservation requires the integration of multidisciplinary science and pro-active management driven by the high community values placed on the Broadwater and its waterways.

References

Alam K, Hossain S, Dalrymple B (2008) WSUD in the Gold Coast. In: New South Wales and Queensland Stormwater Industry Association Conference, 8–11th July 2008, Gold Coast, Australia

Ali A, Lemckert CJ, Zhang H (2009) Numerical study of the hydrodynamics of a very shallow estuarine system – Coombabah lake, Gold Coast, Australia. J Coast Res 56:922–926

Ali A, Lemckert CJ, Dunn RJK (2010) Salt fluxes within a very shallow sub-tropical estuary. J Coast Res 26:436–443

ANZECC/ARMCANZ (2000) Australian and New Zealand guidelines for freshwater and marine water quality, vol. 1. Environment Australia, Canberra

Australian Bureau of Statistics (ABS) (1986) Year book: Australia, Catalogue No. 1301.0

Australian Bureau of Statistics (ABS) (2008) Regional population growth, Australia, Catalogue No. 3218.0

Baird D, Marais JFK, Wooldridge T (1981) The influence of a marina canal system on the ecology of the Kromme Estuary. St Francis Bay S Afr J Zoo 16:21–34

Benfer N, King B, Lemckert C, Zigic S (2010) Modeling the effect of flow structure selection on residence time in an artificial canal system: case study. J Waterw Port Coast Ocean Eng 136:91–96

Blain, Bremner and Williams Pty Ltd (1985) Storm tide statistics – Surfers Paradise Region. Report prepared for the Beach Protection Authority, Queensland. January, 1985, 15 pp

BMT WBM Oceanics (2010) Gold Coast Sustainable Loads Study. Report prepared for Gold Coast City Council, 240 pp

Burton ED, Phillips IR, Hawker DW (2004) Trace metals and nutrients in bottom sediments of the Southport Broadwater, Australia. Mar Pollut Bull 48:378–402

Burton ED, Phillips IR, Hawker DW (2005) Geochemical partitioning of copper, lead, and zinc in benthic, estuarine sediment profiles. J Environ Qual 34:263–273

Burton ED, Sullivan LA, Bush RT, Powell B (2008) Iron-sulfide and trace element behaviour in sediments of Coombabah Lake, southern Moreton Bay (Australia). Mar Poll Bull 56:1353–1358

Connolly RM (2003) Differences in trophodynamics of commercially important fish between artificial waterways and natural coastal wetlands. Estuar Coast Shelf Sci 58:929–936

Cooper JAG, Lemckert CJ (2012) Extreme sea-level rise and adaptation options for coastal resort cities: a qualitative assessment from Gold Coast, Australia. Ocean Coast Manag 64:1–14

Cosser P (1989) Water quality, sediments and the macroinvertebrate community of residential canal estates in south-east Queensland, Australia: a multivariate analysis. Water Res 23:1087–1097

Davies S, Mirfenderesk H, Tomlinson R, Szylkarski S (2009) Hydrodynamic, water quality and sediment transport modelling of estuarine and coastal waters on the Gold Coast, Australia. J Coast Res 56:937–941

Department of Environment and Resource Management (DERM) (2012) Southern Moreton Bay. http://www.derm.gov.au/parks/southern-moreton-bay/about.html. Viewed 10 August 2012

Dunn RJK (2009) Biogeochemical processes of a sub-tropical coastal lagoon (Coombabah Lake, southern Moreton Bay, Australia): with emphasis on organic matter and nutrient dynamics. PhD thesis, Griffith School of Engineering, Griffith University, Australia

Dunn RJK, Teasdale PR, Warnken J, Schleich RR (2003) Evaluation of the diffusive gradient in a thin film technique for monitoring trace metal concentrations in estuarine waters. Environ Sci Technol 37:2794–2800

Dunn RJK, Ali A, Lemckert CJ, Teasdale PR, Welsh DT (2007a) Short-term variability of physico-chemical parameters and the estimated transport of filterable nutrients and chlorophyll-a in the urbanised Coombabah Lake and Coombabah Creek system, southern Moreton Bay, Australia. J Coast Res 50:1062–1068

Dunn RJK, Teasdale PR, Warnken J, Jordan MA, Arthur JM (2007b) Evaluation of the in situ, time-integrated DGT technique by monitoring changes in heavy metal concentrations in estuarine waters. Environ Pollut 148:213–220

Dunn RJK, Lemckert CJ, Teasdale PR, Welsh DT (2007c) Distribution of nutrients in surface and sub-surface sediments of Coombabah Lake, southern Moreton Bay (Australia). Mar Pollut Bull 54:606–614

Dunn RJK, Welsh DT, Teasdale PR, Lee SY, Lemckert CJ, Meziane T (2008) Investigating the distribution and sources of organic matter in surface sediment of Coombabah Lake (Australia) using elemental, isotopic and fatty acid biomarkers. Cont Shelf Res 28:2535–2549

Dunn RJK, Welsh DT, Jordan MA, Teasdale PR, Lemckert CJ (2009) Influence of natural amphipod (*Victoriopisa australiensis*) (Chilton, 1923) population densities on benthic metabolism, nutrient fluxes, denitrification and DNRA in sub-tropical estuarine sediment. Hydrobiologia 628:95–105

Dunn RJK, Catterall K, Hollingsworth A, Kirkpatrick S, Capati G, Hudson S, Khan S, Panther JG, Stuart G, Szylkarski S, Teasdale PR, Tomlinson RB, Welsh DT (2012a) Short-term variability of nutrients and fecal indicator bacteria within the Gold Coast seaway, southern Moreton Bay (Australia). J Coast Res 28:80–88

Dunn RJK, Welsh DT, Jordan MA, Arthur JM, Lemckert CJ, Teasdale PR (2012b) Interactive influences of the marine yabby (*Trypaea australiensis*) and mangrove (*Avicennia marina*) leaf litter on benthic metabolism and nitrogen cycling in sandy estuarine sediment. Hydrobiologia 693:117–129

Dunn RJK, Welsh DT, Jordan MA, Waltham NJ, Lemckert CJ, Teasdale PR (2012c) Benthic metabolism and nitrogen dynamics in a sub-tropical coastal lagoon: microphytobenthos stimulate nitrification and nitrate reduction through photosynthetic oxygen evolution. Estuar Coast Shelf Sci 113:272–282

Environment Australia (2003) Fact sheet Australia's Ramsar sites. Wetlands Section, Environment Australia, Commonwealth of Australia, Canberra

Environmental Protection Agency (EPA) 2005 Shorebird management strategy Moreton Bay, p 10

Erftemeijer PL, Robin Lewis RR (2006) Environmental impacts of dredging on seagrasses: a review. Mar Pollut Bull 52:1553–1572

Eyre B, Ferguson AJP (2009) Denitrification efficiency for defining critical loads of carbon in shallow coastal ecosystems. Hydrobiologia 629:137–146

Eyre BD, Maher D (2011) Mapping ecosystem processes and function across shallow seascapes. Cont Shelf Res 31:S162–S172

Eyre BD, Maher D, Oakes J, Erler DV, Glasby TM (2011a) Differences in benthic metabolism, nutrient fluxes, and denitrification in *Caulerpa taxifolia* communities compared to uninvaded bare sediment and seagrass (*Zostera capricorni*) habitats. Limnol Oceanogr 56:1737–1750

Eyre BD, Ferguson AJP, Webb A, Maher D, Oakes JM (2011b) Denitrification, N-fixation and nitrogen and phosphorus fluxes in different benthic habitats and their contribution to the nitrogen and phosphorus budgets of a shallow oligotrophic sub-tropical coastal system (southern Moreton Bay, Australia). Biogeochemistry 102:111–133

Eyre BD, Ferguson AJP, Webb A, Maher D, Oakes JM (2011c) Metabolism of different benthic habitats and their contribution to the carbon budget of a shallow oligotrophic sub-tropical coastal system (southern Moreton Bay, Australia). Biogeochemistry 102:87–110

Ferguson AJP, Eyre BD, Webb A, Maher D (2004) Sediment biogeochemistry. In Pimpama River Estuary ecological study: chapter 11. Report prepared for Gold Coast Water, SKM, Brisbane

GHD Pty Ltd (2003) Coombabah creek environmental inventory. GHD Pty Ltd, Brisbane
Gold Coast City Council (GCCC) (2002) Report No 47, Gold Coast Harbour Vision 2020 Project: Dredging of the Broadwater, Gold Coast City Council, Bundall
Gold Coast City Council (GCCC) (2010) Gold Coast City local disaster management plan. Gold Coast City Local Disaster Management Group Version 5.1 22 February 2010. http://www.goldcoast.qld.gov.au/attachment/publications/disaster_management_plan.pdf
Guest GA, Connolly RM (2004) Fine scale movement and assimilation of carbon in saltmarsh and mangrove habitat by resident animals. Aquat Ecol 38:599–609
Harper BA, Sobey RJ, Stark KP (1977) Numerical simulation of tropical cyclone storm surge along the Queensland coast part X – Gold Coast. Department of Civil and Systems Engineering, James Cook University, Townsville, p 15
Hollingsworth A, Connolly RM (2006) Feeding by fish visiting inundated subtropical saltmarsh. J Exp Mar Biol Ecol 336:88–98
Hunt S, Lemckert C (2001) Suspended sediment distribution within a partially mixed subtropical estuary. J Coast Res 34:382–387
Hunt S, Lemckert CJ, Schacht C (2006) Location of turbidity maxima with a microtidal estuary and some limitations of laser *in situ* particle sizing. J Coast Res 39:520–525
Johnson TC, Williams SH (1989) From canals to lakes in south-east Queensland (Australia); water quality aspects. Water Sci Technol 21:261–265
Jordan MA, Teasdale PR, Dunn RJK, Lee SY (2008) Modelling copper uptake by *Saccostrea glomerata* with diffusive gradients in a thin film measurements. Environ Chem 5:274–280
Jordan MA, Welsh DT, Dunn RJK, Teasdale PR (2009) Influence of *Trypaea australiensis* population density on benthic metabolism and nitrogen dynamics in sandy estuarine sediment: a microcosm simulation. J Sea Res 61:144–152
Knight JM, Dale PER, Dunn RJK, Broadbent GJ, Lemckert CJ (2008) Patterns of tidal flushing within a mangrove forest: lake Coombabah, south east Queensland Australia. Estuar Coast Shelf Sci 76:580–593
Lee KM, Lee SY, Connolly RM (2011) Short-term response of estuarine sandflat trophodynamics to pulse anthropogenic physical disturbance: support for the intermediate disturbance hypothesis. Estuar Coast Shelf Sci 92:639–648
Lemckert CJ (2006) A conceptual model for designing canal estates to maximise water quality. J Coast Res 39:822–825
Leon LM, Warnken J (2008) Copper and sewage inputs from recreational vessels at popular anchor sites in a semi-enclosed Bay (Qld, Australia): estimates of potential annual loads. Mar Pollut Bull 57:838–845
Maxted JR, Eskin RA, Weisberg SB, Chaillou JC, Kutz FW (1997) The ecological condition of dead-end canals of the Delaware and Maryland coastal bays. Estuaries 20:319–327
McInnes K, Hubbert G, Oliver S, Abbs D (2000) Gold Coast Broadwater study: storm tide return periods and 1974 floodwater modelling. Report prepared for Gold Coast City Council
McLennan M, Sumpton W (2005) The distribution of seagrasses and the viability of seagrass transplanting in the Broadwater, Gold Coast, Queensland. Proc R Soc Qld 112:31–38
Melville AJ, Connolly RM (2003) Spatial analysis of stable isotope data to determine primary sources of nutrition for fish. Oecologia 136:499–507
Melville AJ, Connolly RM (2005) Food webs supporting fish over subtropical mudflats are based on transported organic matter not in situ microalgae. Mar Biol 148:363–371
Mirfenderesk H (2009) Flood emergency management decision support system on the Gold Coast, Australia. Aust J Emerg Manage 24:48–58
Mirfenderesk H, Tomlinson R (2007) Numerical modelling of tidal dynamics and water circulation at the Gold Coast Broadwater, Australia. J Coast Res 50:277–281
Mirfenderesk H, Tomlinson R (2008) Observation and analysis of hydrodynamic parameters in tidal inlets in a predominantly semidiurnal regime. J Coast Res 24:1229–1239
Mirfenderesk H, Tomlinson R, Szylkarski S, Rasmussen EB (2007) Development of a three dimensional model using dye release experiment. In: Proceedings of 30th international conference of coastal engineering, San Diego (United States of America), World Scientific Publications, pp 2118–2129
Mortimer MR (2000) Contaminants in oysters collected from residential canals between Runaway Bay and the Sovereign Islands in January 2000. Report of a study conducted for Gold Coast City Council, January 2000
Mortimer MR, Cox M (1998) Contaminants in oysters and crabs in a canal adjacent to Lae Drive, Runaway Bay, Queensland. Report of a study conducted for Gold Coast City Council, September 1998
Morton RM (1989) Hydrology and fish fauna of canal developments in an intensively modified Australian estuary. Estuar Coast Shelf Sci 28:43–58
Morton RM (1992) Fish assemblages in residential canal developments near the mouth of a subtropical Queensland estuary. Aust J Mar Freshwa Res 43:1359–1371
Moss A, Cox M (1999) Southport Broadwater and adjacent pacific ocean: water quality study 1979–1998. Queensland Environmental Protection Agency, Brisbane, 28p
Oakes JM, Connolly RM, Revill AT (2010) Isotope enrichment in mangrove forests separates microphytobenthos and detritus as carbon sources for animals. Limnol Oceanogr 55:393–402
Pagès A, Teasdale PR, Robertson D, Bennett WW, Schäfer J, Welsh DT (2011) Representative measurement of two-dimensional reactive phosphate distributions and co-distributed iron(II) and sulfide in seagrass sediment porewaters. Chemosphere 85:1256–1261
Poiner IR (1977) Microvariation in the fauna of a sublittoral sand bank, Moreton Bay, Queensland. Aus Ecol 2:297–308
Pratt C, Warnken J, Leeming R, Arthur JM, Grice DI (2007) Detection of intermittent sewage pollution in a subtropical, oligotrophic, semi-enclosed embayment system using sterol signature in sediment. Environ Sci Technol 41:792–802
Preda M, Cox ME (1998) Pimpama River, an example of catchment drainage into Moreton Bay. In: Tibbetts IR, Hall NJ, Dennison WC (eds) Moreton Bay and catchment. School of Marine Science, The University of Queensland, Brisbane, pp 165–172
Robertson D, Welsh DT, Teasdale PR (2009) Investigating 2-dimensional distributions of iron (II) and sulphide in heterogenous sediment using DET and DGT. Environ Chem 6:60–69
Robinson JE, Newell RC, Seiderer LJ, Simpson NM (2005) Impacts of aggregate dredging on sediment composition and associated benthic fauna at an offshore dredge site in the southern North Sea. Mar Environ Res 60:51–68
Ross J (1999) The influence of residential canal estates on estuarine fish assemblages in the Gold Coast regions, Australia. MSc thesis, Department of Zoology, University of Queensland, St. Lucia, Australia
Sennes G, Castelle B, Bertin X, Mirfenderesk H, Tomlinson RB (2007) Modelling of the Gold Coast Seaway tidal inlet. Aus J Coast Res 50:1086–1091
Smith DI (2002) The what, where and how much of flooding in Australia. In: Smith DI, Handmer J (eds) Residential flood insurance the implications for floodplain management policy. Centre for Resource and Environmental Studies, Australian National University, Canberra, pp 73–98
Spilmont N, Meziane T, Seuront L, Welsh DT (2009) Identification of the food sources of sympatric ghost shrimp (*Trypaea australiensis*) and soldier crab (*Mictyris longicarpus*) populations

using a lipid biomarker, dual stable isotope approach. Aus Ecol 34:878–888

Stephenson W (1980) Relationships of the macrobenthos of Moreton Bay to prawns and to abiotic factors. Aus Ecol 5:143–149

Stephenson W, Cook SD (1977) Aggregation of sublittoral macrobenthos in Moreton Bay, Queensland. Aus Ecol 2:419–428

Stuart G, Hollingsworth A, Thomsen F, Szylkarski S, Capati B, Khan S, Teo L, Tomlinson R, Kirkpatrick S (2009) Gold Coast seaway SmartRelease decision support system–optimising recycled water release in a sub tropical estuarine environment. Water Sci Technol 60:2077–2084

Teasdale PR, Welsh DT, Dunn RJK, Robertson D (2009) Investigating nutrient sources and processes in the saltwater creek catchment. Griffith Centre for Coastal Management, research report, Gold Coast, Australia, 113 p

Thomas BE, Connolly RM (2001) Fish use of subtropical saltmarshes in Queensland, Australia: relationships with vegetation, water depth and distance onto the marsh. Mar Ecol Prog Ser 209:275–288

Waltham NJ (2002) Health of the Gold Coast Waterways report. Gold Coast City Council, 181 pp

Waltham NJ (2009) Fish in artificial urban waterways: feeding, ecology and contamination. PhD thesis, Griffith School of Environment, Griffith University, Australia

Waltham NJ, Connolly RM (2006) Trophic strategies of garfish, *Arrhamphus sclerolepis*, in natural coastal wetlands and artificial urban waterways. Mar Bio 148:1135–1141

Waltham NJ, Connolly RM (2007) Artificial waterway design affects fish assemblage in urban estuaries. J Fish Bio 71:1613–1629

Waltham NJ, Connolly RM (2011) Global extent and distribution of artificial, residential waterways in estuaries. Estuar Coast Shelf Sci 94:192–197

Waltham NJ, Teasdale PR, Connolly RM (2011) Contaminants in water, sediment and fish from natural and artificial residential waterways in southern Moreton Bay. J Environ Monit 13:3409–3419

Warnken J, Leon M (2006) Recreational and tourist vessel traffic in south east Queensland's smooth coastal waters Moreton Bay and Gold Coast Broadwater. Sustainable Tourism Cooperative Research Centre (STCRC), 33 p

Warnken J, Dunn RJK, Teasdale PR (2004) Investigation of recreational boats as a source of copper at anchorage sites using time-integrated diffusive gradients in thin film and sediment measurements. Mar Pollut Bull 49:833–843

Webster T, Lemckert C (2002) Sediment resuspension within a microtidal estuary/embayment and the implication to channel management. J Coast Res 36:753–759

Werry JM (2010) Habitat ecology of the bull shark, *Carcharhinus leucas*, on urban coasts in eastern Queensland, Australia. PhD thesis, Australian Rivers Institute and School of Environment, Griffith University Gold Coast, Australia

Werry J, Lee SY (2005) Grapsid crabs mediate link between mangrove litter production and estuarine planktonic food chains. Mar Ecol Prog Ser 293:165–176

Werry JM, Lee SY, Lemckert CJ, Otway NM (2012) Natural or artificial? Habitat-use by the bull shark, *Carcharhinus leucas*. PLoS One 7:e49796

Young PC, Wadley VA (1979) Distribution of shallow-water epibenthic macrofauna in Moreton Bay, Queensland. Aus Mar Biol 53:83–97

Zeller BM (1999) Bull shark *Caracharhinus leucas* Valenciennes in estuarine and nearshore waters of the Gold Coast, Queensland. Department of Primary Industries and Fisheries, Brisbane

Zigic S, King BA, Lemckert C (2002) Mixing between two canals connected by an automated Bi-directional gated structure, Gold Coast, Australia. Estuar Coast Shelf Sci 55:59–66

Zigic S, King B, Lemckert C (2005) Modelling the two-dimensional flow between an estuary and lake connected by a bi-directional hydraulic structure. Estuar Coast Shelf Sci 63:33–41

Hydrodynamics and Sediment Transport in a Macro-tidal Estuary: Darwin Harbour, Australia

F.P. Andutta, X.H. Wang, Li Li, and David Williams

Abstract

Sediment dynamics studies were undertaken for Darwin Harbour (DH), which is a tidal dominated mangrove system in the Northern Territory of Australia. DH is located in a region with extensive mangrove and tidal flat areas, which function as trapping zones of fine cohesive sediment. Transport of sediment was estimated for the dry season, and thus river discharge was negligible. Numerical simulations were also made with two scenarios: (S_1) where the numerical mesh included mangrove and tidal flats, and (S_2) in which the mesh neglected these areas. For the first scenario, the formation of two Estuarine Turbidity Maxima zones (ETM) were verified, and located at the inner and outer harbour. In addition to the formation of ETM zones, for the second scenario increased tidal asymmetry was predicted, which resulted in landward sediment transport. The suspended sediment concentration within these ETM zones was modulated by spring and neap tidal conditions. From our simulations we demonstrated that the sediment transport of small particles, e.g. 2 μm particle size, in DH is driven by flood dominance, which is affected by wet/dry areas such as mangroves and tidal flats. Therefore, mangrove areas of DH may trap fine sediment for long periods, and if the trapped sediment carries pollutants one would expect conditions similar to many European estuaries, where pollutant sediment has been found to be buried for over tens to hundreds of years.

Keywords

Darwin Harbour • Tropical estuary • Dry season • Macro-tides • Sediment transport • Flood dominance • Hydrodynamic modelling • Sediment transport modelling

F.P. Andutta (✉) • X.H. Wang • L. Li
School of Physical, Environmental and Mathematical Sciences,
University of New South Wales at Australian Defence Force Academy
UNSW-ADFA, ACT 2600, Canberra, Australia
e-mail: f.andutta@adfa.edu.au

D. Williams
Australian Institute of Marine Sciences (AIMS),
23 Ellengowan Drive, Brinkin, NT 0811, Australia

E. Wolanski (ed.), *Estuaries of Australia in 2050 and Beyond*, Estuaries of the World,
DOI 10.1007/978-94-007-7019-5_7,

Box 1

Fernando Andutta and colleagues studied Darwin Harbour. They focused on modelling the sediment dynamics studies in order to understand how the estuary will change with developments in the catchment. They demonstrated that the sediment transport of small particles is driven by flood dominance, which is affected by wet/dry areas such as mangroves and tidal flats. Therefore, mangrove areas of Darwin Harbour may trap fine sediment for long periods, and if the trapped sediment carries pollutants one would expect conditions similar to many European estuaries, where polluted sediment has been found to be buried for tens to hundreds of years.

Introduction

The understanding of sediment transport is important to the development and management of harbours and many other sea-structures. The top 15 cities of Australia in terms of largest population are located along the coast (aside from Canberra), and often near estuaries and harbour areas (e.g. Sydney, Perth, Melbourne, Adelaide, Darwin and Townsville). These 15 cities contain over 80 % of Australia's total population in 2012, which means over 17 million people are currently living in areas near estuaries, bays and harbours. The large number of people living in these transitional areas between land and the ocean (e.g. estuaries) results in a significative human impact on coastal aquatic environments (Harvey and Caton 2010). Because of this, the understanding of erosion, deposition and sediment transport processes should be relevant for the majority of Australians. It should be noted that coastal management may have different definitions, which sometimes vary according to different cultures, countries and even the perspective of view of an author. Harvey and Caton (2010) defined coastal management as the management of human activities and the sustainable use of coastal resources in order to reduce adverse impacts on coastal environments in the present and in the future. This definition means a complex integrated approach is necessary, as it is not possible to treat the coastal land and the coastal water as individual ecosystems. Both are interconnected systems, and therefore interconnected management is important.

The development of cities around ports is often associated with the expansion of ports and activities such as oil, coal, and gas exportation (e.g. Gladstone Port, Abbott Point, Hay Point, Port Kembla, Darwin Harbour). These activities result in multiple environmental stresses, such as dredging to facilitate the navigation of larger ships, land reclamation, and changes in the sediment and nutrient runoff to catchment areas (Andutta et al. 2006b; Wang and Pinardi 2002). Treated sewage from coastal cities in Australia is discharged into estuaries, rivers, beaches and coastal waters. As a result, hotspots of water pollution have been observed to be linked to coastal cities, which are releasing industrial wastes, rubbish through wastewater systems, sediments, heavy metals, nutrients etc. (Zann and Kailola 1995). Additionally, some anthropogenic impacts may even reach extremely important marine ecosystems, for example the fine coal dust observed to reach the southern area of the Great Barrier Reef (GBR) (Burns and Brinkman 2011), and was predicted using numerical modelling simulations by Wolanski and Andutta.

The increase in mud concentrations in coastal waters is a worldwide ecological issue that can negatively affect many marine organisms, for example limiting the growth of phytoplankton at the subsurface, and the growth of pearl farms (e.g. pearl farms near Darwin and Broome). Wolanski (2007) observed that many historical sandy coasts around Australia had been replaced by muddy coasts, and this change is envisaged as a permanent degradation (unlikely to be naturally reversed). Recreational and maritime activities can be adversely impacted by processes of sediment resuspension and deposition, and this can lead to economical loss. For example, a possible reduction of coral reefs near Cairns would decrease appeal to eco-tourism, and therefore impact the local economy. Marine sediment may also carry nutrients and pollutants from land sources to coastal waters, transporting these substances through areas where numerous marine species reproduce, e.g. bays and estuaries. An understanding of sediment transport leads to a better comprehension of pollution control, helping to preserve the marine ecosystem through integration with coastal management system (e.g., Frascari et al. 1988).

The spatial and temporal variability of the transport of sediment also has hydrographical implications, such as sediment deposition in navigable channels. For some estuarine systems, dredging maintenance can be really expensive e.g. Yangtze Estuary (Wu et al. 2009; Liu et al. 2011), where over 100 million m^3 of sediment has been dredged per year from 2006 to 2008. The dynamics of sediment transport depend on water circulation, salinity concentration, biological interaction, and sediment type. For estuaries, sediment distribution is often comprised of cohesive sediment (e.g. clay and mud), and non-cohesive sediment (e.g. sand). Cohesive sediments are usually transported in the water column, because they are easily suspended by water currents. Alternatively, non-cohesive sediments are mainly transported along the bottom by the processes of rolling, saltation, and sliding. However, under some circumstances, sediment can be transported along the bed as a bottom fluid sediment layer (Puig et al. 2004).

The study of sediment transport in coastal aquatic systems is usually complex because of complex geometry, many sediment types, and many boundary forcing mechanisms such as tides, wind, river discharge, density-driven currents etc. Sediment transport research has always received attention (e.g. Kessel et al. 2011; Allen et al. 1980), and also has implications on contaminant behaviour. Toxic substances, e.g. metals, tend to associate with fine sediment particles, which are potentially the most mobile sedimentary fraction under normal energy conditions of the system (Taylor and Hudson-Edwards 2008; Thonon 2006). Estuaries often have high sediment concentrations in the water column, e.g. macro tidal estuaries, and the overall sediment transport can be upstream for some particular systems (e.g. Margvelashvili et al. 2006). Some of the many physical processes affecting estuaries are: (1) settling lag and scour lag effects (van Straaten and Kuenen 1958; Postma 1961); (2) shoaling tidal waves that cause an asymmetric distribution of velocity and suspended sediment concentration, known as tidal pumping (Dyer 1986); (3) vertical velocities associated with flocculation and hindering (Pejrup 1988); (4) bed-load transport and bed solidification and liquefaction (Kessel and Kranenburg 1998; Maa and Mehta 1987). To add complexity to the understanding of the hydro-dynamical and morphological changes in many aquatic systems, the combined effect of headlands, rivers, and embayments creates a complicated bathymetry that leads to the formation of many tidal jets within narrow channels, eddies etc.

This manuscript addresses the study of sediment transport within Darwin Harbour (DH), which is an embayment next to Shoal Bay (SB), see Fig. 1a. DH extends from Charles Point to Lee Point. Some previous studies define Darwin Harbor to be the area extending from Charles Point to Gunn Point, which comprises the water area of Shoal Bay. However, a harbour by definition is any natural or artificial place where vessels may seek shelter from stormy weather, and thus Shoal Bay is not part of DH. Harbours differ from bays in that a bay describes a geographical feature, while a harbour is defined by its function. We

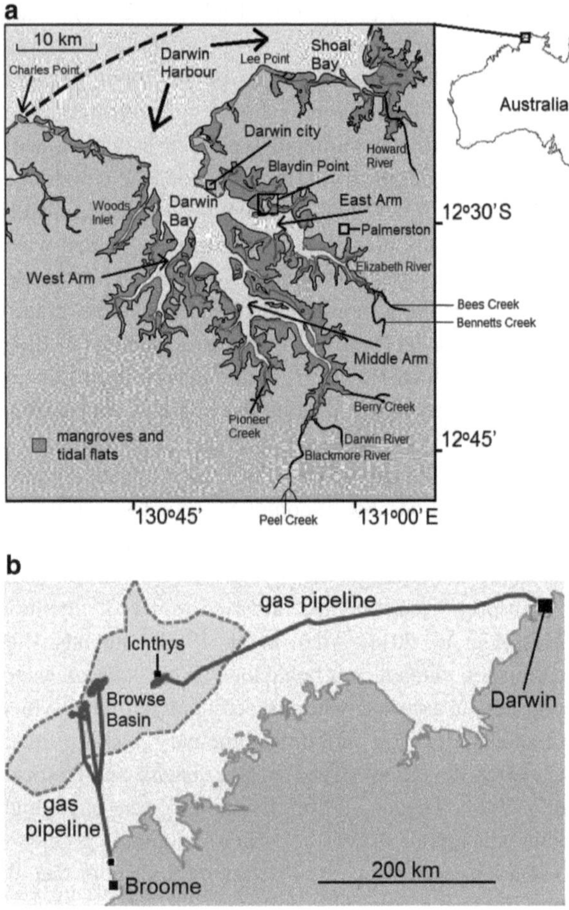

Fig. 1 (**a**) Map of Australia showing the location of DH in the Northern Territory (NT), and indication of Shoal Bay. Areas of mangroves and tidal flats, and indication of Blaydin Point in EA. Adelaide River is located near Shoal Bay on the right side (not shown in this map). (**b**) Browse Basin and indication of Ichthys Field with gas pipeline extending to Darwin Harbour

focused on sediment transport studies and the formation of Estuarine Turbidity Maxima zones (ETM) within DH. Some physical mechanisms affecting sediment transport in DH were investigated, as well as their implications on net sediment flux and distribution of ETM zones. Additionally, the likely future of DH due to the expansion of port activities is discussed. Currently, dredging in DH is present ~16 days per year, resulting in a dredged volume of ~50,000 m^3 from East Arm (EA) (Australian Natural Resource Atlas).

For the East Arm in DH, INPEX Browse Ltd (INPEX) is undertaking the Ichthys Gas Field Development to extract liquefied gas from Browse Basin, see Fig. 1b. The project requires the development of onshore, nearshore and offshore infrastructures. In addition, nearly 15 million cubic metres of mainly hard soil will be dredged in DH by Van Oord Ltd. (Sinclair Knight Merz Pty Ltd. 2011). To obtain approval for this development, stakeholders required an assessment of the potential impacts on many local marine species, e.g. mud crabs, and the likely consequences of elevated suspended sediment concentration in DH (Fig. 1). The project by INPEX was approved in 2012, and an operation to dredge a shipping channel in DH started at the end of 2012, and is expected to take less than 2 years to complete. INPEX is bringing in the largest dredge ship ever seen in Australia to start the dredging operation, which will reduce the previously estimated time of 4 years to only ~15 months. On completion, the harbour will have a deeper channel for liquefied natural gas (LNG) carriers to load gas at a multi-billion dollar processing plant being built at Blaydin Point (in EA).

Although DH is of great economic importance to the Northern Territory (NT), most of the current knowledge about the main driving forces for the local hydrodynamics is due to efforts by numerous researchers (Li et al. 2012; Asia-Pacific Applied Science Associates 2010; Williams 2009; Wolanski et al. 2006; Williams et al. 2006; Ribbe and Holloway 2001). Williams (2009) used a 2D sediment model to assess changes in sediment transport if the sandbar in the EA is removed. Li et al. (2012) used a 3D hydrodynamical model of DH (see Fig. 1), and verified the effect that the mangrove and tidal flat areas have on the tidal asymmetry. It was predicted that a decrease in area of the tidal flats and mangroves would lead to increased tidal asymmetry of flood dominance. Evidently, such changes in water circulation lead to changes in the local redistribution of sediment, e.g. different patterns of erosion and deposition areas. This manuscript also concurs with the study by Li et al. (2012), using the same 3D hydrodynamical model, i.e. Finite Volume Community Ocean Model (FVCOM). Aside from the modelling by Li et al. (2012), previous hydrodynamical models of DH were often vertically integrated (e.g. Asia-Pacific Applied Science Associates 2010). For DH, vertically integrated models have always simulated water circulation properly. This is because tidal currents prevail over small baroclinic currents (Asia-Pacific Applied Science Associates 2010), especially during the dry season. However, suspended sediment concentration varies along the water column, and thus small changes in horizontal currents in the water column might be important for sediment dynamics. Our model is a fully 3D hydro-sediment model using an unstructured mesh, and thus capable of capturing these small current changes within the many layers. The 3D hydrodynamical model was coupled to a sediment model that uses the density induced stratification according to Wang (2002). Other physical mechanisms were also included and will be discussed further in detail. The role that tidal flats and mangrove areas play on the transport of cohesive fine sediment (e.g. 2 μm particle size) was analysed, and some of the physical mechanisms for sediment transport were quantified, e.g. tidal pumping, residual circulation, stokes drift etc.

Darwin Harbour Description

DH is a semi-enclosed estuarine system with extensive mangrove areas (Fig. 2c), in which wind wave activity is sufficiently diminished to allow the development of a harbor and recreational facilities. Economic activities of DH are all located along-side the EA, Fig. 2a, b. Depths in DH range from 0 to 20 m, with a maximum of up to ~40 m in coastal areas. DH is located in the Northern Territory (NT) of Australia, and its surrounding lands are occupied by the cities of Palmerston and Darwin. DH is the embayment next to Shoal Bay (see Fig. 1). The two largest economic sectors in Darwin city and the surrounding areas are the mining and tourism industries (exceeding $2.5 billion per annum), which are currently attracting people from around Australia and overseas to migrate to this region.

Between 2000 and 2001, exploratory research resulted in the discovery of an extremely promising gas and condensate field, which is the Ichthys Field located at Browse Basin. The Ichthys Project will have an initial capacity to produce 8.4 million tonnes of Liquid Natural Gas (LNG) per annum, and 1.6 million tonnes of liquefied petroleum gas (LPG) per annum, as well as approximately 100,000 barrels of condensate per day at peak. After preliminary processing at the offshore central processing facility (CPF), the gas will be transported from the CPF through a subsea pipeline more than 885 km to the onshore LNG processing plant proposed for Blaydin Point on Middle Arm Peninsula, Darwin, Northern Territory. It will be cooled to below minus 161 °C, the point at which the gas becomes a liquid, known as LNG. Nearly AUD $34 billion has been formally opened by the Australian government for the Ichthys liquefied natural gas project in Darwin in May 2012. The Northern Territory Government has approved development of a Village, which will accommodate 3,500 anticipated workers at the peak of onshore construction. The Howard Springs Accommodation Village is being developed to house the fly-in

Fig. 2 (**a**) EA wharf, mangroves *dark red*, extensive intertidal mudflats, (**b**) Sediment plumes, looking from Charles Darwin national park to Darwin city, (**c**) mangroves from Middle Arm

fly-out workforce required to build the Ichthys gas processing facilities at Blaydin Point, Darwin. The Ichthys Project is expected to start production by the end of 2016.

As economic activity increases, so will the population (Fig. 3). For Darwin city (Fig. 2b), the population is predicated to be between ~170 and ~335 thousand people by 2056. Consequently, an increase in natural resource usage and anthropogenic stress on the terrestrial and aquatic environments is almost inevitable. Of all the estuarine systems located near capital cities of Australia, DH (near Darwin city) was the only system classified as largely unmodified by 2002 (Estuary Assessment 2002: Estuaries by Australian Natural Resources Atlas). This condition is, however, likely to change in the next couple of years due to port expansion, and later due to the increasing use of natural resources, which is associated with the predicted population increase (see prediction in Fig. 3).

DH contains the West Arm (WA), Middle Arm (MA) and East Arm (EA) catchments (Fig. 1). The major freshwater inputs for DH are predominantly from the Elizabeth River (for EA), Berry Creek, and Blackmore and Darwin Rivers (for M A). For WA, the fresh water input is considered relatively small compared to the other arms. For the embayment next to DH, i.e. Shoal bay, the fresh water input is from Howard River (Wilson et al. 2004). In DH evaporation usually exceeds rainfall throughout the year, except during the wet season. During the dry season, fresh-water input into DH is negligible and evaporation exceeds river discharge (Avg. Rainfall in Table 1). As a result, in the dry season, salinity concentrations in DH usually become 0.8 psu higher than the adjacent coastal waters (Padovan 1997), and DH becomes a hypersaline system like many others in Australia (e.g. Andutta et al. 2011, 2012, 2013). Michie et al. (1991) reported that from September to October the salinity in DH is typically 35 psu, and the lowest salinity values coincide with the wet season, with values of 5 psu observed in the further upper reaches of MA. From February to October, the evaporation rate varies between 170 and 270 mm, with an average annual evaporation rate of ~2,650 mm. Tropical savannah is the predominant climate of the region, with a mean temperature of ~28 °C (Monthly avg. temperatures in Table 1), which slightly decreases during winter to ~23 °C, and increases during summer to ~32 °C. Water surface temperature during June-July is ~23 °C (winter), ~33 °C during October-November (summer), followed by a small decline of ~4 °C from December to February due to the wet season. Michie et al. (1991) reported that temperatures from the inner harbour to the upstream location in MA showed very little spatial change. Located in a subarid/humid area, DH has a typical rainfall of 1,700 mm year^{-1} (2,500 mm in exceptionally wet years), see Table 1. Runoff typically varies between 100 and 750 mm year^{-1}, with the maximum occurring during the wet season (Wang and Andutta 2012; Milliman and Farnsworth 2011).

DH is forced by semi-diurnal tides, and is classified as a macro-tidal estuary with the form number Nf = 0.32 (criteria of A. Courtier of 1938; Defant 1960). The maximum observed tidal range is ~7.8 m, with mean spring and neap tidal ranges of ~5.5 and ~1.9 m, respectively (Wang and Andutta 2012; Li et al. 2012; Milliman and Farnsworth 2011; Woodroffe et al. 1988; Michie 1987). This tidal range is relatively large compared to the mean depth, and therefore the potential energy from sea level oscillation can easily

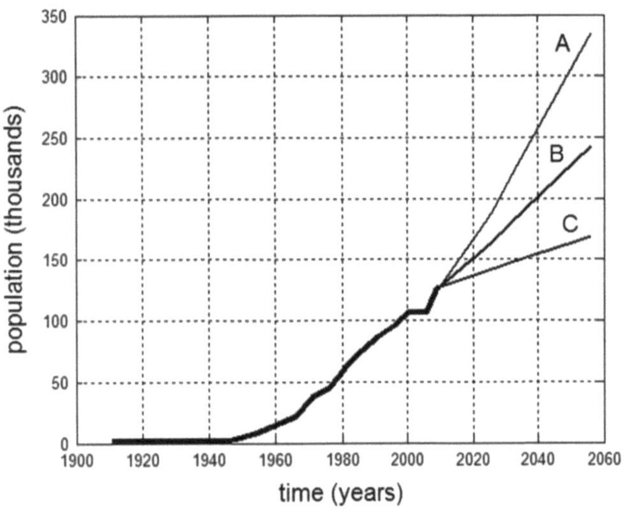

Fig. 3 Historical data of the population of Darwin city from 1911 to 2010 (*thicker line*), and the projections from cases *A*, *B* and *C* cover the period from 2010 to 2056. Different assumptions about level of fertility, mortality, internal migration and overseas migration were taken into account (Source: Australian Bureau of Statistics)

Table 1 Climate data obtained at Darwin Airport station (data from 1941 to 2012) indicating monthly average (Avg.)

Month	Jan	Feb	Mar	Apr	May	Jun	Jul	Aug	Sep	Oct	Nov	Dec	Year
Record high °C	35.6	36.0	36.0	36.7	36.0	34.5	34.8	36.0	37.7	38.9	37.3	37.0	38.9
Avg. high °C	31.8	31.4	31.9	32.7	32.0	30.6	30.5	31.4	32.6	33.2	33.3	32.6	32.0
Avg. low °C	24.8	24.7	24.5	24.0	22.1	19.9	19.3	20.4	23.0	24.9	25.3	25.3	23.2
Record low °C	20.2	17.2	19.2	16.0	13.8	12.1	10.4	13.2	14.3	19.0	19.3	19.8	10.4
Rainfall mm	426.6	374.3	321.3	101.2	21.2	1.9	1.2	5.0	15.7	71.0	140.2	252.0	1,731.6
Avg. rainy days	21.4	20.4	19.6	9.3	2.2	0.6	0.5	0.6	2.3	6.7	12.4	16.8	112.8

Source: Climate statistics for Australian locations, October 2012 diminish vertical stratification. The DH area comprises numerous mangroves and tidal flats. Nearly 5 % of the whole mangrove area in the Northern Territory belongs to DH, i.e. ~274 km^2 (Tien 2006).

Wind conditions for DH do not vary much within a spatial scale of a few tens of kilometres (Asia-Pacific Applied Science Associates 2010). Therefore, homogeneous wind conditions are considered representative and can be applied in numerical simulations. Tropical cyclones may occur in this area during the wet season, and winds are predominantly from the east (in the range of ~160° to ~200°, i.e. winds from NW and SW). During the wet season, ~17 % of the easterly wind speed is in the range of 7.5–10 m s^{-1}. The most frequent wind speed in the wet season is ~9 m s^{-1}; however, characteristically extreme wind conditions are ~18 m s^{-1}. In contrast, during the dry season, SE winds prevail (in the range of ~325° to ~360°, i.e. winds from SE), and ~19 % of the E-SE wind speed is in the range of 7.5–10 m s^{-1}. The most frequent wind speed in the dry season is ~9.3 m s^{-1}, but characteristically extreme wind conditions are ~14 m s^{-1} (Asia-Pacific Applied Science Associates 2010). During intense cyclones the wind can reach speeds of up to ~60–70 m s^{-1}, e.g. cyclone Tracy in 1974.

Wave conditions for DH are reported to have a small effect when compared to its macro-tidal currents. Asia-Pacific Applied Science Associates (2010) provided some numerical simulations for different wave conditions. Typical moderate and high energy wave conditions were applied, from N and NW directions and periods of 5 and 10 s. They concluded that wave energy reduces landwards, and wave bottom stress is usually less than 0.1 N m^2 inside the bay (aside from shallow areas near the bay entrance).

The mean depth at the inner part of Darwin Bay is in the range of ~15 to 20 m. If one considers the average depth to be $\overline{H} = 15$ m, wavelengths λ larger than 30 m would affect bottom stress. The exact formula of the phase speed of gravity (C_w) is $C_w = \left[(g\lambda/2\pi)\tanh(2\pi\overline{H}/\lambda)\right]^{1/2}$ or $C_w = \lambda/T$, where T is the wave period and g is the gravity acceleration. Therefore, the period is expressed by $T = \left[(2\pi\lambda/g)\tanh^{-1}(2\pi\overline{H}/\lambda)\right]^{1/2}$; thus, areas shallower than 10 m would be affected by waves of periods larger than $T \sim (4.3 + 3.5i)$, i.e. absolute value of ~5.6 s. Evidently, bottom stress is just one of the different wave effects on sediment transport. For DH, one would consider the effect from the build-up of pore pressure within the sediment, which causes bed liquefaction and contributes to bed-load transport (Kessel and Kranenburg 1998; Maa and Mehta 1987). However, it is likely that the currents from the macro-tides within DH dominate erosion of sediment from the bottom, and thus overcome the effect of bottom liquefaction. The only optimum condition to enable the build-up of pore pressure in DH would be times of high or low tides coinciding with times of wind-waves occurring at the harbour entrance.

Model Description, Configuration and Calibration

To simulate the hydrodynamics and transport of sediment for DH, the unstructured numerical model FVCOM was applied (Chen et al. 2003). FVCOM is a three-dimensional hydrodynamic model using unstructured, finite element mesh. Two numerical meshes were applied, the first mesh (9,666 horizontal cells) contained both tidal flats and mangrove areas, and the second mesh (3,607 horizontal cells) excluded these areas. For both meshes the horizontal resolution varied between ~20 m and ~3,300 m (Fig. 4). The increased horizontal resolution was applied in the inner harbor, while the lower horizontal resolution was applied in the coastal area. Twenty vertical sigma layers were used. Higher vertical resolution was applied for the layers near the surface and bottom. The bathymetry data was obtained from Australian Institute of Marine Sciences (AIMS). Figure 3 shows that the generated mesh is adequate to simulate the hydrodynamics of DH, because it covers all mangroves and tidal flat areas, which are treated as wet/dry cells with higher bottom friction.

Boundary Forcings

Tidal components were used to force the model at the external open boundary, i.e. the coastal zone. These components were obtained from the TPXO7.2 global model for ocean tides (http://volkov.oce.orst.edu/tides/TPXO7.2.html). The diurnal tidal components applied at the open boundary were K_1, O_1, P_1 and Q_1, and the semi-diurnal tidal components were M_2, S_2, N_2 and K_2. Additionally, the components M_4, MS_4, MN_4, M_f and M_m were used. These tidal components represent over 99 % of sea level variations for all of the DH area (Wang and Andutta 2012; Li et al. 2012). Hourly sea surface elevation data from 1991 to 2010 from the station of the Bureau of Meteorology (BoM, Fig. 3) were analysed to study the principal tidal characteristics of the harbour, and used to validate the model results. Additionally, data of sea surface elevation from AIMS was used to validate the model (Blaydin station, Fig. 4).

The drainage basin of DH covers ~1,833 km^2, and the surface water in low tide is ~863 km^2, resulting in a total surface area of ~2,696 km^2 (Wang and Andutta 2012; Li et al. 2012), which is smaller than the drainage basin from the nearby Adelaide River (~7,600 km^2). The annual

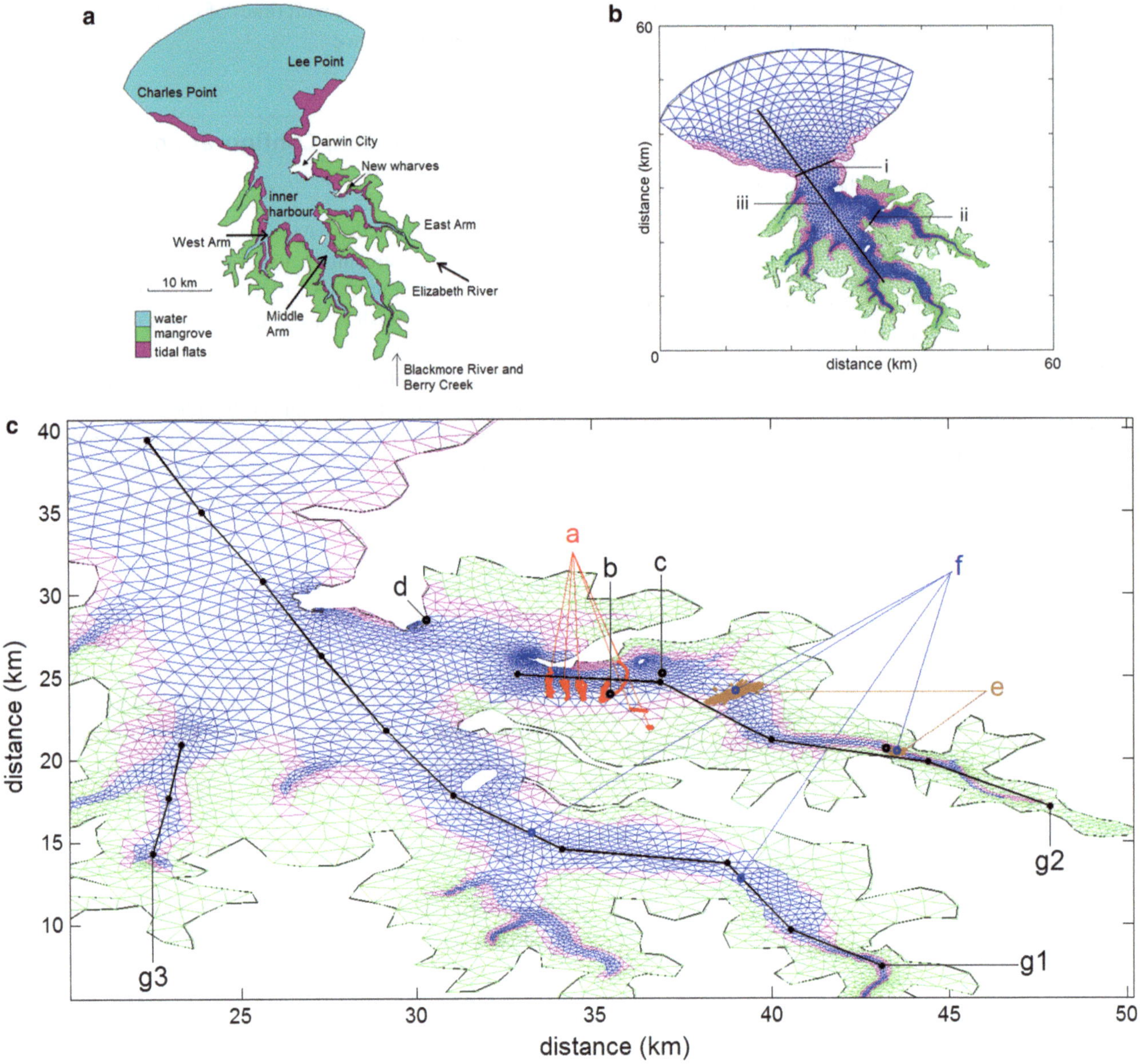

Fig. 4 (**a**) DH domain with indication of water, mangrove and tidal flat areas. Grid elements over mangrove and tidal flat areas are wet/dry elements. (**b**) Numerical grid of DH with resolution from about 20 m to 3.3 km, and colours to indicating water area (*blue*), tidal flats (*pink*) and mangrove areas (*green*). Indication of data extract from transects *i*, *ii* and *iii* from simulations. (**c**) Location of field measurements of tides, salinity, temperature, and water currents. Detailed description is shown in method section in Table 2. Different scales were used for the *vertical* and *horizontal axes*, and view is on a plan surface

discharge from the Adelaide River is ~2 km^3/year (~63 m^3 s^{-1}) (Milliman and Farnsworth 2011). Taking into account a negligible spatial variation of rainfall over the drainage areas of DH and Adelaide River, the annual mean river discharge for DH is roughly estimated using $Q = \alpha 360$ (where $\alpha = 2696/7600$ is the ratio between both catchments), thus the annual mean river flow for DH would be ~22 m^3 s^{-1}. Data obtained from telemetered gauging stations in DH, Shoal Bay and Adelaide River area were evaluated. Flow discharge monitored at most stations within these areas show high correlation. The high correlation of measured river flow indicates the rainfall conditions within these areas are similar. The stations compared were G8150018 (Elizabeth River), G8150322 (Bennetts Creek), G8150036 (Bees Creek), G8150098 (Blackmore River), G8150028 (Berry Creek), G8150321 (Peel Creek), G8150179 (Howard River) for DH (in Fig. 1), and stations G8170085 (Acacia Creek), and G8170020 (Adelaide) for Adelaide River (location not shown in Fig. 1). Evidently, ~22 m^3 s^{-1} is a rough estimate that implies that these drainage basins have similar vegetation distribution, evapotranspiration, soil permeability, ground water etc. Rainfall

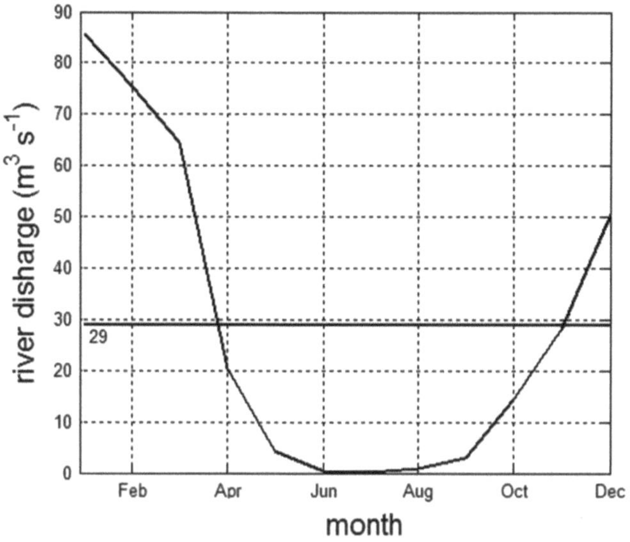

Fig. 5 Estimated monthly mean river flow Q for DH, and the annual mean discharge of 29 $m^3 s^{-1}$

usually varies in space and time, and thus a limited number of rainfall stations over the drainage basin may have an impact on runoff estimates (Sun et al. 2010). Nevertheless, this estimate provides a reasonable mean annual flow for DH, which is likely to differ by an acceptable percentage from the actual DH runoff. A different approach to estimate the river discharge for DH is to use the average runoff and groundwater of 500 mm/year, combined with the drainage basin area of 1,833 km^2, which results in an annual average discharge of ~29 $m^3 s^{-1}$. Note that this estimate is close to the previous estimative of ~22 $m^3 s^{-1}$.

The river discharge in DH is controlled by rainfall; therefore, the climate data obtained for over 70 years is important information that can be used to estimate the average runoff in each month (See Table 1). Consider the annual estimated river discharge from runoff and groundwater to be $Q_e \sim 29$ $m^3 s^{-1}$ (as described previously), where the mean river flow per month, Q_m, is estimated using the factor $\beta = R_m/R_a$ in equation $Q_m = Q_e \beta$, where R_a is the monthly mean rainfall, and R_y is the annual mean rainfall. The estimated monthly average discharge for DH is shown in Fig. 5, which can be used in simulations under typical conditions for DH. From May to October, which is a period that covers the dry season, the mean river flow into DH is smaller than 15 $m^3 s^{-1}$. In contrast, from December to March, which is typically the wet season, the river flow usually exceeds 50 $m^3 s^{-1}$. Although the mean river discharge is estimated to increase to ~85 $m^3 s^{-1}$ in January, for the dimensions of DH, this flow would result in relatively small residual circulation except for the upper reaches of the three arms. In addition, for the main entrance of the Harbour, which is 6,000 m wide and 20 m deep, this flow would result in residual currents of ~10^{-3} m s^{-1}. This indicates that the residual circulation and baroclinic circulation due to river discharge is likely to be more important at the further upstream areas along the EA, MA and WA.

For the internal boundaries of DH, e.g. upstream river zones, there are three main sources of fresh water in the domain (Elizabeth River, Blackmore River and Berry Creek); however, the simulation was for the dry season and thus river discharge was negligible (Wang and Andutta 2012; Li et al. 2012; DHAC 2003; Burford et al. 2008). The monthly estimate of fresh water inflow for DH shows negligible inflow for the dry season (Fig. 4). In the dry season, the baroclinic circulation due to the horizontal salinity gradient is often confined to a few upstream areas in DH (Asia-Pacific Applied Science Associates 2010), and these baroclinic currents are usually less than 3 % of the maximum tidal current intensity (Li et al. 2012). The temperature field was also reported to have little spatial variation (Michie et al. 1991), and to have little seasonal variation (observed from data at BoM station from 01/05/1991 to 31/12/2010). Therefore, salinity and temperature have a negligible effect on water circulation during the dry season.

At the surface, the wind is sometimes an important mechanism that causes sediment resuspension in estuaries by wind-driven currents and wind-driven waves (Mehta 1988). However, as previously discussed for DH, the wind was observed to have a negligible effect compared to tidal currents (Li et al. 2012; Asia-Pacific Applied Science Associates 2010). Tides dominate the transport of sediment with currents of up to ~2.5 m s^{-1}, with typical tidal oscillation between 3.7 and 7.8 m (Wang and Andutta 2012; Li et al. 2012). In summary, the effects from wind-driven currents, wind-driven waves, river discharge, and the heat flux at the sea surface were negligible, resulting in simulations only forced by tides.

Physical and Numerical Parameters

Bottom sediment distribution in DH consists predominantly of small particles, which are often transported through the water column. Therefore, this study only considers the transport of suspended sediment (TSS) and neglects the bed load transport of sediment. The equations by Wang (2002) were used to calculate TSS. Because salinity and temperature were assumed to be constant, only sediment concentration would affect water density. The water density (Eq. 1) was calculated assuming the volumetric relation,

$$\rho = \rho_w + \left(1 - \frac{\rho_w}{\rho_S}\right)C, \tag{1}$$

where C [kg m^{-3}] is the sediment concentration, (ρ_w) [kg m^{-3}] is the water density calculated by the equation of state, and (ρ_S) is the sediment dry density (assumed to be 1,250 kg m^{-3}).

The bottom drag coefficient (C_d) was set to be a function of the water depth for water areas, mangroves and tidal flats (see Eq. 2). The bottom drag coefficient also depends on bottom roughness length, z_0, which is often considered a user-defined free parameter, although there have been some studies showing the space and time dependence of the z_0 parameter (e.g. Cheng et al. 1999; Xu and Wright 1995; Ling 1976; Ling and Untersteiner 1974). z_0 was assumed to be 0.035 m for water and tidal flat areas, and 0.15 m for the mangrove areas, which are reasonable values to be applied for these areas (Straatsma 2009). Higher bottom roughness length was applied in mangrove areas, because of the influence of roots and trees that significantly increase the friction and thus reduce water speed (Mazda et al. 1997). From empirical experiments within mangroves, C_d was observed to vary between 1 and 10, depending upon tidal conditions, mangrove species, mangrove density (i.e. degree of aggregation of mangroves), and patchiness of mangrove distribution. Therefore, a maximum value for C_d was established because calculated values of C_d increase to unrealistic values when water depths are too small, c.a. few centimetres (see Eq. 2).

$$C_d = \min\left\{ \left[\frac{1}{k/(1+AR_f)} \ln\left(\frac{z_b}{z_0}\right) \right]^{-2}, \quad 10 \right\}, \tag{2}$$

where A is an empirical non-dimensional constant; Adams and Weatherly (1981) determined $A = 5.5$ for a sediment-laden oceanic bottom boundary layer. R_f is a flux Richardson number by layer, and in this case for the bottom sigma layer. z_b is the distance to the computational grid point closest to the bed, and K is the Von Kármán constant, c.a. ~0.4 (Telford 1982; Orszag and Patera 1981).

The erosion rate was assumed to be higher than 5×10^{-6} kg m^{-2} s^{-1} for permanent water areas, and assumed to be lower than 1×10^{-7} kg m^{-2} s^{-1} for tidal flats and mangrove areas, because of the influence of roots and trees that inhibit erosion in mangrove areas (Mazda et al. 1997). Equations by Ariathurai and Krone (1976) were used to calculate the vertical sediment flux, E [kg m^{-2} s^{-1}], i.e. erosion and deposition.

$$E = \begin{cases} E_0\left(\dfrac{|\tau_b|}{\tau_c} - 1\right), & if\ |\tau_b| > \tau_c \\ 0, & if\ |\tau_b| = \tau_c\,, \\ C_b W_S\left(\dfrac{|\tau_b|}{\tau_c} - 1\right), & if\ |\tau_b| < \tau_c \end{cases} \tag{3}$$

where E_0 is the erosion coefficient [kg m^{-2} s^{-1}], τ_c [N m^{-2}] is the critical resuspension and deposition stress, and C_b [kg m^{-3}] is the sediment concentration at the model's bottom layer. The critical erosion stress was assumed to be 0.10 N m^{-2} for water areas, and 1.0 N m^{-2} for tidal flats and mangrove areas, while the critical deposition stress for the whole domain was 0.08 N m^{-2}. The settling velocity (Ws) was assumed to be 1×10^{-5} m s^{-1}.

To solve the vertical mixing at the vertical sub-grid scale, the Mellor-Yamada level 2.5 turbulence closure scheme was used (Mellor and Yamada 1982). The extra turbulent mixing due to waves is not yet included in the eddy diffusivity for FVCOM; however, the small effect by waves has been discussed, and previous studies showed that wave influence is negligible within DH (Asia-Pacific Applied Science Associates 2010).

In the past, c.a. before 10 years ago, the coefficients of horizontal diffusivity and viscosity were often treated as user-defined physical parameters. Therefore, to simulate sub-grid scale processes, assumptions were often required to choose the values of the horizontal eddy diffusion and eddy viscosity coefficients, which usually depend on grid size and are likely to vary in time. Many models now apply parameterizations to account for the horizontal viscosity coefficient. The well-known viscosity parameterization by Smagorinsky (1963) is applied in many hydrodynamical studies (e.g. Andutta et al. 2011, 2012, 2013). The horizontal eddy diffusivity, K_h, is often user-defined, and in some cases calculated as a function of the grid-size (e.g. Andutta et al. 2011, 2012, 2013). In this study, the horizontal diffusivity was adjusted to fit observed values of suspended sediment concentration against model results. The remaining numerical and physical parameters are described in more detail by Li et al. (2012) and Wang and Andutta (2012).

Initial Conditions and Simulation Scenarios

For the initial conditions, a homogeneous distribution of salinity and temperature was assumed for the whole domain, with a salinity of 33 psu and temperature of 25 °C, which were constant in time (Wang and Andutta 2012; Li et al. 2012; Michie et al. 1991). Two different scenarios for the simulations were analysed. For the first scenario tidal flats and mangrove areas were applied (S_1), and for the second scenario tidal flats and mangrove areas were excluded from the domain (S_2). These simulations provide an understanding of the independent effects of tidal flats and mangroves in the transport of sediment in DH. Mangrove areas usually cause trapping of sediments (Victor et al. 2004; Wolanski and Spagnol 2000; Wolanski et al. 1998), and also affect asymmetry of tidal currents (Li et al. 2012). The sediment model was run for 40 days, from 1 October 2011 to 9 November

Table 2 Location of the current meter mooring sites, tidal gauge, transects of velocity profiles, CTP, and NTU profiles

Description	Indication in Fig. 3c	Location	Number of transects or profiles	Period of measurements (local time)
ADCP transect	a	Across East Arm	36	30/09/2007; from 15:16:00 to 17:04:00
ADCP profiles	b	East Arm (Blaydin station)	4,175	from 19/06/2009 22:50:00 to 18/07/2009 22:30:00
ADCP profiles	c	East Arm (Hudson station)	4,621	from 01/08/2009 08:50:00 to 02/09/2009 10:00:00
Tides	d	East Arm (BoM station)	847,200	from 01/05/1991 to 31/12/2010
ADCP transect	e	Across East Arm	40	26/10/2011; from 07:11:00 to 18:05:00
CTD profiles	f	Blackmore and Elizabeth Rivers	35	from 26/10/2011 06:46:00 to 27/10/2011 16:53:00
CTD and OBS transect	g1	Along Middle Arm until outer harbor	10	07/11/2012: from 10:46:00 to 13:18:00
CTD and OBS transect	g2	Along East Arm	5	06/11/2012: from 11:40:00 to 12:35:00
CTD and OBS transect	g3	Along West Arm	3	06/11/2012: from 13:11:00 to 13:50:00

The exact location of each obtained measurements is shown in Fig. 3

2011, and the sediment model started 12 h after the beginning of the hydrodynamical model. The first 10 days of simulation were used to generate the conditions for the following 30 days of simulation, i.e. "*hot start*". The calibration and validation of tidal oscillation and tidal currents was made using simulations from different periods, but all comparisons were during the dry season (negligible fresh water inflow), and under the same tidal forcings. The cohesive sediment was considered within the whole domain, with a grain size of 0.002 mm. This fine sediment constitutes the larger part of the suspended sediment for DH (Asia-Pacific Applied Science Associates 2010), and fine sediment commonly extends increasingly on estuary beds (Brenon and Le Hir 1999).

Estimation of Tidal Asymmetry

The skewness parameter γ, Eq. 4, was calculated to verify the effect of mangrove and tidal flat areas on tidal asymmetry. The tidal asymmetry parameter γ allows identification of the major factors controlling tidal asymmetry. For the skewness parameter, the tidal components M_2 and M_4 were verified to control tidal asymmetry (Li et al. 2012), and so the expression to calculate γ was:

$$\gamma_{M_2/M_4} = \frac{\frac{3}{2} a_{M_2}^2 a_{M_4}^2 \sin\left(2\varphi_{M_2} - \varphi_{M_4}\right)}{\left[\frac{1}{2}\left(a_{M_2}^2 + 4a_{M_4}^2\right)\right]^{\frac{3}{2}}} \tag{4}$$

where φ and a are the phases and amplitudes of the astronomical tides M_2 and M_4, respectively.

Model Calibration and Validation

The skill method suggested by Wilmott (1981) was applied to quantify the agreement of velocities and sea level; similar studies have applied and shown the advantages of this quantitative parameter (e.g. Andutta 2006, 2011 and Andutta et al. 2006a). This parameter was used for the final tuning of the user-defined physical parameters (e.g. diffusion coefficient, bottom roughness etc), and to determine which was the most representative turbulent closure method (e.g. method by Mellor and Yamada 1982). The skill parameter is calculated using the equation,

$$Skill = 1 - \frac{\Sigma|X_{model} - X_{obs}|^2}{\Sigma\left(\left|X_{model} - \overline{X}_{obs}\right| + \left|X_{obs} - \overline{X}_{obs}\right|\right)^2}, \tag{5}$$

where X_{obs} and X_{model} are respectively the observed and simulated properties, and $\overline{X}_{obs}$ represents the time averaged value. The skill parameter is a dimensionless quantity. It varies from 1 to zero, with 1 indicating the best fit, and zero indicating a complete disagreement between the observed and theoretical results.

To validate model results, tidal oscillation and water current measurements from different periods during the dry season were used (see Table 2), e.g. data from 2007, 2009, 2011 and 2012. Measurements were obtained at anchored stations, from cross-channel and along-channel transects. The instruments used to obtain measurements were tidal gauges, ADCP (Acoustic Doppler Current Profiler), ADP (Acoustic Doppler Profiler), CTDs (Conductivity, Temperature and Depth), and Optical Backscatter Sensors (OBS). The OBS were used to measure Nephlometer Turbid Units

(NTU), which were converted into suspended sediment concentration SSC.

Simulated tides and water currents were evaluated using the skill parameter, and some comparisons of time-series are shown in Fig. 5. From these figures it is evident that the model satisfactorily reproduces the observations of tidal oscillation and currents. Tidal data obtained at Blaydin, Hudson and BoM stations were used to verify the model results (see Fig. 3). The Nortek ADCP data obtained at the Blaydin and Hudson stations were from a period of over 1 month at 10 min intervals, while data at the BoM station were from a longer period (see Table 2). Figure 5 shows the predicted tides, which agree well with the field data from the Blaydin and BoM stations (positions b and d in Fig. 4c). Skill values of 0.95 and 0.98 were calculated from the comparison between theoretical and observational data of tides. For Hudson station the tides also showed good agreement, with a skill value of 0.97 (position c in Fig. 3, but comparison not shown). The model also performed well in predicting the water currents at Blaydin station near the surface, middle and bottom (Fig. 6b). The skill values over ~0.90 were calculated for Blaydin station from comparison between along-channel and across water velocities. Only the along-channel orientation is shown; the across-channel component of velocity was relatively small and thus is not shown. For Hudson station, the simulated velocity profiles showed good agreement with observations (figure not shown), with the skill value at the different vertical layers calculated to be ~0.90. Transects obtained across EA at locations (a) and (e) are shown in Fig. 3, and measurement periods are described in Table 2. These transects were used to evaluate model performance of tides and the cross-sectional structure of water currents. As expected, the tides showed good agreement; however, the skill value was not calculated, because the transects were made only a few times and the period between each was not consistent. Figure 6c shows that the model results reproduce the currents across EA well, as seen at the transect from a single moment during flood currents.

CTD profiles made along EA, MA and WA were obtained in order to support the assumption of constant values for salinity and temperature during the dry season (four positions indicated by f, and locations at transects g1, g2 and g3 in Fig. 4c). From (f), the average salinity from the two sites along the EA was ~35.7 psu, and the salinity difference between these two locations was less than 0.3 psu. Along the MA the mean salinity from the two sites was ~36.3 psu, and the salinity difference between these two locations was less than 0.5 psu. The average temperatures were 30.1 and 30.6 °C in EA and MA, respectively.

During November 2012 at the end of the dry season, profiles of salinity and temperature were obtained along the three arms (g1, g2 and g3 in Fig. 4). All profiles were obtained during neap tides, during the short period between high tide and peak ebb currents. The salinity profiles were observed to be nearly well-mixed in all locations. The maximum vertical salinity gradients in the East and West Arms were observed at the furthest upstream locations, but did not exceed 1 psu between surface and bottom. Along the MA, the salinity was observed to increase from ~34.2 psu at coastal areas to ~36.2 at the furthest upstream location. For the WA, the average salinity using data from the first two locations close to the arm entrance was ~35 ± 0.2 psu, and ~31.5 psu at the furthest upstream location. For the EA, the average salinity calculated using data from the first four profiles was ~34.2 ± 0.2 psu, and ~31.5 psu at the furthest upstream location. The average temperature of 31.5 ± 0.5 °C was calculated using data from transects g1, g2, and g3.

Tidal amplitudes of the main semi-diurnal and diurnal components (e.g. M_2, S_2, N_2, K_2, K_1, O_1 and M_4) were calculated using the Fourier Transform (Franco and Rock 1971). The observed and predicted amplitudes are shown in Table 3. Observed amplitudes were obtained from harmonic analyses using data from 1992 to 2009. The semi-diurnal components represent nearly 78 % of the total amplitude, showing that DH is a semidiurnal environment. Model results show good agreement between measured and predicted amplitude of the main tidal components for DH. The deviations were calculated for all components (not shown). The largest deviation of 20 % was for M_4, but this component is relatively small when compared to all seven other components from Table 2. Therefore, the model performed well in predicting the amplitude of the most important tidal components for DH.

Model Results

The model results reveal current flows in DH. Figure 7a, b show the model results of vertically averaged current velocities during flood and ebb currents in spring tides. Current speeds increase from the outer harbour to the channel, and then slightly decrease in the inner harbour. Current velocities in the arms are larger than those in the inner harbour. The peak current velocity is about 2.5 m s^{-1} in MA. For the inner harbour and arms, the water flow patterns are in accordance with the shoreline. Current speeds fall to almost zero in the mangrove areas because of the large amount of friction.

Tidal asymmetry in the harbour was calculated using Eq. (4), and shown in Fig. 6c. The skewness parameter distribution shows flood dominance in DH, and the skewness value increases from the outer harbour (0.07) to the inner harbour (0.1) and the arms (0.15).

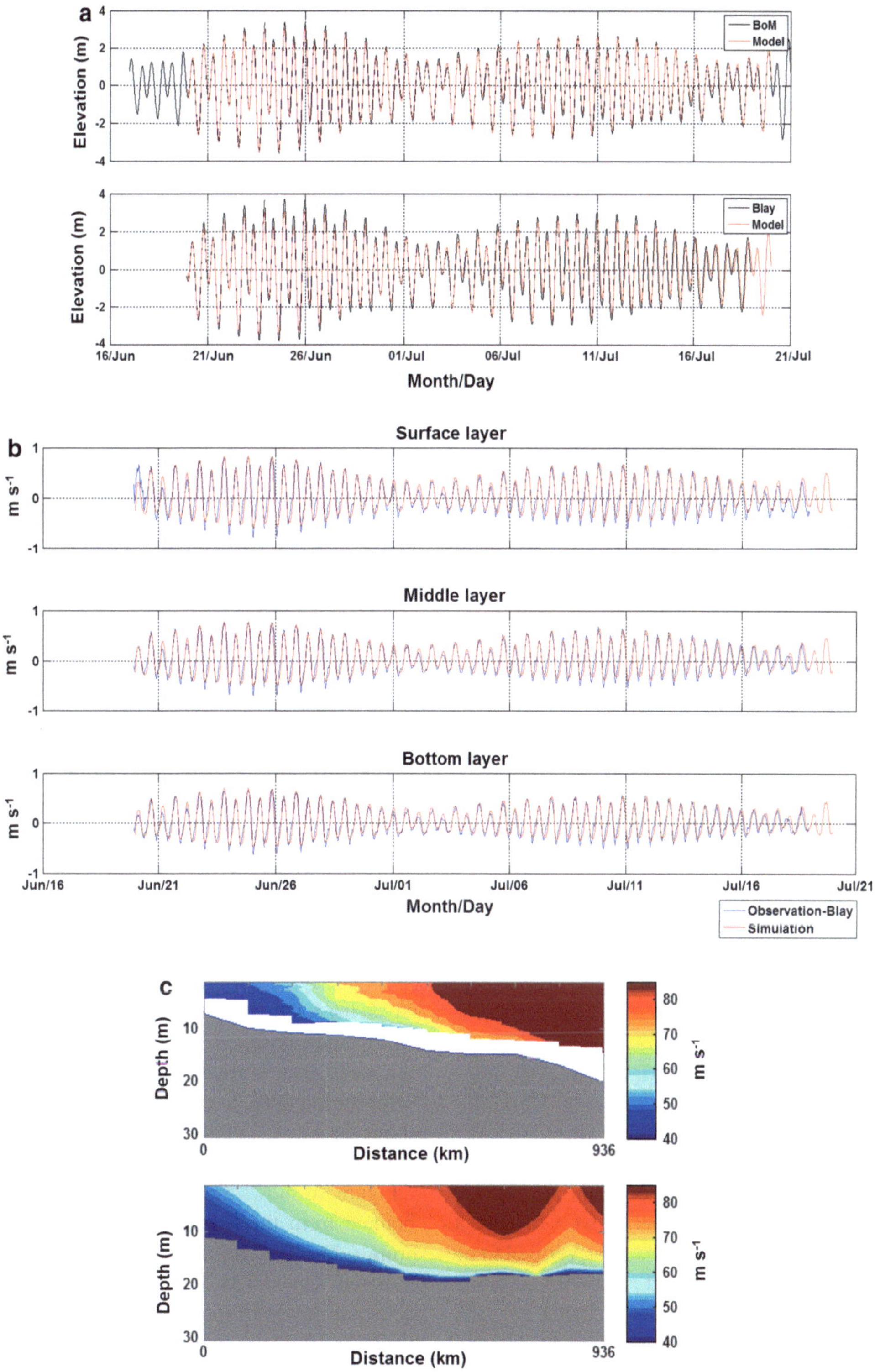

Fig. 6 (**a**) Modelled tides (m), compared to measured tides at stations BoM and Blaydin (positions b and d in Fig. 3c). The skill parameter of ~0.98 was calculated for BoM and Blaydin stations. (**b**) Modelled along-channel velocities (m s^{-1}), compared to observations at station Blaydin. The skill parameter of 0.95 was calculated from comparisons at surface, and 0.97 for the *middle* and *bottom* layers. Across-channel velocities were negligible from observation and simulation (not shown). (**c**) Comparison of observed (*top*) and predicted (*bottom*) along-channel velocities using data from transect shown in Fig. 3c, on 30th of September 2007

Table 3 Comparison between theoretical and observed amplitude of tidal components

Tidal component	Nature	Component name	Period (solar hours)	Observed (m)	Model (m)
M_2	Semi-diurnal	Principal lunar	12.42	1.85	1.70
S_2	Semi-diurnal	Principal solar	12.00	0.96	0.93
N_2	Semi-diurnal	Larger lunar elliptic	12.66	0.35	0.29
K_2	Semi-diurnal	Luni-solar	11.97	0.27	0.25
K_1	Diurnal	Luni-solar diurnal	23.93	0.58	0.53
O_1	Diurnal	Principal lunar diurnal	25.82	0.33	0.30
M_4	Compound	–	6.21	0.05	0.04

Measurements were taken near BoM station

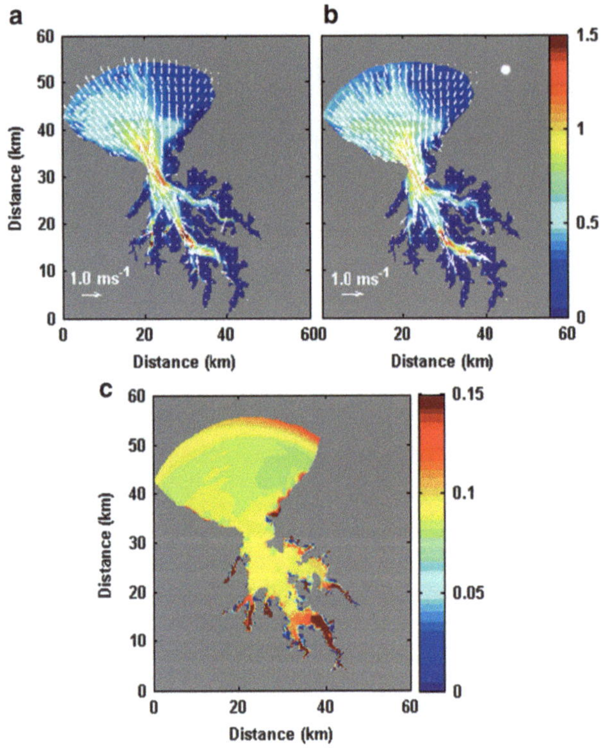

Fig. 7 (**a**) Ebb currents, and (**b**) flood currents during spring tide. (**c**) Tidal asymmetry calculated by the skewness parameter (Eq. 4)

The skewness parameter was also calculated using the observed current velocities at Station Blaydin, γ_{obs}, and is shown in Table 4. The along the channel currents indicate a flood dominance at all depths. In DH, γ_{obs} was verified to be slightly larger than the model predicted γ, ca. ~0.1.

The formation of Estuarine Turbidity Maxima zones is commonly observed in macro-tidal estuaries (e.g. Manning et al. 2010; Uncles and Stephens 2010). This is because strong water currents in macro-tidal estuaries can suspend large amounts of fine sediment from the bottom, and thus SSC increases to high values. Therefore, ETM zones can be formed even under conditions of low sediment input from upstream locations. ETM zones have sediment concentrations in the water column that are typically one or two orders of magnitude higher than upstream and coastal areas (Nichols and Biggs 1985). The understanding of ETM zones is important for the local ecosystem because some toxic substances, e.g. metals, tend to attach to fine sediment particles, which are often the most mobile sedimentary fraction in estuaries (Doxaran et al. 2009; Taylor and Hudson-Edwards 2008; Thonon 2006). ETM formation is driven by different processes such as tidal currents, tidal asymmetry, salinity stratification, density driven currents, waves, water density stratification due to salinity, temperature and SSC (Manning et al. 2010; Uncles and Stephens). Additionally, bathymetry modulates water circulation and therefore may affect the location and formation of ETM zones (Brenon and Le Hir 1999).

Figure 8a shows some snapshots of the vertical distribution of SSC during spring tides along the MA and towards the coastal zone (see transect iii indication in Fig. 4b). Two ETM zones can be observed from the simulation results, with the upstream ETM zone having SSC up to (~0.3 kg m^{-3}), which is nearly double the suspended sediment concentration of the ETM zone along the entrance channel (~0.2 kg m^{-3}). The ETM zone near the channel moves nearly 10 km between low and high tides. The higher SSC in the ETM zone along the MA is predicted to remain in this channel; however, a decrease in SSC is shown during high tides. The vertical stratification of SSC is negligible during spring tides. In contrast, during neap tides the vertical mixing decreases (figure not shown), and thus SSC showed some stratification through the water column.

Figure 8b shows the SSC seen at the sigma bottom layer, obtained from the simulation considering scenario S_1, for spring tides. The formation of two ETM zones in DH was predicted. One ETM zone is formed along the MA, and another is observed in coastal areas near DH entrance. The formation of these two ETM zones is due to the strong bottom stress by the tidal currents, with water speeds that easily exceed 1 m s^{-1}. The ETM zone formed in the MA shows higher SSC than at the harbour entrance (nearly double). This high SSC is caused by the combined effect of strong currents due to the shoaling effect within the MA, and the availability of fine sediment particles. In spring tides, the SSC along the MA reaches values as high as ~0.3 kg m^{-3}, and values of ~0.2 kg m^{-3} at the channel's

Table 4 The skewness parameter γ of observed currents at Station Blaydin

Vertical layer and (depth in metres)	Main velocity skewness
1 (1.8)	0.16
2 (2.8)	0.12
3 (3.8)	0.11
4 (4.8)	0.11
5 (5.8)	0.12
6 (6.8)	0.13
Average 1–6 layers	0.12

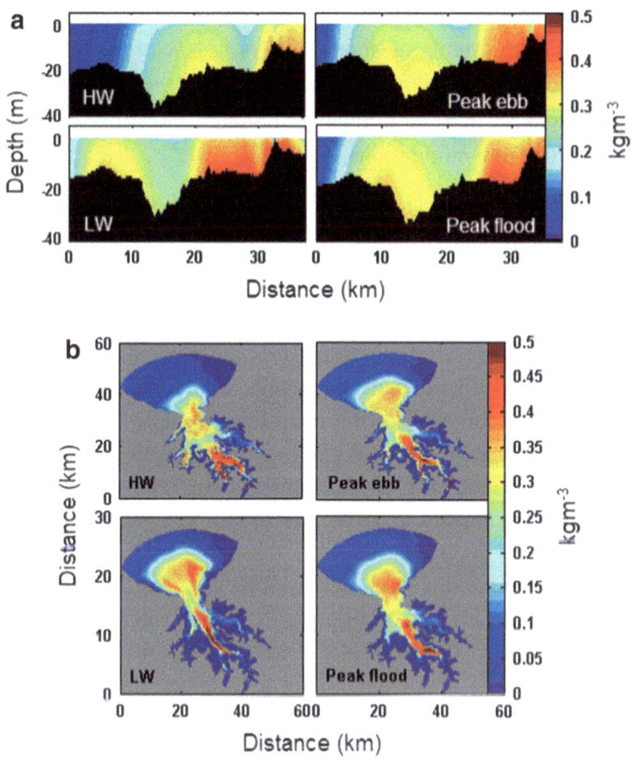

Fig. 8 (**a**) Vertical distribution of suspended sediment concentration (SSC) from transect (iii) in spring tidal cycle (see Fig. 3b). *Left* and *right side* of figure denote respectively coastal and upstream area. *HW* and *LW* refer to high water and low water, respectively. (**b**) *Bottom* distribution of SSC (kg m^{-3}) in spring tidal cycle. *HW* and *LW* refer to high water and low water, respectively

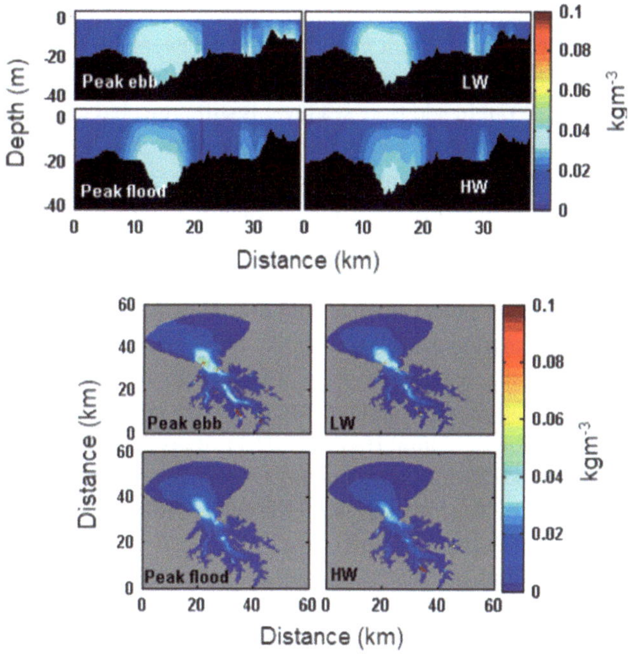

Fig. 9 (**a**) Vertical distribution of suspended sediment concentration (SSC) from transect (iii) in neap tidal cycle (see Fig. 3b). *Left* and *right* side of figure denote coastal and upstream areas, respectively. *HW* and *LW* refer to high water and low water, respectively. (**b**) *Bottom* distribution of SSC (kg m^{-3}) in neap tidal cycle. *HW* and *LW* refer to high water and low water, respectively

main entrance. One should note that if the sediment flux was to change such that these two zones merged into one, an ETM zone of much higher SSC would be formed.

Results from neap tides (Fig. 9a, b) found that the associated less energetic tidal currents erode a much smaller amount of sediment within DH. Moreover, the weaker tidal currents along the MA during neap tides are not able to form the ETM zone of high SSC. The ETM zone within DH shows a SSC lower than ~0.1 kg m^{-3}. During a semi-diurnal tidal cycle, the patches of ETM zones move over 10 km seawards and landwards by ebb and flood currents, respectively.

Future Implications for the Marine Ecosystem of Darwin Harbour

Many estuaries in Europe have sequestered pollutants on the fine sediment, and some of this sediment-bound pollution is observed to date from around the time of the industrial revolution (Den Besten et al. 2003; Löser et al. 2001; Clark et al. 1997). In the past few years, DH has undergone considerable developments because of natural resources from the Ichthys Gas Field (Asia-Pacific Applied Science Associates 2010). Intensive dredging activities are currently taking place, but little is known about threshold limits of high SSC that the DH marine ecosystem can sustain. Some other muddy, macro-tidal and mangrove fringed harbours in Asia (e.g. Ho Chi Minh City and Jakarta), are observed to have high concentrations of pollutants buried within their mud (Rochyatun and Rozak 2008; Minh et al. 2007; Hong et al. 2000; Williams et al. 2000). One would fear a similar future for the DH marine ecosystem, because a large pollution event may result in a considerable portion of pollutants to be buried in the mud. If this was to happen, the length of time the pollutant would be trapped within mangrove areas, and therefore affecting marine species such as the mud crab, local birds, and fishes would be of concern. There is some fear that climate change will result in sea level rise (IPCC 2007; McInnes et al. 2003), and thus cause changes to local hydrodynamics and sediment transport. Sea level has risen globally nearly 1.7 mm year^{-1} during the twentieth century, caused by thermal expansion of melting glaciers and ice caps (IPCC 2007). In addition, sea level is still projected to rise at a larger rate than it has in previous decades. Not all areas, however, have the same sea level rising rate, and some areas are even reported to have a decreasing sea level. These sea level changes raise concern about the possibility of the sediment-trapped pollutants being released back into the water in European estuaries, and thus negatively impacting the local marine ecosystem. Unlike the estuaries in Europe, DH has only undergone most of its development recently, and its future development seems to be connected with exploitation of natural resources. DH is considered a largely unmodified marine environment (Estuary Assessment 2002: Estuaries by Australian Natural Resources Atlas). Therefore, it is too early to infer the hypothesis that similar consequences to that of European estuaries would also occur in DH.

From our results we demonstrate that the sediment transport of small particles in DH is driven by flood dominance, which is therefore affected by wet/dry areas such as mangroves and tidal flats. Therefore, mangrove areas of DH may function as a sediment trap, and if the trapped sediment carries pollutants one would expect conditions similar to many European estuaries. Additionally, the reclamation of mangrove and tidal flat areas may increase tidal asymmetry (Wang and Andutta 2012; Li et al. 2012), increasing flood dominance, and subsequently increasing the rate of landwards sediment transport. All port developments in Darwin are located in EA; because of this, land reclamation in the next few decades would occur along EA. Nevertheless, if land reclamation was to happen for the EA area, one would propose two scenarios: the increased flood dominance with reduced mangrove areas in EA would result in, (a) waters of DH having higher SSC because of reduced deposition areas, (b) increased sediment deposition rates in the remaining mangrove areas (i.e. along MA and WA), (c) the combined effect from (a) and (b).

Unlike European estuaries where boundary forcings do not include cyclone events, DH is occasionally subject to cyclonic activities in the wet season (Ramsay et al. 2008; Hastings 1990; Nicholls, et al. 1998; Nicholls 1984), which are often followed by intense floods. However, these cyclonic activities are capable of flushing only a small fraction of the sediment from within the mangrove areas; therefore, if mud pollution occurs in the mangroves in DH, it is likely to be permanent like the coastal wetlands in Europe and Asia. Although these cyclonic events are projected to increase in intensity in Australia (McInnes et al. 2003), mud pollution in mangroves is predominantly relieved by bioturbation, mainly by crabs. There are no data on this process for DH.

Conclusions

The water circulation and sea level oscillation in DH was accurately simulated by the model, and our water circulation results concur with other field studies and numerical simulations using structured and unstructured models (e.g. Li et al. 2012; Williams 2009; Williams et al. 2006). Our model results would be adequate for simulating hydrodynamics in harbours and predicting suspended sediment transport patterns. Currents speeds increase from the outer harbour to the channel, and then decrease in the inner harbour. Peak currents of about ~2.5 m s^{-1} occur in the MA. The sediment transport pattern is revealed comprehensively, and areas of high suspended sediment concentration were observed from simulations.

This study shows that for scenario S_1, two Estuarine Turbidity Maxima (ETM) zones are expected to form in DH during spring tides, one in the MA and another near the entrance of the bay. These ETM zones are transported by flood currents (landward direction) and ebb currents (seaward direction). During spring tides, the vertical structure of SSC is well-mixed near the peaks of the flood and ebb currents. In contrast, during neap tides the two ETM zones vanish and a small vertical gradient of SSC is predicted near high and low tides. From scenario S_2, the ETM zone along the MA disappears and the maximum SSC within DH is considerably reduced.

Our simulations used a variable grid-size model that allowed for a high resolution prediction of water circulation within mangrove areas, tidal flats and narrow channels, as well as wet/dry grid elements for all mangrove and tidal flat areas. Our measurements have shown that salinity and temperature fields can be treated as uniform during the dry season, because observed values showed little time and space variation of these parameters during the dry season, and the minor density-driven currents are limited to the upper reaches of the rivers of DH.

In the future, DH is likely to accumulate polluted sediment. Polluted fine sediment that predicted to be trapped within mangrove areas, and only a small fraction is expected to be flushed out during cyclone events. Thus this fine sediment may remain trapped for many years. Because cyclone activities are not effective mechanisms to flush out fine sediment from DH, the likely result is that similar conditions to many European estuaries, where pollutant sediment has been found to be buried since the industrial revolution. Additionally, the increased tidal asymmetry would cause higher rates of deposition within mangroves and other areas of the harbour, and would also increase the average suspended sediment concentration of DH. Therefore, the trapping of polluted sediment within mangrove areas combined with increased suspended sediment concentration in the estuarine waters would negatively impact marine species. Additionally, if sediment pollution affects the mud crabs and many other local marine species that are responsible for local bioturbation, trapping of polluted sediment would increase further.

Acknowledgments F. P. Andutta, X. H. Wang, Li Li and D. Williams were supported by a 2011 Australian Research Council/Linkage Project – LP110100652. This work was also supported by the National Computational Infrastructure National Facility at the Australian National University. This is a publication of the Sino-Australian Research Centre for Coastal Management, manuscript number 13.

References

Adams CE Jr, Weatherly GL (1981) Some effects of suspended sediment stratification on an oceanic bottom boundary layer. J Geophys Res 86:C54161

Allen GP, Salomon JC, Bassoullet P, Du Penhoat Y, de Grandpré C (1980) Effects of tides on mixing and suspended sediment transport in macrotidal estuaries. Sediment Geol 26(1–3):69–90

Andutta FP (2006) Experiments and numerical model with application to the tropical estuary of the Curimataú River, RN. MSc thesis, IO-USP, Oceanography Institute of the University of São Paulo, 122 p

Andutta FP (2011) The estuarine system of the Caravelas and Peruípe rivers (Bahia): observations, simulations, residence time, and advective and diffusive processes. PhD thesis, IO-USP, Oceanography Institute of the University of São Paulo, 121 p

Andutta FP, Miranda LB, Castro BM, Fontes RFC (2006a) Numerical simulation of the hydrodynamic in the Curimataú Estuary, RN Brazil. In: III Simpósio Brasileiro de Oceanografia, 2008, São Paulo. Oceanografia e Mudanças Globais, Instituto Oceanográfico da USP, São Paulo, pp 545–558

Andutta FP, Santos MRM, Miranda LB, Ancona CM, Saldaña-Corrêa FMP, Gianesella SMF (2006b) Monthly variations of physical and chemical properties observed in Santos Bay and Santos Estuarine channel. In: Proceedings of the Environmental and Health World Congress EHWC'2006, Santos, pp 80–84

Andutta FP, Ridd PV, Wolanski E (2011) Dynamics of hypersaline waters in the Great Barrier Reef. Estuar Coast Shelf Sci 94:299–305

Andutta FP, Kingsford MJ, Wolanski E (2012) 'Sticky water' enables the retention of larvae in a reef mosaic. Estuar Coast Shelf Sci 101:54–63

Andutta FP, Ridd PV, Wolanski E (2013) Age and the flushing time in the Great Barrier Reef coastal waters. Cont Shelf Res 53:11–19

Ariathurai R, Krone RB (1976) Finite element model for cohesive sediment transport. J Hydraul Div ASCE 102(HY3):323–328

Asia-Pacific Applied Science Associates (2010) Ichths gas field development project. Description and validation of hydrodynamic and wave models for discharges, spills, geomorphology and dredge spoil disposal ground selection. Report prepared for INPEX Browse, Ltd. Perth, 44 p

Brenon I, Le Hir P (1999) Modelling the turbidity maximum in the Seine estuary (France): identification of formation processes. Estuar Coast Shelf Sci 49(4):525–544

Burford M, Alongi D, McKinnin A, Trott L (2008) Primary production and nutrients in a tropical macrotidal estuary, Darwin Harbour, Australia. Estuar Coast Shelf Sci 79:440–448

Burns KA, Brinkman DL (2011) Organic biomarkers to describe the major carbon inputs and cycling of organic matter in the central Great Barrier Reef region. Estuar Coast Shelf Sci 93(2):132–141

Chen C, Liu H, Beardsley RC (2003) An unstructured grid, finite-volume, three-dimensional, primitive equations ocean model: application to coastal ocean and estuaries. J Atmos Ocean Technol 20:159–186

Cheng RT, Ling C-H, Gartner JW (1999) Estimates of bottom roughness length and bottom shear stress in South San Francisco Bay, California. J Geophys Res 104(C4):7715–7728

Clark JS, Cachier H, Goldhammer JG, Stocks BJ (1997) Sediment records of mass burning and global change. Springer, Berlin/New York, 489 p

Darwin Harbour Advisory Committee – DHAC (2003) Darwin harbour regional plan of management. Department of Infrastructure, Planning and Environment, Darwin

Defant A (1960) Physical oceanography, vol 2. Pergamon Press, New York, 598 p

Den Besten P, de Deckere E, Babut M, Power B, DelValls A, Zago C, Oen A, Heises S (2003) Biological effects-based sediment quality in ecological risk assessment for European waters. J Soil Sediment 3(3):144–162

Doxaran D, Froidefond J-M, Castaing P, Babin M (2009) Dynamics of the turbidity maximum zone in a macrotidal estuary (the Gironde, France): observations from field and MODIS satellite data. Estuar Coast Shelf Sci 81(3):321–332

Dyer KR (1986) Coastal and estuarine sediment dynamics. Wiley, Chichester, 342 p

Franco AS, Rock NJ (1971) The Fast Fourier Transform and its application to tidal oscillations. Boletim do Instituto Oceanográfico 20 (1):145–199. ISSN 0373–5524

Frascari F, Frignani M, Guerzoni S, Ravaioli M (1988) Sediments and pollution in the Northern Adriatic Sea. Ann N Y Acad Sci 534:1000–1020

Harvey N, Caton B (2010) Coastal management in Australia. University of Adelaide Press, Adelaide, 342 p

Hastings PA (1990) Southern oscillation influences on tropical cyclone activity in the Australian/south-west Pacific region. Int J Climatol 10(3):291–298

Hong LCD, Becker-van Slooten K, Sauvain J-J, Minh TL, Tarradellas J (2000) Toxicity of sediments from the Ho Chi Minh City canals and Saigon River, Viet Nam. Environ Toxicol 15(5):469–475

IPCC (2007) In: Solomon S, Qin D, Manning M, Chen Z, Marquis M, Averyt KB, Tignor M, Miller HL (eds) Climate change 2007: the physical science basis. Contribution of working group I to the fourth assessment report of the intergovernmental panel on climate change. Cambridge University Press, Cambridge, 996 p

Li L, Wang XH, Williams D, Sidhu H, Song D (2012) Numerical study of the effects of mangrove areas and tidal flats on tides: a case of study of Darwin Harbour, Australia. J Geophys Res 117:C066011

Ling CH (1976) On the calculation of the surface shear stress using the profile method. J Geophys Res 81:2581–2582

Ling CH, Untersteiner N (1974) On the calculation of the roughness parameter of sea ice. J Geophys Res 79:4112–4114

Liu G, Zhu J, Wang Y, Wu H, Wu J (2011) Tripod measured residual currents and sediment flux: impacts on the silting of the Deepwater Navigation Channel in the Changjiang Estuary. Estuar Coast Shelf Sci 93:192–201

Löser C, Zehnsdorf A, Görsch K, Seidel H (2001) Remediation of heavy metal polluted sediment in the solid bed: comparison of abiotic and microbial leaching. Environ Geol 40(4–5):643–650

Maa P-Y, Mehta AJ (1987) Mud erosion by waves: a laboratory study. Cont Shelf Res 7(11/12):1269–1284

Manning AJ, Langston WJ, Jonas PJC (2010) A review of sediment dynamics in the Severn Estuary: influence of flocculation. Mar Pollut Bull 61(1–3):37–51

Margvelashvili N, Herzfeld M, Webster IT (2006) Modelling of fine-sediment transport in the Fitzroy Estuary and Keppel Bay. Technical report 39. Cooperative Research Centre for Coastal Zone, Estuary and Waterway Management, Brisbane, 38 p. ISBN 1 921017 76 7 (online)
Mazda Y, Magi M, Kogo M, Hong PN (1997) Mangroves as a coastal protection from waves in the Tong King delta, Vietnam. Mangrove Salt Marshes 1(2):127–135. doi:10.1023/a:1009928003700
McInnes KL, Walsh KJE, Hubbert GD, Beer T (2003) Impact of sea-level rise and storm surges on a coastal community. Nat Hazard 30:187–207
Mehta AJ (1988) Laboratory studies on cohesive sediment deposition and erosion. In: Dronkers J, Leussen WV (eds) Physical processes in estuaries. Springer, Berlin, pp 427–445
Mellor GL, Yamada T (1982) Development of a turbulence closure model for geophysical fluid problems. Rev Geophys 20 (4):851–875. doi:10.1029/RG020i004p00851
Michie MG (1987) Distribution of foraminifera in a macrotidal tropical estuary: Port Darwin, Northern Territory of Australia. Aust J Mar Freshw Res 38:249–259
Michie MG, Grey DL, Griffin RK (1991) Sea water salinity and temperature measurements in the Darwin Harbour area 1972–1978. Department of Primary Industries and Fisheries, Darwin
Milliman JD, Farnsworth KL (2011) River discharge to the coastal ocean: a global synthesis. Cambridge University Press, Cambridge, 392 p
Minh NH, Minh TB, Iwata H, Kajiwara N, Kunisue T, Takahashi S, Viet PH, Tuyen BC, Tanabe S (2007) Persistent organic pollutants in sediments from Sai Gon-Dong Nai River basin, Vietnam: levels and temporal trends. Arch Environ Contam Toxicol 52(4):458–465
Nicholls N (1984) The Southern Oscillation, sea-surface-temperature, and interannual fluctuations in Australian tropical cyclone activity. Int J Climatol 4(6):661–670
Nicholls N, Landsea C, Gill J (1998) Recent trends in Australian region tropical cyclone activity. Meteorol Atmos Phys 65(3–4):197–205
Nichols MN, Biggs RB (1985) Estuaries. In: Davis RA Jr (ed) Coastal sedimentary environments. Springer, New York, pp 77–186
Orszag SA, Patera AT (1981) Calculation of Von Kármán's constant for turbulent channel flow. Phys Rev Lett 47:832–835
Padovan AV (1997) The water quality of Darwin Harbour: October 1990 – November 1991. Report No. 34/1997D, Water Quality Branch, Water Resources Division, Department of Lands, Planning and Environment, NT Government, Darwin, December 1997
Pejrup M (1988) Flocculated suspended sediment in a microtidal environment. Sediment Geol 57:249–256
Postma H (1961) Transport and accumulation of suspended matter in the Dutch Wadden Sea. Neth J Sea Res 1:148–190
Puig P, Ogston AS, Mulllenbach BL, Parsons JD, Sternberg RW (2004) Storm-induced sediment gravity flows at the head of the Eel submarine canyon, northern California margin. J Geophys Res 109:C03019
Ramsay HA, Leslie LM, Lamb PJ, Richman MB, Leplastrier M (2008) Interannual variability of tropical cyclones in the Australian region: role of large-scale environment. J Climate 21(5):1083–1103
Ribbe J, Holloway PE (2001) A model of suspended sediment transport by internal tides. Cont Shelf Res 21:395–422
Rochyatun E, Rozak A (2008) The distribution of heavy metals in sediment of Jakarta Bay. Mar Res Indones 33(1):101–107
Sinclair Knight Merz Pty Ltd (2011) Ichthys gas field development project: assessment of potential impacts on mud crabs in Darwin Harbour. Report prepared by Sinclair Knight Merz Pty Limited, Perth, for INPEX Browse, Ltd., Perth
Smagorinsky I (1963) General circulation experiment with primitive equations. Mon Weather Rev 91:91–164
Straatsma MW (2009) 3D float tracking: in situ floodplain roughness estimation. Hydrol Process Int J 23(2):201–212
Sun WC, Ishidaira H, Bastola S (2010) Towards improving river discharge estimation in ungauged basins: calibration of rainfall-runoff models based on satellite observations of river flow width at basin outlet. Hydrol Earth Syst Sci 14:2011–2022
Taylor MP, Hudson-Edwards KA (2008) The dispersal and storage of sediment-associated metals in an arid river system: the Leichhardt River, Mount Isa, Queensland, Australia. Environ Pollut 152 (1):193–204
Telford JW (1982) A theoretical value for von Karman's constant. Pure Appl Geophys 120:648–661
Thonon I (2006) Deposition of sediment and associated heavy metals on floodplains, Geographical studies 337. University of Utrecht, Utrecht, 174 p
Tien AT (2006) Influence of deep aquifer springs on dry season stream water quality in Darwin Rural Area, Water monitoring branch natural resource management division report 6/2006D. Water Monitoring Branch, Department of Natural Resources, Environment and the Arts, Darwin, 54 p
Uncles RJ, Stephens JA (2010) Turbidity and sediment transport in a muddy sub-estuary. Estuar Coast Shelf Sci 87(2):213–224
van Kessel T, Kranenburg C (1998) Wave-induced: liquefaction and transport of mud on inclined beds. In: Dronkers J et al (ed) (1998) Physics of estuaries and coastal seas: proceedings of the 8th international biennial conference on physics of estuaries and coastal seas, The Hague, 9–12 Sept 1996, pp 209–215
van Kessel T, Vanlede J, de Kok J (2011) Development of a mud transport model for the Scheldt estuary. Cont Shelf Res 31(10, Supplement 1):S165–S181
van Straaten L, Kuenen P (1958) Tidal action as a cause of clay accumulation. J Sediment Petrol 28:406–413
Victor S, Golbuu Y, Wolanski E, Richmond RH (2004) Fine sediment trapping in two mangrove-fringed estuaries exposed to contrasting land-use intensity, Palau, Micronesia. Wetl Ecol Manag 12:277–283
Wang XH (2002) Tide-induced sediment resuspension and the bottom boundary layer in an idealized estuary with a muddy bed. J Phys Oceanogr 32:3113–3131
Wang XH, Andutta FP (2012) Sediment transport dynamics in ports, estuaries and other coastal environments. In: Manning AJ (ed) Sediment transport. INTECH. ISBN 980-953-307-557-5, pp. 3–35. http://dx.doi.org/10.5772/51022
Wang XH, Pinardi N (2002) Modeling the dynamics of sediment transport and resuspension in the northern Adriatic Sea. J Geophys Res 107(C12):3225. doi:10.1029/2001JC001303
Williams D (2009) Part 1: hydrodynamics and sediment transport. In: Dredging of sand from Darwin Harbour, hydrographic and marine life, Australian Institute of Marine Science, Arafura Timor Research Facility, Brinkin, pp 1–33
Williams TM, Rees JG, Setiapermana D (2000) Metals and trace organic compounds in sediments and waters of Jakarta Bay and the Pulau Seribu Complex, Indonesia. Mar Pollut Bull 40(3):277–285
Williams D, Wolanski E, Spagnol S (2006) Hydrodynamics of Darwin Harbour. In: Wolanski E (ed) The environment in Asia Pacific harbours. Springer, Dordrecht, pp 461–476
Willmott CJ (1981) On the validation models. Phys Geogr 2:184–194
Wilson D, Padovan A, Townsend S (2004) The water quality of spring and neap tidal cycles in the middle arm of Darwin Harbour during the dry season. Water Monitoring Branch, Natural Resource Management Division, Department of Infrastructure, Planning and Environment, Darwin
Wolanski E (2007) Estuarine ecohydrology. Elsevier, Amsterdam, 157 p
Wolanski E, Spagnol S (2000) Environmental degradation by mud in tropical estuaries. Reg Environ Change 1(3–4):152–162
Wolanski E, Gibbs R, Spagnol S, King B, Brunskill G (1998) Inorganic sediment budget in the mangrove-fringed Fly River delta, Papua New Guinea. Mangrove Salt Marshes 2:85–98

Wolanski E, McKinnon D, Alongi D, Spagnol S, Williams D (2006) An ecohydrology model of Darwin Harbour. In: Wolanski E (ed) The environment in Asia Pacific harbours. Springer, Dordrecht, pp 477–488, 497 p

Woodroffe CD, Bardsley KN, Ward PJ, Hanley JR (1988) Production of mangrove litter in a macrotidal embayment, Darwin Harbour, N. T., Australia. Estuar Coast Shelf Sci 26(6):581–598

Wu J, Wang Y, Cheng H (2009) Bedforms and bed material transport pathways in the Changjiang (Yangtze) Estuary. Geomorphology 104:175–184

Xu JP, Wright LD (1995) Tests of bed roughness models using field data from the middle Atlantic bight. Cont Shelf Res 15:1409–1434

Zann LP, Kailola P (1995) State of the marine environment report. Technical Annex 1. AgPS, Canberra

The Ord River Estuary: A Regulated Wet-Dry Tropical River System

Barbara J. Robson, Peter C. Gehrke, Michele A. Burford, Ian T. Webster, Andy T. Revill, and Duncan W. Palmer

Abstract

The lower Ord River is a wet-dry tropical river functioning as a perennial dry tropical river as a result of regulation. It is currently one of the few heavily regulated rivers in Australia's tropical north, providing water for hydroelectrical production and irrigation. Current plans call for an increase in the area of irrigated land surrounding the lower Ord River and its estuary. The estuary is highly turbid and subject to very strong tides. It can be conceptualised as five connected physical zones – the Riverine Zone, the Tidal Freshwater Zone, the Transitional (Maximum Turbidity) Zone, the Estuary Mouth, and the Tidal Creeks and Flats Zone – distinguished by geomorphology, flow and tidal influence. Each of these physical zones functions as a distinct biogeochemical and ecological functional zone. Here, we describe how these zones function, how they interact, and how the estuary as a whole may respond to the changes expected in the mid-term future.

Keywords

Macrotidal estuary • Dry tropics • River regulation • Kununurra Diversion Dam • Ord Irrigation Area • Lake Argyle • Ecological functional zones

B.J. Robson (✉) • I.T. Webster
CSIRO Land and Water, GP.O. Box 1666, Canberra ACT 2601, Australia
e-mail: Barbara.Robson@csiro.au; Ian.Webster@csiro.au

P.C. Gehrke
SMEC Australia Pty Ltd, 154 Melbourne Street, South Brisbane 4101, Australia
e-mail: Peter.Gehrke@smec.com

M.A. Burford
Australian Rivers Institute, Griffith University, Kessels Rd, Nathan Q 4111, Australia
e-mail: m.burford@griffith.edu.au

A.T. Revill
CSIRO Marine and Atmospheric Research, Hobart, Australia
e-mail: Andy.Revill@csiro.au

D.W. Palmer
Department of Water, Government of Western Australia, P.O. Box 625, Kununurra WA 6743, Australia
e-mail: Duncan.Palmer@water.wa.gov.au

E. Wolanski (ed.), *Estuaries of Australia in 2050 and Beyond*, Estuaries of the World,
DOI 10.1007/978-94-007-7019-5_8, © CSIRO 2013

Box 1

Barbara Robson and colleagues studied the lower Ord River, which is a wet-dry tropical river now functioning as a perennial dry tropical river as a result of flow regulation by a large dam. The estuary is highly turbid and subject to very strong tides. It can be conceptualised as five connected physical zones – the Riverine Zone, the Tidal Freshwater Zone, the Transitional (Maximum Turbidity) Zone, the Estuary Mouth, and the Tidal Creeks and Flats Zone – distinguished by geomorphology, flow and tidal influence. Each of these physical zones functions as a distinct biogeochemical and ecological functional zone. They describe how these zones function, how they interact, and how the estuary as a whole may respond to the changes expected in the mid-term future for increasing agricultural development, increasing demand for electricity generation, and climate change. The results for the estuary are expected to be minor, but are highly uncertain.

Introduction

Many of Australia's tropical estuaries are in very remote locations and are subject to large tides and extreme variations in flow, which combine to make them challenging environments for measurement and monitoring. Distant from large population centres and subject to less intense development pressures than their temperate counterparts (Douglas et al. 2005; Bunn et al. 2006), these systems have historically attracted little political interest. In consequence, tropical estuaries have been relatively little-studied (NGIS Australia 2004), though research over the past few years has considerably advanced the state of knowledge.

Australia's temperate and subtropical water resources are now widely considered to be at the limit of exploitation. Attention is therefore turning to the development potential of tropical rivers and their catchments. Tropical rivers and groundwater account for roughly 70 % of Australia's total fresh-water resources (Camkin et al. 2008). The vast majority of flow in these systems, however, occurs in intermittent peak rainfall events and is not readily available for capture and irrigation (Petheram et al. 2008).

The Ord River is one of the few heavily regulated tropical rivers in Australia. The hydrological, biogeochemical and ecological functioning of the estuary has been strongly impacted by flow regulation (Burford et al. 2011). A close study of this system may provide insight into the future of other tropical estuaries in Australia as land use and water management continue to change.

Context, Climate and History

The Ord River is located in the north of Western Australia (Fig. 1), with a catchment straddling the border of W.A. and Australia's Northern Territory. The lower river and estuary have a combined length of over 100 km (Parslow et al. 2003) and constitute an important recreational and conservation asset, supporting commercial and recreational fishing and some ecotourism. The system, with its associated wetlands and flood plains, is of major cultural significance to East Kimberley indigenous communities (Head 1999). The estuary is considered an important nursery habitat for Bonaparte Gulf prawn stocks (Head 1999) and also provides a diverse array of habitats for other fisheries species of commercial and recreational importance in the river. However, no comprehensive fisheries assessments have been undertaken in the Ord River estuary. Most of the available ecological information about the lower Ord River and Estuary has been published only as reports to or by government departments, so these have been cited where appropriate in this chapter.

Many rivers in northern Australia, including the Ord River, originate in the semi-arid interior before reaching the tropical zone nearer the coast, where they receive most of their flow (Wolanski et al. 2001). Australia's tropical rivers are characterised by episodic flow events associated with storm activity during the wet season (the austral summer). By international standards, these flows are highly variable from year to year (Bunn et al. 2006; Warfe et al. 2011). During the extended dry season, flows in most Australian tropical rivers are low: often zero in the wet-dry tropics, with the exception of a few, such as the Daly River, in which groundwater influxes are sufficient to maintain year-round flow (Webster et al. 2005). Temperatures are relatively high throughout the year.

The inter-annual variability of runoff in the Ord-Bonaparte region is high by international standards, but typical of Australia's tropical north. Mean annual rainfall

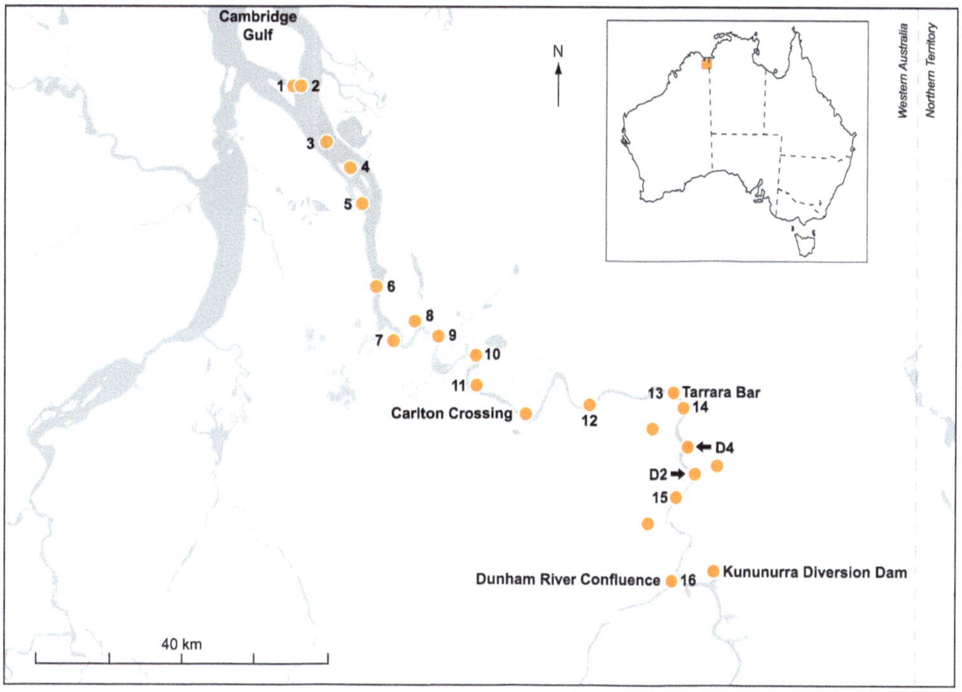

Fig. 1 The lower Ord River and Estuary, showing locations of regular monitoring sites maintained between 2004 and 2008

in the region, at 730 mm, and subsequent runoff are moderate in comparison with other parts of tropical Australia (CSIRO 2009).

Two dams interrupt freshwater flow to the Ord River Estuary. The Kununurra Diversion Dam, built during the 1960s, has a relatively small storage of approximately 100 GL and was too small to have much impact on the timing or magnitude of the annual floods during the wet season. This dam was created to provide irrigation to approximately 140 km^2 of farmland in the catchment, though this agricultural area has had a troubled history, as pests destroyed the originally envisioned cotton crops (Smith 2008), sugar production proved economically unsustainable, and rising groundwater tables and salinisation have been a concern in more recent times. Since the construction of the dam, the groundwater table in the Ord River Irrigation Area has risen by an estimated 10–20 m, though there is some doubt that this is due to irrigation rather than being primarily associated with long-term climate variability (Smith 2008).

Despite these concerns, agricultural and development efforts have continued and diversification of agriculture in the area has increased its resilience and profitability (Smith 2008). In addition to irrigated agriculture, much of the land surrounding the estuary is used for cattle grazing.

A second dam, the Ord River Dam, was constructed upstream of the Kununurra Diversion Dam in the early 1970s to create the largest artificial lake in the southern hemisphere, Lake Argyle. In addition to its irrigation storage function, the Ord River Dam supplies hydropower and water for the Argyle Diamond Mine. Lake Argyle stores 10,760 GL of water at full supply and provides sufficient flood storage to capture all the wet season inflow in most years. To maintain hydroelectric production over the dry season months, water is gradually released through turbines below. Management of the Ord River Dam thus provides a base flow to the estuary throughout the dry season while substantially reducing the frequency and severity of major floods. In so doing, the dam has substantially altered the hydrology and ecology of the lower Ord River and Estuary (Doupe and Pettit 2002; Cluett 2005).

Before the construction of the dams, freshwater flows to the estuary during the dry season (April to December) were negligible, while flows during the 4-month wet season (January to March) were substantial and were characterised by large flood events with peak flows that may have exceeded 30,000 m^3 s^{-1} (Wolanski et al. 2001).

To meet the needs of the hydroelectric scheme, flows of around 50–90 m^3 s^{-1} are now maintained throughout the dry season, while peak wet-season flows are substantially moderated (Wolanski et al. 2001) but still exceed 1,000 m^3 s^{-1}.

The ecological effects of river regulation are more often considered in terms of flow and connectivity (Tetzlaff et al. 2007; Poff and Allan 1995; Pringle 2001; Ward and Stanford 1995) than in terms of overall system productivity and metabolism. Changes that affect flow and connectivity, however, invariably also alter sediment and nutrient loads and concentrations, and may alter salinity and the baroclinic dynamics of estuaries. The environmental impact of these changes on estuaries and adjacent coastal waters has been an issue of increasing concern and interest in recent years.

Fig. 2 *Top left*: Vegetation surrounding the Tidal Freshwater Zone of the Ord River estuary. *Top right*: A saltwater crocodile in the highly turbid water of the Transitional Zone. *Bottom left*: the erosional banks of the Transitional Zone at low tide. *Bottom right*: the wide Estuary Mouth Zone. Photos by Ian Webster

Responses are often complex and depend not only on the magnitude of flows and loads, but also their timing and degree of variability (Parslow et al. 2003).

Hydrological Typology

The lower Ord River (Fig. 2) sits outside the flow-based typology for Australian tropical rivers developed by Moliere et al. (2009) due to regulation of what would previously have been a dryland tropical river. In the classification of Kennard et al. (2010), the river downstream of the dams has shifted from class 10 (predictable summer highly intermittent flows) to class 1 – stable baseflow. In other words, it is perennial.

Latrubesse et al. (2005) reviewed tropical rivers around the world, but focused particularly on the perennial rivers in the wet and wet-dry tropics in Asia, Africa and South America. When compared with these systems, the Ord River, situated in the dry tropics, is clearly a low-flow river. The mean discharge of the Ord River (from a catchment area of approximately 64,000 km^2) is on the order of 200 $m^3\ s^{-1}$, much lower than the mean discharges of any of the river systems reviewed by Latrubesse et al. (2005), which range from 4,000 $m^3\ s^{-1}$ from a 47,300 km^2 catchment (the Caura River, Venuzuala) to 209,000 $m^3\ s^{-1}$ from a 6×10^6 km^2 catchment (the Amazon River). Accordingly, the total annual suspended sediment load carried by the lower Ord River is also much lower (at approximately 130 kt $year^{-1}$ according to our calculations) than in the wet and wet-dry tropical systems included in that review, though this sediment load is still substantial given the smaller volume of water involved.

In terms of seasonality of flow, the lower Ord River in its current, regulated condition, is similar to the unimodal wet-dry tropical rivers reviewed by Latrubesse et al. (2005), with $Q_{monthly}/Q_{mean}$ (where Q represents flow) ranging from 0.33 in the middle of the dry season to 3.0 in March (Fig. 3) – a range comparable to that of the Mekong River (Latrubesse et al. 2005). Under pre-regulation conditions, flow in the lower Ord River dropped to zero during the dry season as is still the case in the upper Ord River (Burford et al. 2011) and in many other rivers in Australia's tropical north.

Considering daily flows, the lower Ord River's Q_{max}/Q_{mean} (30) and Q_{max}/Q_{min} (809) for flows between 1997 and 2007 place it just outside the range represented by semiarid rivers in the dataset presented by Latrubesse et al. (2005).

In summary, the lower Ord River can be considered a dry tropical river that now functions as a semi-arid or wet-dry tropical river, as a consequence of flow regulation. Its estuary is vertically well-mixed and macrotidal, with 7 m spring tides.

Geomorphological Setting

The dominant geological features of the Ord catchment (Wasson et al. 2002) are Precambrian sandstones, shale, basalt and porphyry in the western Kimberley Basin, Precambrian crystalline rocks extending from the Kimberley Basin to the

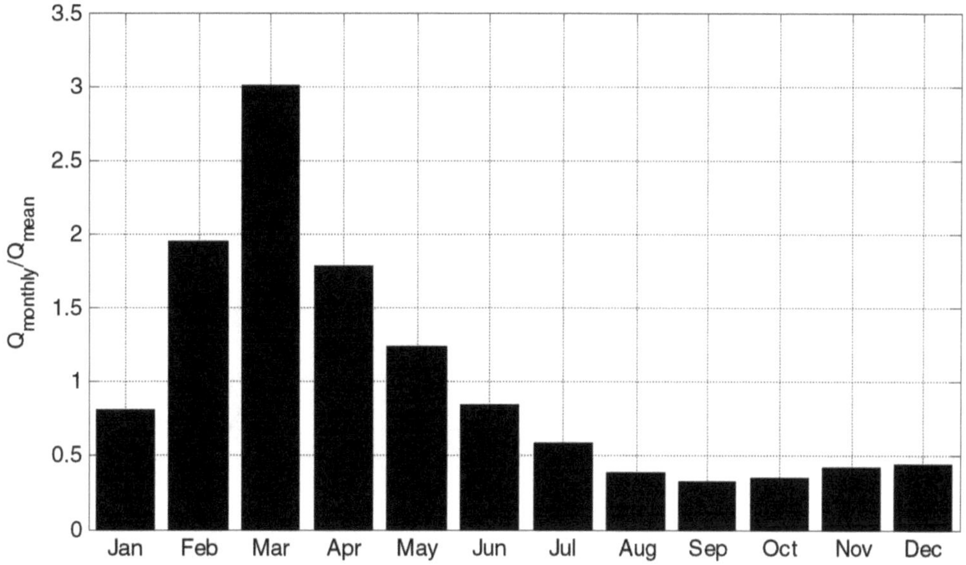

Fig. 3 Seasonality of flow reaching the Ord River estuary, 1998–2007, showing monthly average flow relative to annual average flow. The austral summer runs from December to February, a period that coincides with increased rainfall in the tropics

Halls Creek Fault line, Cambrian basalts to the east of Halls Creek Fault, with areas of uplifted Cambrian sedimentary formations of limestone, siltstone, sandstone and mudstone. Devonian sedimentary rocks toward the centre of the catchment are represented by quartz, sandstone, conglomerate and siltstone. Much of the region is less than 200 m above sea level, with few areas attaining elevations more than 600 m. The landscape to the west of the Halls Creek Fault is rugged, with steep ridges, narrow valleys and sandy alluvial deposits. To the east, the landscape has a lower relief, with wide plains incised by steep-sided, shallow valleys. Approximately 90 million years ago, much of the region was covered by a shallow sea (Bolton et al. 1990). After sea levels receded, older formations were heavily eroded during the Quaternary, about 1.8 million to 10,000 years ago, and sediments were deposited along the river systems, forming broad alluvial plains. The well-drained steep slopes with rock outcrops typically have shallow skeletal soils, with deeper soils largely confined to poorly-drained flatter country (Stewart 1970; Wasson et al. 2002).

It has been estimated that 23.5 × 106 t year^{-1} of sediment enter Lake Argyle and becomes trapped within the lake (Wark 1987; Wasson et al. 2002) The lower Ord River is deprived of sediment immediately downstream of Kununurra Diversion Dam, but receives sediment from erosion of cropping and grazing lands elsewhere in the catchment, and from bank erosion closer to the estuary.

Applying the geomorphological typology of Erskine et al. (2005), the lower Ord River and estuary exhibit a downstream longitudinal sequence of resistant bedrock channels, followed by bedrock-confined reaches; meandering reaches; island- and ridge-anabranching reaches; and co-existent mud-braided and anabranching reaches. Within this geomorphological template, the relative dominance of freshwater and tidal flows superimposes a physical zonation on the ecological patterns and processes.

Five natural zones are apparent (Fig. 3):

1. Riverine zone: characterised by unidirectional flow through two meandering reaches upstream and downstream of the bedrock-constrained reach at Tarrara Bar.
2. Tidal Freshwater Zone (Fig. 2, top left): predominantly freshwater, but water levels rise and fall under the influence of tides. This zone consists of a single meandering reach that becomes strongly depositional in character under the increasing tidal influence downstream.
3. Transitional Zone (the estuarine turbidity maximum zone, Fig. 2, top right and bottom left): dually influenced by freshwater discharge and tidal currents, which result in active eroding banks and mobile islands, creating a complex, dynamic anabranching zone.
4. Estuary Mouth Zone (Fig. 2, bottom right): wide channel with reduced tidal energy resulting in relatively stable mangrove-lined banks and expansive mud flats.
5. Tidal Creeks and Flats Zone: narrow and occasionally deep channels with relatively stable mangrove-lined banks. Characterised by lower turbidity than estuary mouth and transitional zones.

Biogeochemical and Ecological Zones

The physical differences between the five identified zones in the river and estuary create distinct biogeochemical environments and result in distinct zones of ecological function (Fig. 4).

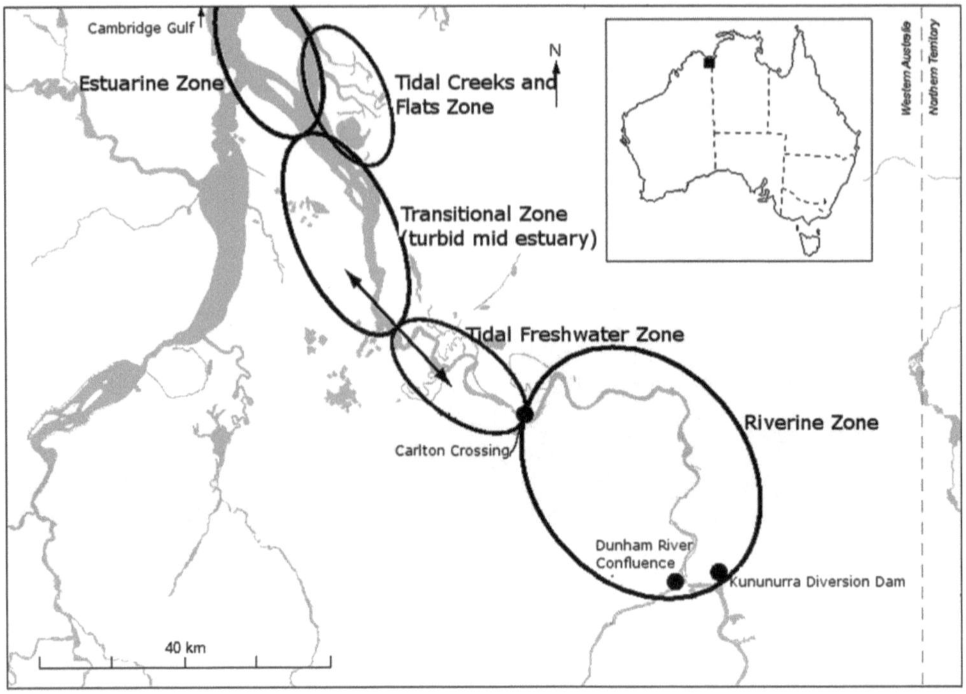

Fig. 4 Approximate extent of five ecological zones in the river-estuary system

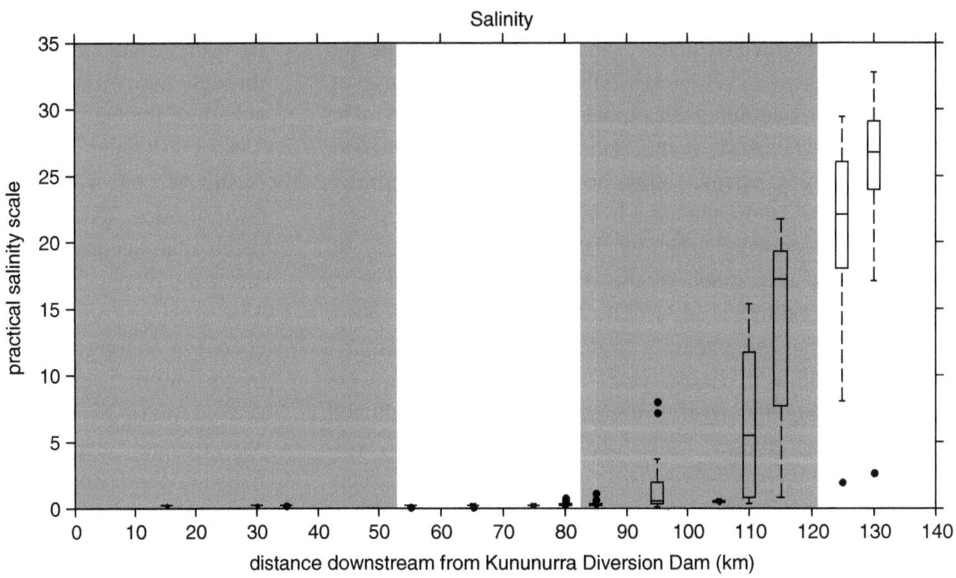

Fig. 5 Box and whisker plot of salinity, showing the mean, upper and lower quartile, and outliers of dry-season (June–December) monthly data, 2002–2006, grouped into 5 km bins. *Shaded* and *unshaded* regions are (from *left*): the Riverine Zone, the Tidal Freshwater Zone, the Transitional Zone and the Estuarine Zone. *Boxes* indicate the median, upper and lower quartiles of dry-season observations within each bin. *Bars* indicated 5 and 95 % intervals, with outliers represented as *dots*.

Figures 5, 6, 7, 8, 9, 10, 11 and 12 summarise the dry-season physical and chemical properties of the main channel of the Ord River Estuary from monthly measurements over a period of 4 years (2002–2006), illustrating how these properties change as a function of distance downstream and how they relate to the bio-physical zones shown in Fig. 4. Sampling procedures and chemical analyses for routine monitoring and more intensive studies of primary production, carbon sources and algal biomass have been described by Robson et al. (2008b) and Burford et al. (2011).

Ecosystem processes during flood conditions, particularly interactions between the river and its floodplains, are not well understood. During flood conditions, residence times in each Zone are greatly reduced, turbidity, suspended sediment and nutrient concentrations are elevated (as is evident in the sediment and nutrient budgets presented here) and benthic sediment material and mudflats are redistributed by strong downstream currents. As in other tropical rivers, little trapping of material occurs during this limited period of highly energetic flow, and short residence times mean that

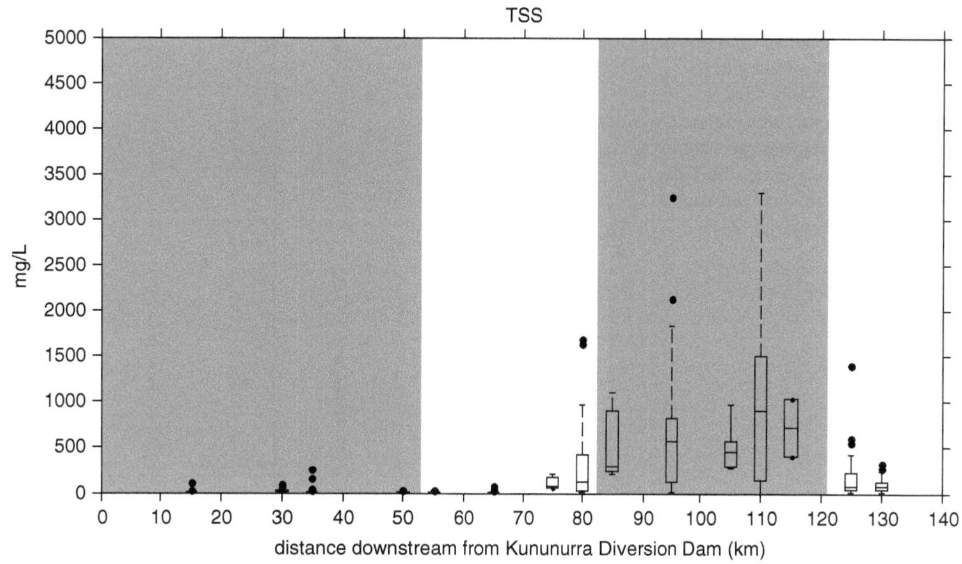

Fig. 6 Box and whisker plot of total suspended solids (TSS), showing the mean, upper and lower quartile, and outliers of dry-season (June–December) monthly data, 2002–2006, grouped into 5 km bins. *Shaded* and *unshaded* regions are (from *left*): the Riverine Zone, the Tidal Freshwater Zone, the Transitional Zone and the Estuarine Zone

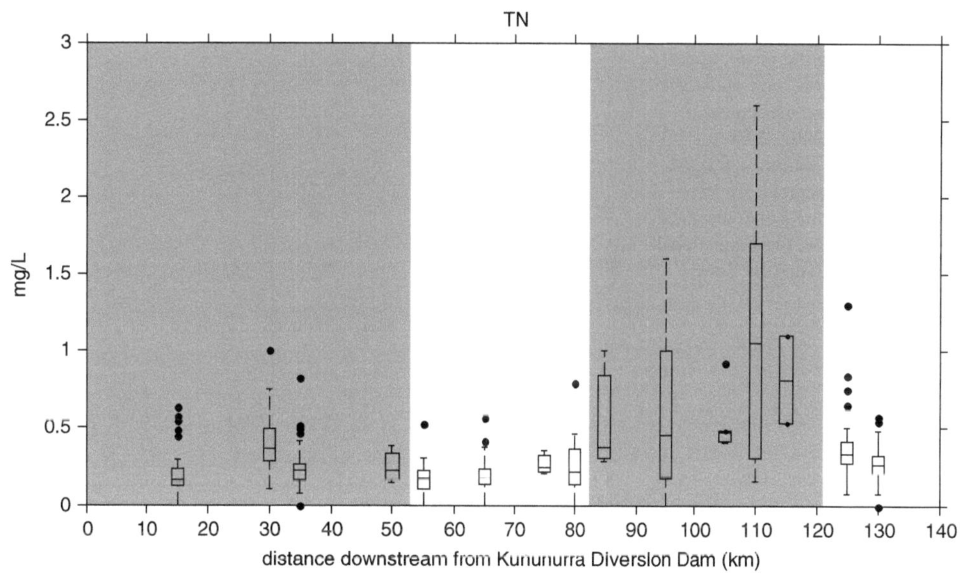

Fig. 7 Box and whisker plot of total nitrogen (TN), showing the mean, upper and lower quartile, and outliers of dry-season (June–December) monthly data, 2002–2006, grouped into 5 km bins. *Shaded* and *unshaded* regions are (from *left*): the Riverine Zone, the Tidal Freshwater Zone, the Transitional Zone and the Estuarine Zone

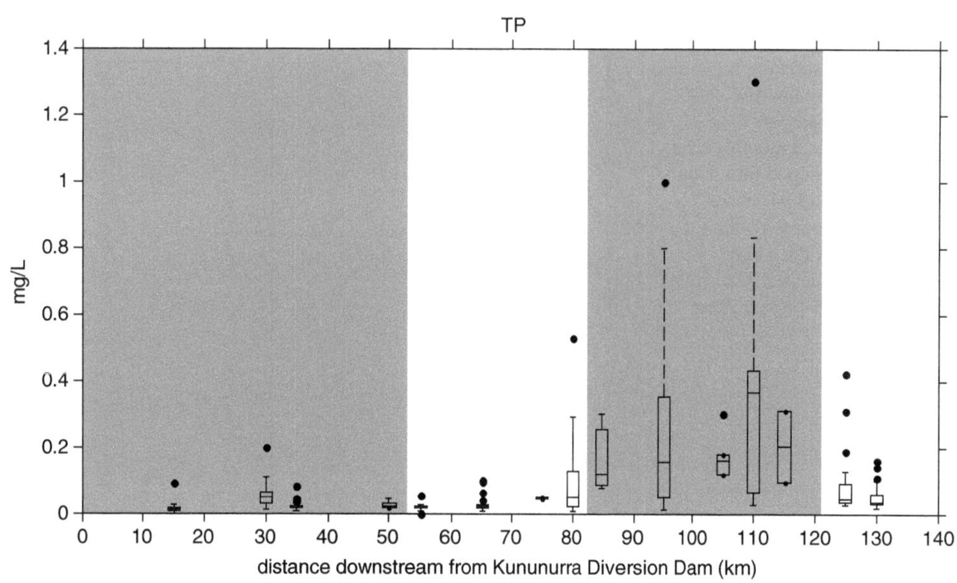

Fig. 8 Box and whisker plot of total phosphorus (TP), showing the mean, upper and lower quartile, and outliers of dry-season (June–December) monthly data, 2002–2006, grouped into 5 km bins. *Shaded* and *unshaded* regions are (from *left*): the Riverine Zone, the Tidal Freshwater Zone, the Transitional Zone and the Estuarine Zone

Fig. 9 Box and whisker plot of dissolved inorganic nitrogen (DIN), showing the mean, upper and lower quartile, and outliers of dry-season (June–December) monthly data, 2002–2006, grouped into 5 km bins. *Shaded* and *unshaded* regions are (from *left*): the Riverine Zone, the Tidal Freshwater Zone, the Transitional Zone and the Estuarine Zone

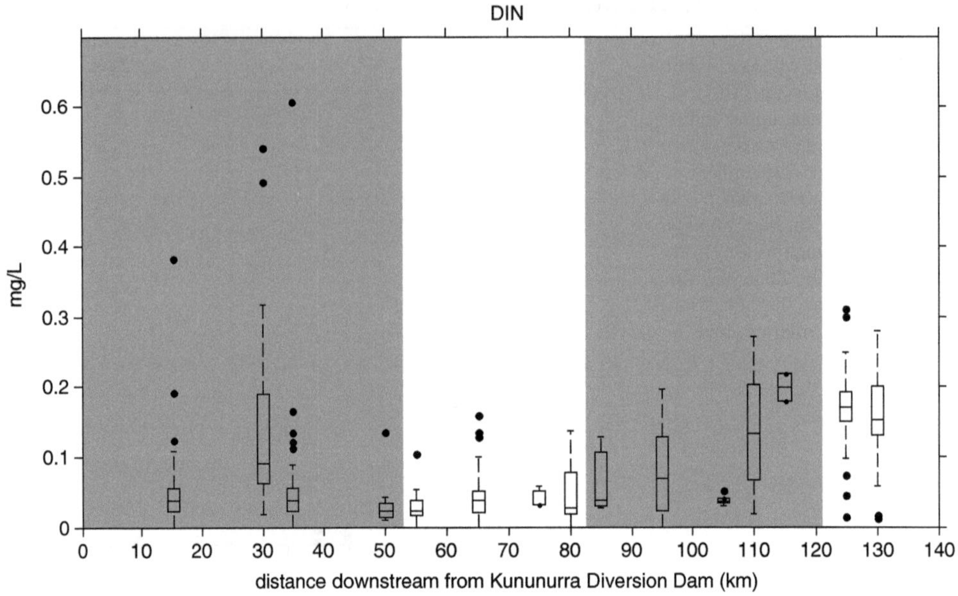

Fig. 10 Box and whisker plot of dissolved inorganic phosphorus (DIP), showing the mean, upper and lower quartile, and outliers of dry-season (June–December) monthly data, 2002–2006, grouped into 5 km bins. *Shaded* and *unshaded* regions are (from *left*): the Riverine Zone, the Tidal Freshwater Zone, the Transitional Zone and the Estuarine Zone

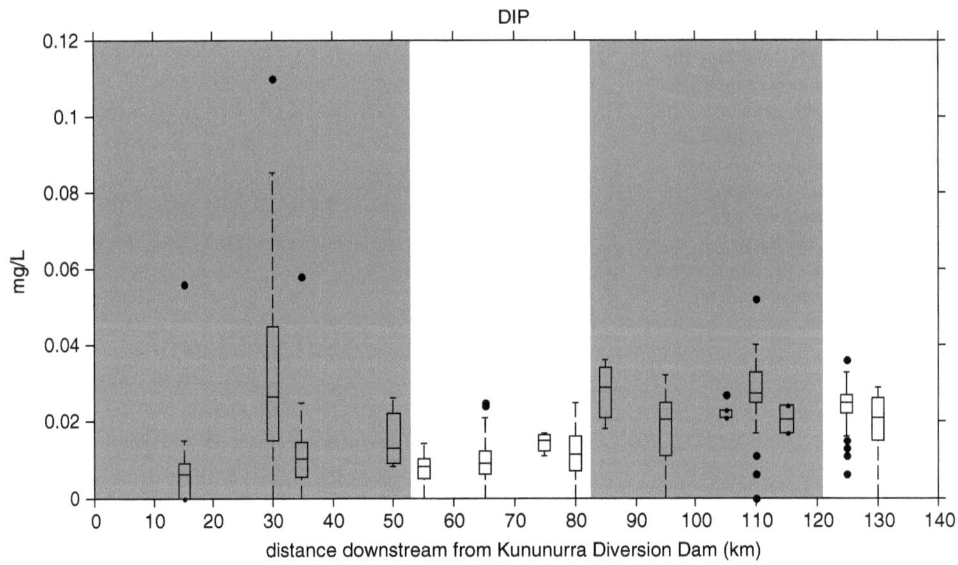

Fig. 11 Box and whisker plot of chlorophyll a, showing the mean, upper and lower quartile, and outliers of dry-season (June–December) monthly data, 2002–2006, grouped into 5 km bins. *Shaded* and *unshaded* regions are (from *left*): the Riverine Zone, the Tidal Freshwater Zone, the Transitional Zone and the Estuarine Zone

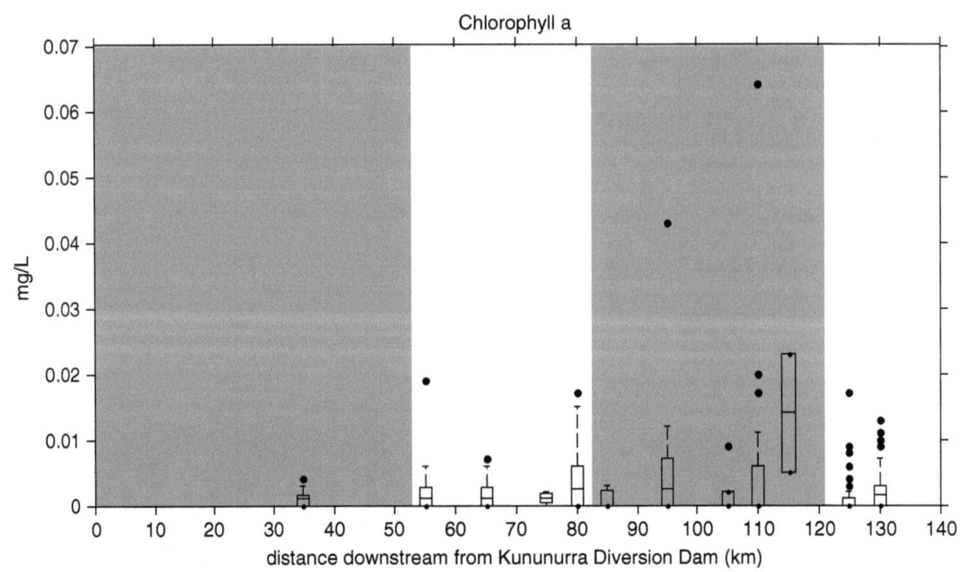

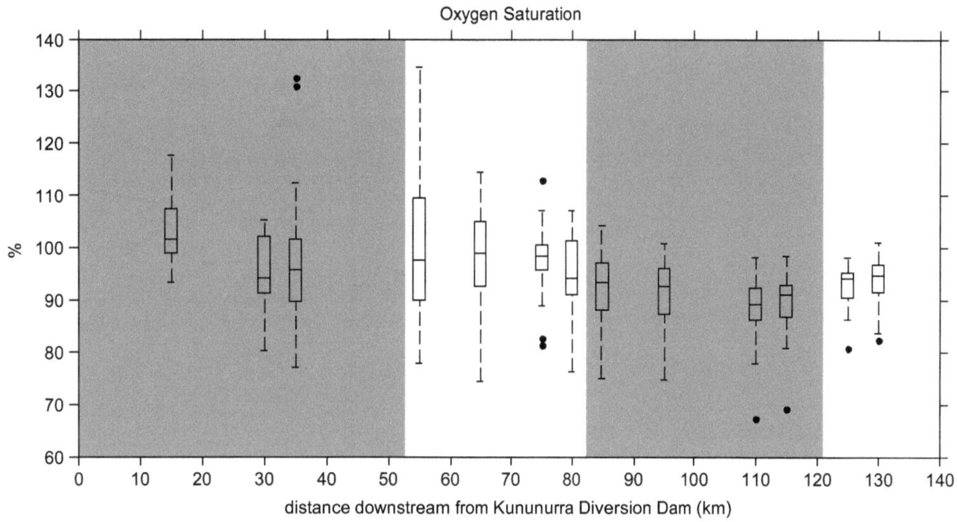

Fig. 12 Box and whisker plot of percent oxygen saturation, showing the mean, upper and lower quartile, and outliers of dry-season (June–December) monthly data, 2002–2006, grouped into 5 km bins. *Shaded* and *unshaded* regions are (from *left*): the Riverine Zone, the Tidal Freshwater Zone, the Transitional Zone and the Estuarine Zone

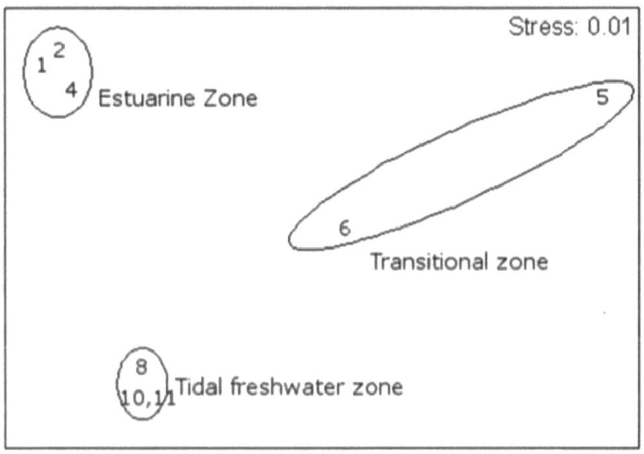

Fig. 13 MDS plot generated from nutrient (TN, TP, DON, DOP, NH3, NOX, FRP, Si, DOC) and chlorophyll a data from 4 years of routine monthly monitoring in the Ord River estuary, using a normalised Euclidean distance measure to compare across sites (site numbers as indicated in Fig. 1). Mantel's Test was performed to correlate the physical parameters, Secchi depth, turbidity and conductivity with the water quality data for each sampling site and time, and to test for relationships between multivariate patterns (BIOENV, RELATE). The Tidal Creeks and Flats Zone (which was not routinely monitored for logistical reasons) and the Riverine Zone (which is not strictly part of the estuary) were not included in the MDS due to insufficient data

biogeochemical processes such as denitrification in the river channel may have little impact (Brodie and Mitchell 2005).

Selected dry-season water quality data (Figs. 5, 6, 7 and 8), and multidimensional scaling (MDS) ordination of monitoring data serves further to confirm a clear separation in terms of water quality between zones (Fig. 13).

The Riverine Zone

Immediately downstream of Kununurra Diversion Dam, the river is characterised by continuously flowing, low-turbidity (~10 NTU) fresh water during the dry season. During wet season flow events, turbidity is much higher due to loads of suspended sediments from the catchment and material scoured from the river bed (Fig. 14).

This zone contains a number of rock bars, shallow riffles and gravel beds, interspersed with relatively deep, long pool sections. Bedrock forms prominent structural features near rock bars, especially upstream of Tarrara Bar (Fig. 1). Dominant riparian habitats for aquatic species include *Melaleuca* trees, freshwater mangrove *Barringtonia* overhangs and snags, *Pandanus aquaticus* clumps, overhangs and snags, *Phragmites* stands, water couch and other littoral grasses, and occasional *Typha* stands. Beds of the macrophyte *Vallisneria* become established in slow-flowing areas during low-flow periods, but are commonly scoured away during high flows. In standing backwaters protected from flow, beds of the macrophyte, *Myriophyllum* are more common. The substratum is commonly sand, gravel and cobbles.

Recent studies have recorded 99 species of invertebrates in this zone (Water and Rivers Commission 2003; Mayes et al. 2005; Bright 2005), and at least 28 species of fish (Morgan et al. 2004; Water and Rivers Commission 2003). Seven of the thirteen most abundant fish species sampled in this zone attained larger sizes than in nearby unregulated rivers (Water and Rivers Commission 2003).

Other aquatic vertebrates recorded in this zone include both freshwater and estuarine crocodiles, several species of freshwater turtles, and two species of water monitors. Approximately 27 species of fish-eating birds have been recorded within the Ord region, although there does not appear to have been a definitive survey of birds in the region.

During the dry season, total suspended solids, total nitrogen and total phosphorus concentrations in the Riverine

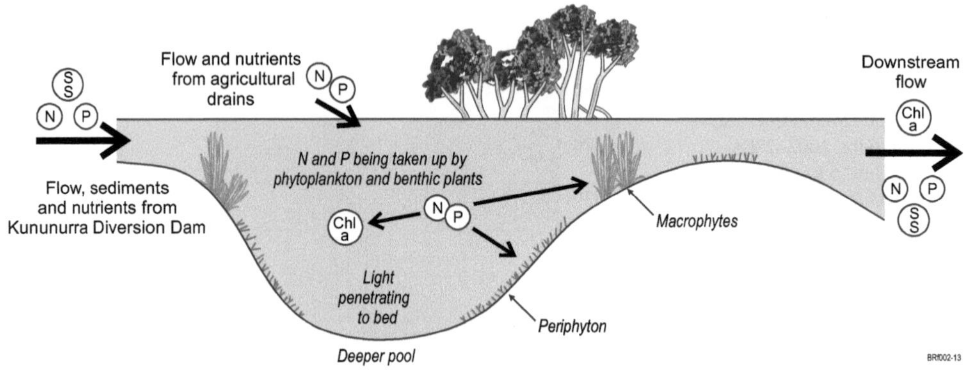

Fig. 14 The Riverine Zone is characterised by clear, fresh water with unidirectional flow and high primary and secondary production

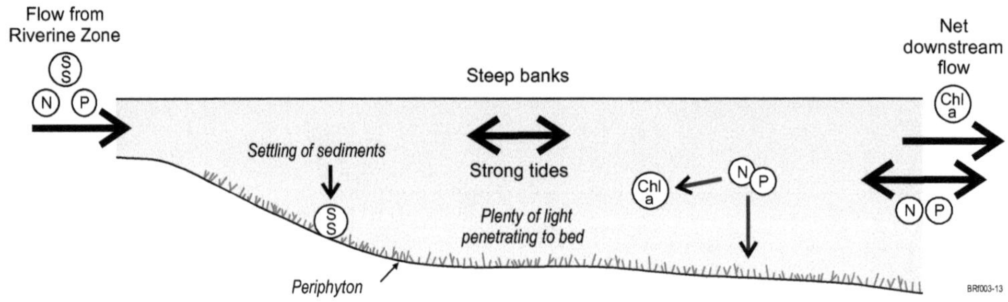

Fig. 15 The Tidal Freshwater Zone has low turbidity and high benthic primary production

Zone during our study were low in comparison with concentrations in many rivers with more heavily developed catchments (Figs. 6, 7 and 8), but dissolved inorganic nutrient concentrations were moderate (Figs. 9 and 10).

The downstream limit of the Riverine Zone is defined by a hydraulic break formed by a rock bar at Carlton Crossing (location shown in Fig. 4) that prevents the incursion of tidal influences upstream.

Dissolved inorganic nitrogen and phosphorus concentrations in water leaving the Riverine Zone during our study (2004–2008) were significantly lower than concentrations in the middle of the Riverine Zone, and we hypothesise that this represents a loss of nutrients from the water column due to uptake of inorganic nutrients by primary producers. Primary production in the Riverine Zone appears to be dominated by benthic algae (with some macrophytes) and limited by the supply of biologically available phosphorus (Burford et al. 2011). Although few chlorophyll measurements are available for this zone, at one site Volkman et al. (2007) measured lower concentrations in the wet season (0.4 μg L^{-1}) compared to dry season samples (1 μg L^{-1}),with samples dominated by diatoms. During our study, oxygen concentrations were around saturation (Fig. 12) and diurnal oxygen curves indicate that this is an area of net autochthonous production. Sediments in this region are relatively low in organic matter content and dominated by terrestrial inputs (Volkman et al. 2007).

Before regulation, flows in the Riverine Zone of the lower Ord River ceased during the dry season and the river functioned as a series of disconnected pools, as unregulated rivers in the area and tributaries of the lower Ord River still do (Trayler et al. 2006). Increased flow volume and stability of low flows appear to have promoted greater habitat diversity and availability during the dry season (Marshall and Storey 2005), greater immigration of predators, and increased net production.

The Tidal Freshwater Zone

Immediately downstream of the Riverine Zone is the first section of the estuary: a freshwater zone that is subject to tidal variations in water levels but is otherwise similar to the Riverine Zone. The extent of fresh water in the estuary varies seasonally according to flow. During the wet season, fresh water has been observed as far downstream as site 5 (Fig. 1) (Parslow et al. 2003), but during the dry season, with regulated flows typically maintained at 50–80 m^3, the extent of this zone is roughly as indicated in Fig. 3. As with the Riverine Zone upstream, turbidity in this zone is <10 NTU during the dry season, but much higher (up to 500 NTU) during the wet season (Fig. 15).

The Tidal Freshwater Zone is characterised by a channel largely confined within banks, with well-developed meanders and secondary high-flow channels. In the upper reaches of this zone, sand banks have been deposited on the inside banks of bends. During high flows, water cuts across these banks creating secondary channels that dry during low flows. Stabilisation of the islands created in this way has created island anabranching in the upstream reaches of this zone. The ridge anabranching that was common in the freshwater meandering zone disappears completely within the first 10 km of tidal influence. The lower two-thirds of this zone exhibits a high frequency of actively eroding mud banks that may be either near-vertical and unvegetated, or slumping, with clumps of grasses binding slumped sections together. At higher bank elevations below the bank-full level, channel benches are evident (Thoms et al. 2004). Bedrock outcrops are absent throughout, except in the immediate vicinity of The Rocks. In the lower two-thirds the substratum becomes predominantly muddy sands, with extensive sand banks becoming exposed at low tide.

Since regulation, this zone has been subject to a progressive sedimentation and narrowing of the low-flow channel (Wolanski et al. 2001) and gradual widening of the channel during high flows that cannot be contained within the narrowed channel. This process appears to have contributed to increased lateral bank erosion and further shallowing of the channel downstream (Gehrke 2009).

Prior to regulation, the Tidal Freshwater Zone probably remained connected throughout the year (unlike the Riverine Zone), but would have been subject to salt water intrusion during the dry season.

Based on the nutrient concentration data from our study, it appears that nutrients reaching this zone have already been subject to biological processing in the Riverine Zone, so concentrations of bioavailable dissolved nitrogen and phosphorus concentrations in the Tidal Freshwater Zone are somewhat lower than they are upstream (Figs. 9 and 10). In other river systems, the magnitude of nutrient depletion will depend on the residence time of water in the Riverine Zone and the existence of additional local nutrient sources in the Tidal Freshwater Zone.

Using a mass balance model to assess nutrient sources and sinks, we estimate 53 tonnes of DIN and 7 tonnes of DIP was removed from the water column in the Riverine and Tidal Freshwater Zones (combined) over the course of the dry season (Fig. 20). Of this, approximately 20 tonnes of nitrogen and 3 tonnes of phosphorus is converted to phytoplankton (calculated on the basis of the observed chlorophyll concentrations (Fig. 11) and assuming a 106:16:1 (i.e. Reynolds) molar C:N:P ratio in phytoplankton). Unfortunately, measurement of benthic algal biomass was not considered in the study design, so we have no direct estimate of benthic algal or macrophyte production or nutrient uptake. From these figures, however, we may speculate that there is a net benthic uptake of up to 4 tonnes of dissolved inorganic phosphorus and 29 tonnes of dissolved inorganic nitrogen (assuming C:N:P = 106:16:1) in the freshwater part of the lower Ord River. Analysis of diurnal oxygen curves suggests that the contribution of benthic production to total photosynthesis in the tidal freshwater zone is greater than these estimates, with phytoplankton contributing less than 10 % to total photosynthesis (Burford et al. 2011). It is possible that, as in the Daly River (Webster et al. 2005), much of the observed benthic photosynthesis occurs without significant nutrient uptake or biomass production. Instead, excess photosynthesis may reduce algal cellular C:N:P ratios and produce low-molecular weight extracellular organic matter (Webster et al. 2005).

Based on measured Secchi depths and the river bathymetry used in hydraulic modelling (Robson et al. 2008b), the illuminated area of channel bed in the Riverine and Tidal Freshwater zones combined under typical dry-season conditions was approximately $4.5 \times 10^6 \text{ m}^2$. Hence, this postulated 4 tonnes DIP uptake over the course of the dry season by benthic plants and algae amounts to an uptake of approximately 3.6 mg P m^{-2} day^{-1}, which would require a benthic photosynthetic rate of at least 12 mmol O_2 m^{-2} day^{-1}.

Presumably as a result of the unstable bed structure, aquatic macrophytes are uncommon in this zone. Benthic algae thus appear to dominate primary production in the Freshwater Zones, and primary production in the two freshwater zones seems to dominate the autochthonous productivity of the river-estuary system as a whole – this is not inconsistent with the River Continuum Concept (Vannote et al. 1980). Volkman et al (2007) reported variable chlorophyll values for this section of the estuary with dry season values ranging from 1 to 2.5 μg L^{-1} and wet season values 0.4–1 μg L^{-1}. Species composition was also highly variable, with diatoms being consistently important but at one site flagellates and chlorophytes were abundant. Sedimentary organic matter in this region begins to show signs of a change from terrestrial inputs dominating to a greater proportion of algal material and while organic carbon concentrations are still low they are greater in the dry season, reflecting the increased input from algal material (Volkman et al 2007).

As is often true, though the ecology of the Riverine Zone has been the subject of several studies, the ecology of the estuary itself has not been intensively studied (Gehrke 2009), though Kay (2004) reports a study of estuarine crocodiles in this zone. The list of fish and other faunal species observed or expected to occur in this zone is very similar to that in the Riverine Zone upstream, though we anticipate that many freshwater species may be less common in the tidal zone.

The Transitional (Maximum Turbidity) Zone

Downstream of the Tidal Freshwater Zone, the Transitional Zone is characterised by a strong lengthwise salinity gradient (Fig. 5), strong semi-diurnal tidal currents and high turbidity (Fig. 16).

Measured tidal velocities exceeded 1 ms^{-1} at mid-tide, with measured tidal ranges of 5–7 m (our measurements). The tidal excursion (distance over which water travels in each direction on a tidal cycle) at spring tide exceeds 20 km (Robson and Webster submitted).

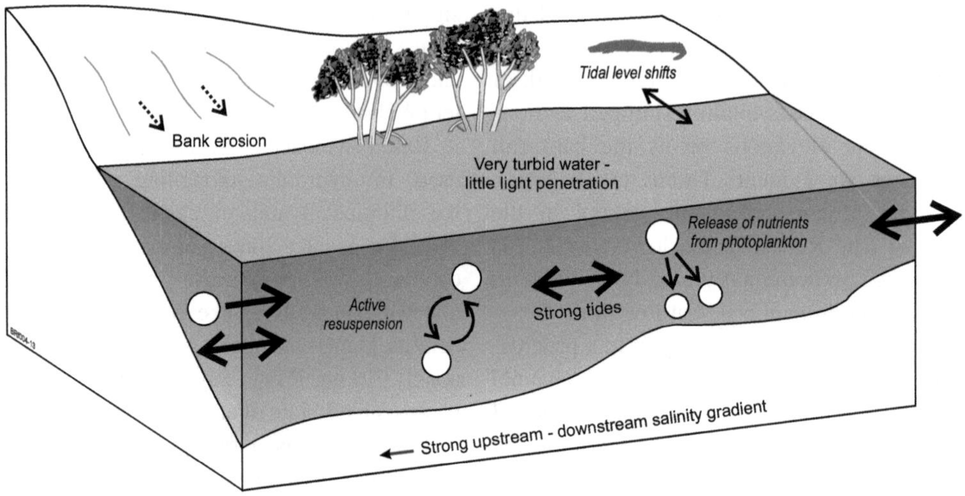

Fig. 16 The Transitional Zone is a high-energy, highly turbid zone with strong a salinity gradient and low primary production

A tidal bore has been evident on occasion, especially during spring tides, when the flood tide moving upstream creates a series of standing waves over shallow sections of the channel (Wolanski et al. 2001).

This high-energy zone is characterised by extensive bank erosion and the presence of a number of low-relief islands formed from sediment deposits that have been consolidated by vegetation. Accordingly, this zone contains a number of mud-braided and anabranching reaches as recognised by Erskine et al. (2005). The material in these islands is continually being reworked by tidal and river flows. Islands in this zone are dynamic and several have been documented to appear or be removed by scouring over several years (e.g. Thom et al. 1975).

This Zone is also characterised by very high and tidally variable turbidity (often >1,000 NTU, the upper limit of measurement of our Hydrolab instruments) and similarly high and variable total suspended solids concentrations (Fig. 6). High concentrations of suspended sediments were maintained by flocculation of colloidal sediment material flowing in from upstream and by tidal pumping of material from downstream (Wolanski et al. 2001), which is kept in suspension by strong tidal currents. Concentrations of particulate nutrients associated with these sediments were high (Figs. 7 and 8), while dissolved inorganic nutrient concentrations (DIN and DIP, Figs. 9 and 10) were also higher than values in the neighbouring zones upstream and downstream. This increase in DIN and DIP in comparison with the Tidal Freshwater Zone is consistent with the senescence and subsequent breakdown of freshwater phytoplankton and benthic algal material that reach this Zone. The strong salinity gradient, rapid salinity changes and low light environment all inhibit primary production.

As in our study, Volkman et al (2007) reported similar chlorophyll concentrations at sites in this zone to those in the tidally influenced freshwater zone, which probably reflects phytoplankton being transported downstream. However, there was no evidence for chlorophyll degradation products, which would be considered unusual in this type of environment unless there were process causing rapid subsequent degradation of those compounds. Diatoms remained the dominant species, with an increasing contribution of cyanobacteria (Volkman et al 2007). Dry season sediment organic matter was dominated by inputs from mangroves, though this was overprinted in the wet season with terrestrially derived material. A contribution from cyanobacteria was also evident during both wet and dry sampling periods (Volkman et al. 2007).

Oxygen concentrations in the Transitional Zone are depressed below saturation (Fig. 12), which we suggest reflects bacterial breakdown of organic material.

Mangroves in this zone have been described in some detail by Thom et al (1975). The mangrove forests are dominated by *Avicenia marina*, with some *Exocoecaria agallocha, Aegialitis annulata* (rare) and *Aegiceras coriculatum* (rare). Other vegetation present along the banks includes *Rhizophora stylosa,* samphires *Sesuvium portulacastrum* and *Suaeda arbusuloides*, and the grass *Sporobolus virginicus*. Steep eroding banks at the river's edge support tall *Avicennia*, mixed with *Ceriops tagal, Sporobolus* and the halophyte *Arthrocnemum* sp.

With the exceptions of Thom et al (1975, mangroves), Loneragan et al. (2002, prawns) and Kay (2004, crocodiles), there have been no comprehensive ecological surveys in this zone. As this zone typically displays estuarine characteristics of tidal flow, fluctuating salinities, and high turbidity, typical freshwater species are generally absent except for individuals carried downstream during high flows and migratory life history stages.

Characteristic species in this zone include estuarine crocodiles, barramundi, bull sharks, popeye mullet and diamond mullet, and mudskippers. Because of the variable

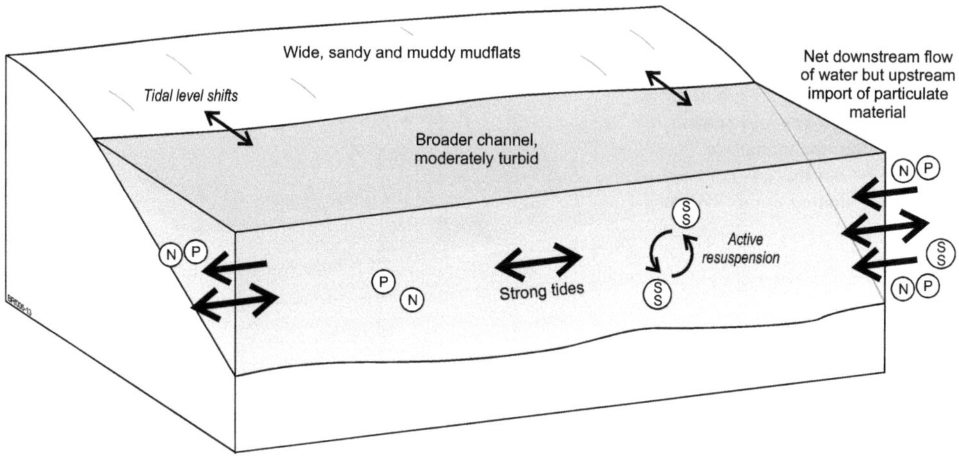

Fig. 17 The Estuary Mouth is a zone of moderate turbidity and tidal energy as the channel broadens and salinity approaches that of seawater. Our understanding of the fauna of this Zone is largely speculative

salinity, species such as commercially important penaeid prawn species are generally absent, whilst small crab species and their burrows are not conspicuous in the vicinity of mangrove stands. The extremely mobile sediments in this zone make it difficult for sessile and burrowing organisms to become established in large numbers.

Juvenile banana prawns (*Penaeus indicus* and *P. merguiensis*) commonly use tidal rivers and estuaries as nursery habitats elsewhere in Cambridge Gulf and the Joseph Bonaparte Gulf region, but the salinities in this zone of the Ord River are too variable and frequently too low to regularly support banana prawn recruitment (Kenyon et al. 2004). Salinities less than 10 ‰ are lethal to *P. indicus* postlarvae and juveniles (Kumlu and Jones 1995). It is believed that the elevated freshwater inflows during the dry season, when salinity would historically have been higher, have contributed to the loss of this zone as a nursery habitat for banana prawns.

The Estuary Mouth Zone

Downstream of the Transitional Zone, the estuary widens out to a broad delta between 2 and 5 km across. In contrast with the steep banks that border the Transitional Zone, wide mudflats line the edge of the channel in the Estuary Mouth Zone. Tidal currents are strong in this zone, but less narrowly confined than in the Transitional Zone. Salinity approaches that of seawater during the dry season (Fig. 5) but falls to less than 4 at times during the wet season. After flood flows recede, it takes several months for tidal currents to re-establish salinities approaching marine values throughout the zone (Fig. 17).

In our measurements, turbidity and particulate concentrations were lower here, though not as low as in the Riverine and Tidal Freshwater Zones (turbidity ~100 NTU, but varying from 0 to >400 NTU depending on the tidal phase). In addition to temporal variation in turbidity over tidal cycles, turbidity is also spatially variable according to the sediment plumes generated by currents as they pass over muddy sediments.

Except for near the edges, the habitat is dominated by sandy and muddy sediment banks of varying depths with no macrophyte or significant algal development. Toward the shore, habitats include a muddy intertidal zone beneath a mangrove zone interspersed with crab burrows and accumulations of mangrove litter and fallen trees.

It was hypothesised (Parslow et al. 2003) that benthic microalgal production on the mudflats might be important in this Zone. Attached microalgae have been found to account for up to 90 % of total primary production in other macrotidal estuaries (Phinney et al. 2004). However, we found no evidence to support this hypothesis for the Ord estuary, with relatively low intertidal mudflat chlorophyll concentrations in the Transition Zone (Burford et al. 2011). Fluctuations in salinity and high mudflat temperatures in summer make this an inhospitable environment for many species, while fluxes of sediment due to tidal action are likely to inhibit algal biomass accumulation. It is also possible that intensive grazing by benthic and intertidal fauna such as crustaceans, worms and mudskippers contributes to the low observed chlorophyll concentrations on mudflats.

Mangroves are the dominant form of riparian vegetation in this zone, forming a narrow band along the upper intertidal zone and islands in the lower reaches of this zone.

Juvenile banana prawns have been recorded in low numbers in the lower estuarine zone, where salinities above ten are common during the dry season (Kenyon et al. 2004). In addition to salinity, Kenyon et al. (2004) suggested that banana prawn numbers might be low because of either poor larval supply, or because of the high turbidity. Considering the supply of banana prawn larvae to other regions in Cambridge Gulf, and the effective upstream tidal transport within the Ord estuary, it would be unusual for such a large area to have a poor supply of larvae.

The large expanses of relatively homogeneous bottom sediments in the open estuary provide limited variety of habitat

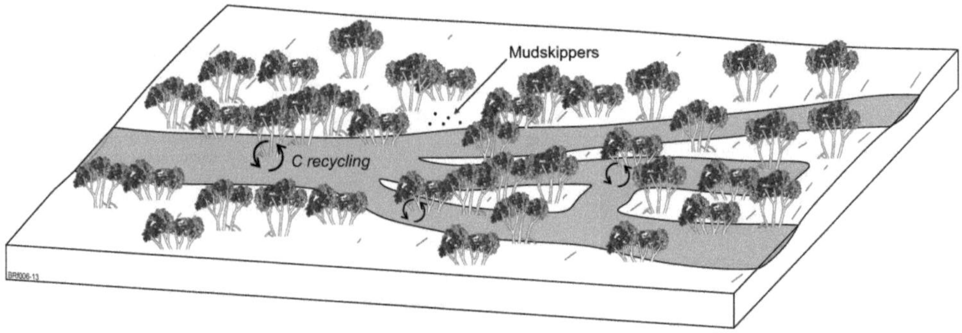

Fig. 18 Hydrodynamically complex, the Tidal Creeks and Flats Zone is home to a rich mangrove ecosystem. Carbon fixed by mangroves may remain largely within the mangrove community, but this has not been intensively studied in this system

for estuarine fauna. Epifauna and infauna have not been sampled, and little is known about fish in open waters. Schooling species such as mullets and herring, and species migrating between the estuary and freshwaters must pass through this zone, but their distribution between open waters and the more protected banks and fringing mangroves is unknown.

The Tidal Creeks and Flats Zone

A system of tidal creeks and mudflats extends to the east of the estuarine zone, supporting mangrove habitat. The extent of this zone varies considerably from one river system to the next, but tidal creeks in this type of estuary generally provide a substantial area of intertidal mudflat that exceeds the mudflat area of the main channel (Fig. 18).

Numerous gutters and narrow creeks penetrate the mangrove forest, creating a complex network of branching channels with complex hydrodynamics. Some channels are heavily shaded and impassable to boats, while other sections have open canopies, mangrove root masses, and open high-tidal flats beyond the mangrove forest, with relatively steeply sloping intertidal banks at the water's edge. Occasional submerged rock bars and rock rubble patches occur along the main creek channel. These habitats combine to provide the most diverse and complex habitat template in the entire marine catchment basin. Much less bank erosion is evident in this area than along the main estuarine channel.

Except for periods of heavy local rainfall which result in intense freshwater runoff, water quality in these mangrove creeks is largely influenced by the quality of water entering from the main estuary, and the settling and resuspension of particles by tidal flow. Observations during February 2007 suggest that during the low flow season these creeks provide a depositional and filtering environment that acts as a sink for materials entering from the main estuary.

No comprehensive ecological survey has been completed in this zone. Mangrove species are similar to those described for the mangrove fringe in the Estuary Mouth Zone, including *Sonneratia alba*, *Avicennia marina*, *Aegiceras corniculatum*, *Bruguiera parviflora*, *Rhizophora stylosa*, *Ceriops tagal* and *Aegialitis annulata*. *Sporobolus virginicus* grassland and samphire grow on the tidal mudflats behind the mangroves. Other species of mangrove recorded occasionally include *Xylocarpus moluccensis*, *Excoecaria agallocha* and *Camptostemon schultzii* (Department of Conservation and Land Management 2003). No seagrasses or macroalgae were observed during February 2007. Intertidal banks are relatively steep, offering limited habitat area for intertidal microalgal production.

In terms of its biogeochemistry and productivity, the Tidal Creeks and Flats Zone has otherwise not yet been studied in detail in the Ord River, but has been studied in other tropical river systems. Volkman et al. (2007) studied sources of organic matter in the Ord River Estuary and found evidence of substantial contributions to sediment organic matter in this Zone from mangroves. The rich organic material supply in mangrove systems may support a high biomass of meiofauna, though the evidence is equivocal, possibly due to the low nutritional quality of mangrove biomass (Coull 2009). Meiofaunal biomass has not been measured or estimated in the Ord River Estuary.

Several species of fiddler crabs (*Uca* spp.) and grapsid crabs inhabit the intertidal banks in this zone, reaching high levels of abundance. Elsewhere, fiddler crabs provide a number of ecosystem functions (Lee 1998). Their dense burrows allow water exchange and oxygenation of the sediments, and mobilise buried organic material and nutrients, promoting the development of a rich mangrove meiofauna. These crabs feed on organic matter deposits on mud banks exposed at low tide, including meiofauna, microphytobenthos and mangrove detritus (Guest and Connolly 2004). After mating, female crabs release their larvae into the water column at night during spring tides, when the larvae are consumed in large quantities by zooplanktivorous fish (Robertson et al. 1988; Sheaves and Molony 2000). Small mangrove crabs may also contribute substantially to the diet of fish in mangrove systems, though the evidence for this is not strong (Lee 2009). Birds such as herons and egrets also appear to consume large numbers of mangrove crabs.

Like crabs, mudskippers feed on surface deposits on the intertidal banks. Other species observed during February

2007 include garfish (*Arrhamphus sclerolepis*), longtom (*Strongylura* sp.). spotted scat (*Scatophagus argus*), seven-spotted archerfish (*Toxotes chatareus*), several species of mullet, and herring. Large unidentified predacious fish were observed chasing mullet approximately 30–40 cm long, which jumped out of the water onto the bank to evade capture.

Crocodiles densities have not been quantified in this region, but marker buoys attached to equipment deployed overnight bore several large tooth indentations when collected the following morning, confirming the presence of large crocodiles. Raptors are also anecdotally observed.

Odum and Heald (1972) suggested that mangroves and mangrove detritus may play a dominant role in the production of tropical rivers, and this view has been widely accepted. More recent work, however, has found equivocal evidence for this claim. Though Manson et al. (2005) found evidence for correlation between the extent of mangroves associated with estuaries in north-eastern Australia and catch per unit effort in several key fisheries, Sheaves et al. (2012) more recently found no such relationship in the case of banana prawn fishery productivity. Burford et al. (2008), studying the Darwin Harbour, Alongi (2001), studying the nearby King Sound, and Ford et al. (2005), studying the Fitzroy Estuary (Queensland) all found that much of the carbon production by mangrove forests remained within the forest or was respired, providing little subsidy to production in the tidal creeks or open estuary channel. Water column productivity in the tidal creeks of these systems was driven by algal production rather than mangrove detritus. We may speculate that the same is true in the Ord River Estuary: though important habitats in their own right, mangrove forests may not be essential to the productivity of the estuary itself.

Interactions Between the Five Zones

We have described five physically and biogeochemically distinct functional zones in the lower Ord River and Estuary. Obviously, these five zones do not function independently. The downstream zones rely on water, carbon and nutrients delivered from upstream, and are therefore affected by biogeochemical processing of these materials in each upstream zone. The location of the salinity gradient and hence that of the Transitional (Maximum Turbidity) Zone is a function of flow and varies in response to changes in flow over time (Parslow et al. 2003).

Less obviously, processes in downstream zones can also affect conditions upstream, through tidal pumping in the estuary, as well as through the upstream migration of fish and other wildlife.

Conceptually, then, we consider each of the functional zones in terms of physical and biogeochemical function.

Figures 19 and 20 summarise wet-season and dry-season budgets for flow, sediments, total nitrogen (TN), total phosphorus (TP), dissolved inorganic nitrogen (DIN) and dissolved inorganic phosphorus (DIP) for the lower Ord River and estuary.

Our method of calculation of materials budgets is described in the appendix to this chapter. These budgets are presented with high uncertainty due to the limited accuracy of nutrient load estimates, but can be considered indicative of wet- and dry-season processes in the system.

Freshwater Inflows

During the wet season, heavy loads of suspended sediments (90 kT in our budget, calculated for a year of median rainfall) and nutrients (472 tonnes of TN and 76 tonnes of TP) and are washed into the lower Ord River. These come primarily from the catchment downstream of the large Ord River Dam, with the greatest contributions coming from the area between the two dams (~1,000 km^2) and Dunham River catchment (~4,200 km^2). Releases from the outlet works of the Ord River Dam and spillage from Lake Argyle are occasionally turbid and do contribute some sediments and nutrients to the lower Ord catchment, however Lake Argyle traps the great majority of sediments and nutrients from upstream (Magilligan et al. 2006). Although not quantified, it is clear that sediment and nutrient inputs from the upstream catchment are now much lower than historical loads (prior to construction of the Ord River Dam), which has led to infilling of Lake Argyle and of the channel upstream of the dam (Dixon and Palmer 2010).

When the Dunham River is in flow, it contributes a smaller flow volume but accounts for a greater proportion of total nitrogen, phosphorus and total suspended solids loads than flows passing over Kununurra Diversion Dam (KDD). The Dunham River ceases to flow entirely during the dry season (Fig. 20), as did the Ord River itself prior to regulation.

Release of water from the Ord River Dam and hence flows through Kununurra Diversion Dam continues throughout the dry season to support hydroelectric production.

Return flow from irrigation drains represents a very small component of total flows, but accounts for a disproportionately large fraction of nutrient inputs, including approximately a third of the dissolved inorganic nitrogen and phosphorus entering the lower river during the dry season. Return irrigation flows are not an important contributor to nutrient loads during the wet season, when strong flows from Dunham River are recorded.

Processing of Sediments and Nutrients in the Freshwater Zones

During the wet season, residence time in the Riverine and Tidal Freshwater Zones is low (<1 day while Lake Argyle

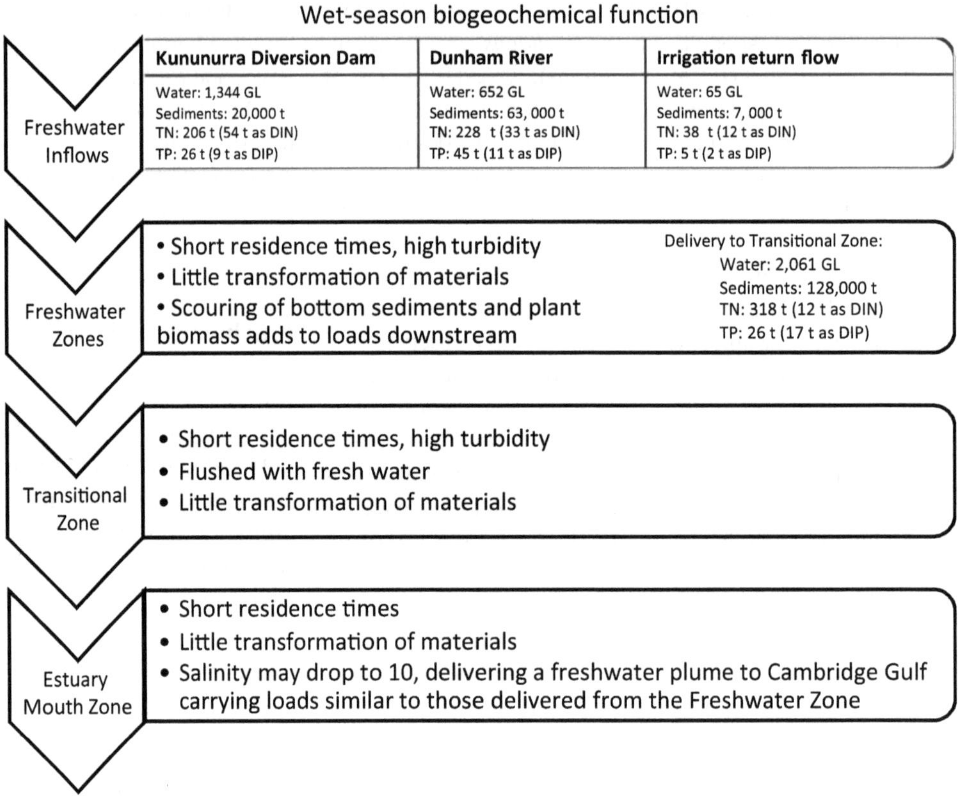

Fig. 19 Estimated wet-season inputs and outputs of water, sediments and nutrients in lower Ord River during a typical (median flow) year. Values are as calculated for the period from February to May, inclusive, using modelled flows past Kununurra Diversion Dam for a year of median rainfall (based on records over the last century). Estimated flow volumes and associated wet-season nutrient and sediment loads for an unusually wet year are three to five times those shown here

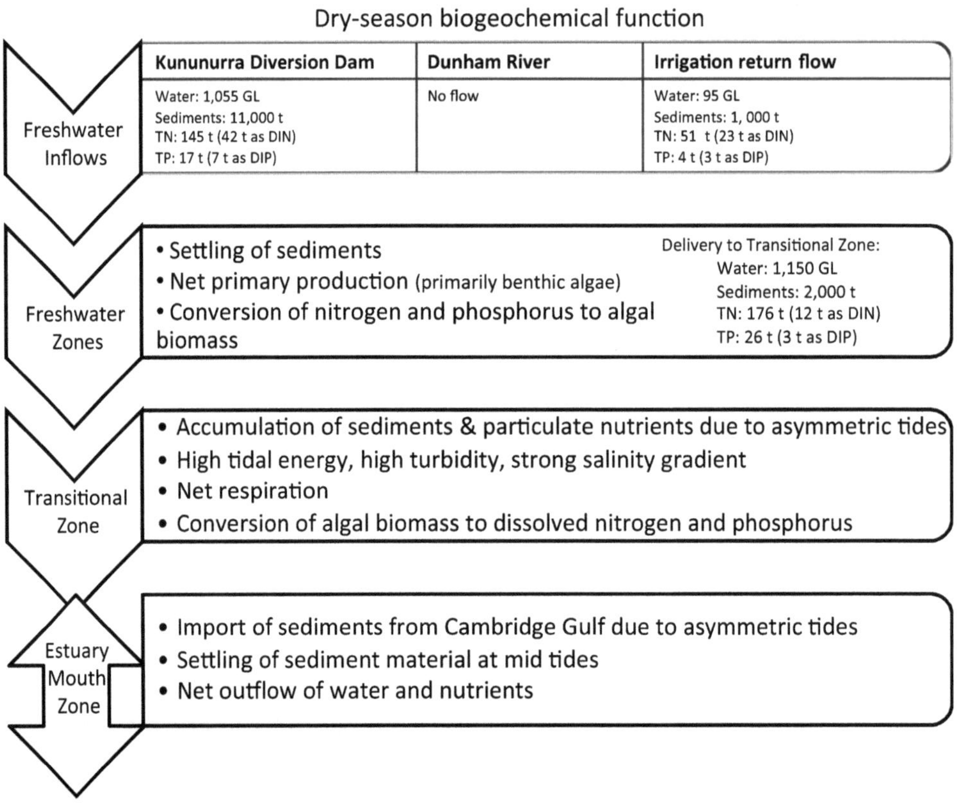

Fig. 20 Estimated dry season inputs and outputs of water, sediments and nutrients in lower Ord River during a typical (median-flow) year. Values are as calculated for the low-flow period (i.e. from June to January), inclusive, using modelled flows past Kununurra Diversion Dam for a year of median rainfall (based on records over the last century)

was spilling) and we infer, from the results of our nutrient budget, that little biogeochemical transformation took place in the freshwater zones during this time. The sediment bed in the Riverine Zone alternates between rocky and sandy sections, providing little fine material that might be resuspended by high flows. In the Tidal Freshwater Zone, by contrast, scouring of the bed adds to the sediment load delivered downstream.

During the dry season, the residence time of water in the Riverine and Tidal Freshwater Zones is three to 5 days.[1] Slower flows allow settling of sediment material and uptake of dissolved nitrogen and phosphorus by primary producers in both the Riverine and Tidal Freshwater Zones. Although the total nitrogen and phosphorus loads reaching the Transitional Zone are similar to the nitrogen and phosphorus loads entering the Riverine Zone, only 18 % of dissolved inorganic nitrogen and 30 % of dissolved inorganic phosphorus loads from tributaries reaches the Transitional Zone. The remainder is converted into macrophyte and algal biomass within the Riverine Zone, and this is reflected in higher chlorophyll *a* concentrations in the Transitional Zone (Fig. 11).

Transformations in the Transitional and Estuary Mouth Zones

During the wet season, heavy loads of suspended sediments and particulate nutrients are washed through the estuary and into the Cambridge Gulf. During the dry season, strong asymmetric tides pump sediments upstream from the Estuary Mouth to the Transitional Zone (Wolanski et al. 2001), creating a counter-intuitive net influx of particulate (and hence total) material from the seaward boundary into the estuary.

The high TSS, TN and TP concentrations observed in the Transitional Zone are maintained by a combination of this tidal pumping effect and active resuspension of benthic particulate matter by tidal currents.

Dissolved inorganic nitrogen and phosphorus in this zone are probably released from decaying algal material (based on the observed lower oxygen concentrations in this zone, Fig. 12). The high turbidity, increased turbulence and strong salinity gradient probably make this Zone inhospitable to freshwater phytoplankton (Burford et al. 2011). Oxygen measurements in this zone were typically made later in the day than measurements upstream (around 13:00 versus around typically just after 08:00 h at the upstream end of the tidal freshwater zone), so Fig. 12 probably underemphasises the degree of oxygen depression in this zone. During our study the Transitional Zone was thus a source of DIN and DIP and a sink for chlorophyll *a*.

[1] Residence times were calculated using a one-dimensional hydraulic model of the river, which is described by Robson and Webster submitted.

A Changing Estuary

In this chapter, we have presented a description of the history, physical, biogeochemical and ecological function of a previously little-studied system, the lower Ord River Estuary, and shown how the physical environment underpins its biogeochemical and ecological function. Here, we will make a few brief observations regarding how the estuary has changed and how it may change in future.

The Past

The most obvious change to the Ord River estuary over the past 100 years has been the regulation of the river, converting a seasonally-flowing wet-dry tropical river to a perennially flowing system. This has substantially altered the hydrology, morphology and ecology of the system, contributing to infilling of the channel, erosion of banks, downstream movement of the salinity gradient and maintenance of water levels upstream of the influence of seawater.

These changes have had some positive outcomes: as well as the values placed on a perennially-flowing river, construction of the two dams created two Ramsar-listed wetlands, Lakes Argyle and Kununurra (Department of Water 2012b).

While it is likely that the introduction of dry-season freshwater flows has increased overall primary production in the freshwater zones of the lower river, Warfe et al. (2011) speculate that both primary production and production of prawns in the estuary have diminished as a result of changes in salinity.

More subtly, agricultural development of the catchment has influenced sediment and nutrient loads reaching the estuary. Because these changes have not been monitored, we can speculate on the effects only through modelling and by comparison with other river-estuary systems in the region.

Wolanski et al. (2001, 2004) modelled sediment dynamics and morphology in the Ord River estuary, finding that the system is still adjusting morphologically to the effects of the construction of the dams, with sedimentation leading to rapid infilling of the channel. Burford et al. (2011) considered the likely effects of flow regulation on the biogeochemistry of the system in more qualitative terms, arguing that regulation has probably resulted in a reduction in primary production in the estuary due to a reduction in nutrient export associated with peak flows.

Brodie and Mitchell (2005) argue that Australian tropical rivers in the past were probably characterised by low to moderate suspended sediment concentrations and low concentrations of dissolved inorganic nutrients during flow events. They speculate that grazing in the last 50–200 years has increased suspended sediment and particulate nutrient loads to present levels, while fertilised agriculture in some catchments has probably increased dissolved nutrient loads. Unfortunately, historical records of water quality in these systems are very limited.

Nutrient concentrations in the Ord River (including the Tidal Freshwater Zone of the estuary), though low in comparison with the major rivers of temperate Australia (Robson et al. 2008a; Hearn and Robson 2000), were found to be higher than in unregulated perennially flowing rivers in the wet-dry tropics of Australia's north, such as the groundwater-fed Daly River (Webster et al. 2005). Production of algal biomass is correspondingly higher in the Ord River. This difference can be largely attributed to the dry-season contribution of irrigation return flows, which, as demonstrated in the nutrient budgets, were enriched in inorganic nitrogen and phosphorus, presumably derived from fertilisers.

The Future

Changes affecting the Ord River estuary in the coming decades are likely to fall into three categories: (1) increasing agricultural development; (2) increasing demand for electricity generation; and (3) climate change.

The area of irrigated land in the Western Australian side of Ord River catchment is currently being increased from 140 to 470 km^2 (Fig. 21) to allow increased production of diverse crops, including field crops, horticulture, a sandalwood forestry and irrigated pasture. Water extraction for irrigation purposes is expected to double over the next few years to meet these demands (Department of Water 2012a, b). This water will be taken from Lake Kununurra and so will not directly affect flow in the lower river.

This intensification of agriculture is likely to be accompanied by increases in nutrient and sediment generation from the catchment, though it is anticipated that this will be relatively minor as irrigation return flows will be diverted from the river.

In the near future (2012–2018), a moderate increase in demand for hydroelectricity is anticipated (Department of Water 2012a). This will require dry season flows to be maintained where possible at the upper end of the current range (i.e. around 90 m^3 s^{-1}) during this period. In the mid-term future, demand for electrical production from the Ord River Dam may return to current levels.

The Western Australian Department of Water (2012b) has modelled the impacts of the planned 2012–2018 increases in irrigation and hydroelectric production on flows reaching the lower river and estuary. They propose management of the dam to ensure an environmental water provision (EWP) that maintains a dry-season base flow to the estuary in the range 32 to 60 m^3 s^{-1} in order to maintain the current environmental values of the system.

The anticipated effects of current development plans on hydrodynamics, water quality primary production and trophic ecology in the lower river and estuary have also been modelled (Robson et al. 2008b; Department of Water 2012b) and are believed to be relatively minor, though the results for the estuary in particular are highly uncertain (Robson and Webster submitted).

If nutrient loads to the lower river increase significantly as a result of ongoing changes in catchment land use, algal production and turbidity in the Riverine Zone are likely to increase at the expense of benthic primary production and the habitat that benthic plants provide (Robson et al. 2008b; Robson and Webster submitted). Reductions in dry-season flow are likely to result in an upstream movement of the Transitional Zone and an increase in turbidity within this zone (Parslow et al. 2003), though again, with the currently planned changes, this effect is unlikely to be dramatic (<5 km shift in the salinity gradient and perhaps a 20 % increase in maximum turbidity during drier years in this, already highly turbid, zone of the estuary).

The impacts of these changes on nutrient dynamics and ecological function in the estuary are more difficult to predict (Robson and Webster submitted), however any changes in physical habitat and biogeochemical function the Transitional Zone and Estuary Mouth Zone are likely to be of particular significance to commercial prawn and recreational fisheries in the Ord River.

In the longer term, it is anticipated that irrigated agriculture will expand further, to new parts of the catchment in both Western Australia and the Northern Territory (Department of Water 2012b). In making these plans, the government has referred to the science reported here and elsewhere, considering ecological water requirements of the river, as well as social, cultural and economic considerations. Minimum dry-season flows have been established to maintain suitable habitat for benthic algae, macrophytes and fish (Department of Water 2012b).

Proposals for extraction of energy from tides may also re-emerge in the long-term future. A tidal power plant would certainly have a major impact on the estuary, extracting considerable physical energy from the system and reshaping the estuary in consequence.

Finally, climate change will certainly affect the estuary in coming decades. Regional climate models are split between predicting an increase in runoff for this particular catchment, predicting little change, or a decrease in runoff (Petheram et al. 2008). Changes in runoff will inevitably be associated with changes in nutrient and sediment loads from the

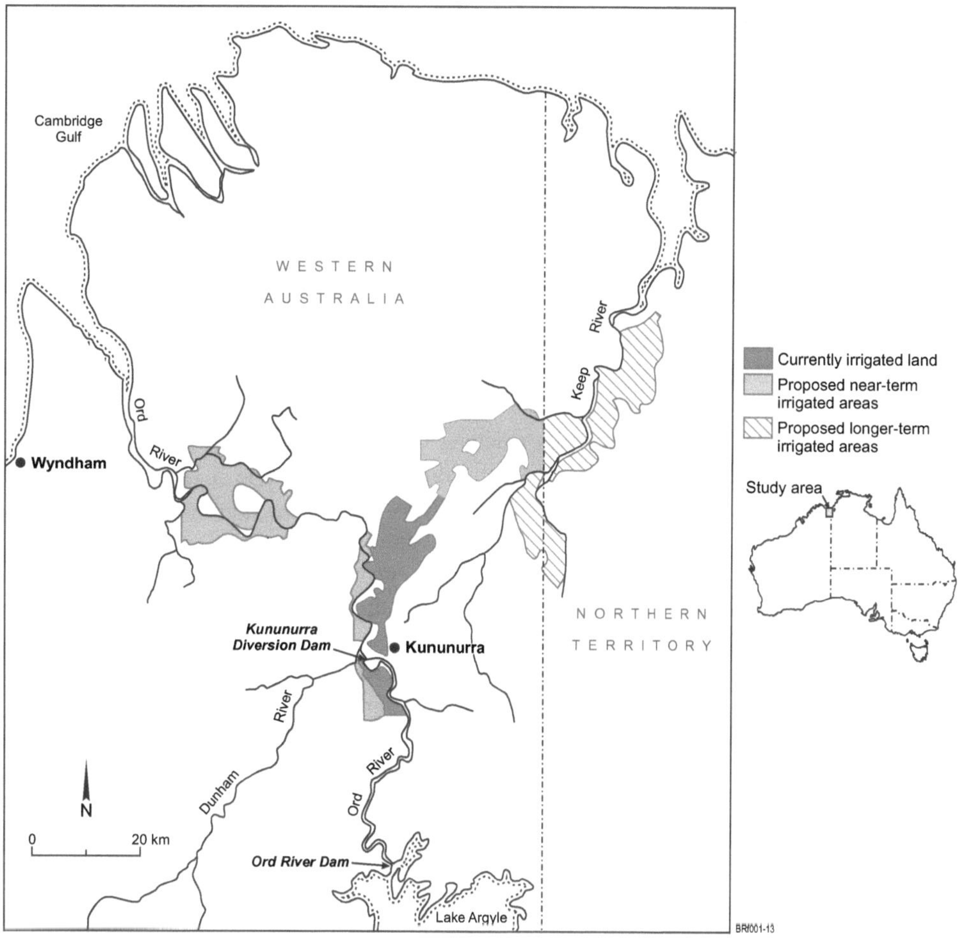

Fig. 21 Location of current and planned future irrigation areas around the lower Ord River (After Department of Water 2006)

catchment. Even if runoff changes relatively little, climate change will affect the estuary through an increase in sea-level, increase in air and water temperatures, and acidification associated with increasing atmospheric carbon dioxide concentrations. The likely effects of these changes have not been evaluated.

Concluding Remarks

The ecological character of the lower Ord and estuary have changed substantially following regulation as a result of reduced sediment supply, perennial dry season flows and increased dissolved nutrient loads associated with changes in catchment land use. The estuary is substantially different from its neighbours in the region as a result of these changes, but retains strong conservational, economic and cultural value.

The implications of these changes are manifest differently in different zones of the river and estuary and it is important to consider these zones separately and together when considering how the system is likely to respond to ongoing and future changes.

The five-zone conceptual model presented in this chapter provides a physical template for understanding potential changes in other macrotidal tropical estuaries subject to increasing pressure from catchment development and water resource development. Most ecological studies consider only freshwater OR estuarine system components, resulting in the interacting zones being under-represented in the literature and in our conceptual understanding. This template provides a tool for a more comprehensive approach to considering potential impacts of development (and other changes) on coastal ecosystems.

Acknowledgments Much of the data and analysis presented in this chapter is the product of original research, funded by the Australian Government, Government of Western Australia, and Rangelands NRM Coordinating Group and is gratefully acknowledged. In-kind support was provided by CSIRO, Griffith University and the Department of Water (DoW), Government of Western Australia.

The support of the Rangelands NRM Coordinating Group and its co-ordinator, Liz Brown, in linking us with interested stakeholders is gratefully acknowledged.

Thanks are due to Scott Goodson (DoW) for field support and for introducing us to the Miriuwung-Gajerrong community, the Traditional Owners of the region. Thanks to Ian Loh (DoW) for his comments regarding the history and planned future of water allocation in the river, and for working with us to model future scenarios.

Particular thanks are due to John Parslow, Nugzar Margvelashvili and John Volkman (all CSIRO) for their contributions to the initial study design, 2002–2003 fieldwork, and subsequent discussions. DoW staff also assisted with field work.

Technical support in laboratory analyses was provided by Rebecca Esmay and Stephane Armand (CSIRO).

Finally, thanks to Andrew and Kate McEwen and their team for providing a great field operation work area and accommodating our ever-changing needs.

Appendix: Calculation of Suspended Sediment and Nutrient Budgets

Suspended sediment and nutrient loads reaching the lower river were calculated by multiplying daily flows from Kununurra Diversion Dam by estimated nutrient and sediment concentrations at the furthest upstream monitoring site, then adding similarly calculated loads from Dunham River and from the small volume of irrigation return flow entering the river a few kilometres downstream, at D2, D4 and D6 (Fig. 1).

Monthly water quality samples are unfortunately not sufficient to allow precise estimates of daily suspended sediment and nutrient concentrations. Given wide variations in flow, simple interpolation between observations is not satisfactory. Daily suspended sediment and nutrient concentration estimates were therefore derived, where possible, from empirical relationships between available water quality observations and observed flows. Logistical considerations and the unpredictability of rainfall events prohibited a precise definition of flow-concentration relationships in the river, but in most cases, there were sufficient data from the observation program to define some approximate relationships.

Concentrations of particulate phosphorus (PP), and total suspended sediments (TSS) in Dunham River correlated reasonably well with flow in Dunham River (r2 = 0.65 for PP, r2 = 0.69 for TSS with 32 degrees of freedom). The correlation between flow and particulate nitrogen (PN) was poor (r2 = 0.26), however better data were not available. Daily estimates of PN as well as PP and TSS in Dunham River were therefore obtained from quadratic regressions fit to the available flow and concentration data.

Water quality in outflow from Kununurra Diversion Dam (KDD) is not directly monitored, but was assumed to match water quality just downstream of the dam, at site 15 (Ivanhoe Crossing, Fig. 1). Gaps in the monitoring record at this site were filled with data from the nearest downstream site – usually site 13 (Tarrara Bar).

Suspended sediment and nutrient concentrations downstream of the dam showed no clear relationship with flows from the dam. Similarly, suspended sediment and nutrient concentrations in the irrigation drains (D2, D4 and D6, Fig. 1) did not correlate with flows in these drains. Correlations were found, however, between TSS, PN and PP at these sites and flow measured in Dunham River. PN, PP and TSS concentrations in all tributaries were thus specified as fitted quadratic functions of flow in the Dunham River.

For concentrations just downstream of Kununurra Diversion Dam, these functions provided an acceptable fit (r2 for PP = 0.71; r2 for TSS = 0.74 and r2 for PN = 0.30, with 32 degrees of freedom). The strength of the relationships between water quality and flow in the irrigation drains (D2, D4 and D6) was weak, however the contribution of these drains to the total suspended sediment and total nutrient loads is relatively small, so the high error in this approximation does not contribute greatly to the uncertainty of the overall nutrient and sediment budgets. For concentrations in D4 vs. Dunham River flow, r2 for PP = 0.57; r2 for TSS in D4 = 0.27 and r2 for PN = 0.15, with 22 degrees of freedom. For concentrations in D2 vs. Dunham River flow, r2 for PP = 0.22, for TSS = 0.13, and for PN = 0.19, with 21 degrees of freedom.

No correlation was found between concentrations of dissolved organic or dissolved inorganic nitrogen and phosphorus, and flow. Hence, dissolved nutrient concentrations (DIN and DIP) for each tributary were set to a constant value equal to the mean concentration observed in that tributary.

A numerical model of flows, sedimentation and biogeochemical transformations (Robson et al. 2008b; Robson and Webster submitted), calibrated against observed water quality in the lower river and estuary, was used to determine the quantity of suspended sediments and nutrients reaching the estuary, given the calculated loads reaching the lower river.

Sources and sinks within the river were calculated by taking the difference between the calculated loads entering the river from tributaries and drains and the modelled loads delivered to the estuary.

References

Alongi DM (2001) The influence of mangrove biomass and production on biogeochemical processes in tropical macrotidal coastal settings. Org Sediment Interact 21:223–241

Bolton BR, Berents HW, Frakes LA (1990) Groote Eylandt manganese deposit. In: Hughes FE (ed) Geology of the mineral deposits of Australia and Papua New Guinea, vol 14, Monograph series/ Australasian institute of mining and metallurgy. Australasian Institute of Mining and Metallurgy, Parkville, pp 1575–1579

Bright F (2005) Lake Kununurra redclaw trapping trial and rule recommendations. Report to W.A. Department of Fisheries, Perth

Brodie JE, Mitchell AW (2005) Nutrients in Australian tropical rivers: changes with agricultural development and implications for receiving environments. Mar Freshw Res 56(3):279–302. doi:10.1071/ Mf04081

Bunn SE, Thoms MC, Hamilton SK, Capon SJ (2006) Flow variability in dryland rivers: boom, bust and the bits in between. River Res Appl 22(2):179–186. doi:10.1002/Rra.904

Burford MA, Alongi DM, McKinnon AD, Trott LA (2008) Primary production and nutrients in a tropical macrotidal estuary, Darwin

Harbour, Australia. Estuar Coast Shelf Sci 79(3):440–448. doi:10.1016/j.ecss.2008.04.018

Burford MA, Revill AT, Palmer DW, Clementson L, Robson BJ, Webster IT (2011) River regulation alters drivers of primary productivity along a tropical river-estuary system. Mar Freshw Res 62 (2):141–151. doi:10.1071/Mf10224

Camkin JK, Bristow KL, Petheram C, Paydar Z, Cook FJ, Story J (2008) Designs for the future: the role of sustainable irrigation in northern Australia. In: Sustainable irrigation management, technologies and policies II, vol 112. WIT, Southampton, pp 293–302

Cluett L (2005) The role of flooding in morphological changes in the regulated Lower Ord River in tropical northwestern Australia. River Res Appl 21(2–3):215–227. doi:10.1002/Rra.842

Coull BC (2009) Role of meiofauna in estuarine soft-bottom habitats*. Aust J Ecol 24(4):327–343

CSIRO (2009) Water in the Ord-Bonaparte region. In: CSIRO (ed) Water in the Timor Sea drainage division. A report to the Australian Government from the CSIRO Northern Australia Sustainable Yields Project. CSIRO Water for a Healthy Country Flagship, Canberra, pp 183–272

Department of Conservation & Land Management (2003) Information Sheet on Ramsar wetlands (Ord River Floodplain). Department of Conservation and Land Management, Perth

Department of Water (2006) Ord river water management plan, Water resource allocation planning series. Government of Western Australia, Perth

Department of Water (2012a) Ord surface water allocation plan: for public comment. Water resource allocation planning series. Department of Water, vol report no 48. Government of Western Australia, Perth

Department of Water (2012b) Ord surface water allocation plan methods report : background information and description of methods for the Ord surface water allocation plan. Government of Western Australia, Perth

Dixon RNM, Palmer DW (2010) Lake Argyle sedimentation – 2006 survey. Salinity and land use impacts series. Department of Water, Perth

Douglas MM, Bunn SE, Davies PM (2005) River and wetland food webs in Australia's wet-dry tropics: general principles and implications for management. Mar Freshw Res 56(3):329–342. doi:10.1071/Mf04084

Doupe RG, Pettit NE (2002) Ecological perspectives on regulation and water allocation for the Ord River, Western Australia. River Res Appl 18(3):307–320. doi:10.1002/Rra.676

Erskine WD, Saynor MJ, Erskine L, Evans KG, Moliere DR (2005) A preliminary typology of Australian tropical rivers and implications for fish community ecology. Mar Freshw Res 56(3):253–267. doi:10.1071/Mf04078

Ford P, Tillman P, Robson B, Webster IT (2005) Organic carbon deliveries and their flow related dynamics in the Fitzroy estuary. Mar Pollut Bull 51(1–4):119–127. doi:10.1016/j.marpolbul.2004.10.019

Gehrke P (2009) Ecological patterns and processes in the lower Ord River and Estuary. CSIRO Water for a Healthy Country National Research Flagship, Canberra

Guest MA, Connolly RM (2004) Fine-scale movement and assimilation of carbon in saltmarsh and mangrove habitat by resident animals. Aquat Ecol 38(4):599–609. doi:10.1007/s10452-004-0442-1

Head L (1999) The northern myth revisited? Aborigines, environment and agriculture in the Ord River Irrigation Scheme, stages one and two. Aust Geogr 30(2):141–158

Hearn CJ, Robson BJ (2000) Modelling a bottom diurnal boundary layer and its control of massive alga blooms in an estuary. Appl Math Model 24(11):843–859

Kay WR (2004) Movements and home ranges of radio-tracked Crocodylus porosus in the Cambridge Gulf region of Western Australia. Wildl Res 31(5):495–508. doi:10.1071/Wr04037

Kennard MJ, Pusey BJ, Olden JD, MacKay SJ, Stein JL, Marsh N (2010) Classification of natural flow regimes in Australia to support environmental flow management. Freshw Biol 55(1):171–193. doi:10.1111/j.1365-2427.2009.02307.x

Kenyon RA, Loneragan NR, Manson FJ, Vance DJ, Venables WN (2004) Allopatric distribution of juvenile red-legged banana prawns (Penaeus indicus H. Milne Edwards, 1837) and juvenile white banana prawns (Penaeus merguiensis De Man, 1888), and inferred extensive migration, in the Joseph Bonaparte Gulf, northwest Australia. J Exp Mar Biol Ecol 309(1):79–108. doi:10.1016/j.jembe.2004.03.012

Kumlu M, Jones DA (1995) Salinity tolerance of hatchery-reared postlarvae of Penaeus-Indicus Milne-Edwards, H. originating from India. Aquaculture 130(2–3):287–296. doi:10.1016/0044-8486(94)00319-J

Latrubesse EM, Stevaux JC, Sinha R (2005) Tropical rivers. Geomorphology 70(3–4):187–206. doi:10.1016/j.geomorph.2005.02.005

Lee S (1998) Ecological role of grapsid crabs in mangrove ecosystems: a review. Mar Freshw Res 49(4):335–343

Lee S (2009) Tropical mangrove ecology: physical and biotic factors influencing ecosystem structure and function. Aust J Ecol 24 (4):355–366

Loneragan N, Fisheries Research and Development Corporation (Australia), CSIRO Mathematical and Information Sciences, CSIRO Marine Research (2002) The growth, mortality, movements and nursery habitats of red-legged banana prawns (Penaeus indicus) in the Joseph Bonaparte Gulf: project FRDC 97/105. CSIRO Marine Research, Cleveland

Magilligan FJ, Salant NL, Renshaw CE, Nislow KH, Heimsath A, Kaste JM (2006) Evaluating the impacts of impoundment on sediment transport using short-lived fallout radionuclides. In: Sediment dynamics and the hydromorphology of fluvial systems, vol 306. IAHS, Wallingford, pp 159–165

Manson F, Loneragan N, Harch B, Skilleter G, Williams L (2005) A broad-scale analysis of links between coastal fisheries production and mangrove extent: a case-study for northeastern Australia. Fish Res 74(1):69–85

Marshall A, Storey A (2005) Lower Ord River functional habitat mapping: evaluation and field trial of remote sensing platforms and imaging techniques. Report to the Department of Environment, Perth

Mayes PJ, Thompson GG, Withers PC (2005) Diet and foraging behaviour of the semi-aquatic Varanus mertensi (Reptilia: Varanidae). Wildl Res 32(1):67–74. doi:10.1071/Wr04040

Moliere DR, Lowry JBC, Humphrey CL (2009) Classifying the flow regime of data-limited streams in the wet-dry tropical region of Australia. J Hydrol 367(1–2):1–13. doi:10.1016/j.jhydrol.2008.12.015

Morgan DL, Rowland AJ, Gill HS, Doupe RG (2004) The implications of introducing a large piscivore (Lates calcarifer) into a regulated northern Australian river (Lake Kununurra, Western Australia). Lakes Reserv Res Manag 9(3–4):181–193

NGIS Australia (2004) Australia's tropical rivers – data audit. Land and Water Australia, Canberra

Odum WE, Heald EJ (1972) Trophic analyses of an estuarine mangrove community. Bull Mar Sci 22(3):671–738

Parslow J, Margvelashvili N, Palmer D, Revill A, Robson B, Sakov P, Volkman J, Watson R, Webster I (2003) The response of the lower Ord River and estuary to management of catchment flows and sediment and nutrient loads: final science report. CSIRO Marine Research/Land and Water Australia, Hobart

Petheram C, McMahon TA, Peel MC (2008) Flow characteristics of rivers in northern Australia: implications for development. J Hydrol 357(1–2):93–111. doi:10.1016/j.jhydrol.2008.05.008

Phinney DA, Yentsch CS, Phinney DI (2004) Primary productivity of phytoplankton and subtidal microphytobenthos in Cobscook Bay, Maine. Northeast Nat 11:101–122

Poff NL, Allan JD (1995) Functional-organization of stream fish assemblages in relation to hydrological variability. Ecology 76 (2):606–627

Pringle CM (2001) Hydrologic connectivity and the management of biological reserves: a global perspective. Ecol Appl 11(4):981–998

Robertson AI, Dixon P, Daniel PA (1988) Zooplankton dynamics in mangrove and other nearshore habitats in tropical Australia. Mar Ecol Prog Ser 43(1–2):139–150. doi:10.3354/Meps043139

Robson BJ, Webster IT (submitted) Can we model macrotidal estuaries well? A case study modelling impacts of changes to irrigation and flow regulation in a dryland tropical river. Submitted to Environmental Modelling and Software.

Robson BJ, Bukaveckas PA, Hamilton DP (2008a) Modelling and mass balance assessments of nutrient retention in a seasonally-flowing estuary (Swan River Estuary, Western Australia). Estuar Coast Shelf Sci 76(2):282–292. doi:10.1016/j.ecss.2007.07.009

Robson BJ, Burford MA, Gehrke PC, Revill AT, Webster IT, Palmer DW (2008b) Response of the lower Ord River and Estuary to changes in flows and sediment and nutrient loads. Water for a Healthy Country National research Flagship, CSIRO, Canberra

Sheaves M, Molony B (2000) Short-circuit in the mangrove food chain. Mar Ecol Prog Ser 199:97–109. doi:10.3354/Meps199097

Sheaves M, Johnston R, Connolly RM, Baker R (2012) Importance of estuarine mangroves to juvenile banana prawns. Estuar Coast Shelf Sci 114:208–219. doi:10.1016/j.ecss.2012.09.018

Smith AJ (2008) Rainfall and irrigation controls on groundwater rise and salinity risk in the Ord River Irrigation Area, northern Australia. Hydrogeol J 16(6):1159–1175. doi:10.1007/s10040-008-0301-6

Stewart GA (1970) Lands of the Ord-Victoria area, Western Australia and Northern Territory. Commonwealth Scientific and Industrial Research Organization (CSIRO), Melbourne

Tetzlaff D, Soulsby C, Waldron S, Malcolm IA, Bacon PJ, Dunn SM, Lilly A, Youngson AF (2007) Conceptualization of runoff processes using a geographical information system and tracers in a nested mesoscale catchment. Hydrol Process 21(10):1289–1307. doi:10.1002/Hyp.6309

Thom BG, Wright LD, Coleman JM (1975) Mangrove ecology and deltaic-estuarine geomorphology – Cambridge Gulf-Ord River, Western-Australia. J Ecol 63(1):203–222. doi:10.2307/2258851

Thoms M, Hill S, Spry M, XiangYang C, Mount T, Sheldon F (2004) The geomorphology of the Barwon-Darling basin. In: Breckwoldt R, Boden R, Andrew J (eds) The Darling. Murray-Darling Basin Commission, Canberra, pp 68–103

Trayler K, Malseed BE, Braimbridge MJ (2006) Environmental values, flow related issues and objectives for the lower Ord River. Department of Water, Government of Western Australia, Perth

Vannote RL, Minshall GW, Cummins KW, Sedell JR, Cushing CE (1980) River continuum concept. Can J Fish Aquat Sci 37 (1):130–137

Volkman JK, Revill AT, Bonham PI, Clementson LA (2007) Sources of organic matter in sediments from the Ord River in tropical northern Australia. Org Geochem 38(7):1039–1060. doi:10.1016/j.orggeochem.2007.02.017

Ward JV, Stanford JA (1995) Ecological connectivity in alluvial river ecosystems and its disruption by flow regulation. Regul Rivers: Res Manage 11(1):105–119

Warfe DM, Pettit NE, Davies PM, Pusey BJ, Hamilton SK, Kennard MJ, Townsend SA, Bayliss P, Ward DP, Douglas MM, Burford MA, Finn M, Bunn SE, Halliday IA (2011) The 'wet-dry' in the wet-dry tropics drives river ecosystem structure and processes in northern Australia. Freshw Biol 56(11):2169–2195. doi:10.1111/j.1365-2427.2011.02660.x

Wark RJ (1987) Deposition of sediment in Lake Argyle. Western Australia Water Authority, Perth

Wasson RJ, Caitcheon G, Murray AS, McCulloch M, Quade J (2002) Sourcing sediment using multiple tracers in the catchment of Lake Argyle, Northwestern Australia. Environ Manage 29(5):634–646. doi:10.1007/s00267-001-0049-4

Water and Rivers Commission (2003) Productivity and water flow regulation in the Ord River of North-Western Australia. Environmental flows initiative project, Perth (23092)

Webster IT, Rea N, Padovan AV, Dostine P, Townsend SA, Cook S (2005) An analysis of primary production in the Daly River, a relatively unimpacted tropical river in northern Australia. Mar Freshw Res 56(3):303–316

Wolanski E, Moore K, Spagnol S, D'Adamo N, Pattiaratchi C (2001) Rapid, human-induced siltation of the macro-tidal Ord River Estuary, Western Australia. Estuar Coast Shelf Sci 53(5):717–732. doi:10.1006/ecss.2001.0799

Wolanski E, Spagnol S, Williams D (2004) The impact of damming the Ord river on the fine sediment budget in Cambridge gulf, northwestern Australia. J Coast Res 20(3):801–807

South Australia's Large Inverse Estuaries: On the Road to Ruin

Jochen Kämpf

Once there was a low-lying, swampy country covered with numerous lagoons
Disagreements amongst Ancestral Beings belonging to the bird, animal and reptile families caused great concern of the willy-wigtail, emu and kangaroo families.
After a night of prophetic dreams, a giant kangaroo bone was found which proved to be magic.
When the wise and respected kangaroo pointed the bone at the swampy land, the earth opened up and the sea gradually flooded the low land.
This is how the two Peninsulas (i.e. Yorke and Eyre) and (what we now call) Spencer Gulf were formed.

Narrunga Creation Story (Smith (1930))

Abstract

This chapter provides an overview of the past, present and likely future of South Australian gulfs – Spencer Gulf and Gulf St. Vincent. It describes the distinct physical factors shaping these inverse estuaries, their unique ecology, past environmental degradation and future threats. Rather than direct climate-change impacts, the reader will learn that traditional industrialization poses the biggest threat to the gulfs' ecosystem health, despite recent enhanced efforts of protection and conservation of natural habitat.

Keywords

Spencer Gulf • Gulf St. Vincent • Inverse estuary • Ecosystem • Environmental degradation • Future threats

J. Kämpf (✉)
School of the Environment, Flinders University,
PO Box 2100, Adelaide, SA 5001, Australia
e-mail: jochen.kaempf@flinders.edu.au

E. Wolanski (ed.), *Estuaries of Australia in 2050 and Beyond*, Estuaries of the World,
DOI 10.1007/978-94-007-7019-5_9, © Springer Science+Business Media Dordrecht 2014

Box 1

Jochen Kämpf studied South Australia gulfs – Spencer Gulf and Gulf St. Vincent. He describes the distinct physical factors shaping these inverse estuaries, their unique ecology, past environmental degradation and future threats. Rather than direct climate-change impacts, he concludes that traditional industrialization poses the biggest threat to the gulfs' ecosystem health, despite recent enhanced efforts of protection and conservation of natural habitat.

Introduction

Most of the world's estuaries are located in high-rainfall regions of the tropics and polar-front zones where numerous rivers discharge their freshwater loads into the sea. Such semi-enclosed water bodies in which the salt concentration (salinity) is lower than that of the adjacent sea are called *positive estuaries.*

South Australia – the driest Australian state – is exposed to an arid Mediterranean climate and experiences only marginal and highly transient river flows. The annual-mean domination of evaporation over freshwater inflows makes South Australian gulfs – Spencer Gulf and Gulf St. Vincent – a distinctive type of estuary, termed *inverse estuary* or *negative estuary*, in which the salinity increases with increasing distance from the estuary mouth. Such hypersaline estuaries are relatively rare on Earth. Other examples are the Red Sea and the Persian Gulf in the northern Indian Ocean (Kämpf and Sadrinasab 2006), Shark Bay and Exmouth Gulf in Western Australia (Edyvane 2005), and the hypersaline Coorong estuary/lagoon in South Australia (Kämpf and Bell 2013, see chapter "The Murray/Coorong Estuary. Meeting of the Waters?" of this book).

This chapter describes a broad range of topics of relevance to South Australian gulfs, including physical, ecological, cultural, historical and industrial aspects. The prediction of the future of the gulfs' environment and their people is a difficult task and the outcome strongly depends on one's mindset. On a good day, one may think positively that modern-day people (in particular decision-makers) are now wise and educated enough to do everything they can to protect important natural habitat. On another day, it may rather look like a business-as-usual scenario controlled by large corporate companies whose prime goal is to maximize profits at any means. For a country as Australia that holds substantial natural resources of coal, ire ore, and uranium, to name a few, it is reasonable to expect "business-as-usual" exploitation of natural resources and severe degradation of natural habitat.

Climate, Geography and Bathymetry

South Australia is exposed to a semi-arid temperate climate with annual evaporation exceeding precipitation by far. Observed annual evaporation and precipitation rates (at Adelaide airport) are 1,900 mm/year and 450 mm/year, respectively (Kämpf et al. 2010). This excess of evaporation over precipitation together with low river runoff makes both South Australian gulfs – Spencer Gulf and Gulf St. Vincent – hypersaline inverse estuaries (Nunes et al. 1990). The seasonal climate is a Mediterranean type with dry summers and wet winters.

Spencer Gulf and Gulf St. Vincent are Australia's largest estuaries (Fig. 1). The gulfs are sometimes called the *South Australian Seas,* noting that their sizes are similar to that of the Irish Sea. Both gulfs are shallow bodies of water (Fig. 2), less than 50 m deep, with restricted northern regions that are enclosed by sand spits.

Spencer Gulf received its name by explorer Matthew Flinders in 1802 (Flinders 1814). The gulf was also named *Golfe Bonaparte* by Nicholas Baudin at roughly the same time. The area was first explored on land by Edward John Eyre in 1839 and 1840–1841. European settlement of the shores of the Gulf began in the late 1840s.

Spencer Gulf is bordered by Eyre Peninsula in the west and Yorke Peninsula in the east. Spencer Gulf is the larger of the two gulfs. It is a triangular-shaped embayment, 300 km long. The gulf spans a total area of 2.2×10^4 km^2 (2,200,000 ha) and has a mean depth of 24 m. Spencer Gulf has one entrance, about 130 km wide, with a series of islands – the *Gambier Islands.*

The *Sir Joseph Banks Group* is an archipelago of about 20, mostly small, islands, with a collective land area of 1,275 ha. It is an important seabird breeding site. The islands were visited and named by Matthew Flinders in 1802 (Flinders 1814). They consist mainly of a granite base beneath limestone and are usually capped with calcrete or sandy soil. The highest point of about 50 m is on Spilsby Island.

Spencer Gulf is divided into a deeper southern area and a much shallower (<25 m deep) northern area, called *Upper*

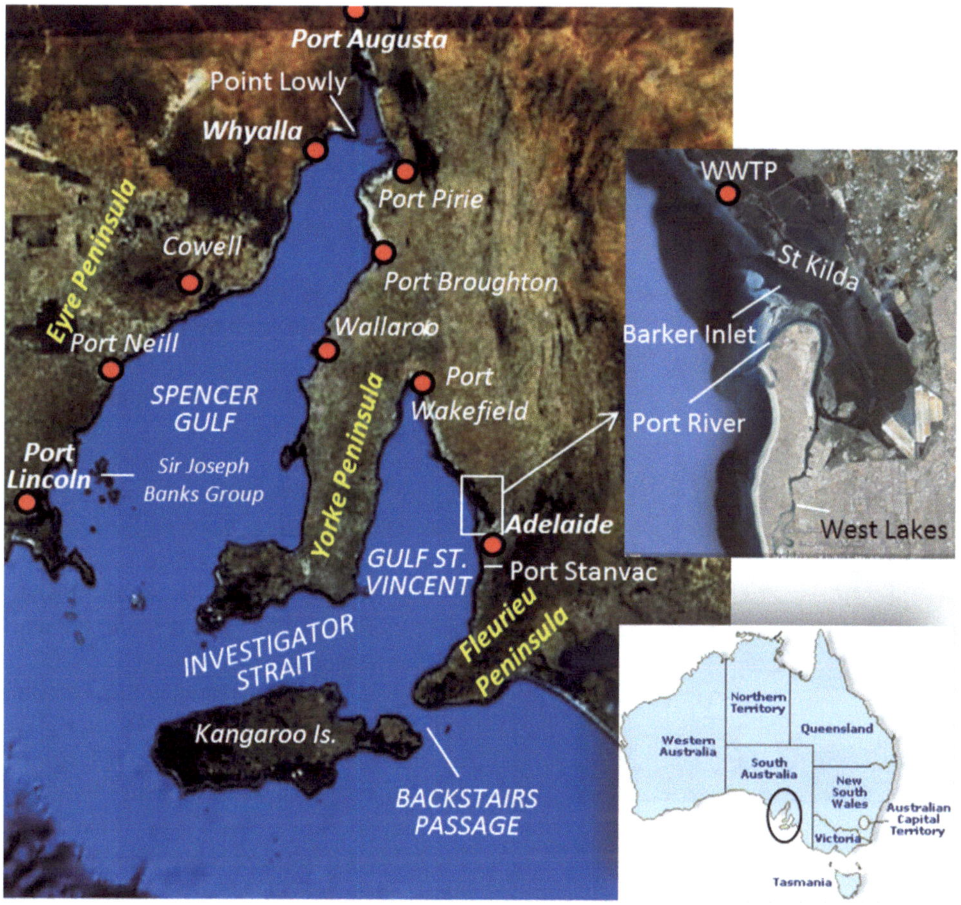

Fig. 1 The location of South Australian gulfs. Major cities and selected regions are *highlighted*. WWTP indicated the location of the Bolivar Wastewater Treatment Plant (Image source: Google Earth)

Spencer Gulf. The main basin of southern Spencer Gulf has an asymmetric topography in cross-section, with a gently sloping side on the west and a steeply sloping side on the east (James et al. 1997). Ward Spit divides the southern and northern areas of the gulf and confines exchange between them to a narrow channel called the *Flinders Channel*, which runs up the center of the northern gulf (Shepherd and Hails 1984; Corlis et al. 2003). Intertidal sand and mud flats flank the channel, along with a series of low terraces and scarps covered in seagrass.

Spencer Gulf receives freshwater runoff only sporadically during floods from numerous creeks and the northward extension of Upper Spencer Gulf which continues as a narrow tidal channel through Port Augusta over a distance of about 10 km (Bye and Harbison 1991).

Gulf St. Vincent was named *Gulph of Saint Vincent* by Matthew Flinders in 1802 (Flinders 1814). Prior to then, it had been known as *Golphe Josephine*. European settlement of the gulf's shores Gulf commenced in 1838.

The gulf is bordered by York Peninsula in the west and Fleurieu Peninsula in the east. Investigator Strait is bordered by Kangaroo Island in the south. Gulf St. Vincent and Investigator Strait together are 210 km in length. Gulf St. Vincent has an area of 6.8×10^3 km^2 (680,000 ha) and a mean depth of 21 m. Investigator Strait has an area of 6.1×10^3 km^2 (610,000 ha) and a mean depth of 34 m. Whereas Spencer Gulf has only one entrance, Gulf St. Vincent is connected to the adjacent shelf via two entrances, Investigator Strait, about 30–45 km wide, and Backstairs Passage, only 14 km wide.

The maximum depth of 41 m in Gulf St. Vincent occurs in the centre of the southern area. Both entrances to the gulf are shallower; Backstairs Passage has a 35 m deep sill and Investigator Strait is <30 m deep due to an north–south trending bar at its midpoint (James et al. 1997; Li et al. 1998). Tidal current sand ridges and scours are present in both entrances (Harris 1994). Northern Gulf St. Vincent is bounded in the south by a latitudinal line through Long Spit

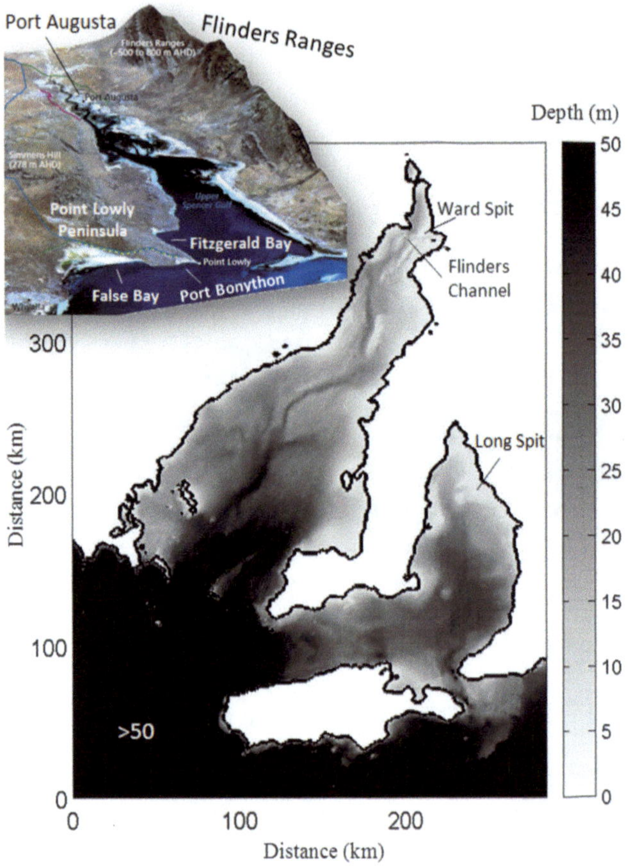

Fig. 2 Bathymetry (m) of South Australian gulfs. The *insert* shows a three-dimensional projection of the Upper Spencer Gulf, modified from BHP Billiton's Supplementary EIS that can be downloaded at http://www.olympicdameis.sa.gov.au/

and has a maximum water depth of 20 m (de Silva Samarasinghe et al. 2003). The northern area has an asymmetric topography with a 20 m deep channel along the western side and gently grading flats on the eastern side (de Silva Samarasinghe 1998).

Most freshwater inflows to the Gulf St. Vincent come from various rivers via the Adelaide Hills from the Torrens Catchment. Major rivers are the Port River, the River Torrens, the Onkaparinga River, and the Myponga River.

The Barker Inlet and Port River region is located about 15 km northwest from the Adelaide city centre (see Fig. 1). The inlet system covers an area of approximately 20 km^2 (2,000 ha). This offshoot from the gulf is a sheltered, marine dominated environment which experiences a maximum tidal range 2.5 m. The inlet is exposed to low and intermittent inflows of freshwater, via ephemeral creeks and stormwater drains, and continuous tidal flushing with hypersaline gulf water. The Port River serves as a shipping channel and extends inland through the historic Inner Harbour of Port Adelaide, to the constructed salt-water West Lakes in the north-western suburbs of Adelaide. Adelaide's port in the Port River is currently South Australia's major export corridor. Both gulfs have numerous smaller bays, each of individual ecological characteristics. A description of marine "treasures" found in some of these bays is outside the scope of this chapter.

Cities and Major Industries

Major cities on the shores of South Australian gulfs are Adelaide (current population around 1.26 Million), Whyalla (~27,000), Port Lincoln (~15,000), Port Augusta (~15,000) and Port Pirie (~14,000). The total population is expected to grow to 2 million by 2050.

Major industries of the Spencer Gulf region are agriculture, tourism, fishing, aquaculture and mining. For more than 100 years, Upper Spencer Gulf has supported a number of heavy industries and urban centres with the steelworks industry located in Whyalla, South Australia's major power plant located in Port Augusta, and the world's largest lead smelter located in Port Pirie.

Marine related industries in Spencer Gulf include a large range of commercial fisheries. The key players of this industry are the tuna fish farming industries located in Port Lincoln and the prawn (trawling) industry. Upper Spencer Gulf supports a particularly important and recreational fishing industry. The region between Whyalla and Port Pirie in particularly productive with a commercial catch of over 6,000 tonnes annually (Knight et al. 2005).

Major industries of the Adelaide region are manufacturing (mainly automotive, electronics and defence), horticulture and viticulture, education, and tourism. Gulf St. Vincent also supports a valuable scalefish industry.

Aboriginal Heritage

Prior to European settlement, five distinct groupings of Aboriginal peoples lived around the shores of South Australia's gulfs. The local aboriginal tribes have lived in the area for approximately 40,000 years.

During initial European contact, the population of Aboriginal people on the Eyre Peninsula is estimated to have been no more than about 2,000 individuals (Berndt 1985). The number of people in each language group varied, as did the size of the territory they occupied. In the northern parts of the peninsula, tribal areas tended to be large and boundaries were not always well defined. Along the coast where water and food resources were more easily accessible, smaller tribal areas had loosely defined boundaries. All Aboriginal groups on the Eyre Peninsula are known to have used a wide variety of native plant and animal (including fish) species for food and other resources.

Little is known about the Nawu (or Naou) people who have lived in the lower Eyre Peninsula. This culture and language is extinct since the 1930s (Tindale 1974). Barngala (or Banggarla) people lived in the upper western part of Spencer Gulf. Barngarla people wore cloaks made from kangaroo skin turned fur inside during winter. In summer, they smeared their bodies with fat and ochre. They hunted both land and marine animals; however they never included oysters and shellfish in their diet. The Barngarla were especially known to "sing" to the sharks and dolphins at Fitzgerald Bay and Point Lowly to help them to drive the fish towards the shore where they could be either caught in the fish traps or speared (Paul Mazourek, 2011, Whyalla Maritime Museum, personal communication). By the 1870s, the majority of Eyre Peninsula aboriginals lived at the fringe camps near the white settlement. The Malkaripangala (a subdivision of Barngarla aboriginal group) virtually disappeared. Nukunu people lived in the northeastern region of Spencer Gulf. Survivors of this tribe now live throughout South Australia including urban centres, with some still living in and around Port Pirie, Port Augusta and down to Port Lincoln.

Narangga (also Narungga) lived on most parts of the Yorke Peninsula (Krichauff 2011). They were a nomadic people who practiced fire-stick farming to flush out wildlife and control vegetation. Their diet also included seafood. Their expertise at fishing was highly skilled. The seal (*mūlta*) was their totems, which highlights the pre-colonial abundance of seals in the region. On the nineteenth century, the southern shores of Yorke Peninsula were explored and exploited by those men who came ashore to hunt kangaroos and seals, to collect timber and fresh water, and to explore undiscovered land. The transient inhabitants of Kangaroo Island owned whaling boats and other small vessels which gave them a great deal of mobility.

The population of the Narangga people at the time of first contact was estimated at 500. This number had halved by 1856 and by 1880 there were less than 100 Aboriginal people of full Narangga descent on Yorke Peninsula (Carmichael and Mudie 1973). Many had died of introduced diseases such as bronchitis, measles, scarlet fever and venereal diseases, or from the lack of plentiful food and water. In the years following colonization, the remaining Narungga people lost much of the use of their language and cultural heritage.

The Kaurna people lived on the eastern side of Gulf St. Vincent on the Fleurieu Peninsula. They were a hunter-gatherer society. At the establishment of South Australia in 1836 the Kaurna population was around 500. The Kaurna population, which may have originally numbered up to 1,000, had been seriously depleted in 1830 due to a smallpox epidemic (Anonymous 1985). Kaurna culture and language was almost completely destroyed within a few decades of the European settlement. However, extensive documentation by early missionaries and other researchers has enabled a modern revival of both language and culture.

Geology and Sedimentology

Spencer Gulf and the Gulf St. Vincent are underlain by two submerged sedimentary basins: the Pirie Basin and the St. Vincent Basin. The two basins formed as a result of tectonic activity during the Tertiary (from 65 to 1.8 million years ago). This activity included episodes of faulting and regional tectonic uplift (Cooper 1985; Leonard 2003). The gulfs occupy active intra-cratonic grabens, and recent seismic activity in the nearby Flinders Ranges indicates tectonic changes continue to affect the structure and morphology of the region (Leonard 2003; Sandiford 2003).

The Pirie and St. Vincent Basins are connected in the north by a narrow onshore trough and contain similar early and middle Tertiary sediments (Alley and Lindsay 1995). The basins are both underlain by Proterozoic-Cambrian basement rocks that outcrop on several offshore islands (Cooper 1985; Fuller et al. 1994). Kangaroo Island is a basement high forming the southern margin of the St. Vincent Basin (Cooper 1985).

The gulfs' region is part of the large area of cool-water carbonate deposition that exists along the southern margin of Australia (James and von der Borch 1991; James et al 1997). In the gulfs, surficial marine sediments display a clear zonation from subtidal to supratidal areas, related to water depth (Gostin et al. 1988). The gulf margins contain nearshore intertidal muddy gastropod-rich sediments, which are inhabited by cyanobacterial mats and mangroves. Shallow water sediments include poorly-sorted muddy bioclastic sand and seagrass, with muddy carbonate sands dominant in water depths greater than 30 m.

Seabed sediments across the region are composed of mixed terrigenous-carbonate material (Gostin et al. 1984). Overall, biogenic carbonate is the dominant sediment component. Biogenic sediments contain bryozoans (B), coralline algae (A), molluscs (M), and foraminifers (F), which are known as the BAMF Carbonate Factory (James et al. 1997).

Terrigenous sediment inputs (derived from erosion of rocks on land) are generally low, except in northern parts of Spencer Gulf where terrigenous components make up 85 % of surface sediments. Such high terrigenous concentrations in these areas are a result of material being eroded from the nearby Flinders Ranges and introduced into the gulf by local drainage (e.g. Fuller et al. 1994). It is also possible that these terrigenous sediments were transported from the shelf into the shallow northern regions during sea level transgression.

Up to 4 m of Holocene surficial sediments have accumulated in northern parts of Spencer Gulf (Hails et al. 1984). Sediments in the area contain the dominant components: gastropods, bivalves, foraminifera, coralline algae and quartz grains. Southern parts of Spencer Gulf have a patchy covering of surficial sediment that is less than 10 m thick. The substrate is exposed on some areas of seabed, revealing Cenozoic limestone (James et al. 1997). Relict carbonate fragments are a significant sediment component in southern Spencer Gulf (Fuller et al. 1994).

In Gulf St. Vincent and Investigator Strait, this biogenic carbonate is generated from two main sources; within the gulf itself and from the continental shelf southwest of Investigator Strait. Terrigenous sediment concentrations are typically low, and inversely proportional to the amount of carbonate (eg James et al. 1997).

The Flinders Channel of Upper Spencer Gulf is covered by a large area of mega-ripples (sandwaves). These ripples are oriented normal to the tidal flow, with wavelengths of 2–20 m and heights of up to 1.3 m. The mega-ripples are made up of reworked local material rather than new material (e.g. Shepherd and Hails 1984). Sediments on the mega-ripple field are composed of well-sorted and medium-grained sand on crests and poorly sorted sands in troughs. Surface sediments collected from sandwaves are often shell-rich, with carbonate concentrations ranging from 25 to 95 % by weight (Shepherd and Hails 1984).

Oceanography

South Australian gulfs are inverse estuaries in which, due to an excess of evaporation over freshwater sources, the salinity increases with increasing distance for the gulfs' entrances. As a consequence, the highest salt concentrations are found near the head of the gulfs. Whereas the ambient ocean salinity is about 36 ppt (roughly equivalent to 36 g per litre), the salt concentration in Upper Spencer Gulf exceeds 43 ppt and can reach up to 50 ppt (Nunes and Lennon 1986). The maximum salinity in upper reaches of Gulf St. Vincent is about 43 ppt (Kämpf et al. 2009). These elevated salinities are a signature of a reduced oceanic influence (also called *flushing*).

The net water circulation in the gulfs and exchange flows with ambient shelf waters are mainly driven by lateral density differences (Kämpf et al. 2009). Density is the weight of seawater per unit volume, being controlled by both temperature and salinity. At the same temperature, saltier water is heavier and, hence, denser than less salty water. In narrow inverse estuaries, the hypersaline dense water flows out as a bottom-attached current and, in return, draws in ambient shelf water in surface layers. In larger estuaries, such as South Australian gulfs, the Coriolis force in the Southern Hemisphere operates such that the outflow is shifted to the left side of the estuary (with respect to the flow direction), whereas the inflow occurs on the other side. Hence, the density-driven circulation in South Australian gulfs is generally a semi-clockwise pattern. In addition, dynamic instabilities of the dense outflows tend to create a field of mesoscale eddies of ~10–20 km in diameter in the gulfs (Lennon et al. 1987) that make the density-driven circulation spatially highly variable.

The exchange circulation of the gulfs with ambient shelf water is subject to pronounced seasonal variations. During the austral summer months (January–March), the gulf waters warm up substantially, and temperature effects tend to compensate salinity effects in terms of density (warm water is less dense than cold water at the same salinity). Horizontal density contrasts in the lower reaches of the gulfs weaken substantially, significantly reducing the gulf-ocean exchange during this period (Petrusevics 1993). To this end, during austral summer zones of sharp temperature contrasts (temperature fronts) establish across the mouth of Spencer Gulf and in Investigator Strait and Backstairs Passage (Fig. 3a). A likely contribution to the establishment of these fronts is summertime upwelling of cold water along the adjacent shelf (Kämpf et al. 2004).

Owing to atmospheric cooling, hypersaline gulf waters attain the highest density during austral winter months (June–July) triggering a strong exchange circulation (Fig. 3b). The seasonal density-driven outflow from Spencer Gulf is well documented (Lennon et al. 1987). Little is known about specific pathways of the outflows from Gulf St. Vincent. Model predictions suggest the existence of two separate outflow branches through Investigator Strait and Backstairs Passage, respectively (see Fig. 3b).

Inflow of ambient sea water, peaking in the winter season, is the dominant means of renewal (flushing) of gulf waters and plays an important role in the gulfs' budgets of salt and dissolved chemicals (Smith and Veeh 1989). Wind-driven effects on the gulf's general circulation play a secondary role (Kämpf et al. 2010).

Tidal currents, being another important part of the gulfs' ecosystem dynamics, are superimposed on density-driven currents. Apart from regular tidal flooding of the intertidal coastal zone, tides involve oscillatory flows moving water back and forth along the shore over relatively short distances of ~2–7 km. South Australian gulfs are subject to co-oscillating tides with tidal ranges of ~1 m near the entrances and increased values toward the heads (Easton 1978). Tidal ranges are markedly amplified in the upper reaches of both gulfs creating extensive intertidal flats and tidal creeks. The tidal range in Upper Spencer Gulf can exceed 4 m (Schluter et al. 1995). A slightly lower tidal range of 3 m is found near the head of Gulf St. Vincent.

Another remarkable feature of the gulfs' tides is that the dominant tidal constituents of the semidiurnal lunar (M2)

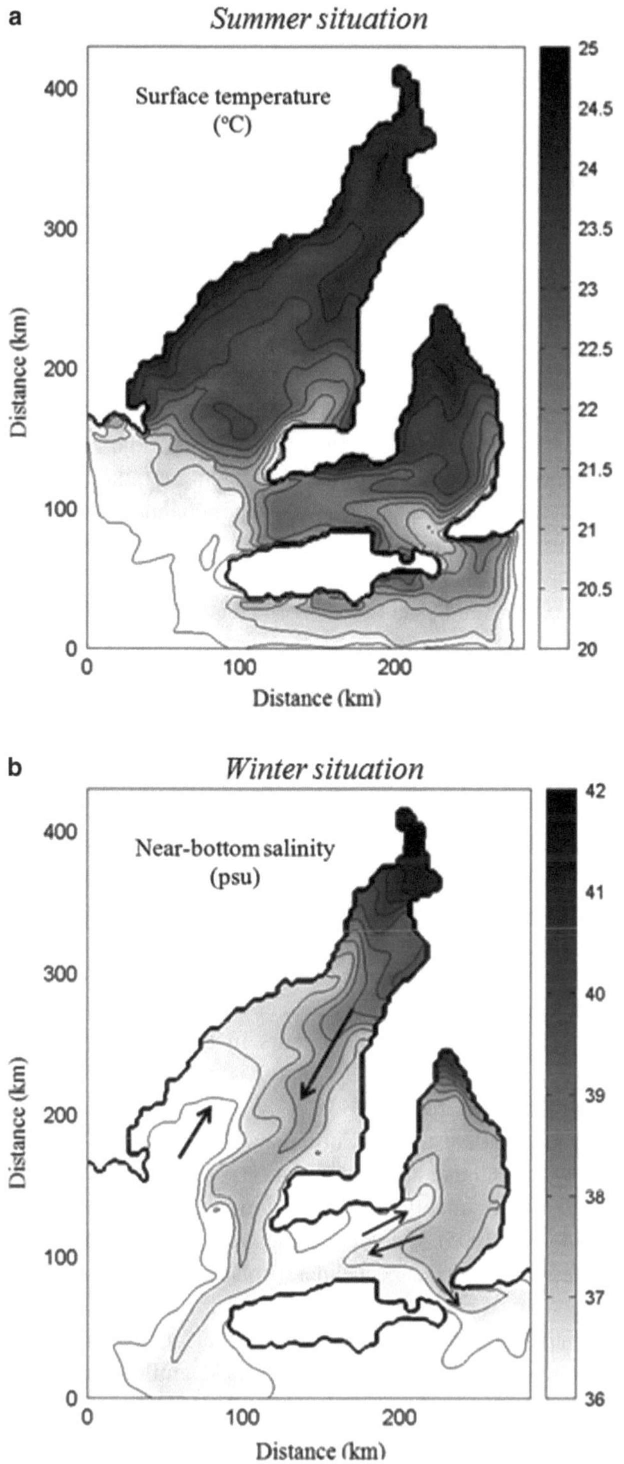

Fig. 3 Distributions of (**a**) sea-surface temperature in austral summer and (**b**) near-bottom salinity in austral winter in South Australian gulfs (From Kämpf et al. 2009)

and solar (S2) tides are almost of the same amplitude. The interference between these constituents implies that tidal currents tend to vanish on a roughly fortnightly basis. During this period, lasting 1–2 days, the so-called neap tide and associated tidal flows almost disappear, which is known as a *dodge tide*, first explained by Chapman (1892). The existence of dodge tides is a critical factor for the management of pollutant discharges, given the lack of tidally induced mixing during these periods. During periods of maximum tidal flows, called *spring tides*, tidal currents are of moderate speed (10–50 cm/s) in most parts of the gulfs with the exception of Backstairs Passage and Upper Spencer Gulf where peak current speeds exceed 1 m/s.

Due to the presence of Kangaroo Island and dissipation effects in shallower water, both gulfs are largely sheltered from vigorous Southern Ocean swell waves (Hemer and Bye 1999). Such waves, although of markedly reduced energy, are still able to create substantial northward sand movement (littoral drift) along the narrow beaches of Adelaide. This beach sand displacement poses a continuous management problem.

The flushing characteristic of South Australian gulfs can be quantified by means of an academic method based on the determination of "water ages". This method uses a calibrated hydrodynamic model, involving all key physical processes, to predict the ageing of water parcels on the basis that (a) all water has an initial age of zero, and that (b) all water entering the gulfs has a starting age of zero. The predicted water age distribution then provides information on which regions of the gulf-wide system are flushed by ambient shelf water on shorter time scales as compared to others. This knowledge is fundamental for a gulf-wide management of marine resources.

The distribution of water age in South Australian gulf is not surprising given that it largely reflects the salinity distribution; that is, higher salinities correspond to older water (Fig. 4). The more relevant finding is rather that the upper reaches of both gulfs are characterized by water ages exceeding a year and are therefore excluded from seasonal flushing of the gulfs' lower reaches. This slow flushing is vital for the maintenance of hypersaline conditions in the upper reaches of the gulfs and it also provides a trapping mechanism for marine larvae that otherwise could be wiped away by the density-driven seasonal flow.

Marine Ecology

South Australian gulfs, like many shallow, sheltered and tidal estuaries, are highly productive ecosystems. South Australian gulfs have a diverse range of habitats and is a globally significant region for temperate biodiversity, exhibiting very high levels of endemism relative to the southern temperate coastline of Australia, which itself has an endemism – or uniqueness of species – of over 85 %, compared to only 15 % in tropical areas such as the Great Barrier Reef (Edyvane 1999a). In particular, temperate

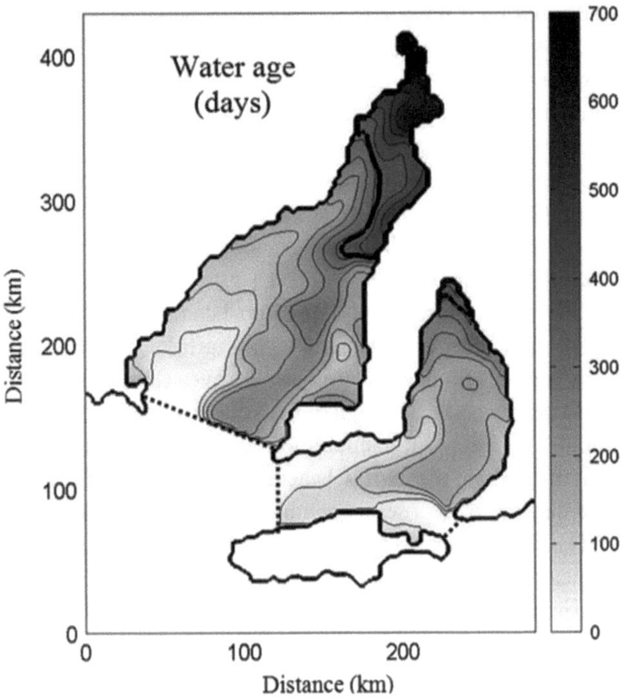

Fig. 4 Water-age distribution in South Australian gulfs (From Kämpf et al. 2009)

Australia has the world's highest level of seagrass diversity (22 species recorded, cf. 15 in tropical waters) and the highest recorded levels of endemism (Edyvane 1999b).

The gulfs' shallow water depths, high tidal ranges and highly saline waters in their northern reaches has resulted in extensive subtidal and supratidal marine wetland environments that support some of the largest areas of temperate seagrasses, mangroves and saltmarshes in Australia (Edyvane 1996).

Seagrasses are the 'grass meadows' of coastal waters. They are flowering plants with roots, making them very distinct from algae. Their extensive root system is essential for the uptake of nutrients and for anchoring the plants firmly into the seabed, which stabilizes the seabed and makes it less vulnerable to sediment erosion. Seagrass meadows are primary producers at the base of the food chain and 'critical' habitats for many fish, crustaceans and other marine animals. At the end of twentieth century, both gulfs had a total of over 500,000 ha of seagrasses, dominated by the *Posidonia australis* (ribbon-weed) species (Fig. 5).

The tidal northern reaches of the gulfs also comprise the largest areas of temperate mangroves and salt marshes in Australia (see Fig. 5). Together, mangrove and saltmarsh communities along the South Australian coast total 82,000 ha, with the largest communities occurring in northern Spencer Gulf (46,000 ha) and Gulf St. Vincent (20,000 ha). Other substantial communities occur in lower Spencer Gulf (6,000 ha) (Edyvane 1999a). Mangroves are dominated by the Grey Mangrove, *Avicennia marina*, one of two mangrove species recorded from temperate Australia.

These highly productive intertidal environments provide key habitat for organisms of both marine and terrestrial origin: e.g., roosting sites for migratory wading birds; feeding and refuge areas for fish (Connolly et al. 1997). Tidal saltmarshes are also an essential hydrological buffer between seaward mangroves and terrestrial ecosystems, regulating salinity and water velocity, and decreasing the suspended sediment load entering the marine environment from land.

Upper Spencer Gulf is a recognised Wetland of National Importance characterised by extensive intertidal mangrove, saltmarsh and sand and mud flats which are linked offshore to extensive seagrass meadows, deeper channels with strong currents, and sandy plain habitats. Sandy and shell grit beaches are also present. Rocky shorelines, headlands and reefs occur mainly on the western coastline, with those near Point Lowly recognised as the world's only know mass spawning aggregation of the Giant Australian Cuttlefish (*Sepia apama*) (Dupavillon and Gillanders 2009).

The western king prawn, *Penaeus latisulcatus*, is a decapod crustacean of the family Penaeidae. *P. latisulcatus* has been reported from the Indo-West Pacific region, the Red Sea, and Arabian Gulf in the west, through Malaysia, Korea and Japan to the north and through Indonesia to New Guinea and Australia to the south (Grey et al. 1983). This species generally only forms high level stocks in areas associated with the hypersaline waters of marine embayments (Kailola et al. 1993), such as Upper Spencer Gulf. The world's largest population of this species occurs in Upper Spencer Gulf, being genetically different to other populations found in Australia (Richardson 1982).

The Giant Australian Cuttlefish (*Sepia apama*) (Fig. 6) is the world's largest sub-population of cuttlefish, growing to 50 cm in mantle length and over 10.5 kg in weight (Reid et al. 2005). Using cells known as chromatophores, the cuttlefish can put on spectacular displays, changing color in an instant act of *camouflage* which protects the species from its predators (e.g. dolphins). Cuttlefish live for 1–2 years only. Recent studies have indicated that these cuttlefish may be a separate sub-population being genetically different to other cuttlefish populations that exist across southern Australia (Bronwyn Gillanders, 2011, personal communication).

Historically, each winter tens of thousands of Giant Australian Cuttlefish aggregated at the inshore reefs in the Point Lowly region to mate and lay their eggs. Many of the cuttlefish die following mating, but there is a proportion of the population that returns again the following year to repeat the process before dying (Hall et al. 2007). If there is recruitment failure in successive years then the entire population

Fig. 5 Images of seagrass (*Posidonia australis*) (Collings et al. 2006), salt marshes (Image credit Peripitus/Wikimedia), and Barker Inlet mangroves at low tide (Image source: http://www.waterwatchadelaide.net.au)

Fig. 6 Images of the Giant Australian Cuttlefish (*Sepia apama*) in Upper Spencer Gulf (Photo credit: Dan Monceaux), the male leafy seadragon (*Phycodurus eques*) in Gulf St. Vincent (Photo credit: Brian Scupham), and the Great White Shark (*Carcharodon carcharias*) (Photo credit: James Bradley, http://cityoftongues.com/writing/great-white-tale/)

could be at risk. The aggregation has been internationally renowned and attracted tourist attention due to the sheer past number of cuttlefish and their elaborate, colourful mating rituals.

Key protected marine species in the region include various species of dolphins, turtles and seals, the Australian Sea Lion (*Neophoca cinerea*), Great White Shark (*Carcharodon carcharias*), various species of seahorses, seadragons including the popular, endemic Leafy Seadragon (*Phycodurus eques*) (Fig. 6), and pipefish. Browne and Smith (2007) have recently discovered a new pipefish species in both Spencer Gulf and Gulf St. Vincent. See Knight and Vainickis (2011) for a full list of protected and endangered species including seabirds, noting that the Giant Australian Cuttlefish has not been listed as a threatened species yet.

Two main species of dolphin inhabit South Australian coastal waters: the bottlenose dolphin, *Tursiops truncatus*; and the common dolphin, *Delphinus delphis* (Long 1996; Long et al. 1997). The bottlenose dolphin is distributed widely in coastal and inshore regions of tropical and temperate waters of the world, and in Australia it is found in all states and the Northern Territory. Male bottlenose dolphins reach physical and sexual maturity at about 15 years of age and have a lifespan of about 43 years (Long 1996). Females calve every 3–6 years and have a gestation period of 12.3 months. A mature adult will consume about 10 kg of fish and squid per day.

According to ecologist Dr Mike Bossley, who has studied the Port River dolphins for over 25 years, Dreaming Stories (legends) of the local Aboriginal Kaurna people indicate that dolphins were abundant in the estuary prior to European colonisation (Mike Bossley, 2008, personal communication). However, historical records indicate that dolphins were rarely sighted for about a century leading up until the 1970s, presumably as a result of pollution and habitat damage. In the mid 1970s a canal estate housing development resulted in clean sea water being piped into the upper reaches of the Barker Inlet substantially improving the water quality. It appears that dolphins began to recolonise the estuary from about that time. The Adelaide Dolphin Sanctuary formally proclaimed in 2005 for the protection of the Port River dolphins. In 2008, the estimated number of dolphins resident in the sanctuary was 30 or more (Anonymous 2008).

Wetlands, Waterbirds and Marine Protected Areas

The Directory of Important Wetlands in Australia (Anonymous 2001) contains a total of 69 South Australian wetland sites. 17 wetlands are classified as intertidal mud, sand or salt flats, of which 16 sites are also classified as marine waters.

In Spencer Gulf there are three coastal wetlands of national significance, namely Upper Spencer Gulf, Franklin Harbour and Tumby Bay. Gulf St. Vincent has a total of five listed coastal wetlands. Two listed wetlands are located in the gulf's upper reaches (Clinton and Wills Creek); the other three listed wetlands are Barker Inlet & St Kilda, Port Gawler & Buckland Park Lake, and the Onkaparinga Estuary, which technically is a sub-estuary of Gulf St. Vincent. In addition, there are three listed wetlands in Investigator Strait on the northeastern side of Kangaroo Island; that is, the American River Wetland System, Busby and Beatrice Inlets (north of Kingscote) and Cygnet Estuary (south of Kingscote).

These coastal wetlands provide important feeding and nesting habitat for many significant seabirds and migratory birds. For instance, the Upper Spencer Gulf has been identified by BirdLife International[1] as an "Important Bird Area" because it supports over 1 % of the world population of Red-necked Stints. Other waders and waterbirds sometimes recorded in significant numbers include Red Knots, Sharp-tailed Sandpipers, Banded Stilts, Pied Oystercatchers, Australian Shovelers and Fairy Terns. Chirruping Wedgebills and Rock Parrots have also been recorded.

Another Important Bird Area is the Sir Joseph Banks Group archipelago in lower Spencer Gulf because it supports over 1 % of the world populations of White-faced Storm Petrels (with up to about 180,000 breeding pairs), Cape Barren Geese (up to about 1,200 individuals), and Black-faced Cormorants (from 3,000 to 5,000 breeding pairs). Other seabirds which breed in the archipelago include Little Penguins, Silver Gulls and Greater Crested Terns. Fairy Terns and Eastern Reef Egrets have also been recorded. The northern and northeastern reaches of Gulf St. Vincent are also listed as an Important Bird Area.

Marine protected areas are nowadays a mainstream management tool for conserving biodiversity and assisting resource management worldwide. Several international, national, and local level initiatives and mechanisms serve to advance marine protected areas as vehicles for promoting the long-term conservation and sustainable use of marine resources and biodiversity (Agardy et al. 2003). The proliferation of these protected areas has been astounding – whereas 20 years ago a scant handful existed, now virtually every coastal country has implemented at least one marine protected area. The first marine protected areas were proclaimed early in the twentieth century. De Silva et al. (1986) listed 430 marine protected areas worldwide created by 1985, but most of those covered relatively small coastal areas. Many more marine protected areas were proclaimed in the last two decades of the twentieth century. By 1995 there were globally at least 1,306 sub-tidal marine protected areas with a median size of 1,584 ha (Kelleher et al. 1995).

In Australia, state/territory laws apply generally to coastal waters (up to three nautical miles from the shore) and Commonwealth laws apply from those waters out to the limit of the Australian fishing zone (200 nautical miles offhore). The Great Australian Bight Marine Park, established in 1995, was South Australia's first Commonwealth marine park. The State's first aquatic state reserves were established earlier in 1971. These reserves, such as Port Noarlunga Reef and Onkaparinga Estuary Aquatic Reserve, covered relatively small regions of local ecologic significance. Despite the great ecologic significance of South Australia's gulfs, particularly their northern reaches, it took until 2007 for the State Government for pass legislation to create a total of 19 marine parks covering 44 % of State waters. Although this degree of spatial coverage sounds impressive, it should be noted that no-fishing, sanctuary zones will cover only 6 % of State waters. The State Government of South Australia has enforced these marine parks and their boundaries only recently (November 2012).

Past Environmental Degradation

Since colonization the marine habitat of South Australian gulfs has been severely degraded through human impacts. Sources of impacts include storm water and waste water discharges, modification of the coast changing sediment drift patterns and destroying salt marshes, commercial and recreational fishing, seabed trawling, introduction of marine pests, destruction from marine debris, shipping traffic and oil spills.

Most of the degradation occurred along the Adelaide metropolitan coast since major urbanization. Up to 70 % canopy-forming algae (kelp forests) were lost (Connell et al. 2008), the seagrass coverage declined from 80 % in 1949 to 28 % in 1993 (Hart 1997), and approximately 25 % of mangroves have been lost, most in the Barker Inlet near the Bolivar wastewater outfall (Connolly 1986; Edyvane 1991). Although saltmarshes have legislative protection under South Australia's Native Vegetation Act (1990), approximately 80 % of tidal saltmarshes and all saltwater teatree (*Melaleuca* spp.), reed beds (Phragmites australis) and native grasses have been lost in the metropolitan region of Gulf St. Vincent since 1954 (see Edyvane 1999a, b).

[1] See http://www.birdlife.org/

Fig. 7 Major discharge of stormwater from the Torrens River into Adelaide's coastal waters on 25 October 2005 (Photo credit: Simon Bryars)

The Adelaide Coastal Water Study (Fox et al. 2007), undertaken in 2003–2005, provided an integrated understanding of the causes of water quality decline, seagrass loss and sediment instability along Adelaide's nearshore coastal waters. The main causes of this decline identified were excessive sediment and nutrient loads of stormwater runoff (Fig. 7) and high nutrient loads of wastewater and industrial discharges. The study recommended substantial reductions of both nitrogen discharges by 75 % and particulate matter loads by 50 % (relative to 2003 levels). Management of highly transient stormwater runoff appears to pose the biggest hurdle for the Environmental Protection Authority to substantially improve water quality in coastal waters (Peter Pfennig, EPA, 2012, personal communication).

Another ongoing environmental issue in Adelaide's waterways is the control of a number of invasive species. Of most concern is the "aggressive" invasive seaweed *Caulerpa taxifolia* in the Port River and West Lakes (see Fig. 1), first detected in 2002.[2] This species has now infested most parts of the Port River and Barker Inlet.

For more than 100 years, Upper Spencer Gulf has supported a number of heavy industries and urban centres (Whyalla, Port Augusta, Port Pirie) discharging wastewater and other pollutants to the marine environment. This includes heavy metals from the lead smelter at Port Pirie, steelworks at Whyalla and power stations near Port Augusta. In the early 1900s mining leases covered 97,000 ha of seagrass meadow area in Spencer Gulf. In the years preceding the First World War, Posidonia fibre was harvested by dredges near Port Broughton for its high cellulose content and used in the manufacture of suits, explosives and household products (Winterbottom 1917). Large clearings can still be detected and the seagrass has not regrown. A major oil spill of the vessel *ERA* near Port Bonython (Point Lowly) in 1992, discharging 300 tonnes of bunker oil, has affected nearly 100 ha of mangroves and 23 ha have died.[3] A recent survey found ten introduced invasive species between Whyalla marina and Fitzgerald Bay (Sabine Dittmann, 2012, personal communication). One of these, a pearl oyster, is now well established in Upper Spencer Gulf and in some cases has replaced native razorfish beds. Habitat changes attributed to prawn trawling have also been documented in Spencer Gulf (Currie et al. 2011).

Toward the end of summer in January/February of 1993 the northeastern coast of Spencer Gulf experienced a major dieback of seagrass affecting nearly 13,000 ha of intertidal and shallow subtidal seagrasses (*Amphibolis antarctica*) (Seddon et al. 2000). The rapid decline was likely caused by natural factors (strong atmospheric heating in conjunction with usually low tides, called negative tides) (Seddon and Cheshire 2001).

In 1998 overfishing of cephalopods (cuttlefish, southern calamary and octopus) in Upper Spencer Gulf has triggered a spatial closure for the take of all cephalopods. Despite this closure, the Giant Australian Cuttlefish breeding aggregation at Point Lowly has been in serious decline for some years (Hall 2012). In the years 1999–2001, the estimated cuttlefish abundance was of the order of 170,000–180,000 animals declining to 75,000 in 2008, and 38,000 in 2011. This season (2012) a local diver reported an alarmingly low number of approximately 6,000 animals, corresponding to only 3.5 % of the 1999 population. Reasons for this dramatic decline, with the Upper Spencer Gulf cuttlefish being on the brink of extinction, are still uncertain. For completeness it should be noted that there have been hydrocarbon leaks from the SANTOS hydrocarbon processing plant into the groundwater aquifer near Port Bonython (Point Lowly) from 2008 for 2–3 years (Anonymous 2010). The extent of groundwater contamination and marine impacts caused by these leaks are still unverified.

Community Efforts

A number of larger community environmental groups were formed in the last decade with the main purpose of conservation and protection of the marine environment. These groups were engaged in invaluable activities including making many formal submissions to city councils and government agencies in marine matters, and organizing numerous community awareness events.

South Australia's first larger marine community group – *Friends of Gulf St. Vincent* – was formed in 2003 under the lead of Ina Patricia (Pat) Harbison, Dr Scoresby Shepherd,

[2] See http://www.pir.sa.gov.au/

[3] See http://www.amsa.gov.au/marine_environment_protection/major_oil_spills_in_australia/

and Dr Ian Kirkegaard. In 2007 this group (together with Dr Simon Bryars and Dr John Jennings) edited the book "Natural history of Gulf St. Vincent" (Shepherd et al. 2008), which provided the first comprehensive multidisciplinary description of the gulf. The mission of this group is to raise community awareness of the need for integrated natural resource management of Gulf St Vincent. In 2009, Dr John Caldecott became the president of Friends of Gulf St. Vincent. In the same year, John launched the *Water Action Coalition*, whose mission is to ensure a sustainable water future of Adelaide.

In 2006, after several years of severe drought conditions, the State Government announced to build a large (100 Gigalitres per annum) seawater desalination plant at Port Stanvac, located 22 km south of Adelaide's city centre. Members of the community and scientists were particularly concerned with the environmental harm caused by the marine discharge of toxic desalination brine. The *Save Our Gulf Coalition* formed in 2008 under the lead of environmentalist Peter Laffan had the mission to stop the desalination plant from being constructed.

In 2010, environmentalists Ruth Trigg and Corrie Vanderhoek formed the *Save Our Gulfs Embassy* combining the work of *Save Our Gulf Coalition*, the *River Lakes & Coorong Action Group* and the *Fresh Water Embassy*, to focus on the two South Australian gulfs.

The Port Stanvac desalination plant received final governmental approval in 2010 and is currently commissioned for full operation. The Government has just (October 2012) announced that the desalination plant may be put on stand-by mode until needed because other water sources are currently cheaper to run Adelaide's water supply.

In 2007, BHP Billiton announced a substantial increase of mining production on the Eyre Pensinula, including open-pit mining, known as the *Olympic Dam Expansion*. In conjunction with this, BHP Billiton announced plans to build a large (100 gigalitres per year) seawater desalination plant at Point Lowly to produce freshwater needed for the mining operations, a deep-water port at Point Lowly and a landing facility in Port Augusta. Draft and supplementary Environmental Impact Statements were submitted in 2009 and 2011, and the Government approval followed promptly in October 2011. A couple of month ago (August 2012) BHP Billiton decided to shelve the Olympic Dam Expansion while asking the State Government for a 4-year extension of the approval conditions.

Local environmentalist Tom Cheesman and Dr Andrew Melvin-Smith formed the *Save Point Lowly* group in Whyalla with the mission to prevent the industrial destruction of the cuttlefish breeding habitat. In 2010 Adelaide's filmmakers Dan Monceaux and Emma Sterling started to document the industrial developments in the Upper Spencer Gulf region in a project entitled "Cuttlefish Country". While the documentary is still work in progress, their *Cuttlefish Country* initiative can be regarded as the first modern Internet-based environmental campaigns in South Australia involving social media, opinion polls, and video clips of interviews with politicians, scientists and locals. Other important environment groups in South Australia are the Wilderness Society, strongly supporting the establishment of marine parks, and the Conservation Council of South Australia (Conservation SA) representing over 50 member groups whose main purpose is conservation and protection of the environment.

Future Threats

The destruction of natural habitat in South Australia will continue to be predominantly driven by "classical" factors such as urban development, wastewater and industrial pollution, stormwater runoff, ship traffic, invasive species, oil spills, trawling and dredging. Population growth in the Adelaide region will have further detrimental environmental consequences on a regional level. In principle, it should be possible to keep these environmental impacts at a minimum.

The by far biggest threat to ecologic health of South Australian gulfs is the planned massive industrialization of the Upper Spencer Gulf region, that the South Australian government classifies as the Heavy Industry Hub.[4] Despite the current delay of BHP Billiton's Olympic Dam expansion and possible extinction risk of the Giant Australian Cuttlefish, South Australia's current government has recently confirmed that this region will remain as the future's main export corridor of mining products.

Major industrial developments in the Spencer Gulf region imply the construction or expansion of several deep-water ports. Approved developments include the BHP Billiton Landing Facility in Port Augusta, the OneSteel (now called Arrium Limited) Harbor expansion in Whyalla, and the Lucky Bay harbor extension (10 km east of Cowell). Proposed projects include the Port Bonython expansion, the Port Spencer construction (20 km southwest of Port Neill), a new wharf at Backy Point Wharf (12 km north of Point Lowly), and modifications of the Port Pirie harbor. Other developments include a Mitsubishi Diesel Storage and Refinery at Point Lowly (approved), ISR uranium mines, south of Whyalla (approved) and the planned Arafura Rare Earths (and uranium) Processing Plant, Whyalla. In addition, a number of seawater desalination plants are proposed. Apart from BHP Billiton's plant at Point Lowly, Breamar Alliance plans to construct a 50 gigalitres per year plants near Port Germein (20 km north of Port Pirie), and Centrex Metals a plant of the same size near Port Spencer. The biggest

[4] See http://www.pir.sa.gov.au/regions/initiatives/usg_heavy_industry_hub

immediate environmental hazard of desalination brine discharges is the development of *dead zones* devoid of dissolved oxygen (Kämpf et al. 2009).

The bizarre circumstance is that the State Government's vision of a "Heavy Industry Hub" in the Upper Spencer Gulf is supposed to coexist with a Wetland of National Significance and the planned Upper Spencer Gulf Marine Park. Whether protection of the precious Upper Spencer Gulf in a zone of heavy industries and ship traffic is possible is more than doubtful. Overall, it is more than likely that nature degradation by traditional industrialization will outweigh that from direct climate change impacts.

Acknowledgements This work has not received financial support. The author thanks Dan Monceaux and Emma Sterling for the provision of background information.

References

Agardy T, Bridgewater P, Crosby MP, Day J, Dayton PK, Kenchington R, Laffoley D, McConney P, Murray PA, Parks JE, Peau L (2003) Dangerous targets: differing perspectives, unresolved issues, and ideological clashes regarding marine protected areas. Aquat Conserv Mar Freshw Ecosyst 13:1–15

Alley NF, Lindsay JM (1995) Tertiary: Pirie basin. In: Drexel JF, Preiss WV (eds) The geology of South Australia, vol 2, The phanerozoic, geological survey of South Australia, Bulletin, 54. Geological Survey of Australia, Adelaide, pp 175–178

Anonymous (1985) Settlers on the hill, a local history of Para hills. City of Salisbury, Salisbury, pp 5

Anonymous (2001) A directory of important wetlands in Australia, 3rd edn. Australian Nature Conservation Agency, Canberra

Anonymous (2008) Adelaide dolphin sanctuary management plan. Department for Environment and Heritage, Adelaide. ISBN 1 921238 07 0

Anonymous (2010) Further contamination at Santos' Port Bonython facility. Media release, Environmental Protection Authority (EPA), Adelaide, 18 Nov 2010. http://www.epa.sa.gov.au/xstd_files/-Corporate/Medianews/media228.pdf. Accessed 14 Jan 2013

Berndt RM (1985) Traditional aborigines. In: Twindale CR, Tyler MJ, Davies M (eds) Natural history of the Eyre Peninsula. Royal Society of South Australia, Adelaide, pp 127–138

Browne RK, Smith K (2007) A new pipefish, *Stigmatopora narinosa* (Syngnathidae) from South Australia. Mem Mus Vic 64:1–6

Bye JAT, Harbison IP (1991) Transfer of inland salts to the marine environment at the head of Spencer Gulf, South Australia. Palaeogeogr Palaeoclimatol Palaeoecol 84(1–4):357–368

Carmichael EJ, Mudie I (1973) History of settlement and development. In: Corbett D (ed) Yorke Peninsula – a natural history. Department of Adult Education University of Adelaide, Adelaide, pp 167–204

Chapman RW (1892) On the dodging tide of South Australia. Aust Assoc Adv Sci 4:261

Collings G, Miller D, O'Loughlin E, Bryars S (2006) Turbidity and reduced light responses of the meadow-forming seagrasses Amphibolis and Posidonia from the Adelaide metropolitan coast. ACWS technical report no. 12 prepared for the Adelaide Coastal Waters Study Steering Committee. South Australian Research and Development Institute (Aquatic Sciences) Publication No. RD01/020817, Adelaide

Connell SD, Russell BD, Turner DJ, Shepherd SA, Kildea T, Miller D, Airoldi L, Cheshire A (2008) Recovering a lost baseline: missing kelp forests from a metropolitan coast. Mar Ecol Prog Ser 360:63–72

Connolly RM (1986) Relation of near-shore benthic flora of the Barker Inlet and northern beaches region to pollution sources – with emphasis on *Ulva* distribution. Department of Environment and Planning, Adelaide

Connolly RM, Dalton A, Bass DA (1997) Fish use of an inundated saltmarsh flat in a temperate Australian estuary. Aust J Ecol 22:222–226

Cooper BJ (1985) The Cainozoic St. Vincent Basin; tectonics, structure, stratigraphy. Spec Publ S Aust Dept Mines Energy 5:35–49

Corlis NJ, Veeh HH, Dighton JC, Herczeg AL (2003) Mixing and evaporation processes in an inverse estuary inferred from [Delta]H and [Delta]O. Cont Shelf Res 23(8):835–846

Currie DR, Dixon CD, Roberts SD, Hooper GE, Sorokin SJ, Ward TM (2011) Relative importance of environmental gradients and historical trawaling effort in determining the composition and distribution of benthic macro-biota in a large inverse estuary. Fish Res 107:184–195

De Silva ME, Gately EM, Desilvestre I (1986) A bibliographic listing of coastal and marine protected areas: a global survey, Woods hole oceanographic institution technical report. Woods Hole Oceanographic Institution, Woods Hole

de Silva Samarasinghe JR (1998) Revisiting Upper Gulf St. Vincent in South Australia: the salt balance and its implications. Estuar Coast Shelf Sci 46(1):51–63

de Silva Samarasinghe JR, Bode L, Mason LB (2003) Modelled response of Gulf St. Vincent (South Australia) to evaporation, heating and winds. Cont Shelf Res 23(14–15):1285–1313

Dupavillon JL, Gillanders BM (2009) Impacts of seawater desalination on the giant Australian cuttlefish Sepia apama in the upper Spencer Gulf, South Australia. Mar Environ Res 67:207–218. doi:10.1016/j.marenvres.2009.02.002

Easton AK (1978) A reappraisal of the tides in Spencer Gulf, South Australia. Aust J Mar Freshw Res 29:467–477

Edyvane KS (1991) Pollution! The death knell of our mangroves? Safic 16:4–7

Edyvane KS (1996) Marine conservation and marine protected areas in South Australia (summary). In: State of the marine environment report for Australia. Technical summary, compiled by Zann L. Great Barrier Reef Marine Park Authority, pp 482–487

Edyvane KS (1999a) Conserving marine biodiversity in South Australia – Part 1 – Background, status and review of approach to marine biodiversity conservation in South Australia, SARDI report number 38, PIRSA, Adelaide

Edyvane KS (1999b) Coastal and marine wetlands in Gulf St. Vincent, South Australia – their loss and degradation. Wetl Ecol Manag 7:83–104

Edyvane KS (2005) Australia, coastal ecology. In: Schwartz M (ed) Encyclopedia of coastal science. Springer, Dordrecht, pp 96–109

Flinders M (1814) In: Nicol G, Nicol W (eds) The voyage to Terra Australis, vol 1. G and W Nicol, London

Fox DR, Batley GE, Blackburn D, Bone Y, Bryars S, Cheshire A, Collings G, Ellis D, Fairweather P, Fallowfield H, Harris G, Henderson B, Kämpf J, Nayar S, Pattiaratchi C, Petrusevics P, Townsend M, Westphalen G, Wilkinson J (2007) Adelaide coastal waters study, final report, vol 1, Summary of Study Findings, CSIRO, June 2007

Fuller MK, Bone Y, Gostin VA, von der Borch CC (1994) Holocene cool-water carbonate and terrigenous sediments from southern Spencer Gulf, South Australia. Aust J Earth Sci 41(4):353–363

Gostin VA, Hails JR, Belperio AP (1984) The sedimentary framework of northern Spencer Gulf, South Australia. Mar Geol 61 (2–4):111–138

Gostin VA, Belperio AP, Cann JH (1988) The Holocene non-tropical coastal and shelf carbonate province of southern Australia. Sediment Geol 60:51–70

Grey DL, Dall W, Baker A (1983) A guide to the Australian penaeid prawns. Northern Territory Government Printing Office, Darwin

Hails JR, Belperio AP, Gostin VA, Sargent GEG (1984) The submarine Quaternary stratigraphy of northern Spencer Gulf, South Australia. Mar Geol 61(2–4):345–372

Hall KC (2012) Estimated abundance and biomass of giant Australian cuttlefish Sepia apama in the spawning aggregation area of northern Spencer Gulf, South Australia. Report prepared for the Olympic Dam EIS Project

Hall KC, Fowler AJ, Geddes MC (2007) Evidence for multiple year classes of the giant Australian cuttlefish *Sepia apama* in northern Spencer Gulf, South Australia. Rev Fish Biol Fish 17:367–384

Harris PT (1994) Comparison of tropical, carbonate and temperate, siliciclastic tidally dominated sedimentary deposits: examples from the Australian continental shelf. Aust J Earth Sci 41:241–254

Hart DGD (1997) Near-shore seagrass change between 1949 and 1996 mapped using digital aerial orthophotography. Metropolitan Adelaide area: Largs Bay – Aldinga, South Australia. A report for the Environmental Protection Authority, Department of Environment and Natural Resources, South Australia and SA Water. Image Data Services, Resource Information Group, Department of Environment and Natural Resources, Adelaide

Hemer MA, Bye JAT (1999) The swell climate of the South Australian Sea. Trans R Soc S Aust 123:107–113

James NP, von der Borch CC (1991) Carbonate shelf edge off southern Australia: a prograding open-platform margin. Geology 19:1005–1008

James NP, Bone Y, Hageman SJ, Feary DA, Gostin VA (1997) Cool-water carbonate sedimentation during the terminal Quaternary sea-level cycle: Lincoln Shelf, southern Australia. In: James NP, Clarke JAD (eds) Cool-water carbonates, SEPM (Society for Sedimentary Geology) special publication no.56. SEPM (Society for Sedimentary Geology), Tulsa, pp 53–75

Kailola PJ, Williams MJ, Stewart PC, Reichelt RE, McNee A, Grieve C (1993) Australian fisheries resources. Bureau of Resource Sciences and the Fisheries Research and Development Corporation, Canberra, p 422

Kämpf J, Bell D (2013) The Murray/Coorong estuary: meeting of the waters? In: Wolanski E (ed) Estuaries of the world. Springer, Dordrecht, pp 31–47

Kämpf J, Sadrinasab M (2006) The circulation of the Persian Gulf: a numerical study. Ocean Sci 2:1–15

Kämpf J, Doubell M, Griffin D, Matthews RL, Ward TM (2004) Evidence of a large seasonal coastal upwelling system along the Southern Shelf of Australia. Geophys Res Lett 31, L09310. doi:10.1029/2003GLO19221

Kämpf J, Brokensha C, Bolton T (2009) Hindcasts of the fate of desalination brine in large inverse estuaries: Spencer Gulf and Gulf St. Vincent, South Australia. Desalination Water Treat 2:325–333

Kämpf J, Payne N, Malthouse P (2010) The marine connectivity in a large inverse estuary. J Coast Res 26(6):1047–1056. doi:10.2112/JCOASTRES-D-10-00043.1

Kelleher G, Bleakley C, Wells S (1995) A global representative system of marine protected areas, 4 vols, GBRMPA/IUCN and World Bank, Washington D.C.

Knight MA, Vainickis AA (2011) Interactions with threatened, endangered or protected species in South Australian managed fisheries – 2008/09, 2009/10, and 2010/11. Report to PIRSA fisheries and aquaculture. SARDI publication no. F2009/000544-2. SARDI research report series no. 593, South Australian Research and Development Institute (Aquatic Sciences), Adelaide

Knight MA, Tsolos A, Doonan AM (2005) South Australian fisheries and aquaculture information and statistics report, SARDI research report series no. 67. SARDI, Adelaide, 79pp

Krichauff S (2011) A journey through Narungga history. Wakefield Press, Adelaide, xi + 203 pp. ISBN 978 1 86254 910 4

Lennon GW, Bowers DG, Nunes RA, Scott BD, Ali M, Boyle J, Wenju C, Herzfeld M, Johansson G, Nield S, Petrusevics P, Stephanson P, Suskin AA, Wijffels SEA (1987) Gravity currents and the release of salt from an inverse estuary. Nature 327:695–697

Leonard M (2003) Small moves towards a big event in South Australia. Aus Geo News 70:4–5

Li Q, James NP, McGowran B, Bone Y, Cann J (1998) Synergetic influence of water masses and Kangaroo Island barrier on foraminiferal distribution, Lincoln and Lacepede shelves, South Australia: a synthesis. Alcheringa 22(1–2):153–176

Long M (1996) Cadmium accumulation and toxicity in the bottlenose dolphin, Tursiops truncatus *and the common dolphin,* Delphinus delphis *in South Australia.* Honours thesis, Department of Botany, The University of Adelaide

Long M, Reid RJ, Kemper C (1997) Cadmium accumulation and toxicity in the bottlenose dolphin *Tursiops truncatus*, the common dolphin *Delphinus delphis*, and some bottlenose prey species in South Australia. Aust Mammal 20:25–33

Nunes RA, Lennon GW (1986) Physical property distributions and seasonal trends in Spencer Gulf, South Australia: an inverse estuary. Aust J Mar Freshw Res 37:39–53

Nunes RA, Lennon GW, Bowers DG (1990) Physical behaviour of a large negative or inverse estuary. Cont Shelf Res 10:277–304

Petrusevics PM (1993) SST fronts in inverse estuaries, South Australia—indicators of reduced gulf-shelf exchange. Aust J Mar Freshw Res 44:305–323

Reid A, Jereb P, Roper CFE (2005) Family Sepiidae. In: Jereb P, Roper CFE (eds) Cephalopods of the world. An annotated and illustrated catalogue of species known to date. Volume 1. Chambered nautiluses and sepioids (Nautilidae, Sepiidae, Sepiolidae, Sepiadariidae, Idiosepiidae and Spirulidae), vol 1, FAO species catalogue for fishery purposes no. 4. FAO, Rome, pp 57–152

Richardson BJ (1982) Geographical distribution of electrophoretically detected protein variation in Australian commercial fisheries. III Western king prawn, *Penaeus latisulcatus* Kishinouye. Aust J Mar Freshw Res 33:933–937

Sandiford M (2003) Neotectonics of southeastern Australia; linking the Quaternary faulting record with seismicity and in situ stress. Special paper – Geological Society of America, 372, pp 107–119. doi:10.1130/0-8137-2372-8.107

Schluter C, Bye JAT, Harbison P (1995) The most vigorous South Australian tide. Trans R Soc S Aust 119(3):123–132

Seddon S, Cheshire AC (2001) Photosynthetic responses of Amphibolis Antarctica and Posidonia australis to temperature and desiccation using chlorophyll fluorescence. Mar Ecol Prog Ser 220:119–130

Seddon S, Connolly RM, Edyvane K (2000) Large-scale seagrass dieback in northern Spencer Gulf, South Australia. Aquat Bot 66:297–310

Shepherd SA, Hails JR (1984) The dynamics of a megaripple field in northern Spencer Gulf, South Australia. Mar Geol 61 (2–4):249–263

Shepherd SA, Bryars S, Kirkegaard I, Harbison P, Jennings JT (eds) (2008) Natural history of Gulf St. Vincent. Royal Society of South Australia, Adelaide, 496pp

Smith WR (1930) How Spencer Gulf came into existence. In: Myths and legends of the Australian Aboriginals. George G Harrap and Co. Ltd., London, pp 168–170

Smith SV, Veeh HH (1989) Mass balance of biogeochemically active MATERIALS (C, N, P) in a hypersaline gulf. Estuar Coast Shelf Sci 29(3):195–215

Tindale NB (1974) Aboriginal tribes of Australia: their terrain, environmental controls, distribution, limits, and proper names. Australian National University Press, Canberra

Winterbottom DC (1917) Marine fibre, Bulletin no. 4. Department of Chemistry. Government Printer, Adelaide

Turbulent Mixing and Sediment Processes in Peri-Urban Estuaries in South-East Queensland (Australia)

Hubert Chanson, Badin Gibbes, and Richard J. Brown

Abstract

An estuary is formed at the mouth of a river where the tides meet a freshwater flow and it may be classified as a function of the salinity distribution and density stratification. An overview of the broad characteristics of the estuaries of South-East Queensland (Australia) is presented herein, where the small peri-urban estuaries may provide an useful indicator of potential changes which might occur in larger systems with growing urbanisation. Small peri-urban estuaries exhibit many key hydrological features and associated ecosystem types of larger estuaries, albeit at smaller scales, often with a greater extent of urban development as a proportion of catchment area. We explore the potential for some smaller peri-urban estuaries to be used as 'natural laboratories' to gain some much needed information on the estuarine processes, although any dynamic similarity is presently limited by a critical absence of in-depth physical investigations in larger estuarine systems. The absence of detailed turbulence and sedimentary data hampers the understanding and modelling of the estuarine zones. The interactions between the various stakeholders are likely to define the vision for the future of South-East Queensland's peri-urban estuaries. This will require a solid understanding of the bio-physical function and capacity of the peri-urban estuaries. Based upon the current knowledge gap, it is recommended that an adaptive trial and error approach be adopted for their future investigation and management strategies.

Keywords

Peri-urban estuaries • Mixing • Dispersion • Sediment processes • Water quality • Ecology • South-East Queensland • Australia

H. Chanson (✉) • B. Gibbes
School Civil Engineering, The University of Queensland, Brisbane, QLD 4072, Australia
e-mail: h.chanson@uq.edu.au http://www.uq.edu.au/~e2hchans/

R.J. Brown
Faculty of Science and Engineering, Queensland University of Technology, 2 George St, Brisbane, QLD 4000, Australia

E. Wolanski (ed.), *Estuaries of Australia in 2050 and Beyond*, Estuaries of the World,
DOI 10.1007/978-94-007-7019-5_10, © Springer Science+Business Media Dordrecht 2014

Box 1

Hubert Chanson and colleagues studied South-East Queensland (SEQ) peri-urban estuaries. These are the many small estuaries that drain highly-urbanised to semi-urbanised small catchments along Moreton Bay. They explore the potential for some smaller peri-urban estuaries to be used as 'natural laboratories' to gain some much needed information on the estuarine processes of the larger estuaries, such as the Brisbane River estuary, which is the dominant estuary in SEQ and which is virtually not studied. They provide detailed turbulence and sedimentary data to help advance science-based models of these estuaries. These models are needed to enable an interaction between the various stakeholders who will define the vision for the future of SEQs peri-urban estuaries.

Based upon the current knowledge gap, they recommend that an adaptive trial and error approach be adopted for the management of peri-urban estuaries.

Introduction

An estuary is formed at the mouth of a river where the tides meet a freshwater flow and where some mixing of freshwater and seawater occurs. Estuaries may be classified as a function of the salinity distribution and density stratification, and the wind, the tides and the river are usually major sources of inputs. Altogether the study of mixing in estuaries is more complicated than in rivers. Estuaries have long been important to the development of communities. Some ancient civilisations thrived in such estuarine systems, such as the lower region of the Tigris and Euphrates Rivers in Mesopotamia, the Nile River delta in Egypt and the Ganges River delta in India. Through some important scientific contributions (Fischer et al. 1979; Dyer 1997; Savenije 2005), the community gained a clearer understanding of the relative sensitivity of estuarine systems and their vulnerability to human and climatic interference. Herein an overview of the broad characteristics of the estuaries of South-East Queensland (Australia) is presented, before exploring the potential for some smaller peri-urban estuaries to be used as 'natural laboratories' to gain some much needed information on the estuarine processes. In the Australian context, small peri-urban estuaries are an emerging concept of estuaries that are increasing in number with the increasing urbanisation of the continent. Small peri-urban estuaries are an interesting sub-class of estuary in their own right, with many unique characteristics making them potentially useful environmental sentinels for larger estuarine systems. They exhibit many key hydrological features (e.g. tidal range, stratification, salt-fresh transition, turbulent mixing processes) and associated ecosystem types (e.g. mangroves, salt-marsh, riparian forests, seagrass, benthic communities) of larger estuaries, albeit at smaller spatial and temporal scales, often with a greater extent of urban development as a proportion of catchment area. The small peri-urban estuaries may provide an useful indicator of potential changes which might occur in larger estuarine systems as the trend of urbanisation grows. For the research community, the smaller spatial scales of such systems have significant advantages in terms of the logistics of field measurement programs, and offer the management organisations with some opportunity to more effectively experiment with management approaches and a reduced number of stakeholders. Furthermore it can be argued that, while adaptive management approaches are not well suited to larger environmental systems due to the lack of controllability of the system (Allen and Gunderson 2011), an adaptive management approach might be successfully applied to smaller systems with a higher level of controllability. In these smaller systems, both uncertainty and controllability are high and there is a potential for learning how the system can be manipulated.

Turbulent Mixing

In natural estuaries, turbulent mixing is one of the most important and challenging processes to investigate. Turbulent mixing exerts a controlling influence on key estuarine processes including sediment transport, storm-water runoff and associated chemical and sediment dynamics during flood events, the release of nutrient-rich wastewater into ecosystems and the exchange of chemicals between benthic and surface water systems. Why? The Reynolds number associated with estuarine flows is typically within the range of 10^6–10^7 and more. The flow is turbulent and characterised by an unpredictable behaviour, a broad spectrum of length and time scales, and its strong mixing properties: "turbulence is a three-dimensional time-dependent motion in which vortex stretching causes velocity fluctuations to spread to all wavelengths between a minimum determined by viscous forces and a maximum determined by the boundary conditions of the flow" (Bradshaw 1971, p. 17). Turbulent flows have a great mixing potential involving a wide range of vortice length scales (Tennekes and Lumley 1972; Hinze 1975). Importantly the velocity field does not map directly to the scalar field (e.g. concentration, temperature). The range of length scales is significantly different at the lower end. The turbulent length scales are bound by the Kolmogorov scale whereas the scalar length scales are bound by the Bachelor scale, which is significantly smaller than the Kolmogorov scale for liquid flows such as that in estuaries (Appendix I). This lower bound in terms of length scales is linked with the viscous dissipation process when the energy of the micro-scale turbulence is converted to heat. Appendix I presents a brief summary of the Bachelor and Kolmogorov scale calculations in estuarine zones, highlighting key differences between small and large estuaries. The turbulent and scalar length scales have practical significance for dispersion and micro scale mixing which is relevant for nutrient uptake of some estuarine and marine organisms (Batchelor 1959; Bilger and Atkinson 1992).

Although the turbulence is a pseudo-random process, the small departures from a Gaussian probability distribution constitute some key features. Further the measured data include usually the spatial distribution of Reynolds stresses, the rates at which the individual Reynolds stresses are produced, destroyed or transported from one point in space to another, the contribution of different sizes of eddy to the Reynolds stresses, and the contribution of different sizes of eddy to the rates mentioned above and to the rate at which Reynolds stresses are transferred from one range of eddy size to another (Bradshaw 1976). Turbulence in natural estuaries is neither homogeneous nor isotropic. A characterisation of turbulence must be based upon long-duration measurements at high frequency to characterise the small eddies and the viscous dissipation process, as well as the largest vortical structures to capture the random nature of the flow and its deviations from Gaussian statistical properties (Chanson 2009). The estuarine flow conditions and boundary conditions may vary significantly with the falling or rising tide. In shallow-water estuaries and inlets, the shape of the channel cross-section changes drastically with the tides. Further the stratification of the water column may hinder vertical diffusion during some wet weather periods. Simply it is far from simple to characterise in-depth the estuarine flow turbulence, and an understanding of turbulence in natural estuaries is particularly important for the accurate prediction of the fate of scalars (chemicals) that might be important for water quality. The challenges associated with field measurements are far from trivial in both large and small estuaries (Lewis 1997).

Sediment Processes

Waters flowing in rivers and estuaries have the capacity to scour channel beds, to carry particles and to deposit sediment materials, hence changing the bed morphology (Graf 1971; Chanson 1999). This phenomenon, called sediment transport, is of great economic significance. For example, to assess the risks of scouring of river banks and bridge piers; to estimate the siltation of a river mouth; to predict the possible bed form changes in estuaries and impact on navigation. Sediment transport also has a significant impact on water quality and ecosystem health through both the direct influence of suspended sediment on the light regime in the water column and the modification of benthic habitats due to erosion and deposition processes. Further many nutrients and chemicals of concern in relation to water quality and human health are often bound to sediment particles or have significant interactions with sediments.

The transported sediments are called the sediment load and some distinction is made between the bed load and the suspended load. The bed load motion characterises sediment grains rolling along the bed while the suspended load refers to grains maintained in suspension by turbulence. While the distinction is often arbitrary when both loads are of the same material, the suspended load can be considerable in fine-particle systems, as observed in the Brisbane River during the January 2011 flood (Event Monitoring Group 2011; Brown and Chanson 2012). The transport of suspended matter occurs by a combination of advective turbulent diffusion and convection (Nielsen 1992; Chanson 1999). Advective diffusion characterises the random motion and mixing of particles through the water depth superimposed to the longitudinal flow motion. Sediment motion by convection may be simplified as the entrainment of particles by very-large scale eddies: e.g. in a sharp river bend.

Outline of Contribution

Herein some particular attention is given to the potential use of small peri-urban estuaries to better understand the turbulent mixing properties that exert a controlling influence on sediment dynamics and water quality and ecosystem health. The challenges associated with up-scaling are discussed with a focus on larger estuarine systems. The estuaries of the future cannot be managed without basic understanding of the physical processes, particularly to support predictive models including computational fluid dynamics (CFD) modelling. However at present there are gaps in both knowledge and data for these systems. As outlined below, developing this understanding and addressing the current knowledge gap will be essential to explore the potential future states of estuaries, a process that is likely to place increasing emphasis on the development and use of predictive models.

Site and Geomorphological and Hydrological Settings

South East Queensland is located in the sub-tropics. The weather is characterized by wet and hot summers, and dry and mild winters. The region is home to over three million people (QOESR 2011) and has undergone significant land use changes with less than 40 % of the catchment now classified as pristine (Catterall et al. 1996). Herein we focus on the South-East Queensland estuaries between Bribie Island and the Gold Coast Seaway (Fig. 1). The coastline includes the estuaries of a few large rivers (Brisbane, Logan/Albert, Nerang, Pine, Caboolture) and of a large number of small, sometimes ephemeral streams, all discharging into Moreton Bay. The combined catchment area discharging into the Bay is 21,220 km^2 (Dennison and Abal 1999). A key feature linking all of the estuaries of South-East Queensland is their common downstream receiving environment: the Moreton Bay. From the Pumicestone Passage in the North to the Gold Coast Broadwater in the South, there are 20 estuaries connecting to Moreton Bay (Figs. 1 and 2). These estuaries are ephemeral over geologic time (Neil 1998). Infilling by tidal deltas on the eastern side of Moreton Bay and by river/estuary deltas to the West combined with fluctuations in sea level have caused Moreton Bay and its estuaries to transition between terrestrial and coastal dominated environments over geological timeframes. This pattern of ongoing transition is likely to continue with the potential for the current in-filling phase accelerated by land clearing and other anthropogenic activities that cause or accelerate discharges of sediment and chemicals to the region's estuaries and Moreton Bay (Neil 1998). It can be argued that the small peri-urban estuaries of South-East Queensland exhibit a greater degree of catchment modification in proportion to their surface area than and provide an indication of the effects of such land transformation for the region's larger estuarine systems.

An overview of basic geomorphological characteristics of South-East Queensland estuaries is provided in Table 1. The estuaries can be categorised by their catchment area into either small, medium and large estuarine systems. The region is dominated by small- (eight with catchment area <100 km^2) and medium- (nine with catchment area 100–1,000 km^2) sized estuaries. Notably the region also contains two large estuaries (Brisbane and Logan-Albert Rivers, catchment area $>1,000$ km^2) which exert a significant influence on the sediment and water quality characteristics of Moreton Bay during large rainfall events (Davies and Eyre 1998; DERM 2011). The South-East Queensland includes 4 major water storages: Lake Samsonvale on the North Pine River, Lake Somerset and Wivenhoe Reservoir on the Brisbane River, and Advancetown Lake on the Nerang River. There are a few further smaller storages, including Lake Kurwongbah, Enoggera Reservoir, Tingalpa Reservoir, Little Nerang Dam and Wyaralong Dam which have a smaller catchment area and storage capacity.

The estuaries of South-East Queensland are classified as wet and dry tropical/subtropical estuaries. The estuarine zones are typically partially mixed, although they tend to be partially stratified during ebb tides and could become stratified after some rainstorm events. Although each estuary is distinctly unique, since topography, river inflow and tidal forcing influence the shape and mixing that occur locally, the vast majority of estuaries of South-East Queensland were found to be a drowned river valley (coastal plain) type (Dyer 1973; Digby et al. 1998). The main topographical features of these coastal plain estuaries are shallow waters with large width to depth ratio, cross-sections which deepen and widen towards the mouth, a small freshwater inflow to tidal prism volume ratio, large variations of sediment type and size, a surrounding of extensive mud flats, and a sinuous central channel. All estuaries experience a mixed semi-diurnal tide with mean tidal ranges from 1.3 to 1.8 m depending on the estuary configuration (Digby et al. 1998). At the Brisbane River mouth, the tidal range is about 0.7 to 2.7 m: the mean neap tidal range is 1.0 m and the mean spring tidal range is 1.8 m. The predominant tidal constituents are the M_2, S_2 and K_1 components which have tidal periods of 12.42, 12.00 and 23.93 h respectively. Diurnal inequalities are observed in Moreton Bay under both spring and neap tidal conditions. A diurnal inequality occurs when the two tidal cycles that occur within the 25 h period of the semi-diurnal tide have different tidal amplitudes and periods.

All the estuaries draining into Moreton Bay experience a similar climate and hydrological regime (Table 2), although the annual averaged rainfall data show an East–West trend with the smaller coastal catchment experiencing higher annual average rainfall totals (1,600–2,000 mm/year) than the western edges of the larger catchments ($\approx$1,000 mm/year) (BOM 2009) (Table 2). The estuaries are characterised by short-lived, episodic, high freshwater inflows during the wet season, and very little or no flow during the dry season. During flood periods, an estuary is flushed to the mouth with freshwater. After flushing, the estuary may change from fully flushed to partially mixed (stratified) back to vertically homogeneous within a few days to a few weeks after the end of the high flow event, depending upon the flood event and river system. The seasonal, event-driven hydrology of the region causes the estuaries to operate in two distinct modes: tidally dominated periods and event dominated periods (Fig. 3).

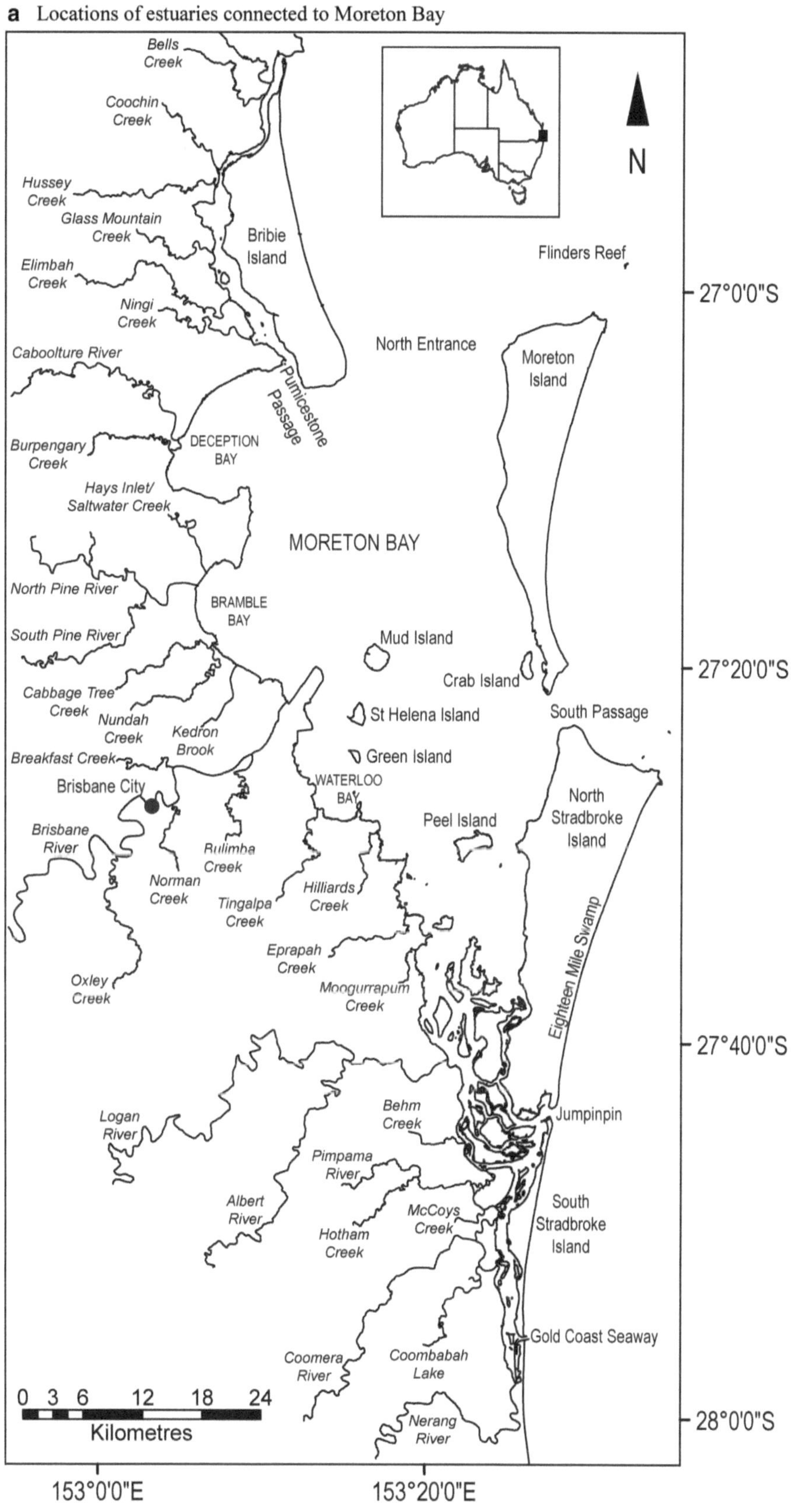

Fig. 1 Estuaries of South-East Queensland and Moreton Bay. (**a**) Locations of estuaries connected to Moreton Bay. (**b**) Sketch of Moreton Bay catchments and main water storages looking West

b Sketch of Moreton Bay catchments and main water storages looking West

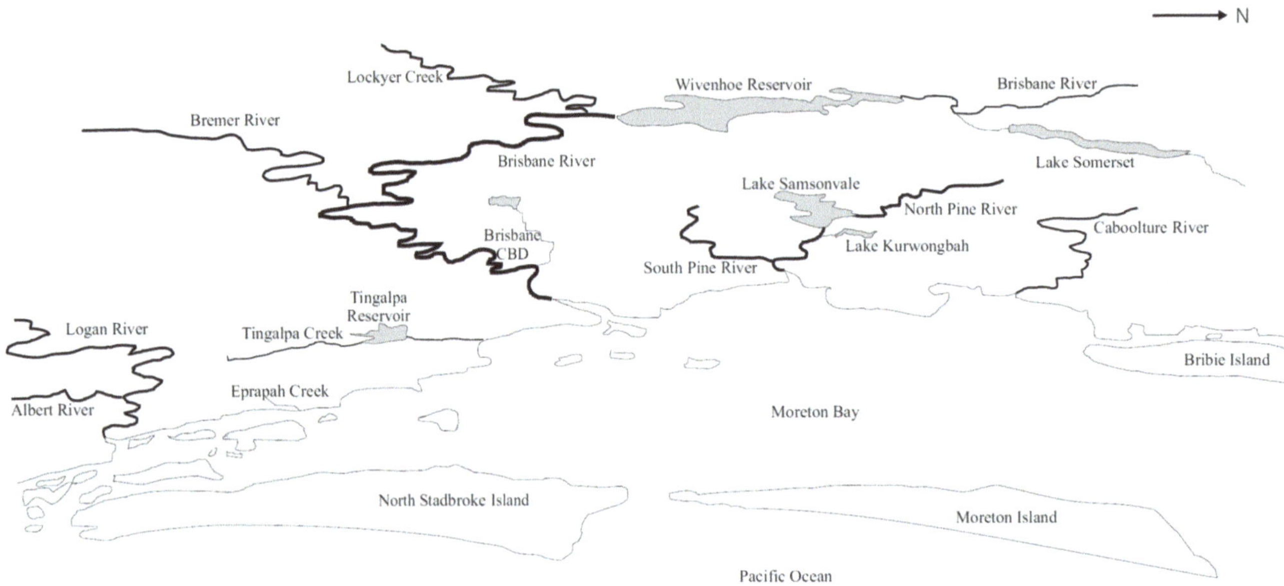

Fig. 1 (continued)

Fig. 2 Photographs of peri-urban estuaries of South-East Queensland – (**a**) Brisbane River in Brisbane (Courtesy of Brisbane Marketing). (**b**) Nudgee Creek on 2 May 2010. (**c**) Trawler in Cabbage Tree Creek on 12 Feb. 2003. (**d**) Field measurements in the upper estuarine zone (AMTD 3.1 km) of Eprapah Creek in June 2006

Figure 3 shows some salinity contours in a large river (Brisbane River) and a small system (Eprapah Creek) as functions of the average middle thread distance (AMTD) measured from the river mouth. Figure 3 highlights the contrasted salinity distributions during drought and shortly after a major event. This dual mode has shaped the sediment processing, water quality and ecosystem health dynamics of the region's estuaries. While some medium to large estuaries

Table 1 Summary of physical classification of South-East Queensland estuaries (After Digby et al. 1998)

Estuary Name	Latitude [°South]	Longitude [°East]	Classification[b]	Catchment area [km^2]	Water area [km^2]	Perimeter [km]	Maximum length [km]	Maximum width [km]	Entrance width [km]
Pumicestone Passage	−27.08	153.151	TD	702	49.88	154.0	36.16	2.8	2.27
Caboolture River	−27.15	153.044	RD	354	1.77	20	7.85	0.36	0.37
Burpengary Creek	−27.16	153.040	TD	108	0.41	8.98	2.92	0.62	0.62
[a]Hays Inelt/Saltwater Creek	−27.26	153.071	TD	74.35	2.67	11.11	3.86	1.25	1.25
Pine River	−27.28	153.063	TD	806	4.043	44.41	12.61	0.66	0.51
Nundah/Cabbage Tree Creek	−27.33	153.088	TD	131	0.41	15.02	3.11	0.17	0.14
Nudgee Creek	−27.34	153.094	TD	1.7	0.09	6.224	3.32	0.13	0.13
Brisbane Airport Floodway/ Kedron Brook	−27.35	153.111	TD	40	0.84	13.71	6.37	0.36	0.25
Brisbane River	−27.37	153.166	RD	13,643	18.67	123.54	45.88	1.34	1.75
Tingalpa Creek	−27.47	153.200	RD	150	0.9	11.96	5.13	0.6	4.1
Hilliards Creek	−27.49	153.266	TD	62	0.35	4.93	1.64	0.52	0.16
Eprapah Creek	−27.56	153.294	TD	31	0.08	3.62	1.61	0.14	0.042
[a]Moogurrapum Creek	−27.59	153.302	TD	15.1	0.06	4.17	2.05	0.13	0.13
Logan-Albert River	−27.69	153.349	RD	3,822	5.01	53.71	21.81	0.79	0.32
[a]Behm Creek	−27.76	153.360	TD	29.81	0.25	13.54	6.75	0.14	0.14
Pimpama River	−27.82	153.396	RD	[a]171	1.76	20.92	7.64	0.36	0.21
[a]McCoys Creek	−27.82	153.378	TD	14.2	0.17	10.11	4.94	0.16	0.16
Coomera River	−27.83	153.396	RD	[a]489	3.62	45.29	16.68	0.36	0.23
Coombabah Lake	−27.87	153.400	TD	[a]44	4.647	34.62	9.37	1.47	0.47
Nerang River	−27.98	153.424	RD	[a]498	4.0	47.31	20.71	0.48	0.3

Notes:
[a]Indicates data not included in Digby et al. (1998). Data based on analysis of a combination of geo-referenced aerial photography, topographic maps and digital elevation models
[b]Classification refers to tidally dominated (TD) or river dominated (RD) estuary

Table 2 Average hydrological conditions of the Brisbane River estuary and Eprapah Creek

Hydrological characteristic	Units	Brisbane River mouth	Brisbane River upper estuary	Eprapah Creek
Station name		Brisbane Airport	Ipswich	Redlands
Station reference number		040842	040101	040265
Period		1994–2012	1913–1994	1953–2010
Air temperature at 09:00	Celsius	21.7	20.7	21.4
Average humidity at 09:00	%	65	66	69
Average wind speed at 09:00	km/h	15.4	5.6	8.4
Average yearly rainfall	mm	1013.5	877.8	1269.7
Maximum monthly rainfall	mm	577.2	780	909.7
Maximum daily rainfall	mm	168.4	340	241
Average clear days	days/year	133.4	84.8	82.1
Average number of cloudy days	days/year	98.9	76	59.9

Reference: Australian Bureau of Meteorology
Notes: Brisbane Airport is located next to the Brisbane River mouth; Ipswich is located less than 7 km from the Brisbane River upper estuarine zone; Eprapah Creek catchment is located less than 10 km from the Redlands meteorological station

listed in Table 1 are characterised as river-dominated estuaries, the riverine influence on hydrodynamics and water quality occurs during large rainfall events of relatively short duration. With their relatively small catchment area, whilst being tide dominated for much of the year, the majority of the smaller estuaries, can experience significant catchment flows over relatively short time scales (hours) during rainfall events. These events can significantly alter the channel hydrodynamic, sediment dynamics, water quality and ecosystem health characteristics of these small estuaries for short periods (<48 h) (Fig. 3b). This rapid response, the small spatial scales and associated logistical ease of operating in such systems makes them attractive for use as a 'natural laboratory' to investigate the influence of rain events on estuarine processes.

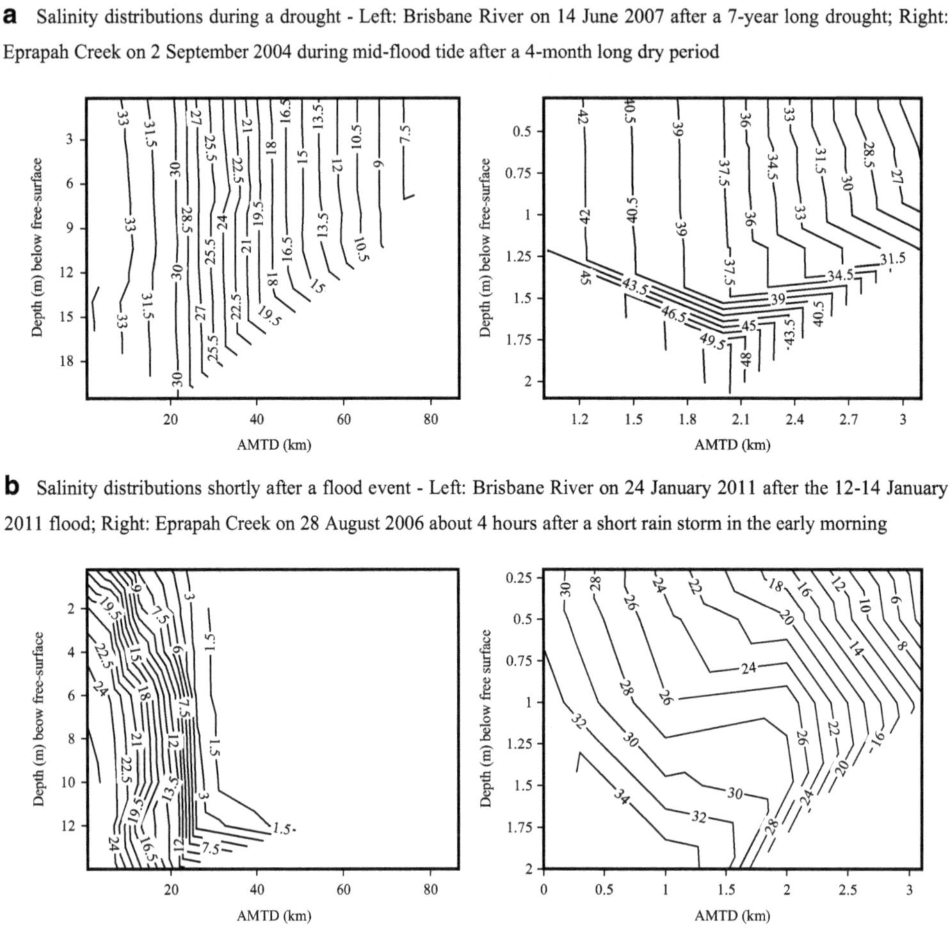

Fig. 3 Salinity distributions in the Brisbane River (*Left*) and Eprapah Creek (*Right*) during wet events and droughts as functions of the average middle thread distance (AMTD), measured from the river mouth. (**a**) Salinity distributions during a drought – *Left*: Brisbane River on 14 June 2007 after a 7-year long drought; *Right*: Eprapah Creek on 2 September 2004 during mid-flood tide after a 4-month long dry period. (**b**) Salinity distributions shortly after a flood event – *Left*: Brisbane River on 24 January 2011 after the 12–14 January 2011 flood; *Right*: Eprapah Creek on 28 August 2006 about 4 h after a short rain storm in the early morning

In many ways, the small estuarine systems have the potential to provide information for a better understanding in medium and large estuarine processes, provided that appropriate methods are available to upscale the physical data.

A more detailed scrutiny of this idea of small estuaries as 'natural laboratories' is presented by exploring the characteristics of the Eprapah Creek estuary and the up-scaling to the larger Brisbane River estuary. The Eprapah Creek estuary is located in Victoria Point, Redland Bay (Table 1, Fig. 2d). The catchment area is 39 km^2 and Eprapah Creek flows eastwards, emptying into Moreton Bay North-West of Victoria Point. The waterway is 12.6 km long and about 4 km of the creek is tidal. Eprapah Creek has two small tributaries, Little Eprapah Creek and Sandy Creek, located in the west of the catchment and discharging into the main channel at the middle of the catchment (Redlands Shire Council 2012). The Brisbane River estuary extends from the river mouth approximately 86.6 km upstream to Colleges Crossing as well as the 22 km of the Bremer River upstream from its junction with the Brisbane River (Table 1, Fig. 2a). Many small tributaries enter this complex tidal estuary including Bulimba, Oxley, Norman and Breakfast Creeks. These tributaries drain urban, industrial and semi-rural catchments (Connell and Miller 1998). Both Eprapah Creek and the Brisbane River lower estuary were adversely affected by severe pollution in the late 1990s (Appendix II). A summary of a landmark court case is presented in Appendix II.

Turbulent Mixing and Sediments, Water Quality and Ecology

For the last decade, a series of turbulence, suspended sediment and water quality measurements were conducted in the estuarine zone of Eprapah Creek (Fig. 2d). The physical

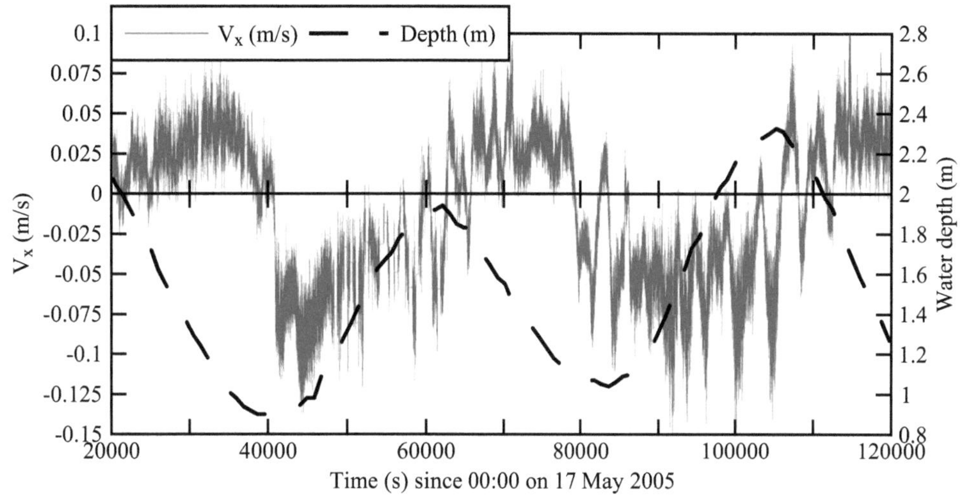

Fig. 4 Water depth and longitudinal velocity in the mid-estuarine zone (AMTD 2.1 km) of Eprapah Creek at 0.2 m above the bed under neap tide conditions on 17 May 2005 – The velocity data were sampled continuously at 25 Hz

studies were performed with state-of-the-art instrumentation to characterise the spatial and temporal variations in mixing properties as functions the tidal and hydrological conditions, including during tide dominated periods and rainfall events. The continuous turbulent velocity sampling at high frequency allows a detailed characterisation of the turbulence field in estuarine systems and its variations during the tidal cycle. Figures 4 and 5 illustrate some results in terms of the water depth, velocity and suspended sediment flux. A brief summary follows.

The bulk flow parameters vary in time with periods comparable to tidal cycles and other large-scale processes. The turbulent properties depend upon the instantaneous local flow properties; they are little affected by the flow history, but their structure and temporal variability are influenced by a variety of parameters including the tidal conditions and bathymetry. A striking feature of the data sets is the large fluctuations in all turbulent properties and suspended sediment flux during the tidal cycle including during slack periods, with some basic differences between neap and spring tide turbulence (Figs. 4 and 5) (Trevethan et al. 2008a). The upper estuarine region of this elongated tidal creek is drastically less mixed than the lower zone during tide dominated periods with some adverse impact on the water quality and ecological indicators (Trevethan et al. 2007). During rainfall events, the estuarine processes are dominated by the significant flushing associated with a strong vertical stratification of the water column, while the depth-averaged salinity data exhibit a dome-shaped intrusion curve (Chanson 2008). The field observations show some significant three-dimensional effects associated with strong secondary currents including transverse shear events (Trevethan et al. 2008b). Short-lived and highly energetic turbulent events, called bursting, play a major role in terms of sediment scour, scalar transport and accretion as well as contaminant mixing and dispersion (Trevethan and Chanson 2010).

Overall the turbulent flow properties are highly fluctuating and a large number of parameters are required simultaneously to characterise the turbulent mixing and its properties' variations with time. In plain terms, the turbulent mixing "does not slack". The mixing properties are not constant and differ between fluid, scalar and sediments. This result has fundamental implications in terms of predictive models: current numerical data are outdated and the predictive models are outclassed by recent development in computational fluid dynamics (CFD) albeit their implementation is not trivial. The mixing properties should not be assumed constant in a shallow estuary, and some similar findings are reported in a number of shallow estuarine systems of Australia and Japan with comparable hydrological and tidal conditions (Chanson and Trevethan 2010).

Detailed data similar to those presented in Figs. 4 and 5 are unavailable for most estuaries of South East Queensland, especially the larger estuaries. There is a critical need for further expert monitoring during non flood events as well as during major flood events. This situation is highlighted by a lack of basic data on flow and mixing for the estuary of the Brisbane River. The most recent data was collected in 1998, consisting of limited drogue and tracer experiments (McAllister and Patterson 1999). These data showed a reasonably long tidal excursion (4–8 km per tidal cycle) and significant mixing. To the authors' knowledge, there has been no detailed physical investigation in the estuarine zone, with an acute absence of high frequency long-duration data sets. Collection of such data in a large estuarine system like the Brisbane River is challenging, especially during flood events with a high risk of equipment loss (Brown and Chanson 2012). Without such data, it is nearly impossible to up-scale detailed physical data collected in small systems to larger estuaries without adverse scaling effects. As a result the ability to understand the underlying hydrodynamic processes driving higher order processes such as sediment transport, mixing of chemicals and ecosystem processes will continue to be limited.

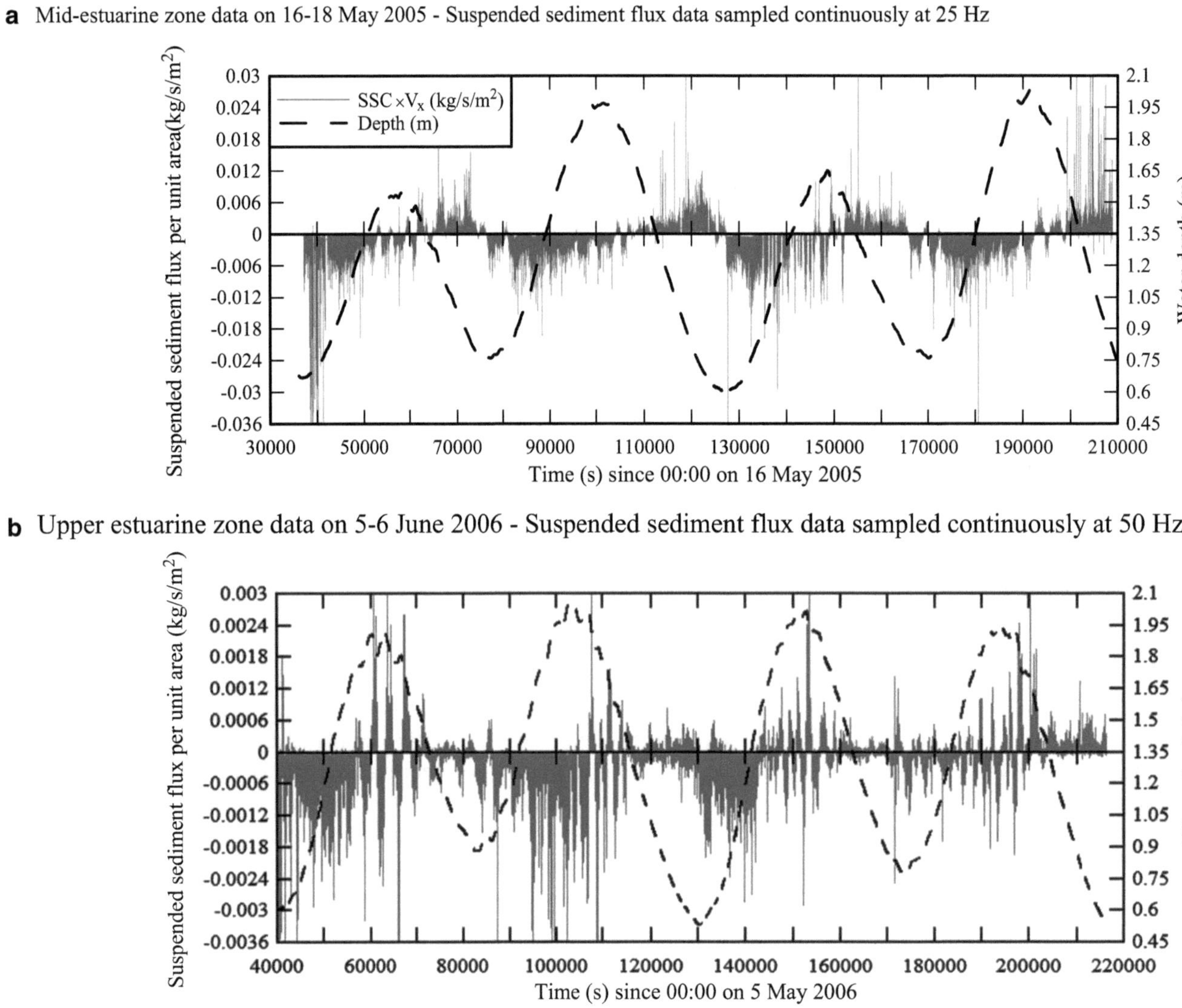

Fig. 5 Water depth and suspended sediment flux per unit area in the mid-estuarine zone (AMTD 2.1 km) and upper estuarine zone (AMTD 3.1 km) of Eprapah Creek at 0.2 m above the bed under neap tide conditions – Note the differences in *vertical* axes scaling between Fig. 5a, b. (**a**) Mid-estuarine zone data on 16–18 May 2005 – Suspended sediment flux data sampled continuously at 25 Hz. (**b**) Upper estuarine zone data on 5–6 June 2006 – Suspended sediment flux data sampled continuously at 50 Hz

Anthropological Influences: Resources, Pressures, Impacts, and Remediation

The condition of South-East Queensland estuaries has been classified as 'modified' or 'extensively modified' because of the impacts of sewage treatment plant discharges, dams and weirs, wetland loss, urbanisation, dredging and entrance modification (Digby et al. 1998). These impacts are a consequence of catchment modifications which have occurred in four distinct phases: (1) natural catchment condition, (2) Aboriginal landscape modification, (3) European agricultural development and associated vegetation clearing, and (4) catchment urbanisation. Prior to European influence, there was some evidence of landscape modification by the local Aboriginal people. Forest burning or 'firestick farming' practices are thought to have led to increased erosion rates and associated sediment delivery to waterways (Hall 1990; Neil 1998). Quantifying the extent of this landscape modification is challenging; however the extent of modification from a 'natural' state is thought to be in the order of 10 % (Neil 1998). Following European settlement about 200 years ago, the landscape was modified by a combination of timber operations and vegetation removal to develop livestock grazing land. This led to a further increase in sediment and nutrient loading to the estuaries, as well as some modifications of the hydrology and hydrodynamics. It is estimated that the catchment sediment yield likely increased by a factor of 2–5 (Neil 1998). Following land transformations associated with agricultural use, a progressive urbanisation of the catchments took place. This influenced the region's estuaries in a range of ways including (a) increased wastewater discharges, (b) altered hydrological performances with more rapid transition of runoff, coupled with increased sediment and chemical runoff, (c) altered regional hydrodynamics and sediment transport processes as a result of construction of large-scale dams for water supply and flood mitigation purposes, and (d) channel modification to support growing fishing and shipping industries as well as a significant recreational boating activities.

Initial management interventions to improve the estuarine ecosystem health generally focused on upgrades to sewage treatment plants (STPs) and industrial point source discharges (SEQHWP 2007). These actions reduced nutrient loads, especially nitrogen, released to estuarine waterway, and in turn resulted in a decline in the occurrence of phytoplankton blooms in many estuaries (SEQHWP 2007). A reduction of the rates of sediment and nutrient (nitrogen and phosphorous) transport from upland catchments to the region's estuaries has been the focus of more recent management interventions (SEQHWP 2007). The retention and restoration of vegetation in riparian zones across large parts of the catchment has been identified as an important component of the management response because of their sediment and nutrient trapping capacity (SEQHWP 2007). The large spatial scales, capital cost and long time frames for realisation of benefits makes such riparian management action challenging. Another intervention was the establishment of the Moreton Bay Marine Park in 1993 to protect the unique values and high biodiversity of the Bay and its associated estuaries. The marine park covers 3,400 km^2, stretching 125 km from Caloundra to the Gold Coast and encompassing most tidal areas of the Bay, including many river estuaries. The landward boundary is generally the line of highest astronomical tide. The majority of the region's estuaries are located within the general use zones: i.e., zones in which boating and both recreational and commercial fishing are permitted (QDERM 2010) – these general use areas having a different level of protection than those designated for habitat protection, conservation and national marine park zones.

A subtle yet distinct difference between the Brisbane River and Eprapah Creek is observed in terms of urbanisation. Within the Brisbane River catchment, urbanisation has been predominantly focused on the coastal/estuarine flood plain regions with urbanisation of riparian zones being a dominant feature (Fig. 2a). In the Eprapah Creek catchment, much of the lower coastal region has been maintained in a vegetated state through the creation of conservation areas. There is little urbanisation in the riparian areas of the Eprapah Creek estuary (Fig. 2d) – a key feature of many peri-urban estuaries (Fig. 2b). This condition (i.e. urban development in the catchment coupled with maintenance of functioning near-natural riparian zones) is often identified as the desirable future state of a catchment landscape for the improvement of water quality (SEQHWP 2007). Both catchments are characterised by a mix of urban and agricultural land uses in the upper catchment areas. On a regional scale it is estimated that between 30 and 65 % of pre-European vegetation cover has been removed by agricultural and urban/industrial development activities (Catterall et al. 1996; DERM 2010). In the Eprapah Creek catchment it is thought that approximately 40 % pre-European vegetation has been cleared for agricultural and urban development (Redlands Shire Council 2012). Another key difference in anthropological influences of the Brisbane River and Eprapah Creek is the extent of dredging. The Brisbane River has undergone significant dredging over an extended period of time (Dobson 1990) while this has been much more restricted at Eprapah Creek. The environment of the Brisbane River was significantly altered by channel dredging which extended the tidal zone from 16 km to in excess of 85 km upstream (Holland et al. 2002). Large increases in flow velocity and turbidity levels, and consequent changes in the fauna and flora, have occurred. The river has had a number of dredging phases, commencing with the opening of the Brisbane River bar in 1862 and dredging to allow shipping to access the dry dock and working docks, then located at the site of the current Southbank Parklands opposite of the Central Business District (CBD) (McLeod 1978). The modern Port of Brisbane Pty Ltd (PBPL) is now responsible for all dredging in the Brisbane River, from Point Cartwright in the north to Hamilton, about 15 km upstream of the river mouth. This includes 90 km of shipping channel and the dredging occurs to a maximum depth of 16.5 m below mean sea level (MSL). Dredging in the Brisbane River and Moreton Bay is now continuously undertaken for: (a) the maintenance of shipping channels servicing the Port of Brisbane, and (b) the reclamation of land using the dredge spoil. At Eprapah Creek, dredging occurred in the last decades at a much smaller scale. Privately owned marinas developed two channels approximately 200 m long, 15 m wide and 2 m deep, connecting the shipping yards to the main channel about 1 km upstream of the river mouth. To the best of the authors' knowledge, the dredging was not carried further into the main channel, although accurate records of such activity are difficult to trace. Personal communications with the current shipyard operators confirmed that both channels have experienced noticeable silting up in recent years to a point where they are serviceable only during high tide, impacting on the operation of the marinas.

The conflict of use between the marinas serving a sector of the community, and the environmental and community access aspects of Eprapah Creek is a microcosm of many issues for the wider management of South East Queensland

Fig. 6 Mixing and dispersion experiments in the wake of an outboard motor in Eprapah Creek (AMTD 2 km)

estuaries. Navigation is a typical example of human interaction with estuaries (Fig. 2a, c). The activity may be for recreational, primary production and transportation. Adverse effects of navigation are especially important when the system has low flushing potential. Impacts of navigation are ubiquitous, often accepted uncritically until serious impacts occur. These include noise, wave erosion of banks and wake/propeller emission of chemicals. The latter includes the emissions of inboard and outboard engines, emissions of oils, antifouling and waste disposal (Kelly et al 2004). The small scale mixing caused by navigation was recently tested by measuring the mixing and dispersion from an outboard motor in a small peri-urban waterway (Eprapah Creek) (Fig. 6). Organic dye was used as a surrogate for exhaust emissions, and dye concentrations were measured with an array of concentration probes stationed in the creek. The results highlighted very significant mixing in-homogeneity, challenging the many conventional modelling approaches.

Summary and Discussion: What Sort of Peri-Urban Estuary Do We Want for 2050 and Beyond?

Defining a future vision for an estuary is a vital and complex task. In the absence of clearly defined description of the desired future state of the system, it is challenging to identify the types of actions required to reach the future state. Key indicators are required to quantify the estuary state and to allow progress towards achieving a given vision. A common approach includes bio-physical measures: e.g., targets for dissolved concentrations of chemicals, suspended sediment concentrations, bio-diversity measurements, measures of the spatial area of a given ecosystem type. While such bio-physical indicators form the basis of 'visions' outlined in many natural resource management plans, there is a growing recognition of the importance of developing system-specific social (e.g. length of access time, number of visits, number of complaints) and economic (e.g. fisheries productivity, revenue from tourism) indicators. As an illustration, Fig. 2a presents some recreational activity on the Brisbane River, while Fig. 2b, c show respectively some residential development and fishing activity in two smaller estuaries. In South-East Queensland, a range of planning and management processes have been undertaken to scope the desired future condition of the region's estuaries. A central element has been the definition of resource condition targets (RCTs) for each estuarine system. The RCTs use a combination of environmental values and specific water quality objectives to classify the current and future state of the waterways, as documented in the South East Queensland Healthy Waterways Strategy 2007–2012 (SEQHWP 2007). This document sets a 2026 timeline and the current set of targets is designed to halt the current decline in water quality and ecosystem health. While this is an important first step in any natural resource management process, the relatively short timeframe (15–20 years) prevents largely the examination of possible long-term (>20 years) future states. A further interesting theme of past commentary on management approaches was the idea of management of 100–1,000 year planning horizons (Davie et al. 1990; Tibbetts et al. 1998). While there are very practical reasons for using the current short timeframes, the process of framing and examining potential future states over longer time periods is likely to be informative. This is particularly true in light of the design life of most infrastructure associated with the management actions adopted to halt the decline of the waterway state: e.g., wastewater treatment plant upgrades, incorporation of water sensitive urban design elements in stormwater drainage systems, restoration of riparian zones. These have a 25–100 year design life. It can be argued that any investment decision in relation to estuary management actions should consider a time-line that at least encompasses the entire lifecycle of the associated infrastructure. If a longer time-line is adopted in relation to the future of the estuaries, the desired outcomes might move beyond 'halting the decline' to a number of different potential future states ranging from some sub-optimal condition to the 'best attainable condition' for the system (Fig. 7). Figure 7 presents a conceptual diagram illustrating different phases of a natural resource management cycle. The future condition depends on the relationship between the system's minimal viable condition and condition at which the system is stabilised. If some minimum viable level is not crossed, the system can regenerate to a range of states (outcomes A, B, C) representing maintenance of the stable condition (outcome C), restoration of the reference condition (outcome B) and even a state of enhanced resource condition compared to the reference condition (outcome A). If the minimum viable level is crossed (e.g. a threshold level is reached), the system may enter an alternative stable state (outcome D).

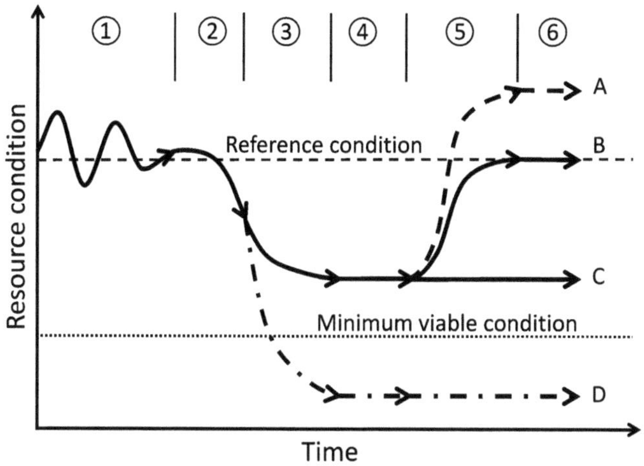

Fig. 7 Conceptual diagram illustrating the different phases of a natural resource management cycle: ① Historical condition with natural fluctuations (often used to define a historical reference condition);② Observation of resource condition decline; ③ Management intervention to halt the decline; ④ System stabilisation; ⑤ Regeneration (augmented regeneration or natural system resilience); and ⑥ Future condition

There are some notable examples of the potential for both planned and unplanned ecosystem restoration to levels equivalent or exceeding those of pre-development conditions. The history of the Sumitomo Copper Mine and associated forestry operations in Besshi, Japan provides an illustration of a long timeframe mining operation commenced around 1690 and completed in 1973, investment in technological advancements and environmental regulation (Nishimura 1989). The transition to different forms of land use, establishment of conservation zones and ecosystem resilience transformed a once barren landscape devoid of most vegetation into a vibrant forest ecosystem that has been sustainably managed for wood production over the past 100 years (Aomame 2007) (outcome A, Fig. 7). The Landes Forest in South-western France is another example of a large-scale ecosystem restoration project that commenced in the mid-nineteenth century and sought to establish a large-scale (~10,000 km^2) pine forest to address severe soil erosion issues developed from centuries of pastoral activities (IFN 2003). In France, another large-scale project was the 'Restauration des Terrains en Montagne' (RTM) conducted in mountain areas during the nineteenth century (~3,000 km^2), with very successful outcomes in terms of drastic soil erosion reduction (Brugnot and Cassayre 2002; Antoine et al. 1995), and later emulated effectively in Japan (Nakao 1993). The challenges associated with more recent large-scale ecosystem restoration efforts are daunting particularly with the issue of transformation from a degraded to a 'netpositive' state. They have been documented for a number of systems including the California Bay Delta, Chesapeake Bay and Mississippi River (Doyle and Drew 2008). The more recent restoration efforts have not had the benefit of longer timeframes associated with the preceding examples, perhaps suggesting that the combination of long timeframes, the transition to different forms of land use, establishment of conservation zones and ecosystem resilience are particularly important factors to consider.

While a 'best attainable' or optimal condition will require some definition in terms of bio-physical conditions, it will also be influenced by socio-economic factors, particularly a level of investments in terms of both economic provisions and management/lifestyle/cultural changes, that the community must be prepared to contribute. Because of their spatial characteristics and associated features, small peri-urban estuary systems offer a unique opportunity to experiment with the various management options to achieve a range of future states: e.g. catchment land use management, stormwater management, fisheries management, management of recreational activities, morphological modification. The inherent bio-physical limits will largely define the optimal state of the system with any given state/condition subsequently refined by subsequent socio-economic considerations. The future vision for the region's estuaries in the bio-physical system might be defined in terms of bio-diversity enhancement, improved ecosystem processing or stronger system resilience. In a socio-economic context a net positive gain might encompass elements such as greater fisheries productivity and enhanced recreational and/or aesthetic values. Conversely the current degradation of South-East Queensland's peri-urban estuaries may have resulted in the crossing of a threshold line that will cause the system to enter an alternative state of lower resource condition from which recovery to a pre-development reference state is not achievable (outcome D, Fig. 7). The development of more detailed visions of the future of South-East Queensland's peri-urban estuaries will involve interactions between the various stakeholders: i.e., community, industry, government agencies and the research community. This process will require a solid understanding of the bio-physical function and capacity of these periurban systems to identify the range of bio-physically feasible scenarios, and there is a critical need to quantify the various factors that will ensure the long term sustainability of the estuary's intended use.

Our current state of knowledge of peri-urban estuaries, and all estuaries more generally, presents a challenge to describe and quantify the key biophysical processes operating in these systems in sufficient detail to accurately characterise the system resilience, minimum viable condition and critical threshold levels. A desirable position does warrant sufficient information and tools to answer adequately key questions about the potential bio-physical states in which these peri-urban estuaries can exist. The understanding and predictive capacity is a necessary starting point for an effective long-term planning process that is able to attract investment to allow the desired future state of these systems to be reached.

Clearly an improved understanding of turbulent mixing and sediment dynamics is needed to provide the basis of improved understanding of higher order estuarine processes such as the transport and fate of chemicals, both natural chemical cycling and pollutant chemicals, as well as the structure and function of biological communities. The latters are often strongly influenced by the light, salinity and dissolved oxygen environments which are in turn directly related to turbulent mixing and sediment dynamics. Such an understanding will necessitate an investment in high quality process measurements (e.g. Fig. 2d) to support the development of useful predictive modelling tools. Given the current trends towards a science- and evidence-based approach to management, there will be an increased reliance on predictive models to better understand and explore the range of possible future states of the region's estuaries. These models must be based on accurate simulation of the hydrodynamics of surface water flows within an estuary. Current approaches to the simulation of estuarine flows typically adopt a very simplified representation of the three-dimensional (3D) flow dynamics. The approaches are based on the equations of momentum, continuity and conservation of heat and salt. At present they usually employ the Boussinesq approximation, neglecting the non-hydrostatic pressure terms, and in some instances replacing the standard vertical turbulent diffusion equation with a simplified mixed layer model. These approaches typically use a time averaged turbulent closure scheme (e.g. eddy viscosity model, Reynolds stress model) to simulate the turbulent processes. The recent measurements in Eprapah Creek system suggest that the simplistic representation of turbulent mixing, and particularly vertical mixing, in these types of models does not provide the level of details required for long term predictions of mixing and sediment dynamics as well as the higher order estuarine processes, without significant undesirable implications. A shift toward computational fluid dynamics (CFD) using the direct Navier–Stokes (DNS) and large eddy simulation (LES) approaches would offer massive advantages in terms of simulation accuracy. Current disadvantages of these modelling approaches for estuarine systems include the computation resources required to complete a simulation and the level of information necessary to describe the system boundary conditions: e.g., morphology and small scale bed roughness, flow vectors and distribution of sediment and dissolved chemicals. For example, a DNS approach requires the model domain to be discretised by a grid with sufficient resolution to capture the length scales associated with the key estuarine processes, bound by the Bachelor scale in the order of 10^{-5} m (Appendix I). Using a rough approach (Nezu and Nakagawa 1993), a model domain with a grid small enough to capture processes at the Bachelor scale would require about 10^{20} and 10^{23} mesh points for Eprapah Creek and the Brisbane River respectively, numbers that are currently unfeasible for routine simulation. The number of operations required for DNS is proportional to $Re^{9/4}$ where Re is the Reynolds number (Lesieur 1997), while the number of operations for LES approach scales as $Re^{3/2}$: for example, Re ~ 10^6 and 10^7 for respectively Eprapah Creek and Brisbane River estuarine zones during dry periods. Altogether the LES approach may be comparatively more efficient for the larger estuarine system models. However both DNS and LES approaches can only be used to investigate turbulence processes in simple geometries with flows corresponding to relatively low Reynolds numbers (~10^5 for DNS) today (Reynolds 1990). The adoption of simplified CFD approaches, employing various non-dynamic turbulent closures, would reduce the need for such fine scale mesh geometries but would also introduce the same issues experienced by the current 3D models.

As concluding remarks, we explored the potential for some smaller peri-urban estuaries to be used as 'natural laboratories' to gain some much needed information on the estuarine processes. While these small estuaries offer significantly advantages in terms of logistics of field measurement programs, the dynamic similarity is presently limited by a critical and accute absence of detailed physical investigations in larger estuarine systems during non flood events as well as during major flood events. Nonetheless it is suggested that the interactions between the various stakeholders (i.e. community, industry, government agencies, research institutions) are likely to define the vision for the future of South-East Queensland's peri-urban estuaries. A longer term view must be adopted, including some systematic in-depth physical data collection in larger estuaries. In a broader context, there is a trend towards public management intervention to better manage the estuarine systems of South-East Queensland. While this trend has a basis in the region's broader legal framework and has some benefits, there are other community-based and privately-based approaches that might also be employed.

Acknowledgements The authors acknowledge the Queensland Government Department of Natural Resources and Mines for access to some data, as well as Dr Alistair Grinham (The University of Queensland) for assistance with some illustration. H. Chanson acknowledges the support of the Australian Research Council (Grants LP0347242 & LP110100431).

Appendices

Appendix I: Kolmogorov and Batchelor Scales

Motions in a turbulent flow exist over a broad range of length and time scales (Roberts and Webster 2002). The length scales are linked to the motion of fluctuating eddies in turbulent flows. The largest scales are bounded by the geometric dimensions of the flow, for instance the depth and width of the channel. The large scales are referred to as the integral length and time scales. Observations indicate that eddies lose most kinetic energy after one or two overturns. The rate of energy transferred from the largest eddies is proportional to their energy times their rotational frequency. The kinetic energy is proportional to the velocity squared, in this case the fluctuating velocity v, that is ascribed by the velocity standard deviation. The rotational frequency is proportional to the standard deviation of the velocity divided by the integral length scale. Thus, the rate of dissipation ε is of the order:

$$\varepsilon \sim v^3/l$$

where l is the integral length scale. The rate of dissipation is independent of the viscosity of the fluid and only depends on the large-scale motion. In contrast, the scale at which the dissipation occurs is strongly dependent on the fluid viscosity. These arguments allow an estimate of this dissipation scale, known as the Kolmogorov microscale η, by combining the dissipation rate and kinematic viscosity ν based upon dimensional considerations:

$$\eta \sim \left(\nu^3/\varepsilon\right)^{1/4}$$

Similarly, the time and velocity scales of the smallest eddies may be derived:

$$\tau \sim (\nu/\varepsilon)^{1/2}$$

$$u \sim (\nu\varepsilon)^{1/4}$$

An analogous length scale may be introduced for the range over which molecular diffusion acts on a scalar quantity. This length scale is referred to as the Batchelor scale L_B and it is proportional to the square root of the ratio of the molecular diffusivity D_m to the strain rate γ of the smallest velocity scales:

$$L_B \sim (D_m/\gamma)^{1/2}$$

The strain rate γ of the smallest scales is proportional to the ratio of Kolmogorov velocity to length scales:

$$\gamma \sim u/\eta \sim \varepsilon^{1/2}/\nu$$

Thus, the Batchelor length scale L_B can be recast into a form that includes both the molecular diffusivity of the scalar and kinematic viscosity of the fluid:

$$L_B \sim \left(\nu^2 D_m{}^2/\varepsilon\right)^{1/4}$$

A further dimensionless number is the Schmidt number Sc defined as the square of the ratio of Kolmogorov to Batchelor length scales:

$$Sc = \eta/L_B \approx \nu/D_m$$

In the estuarine zone of Eprapah Creek, a typical mean velocity is 0.2 m/s with a fluctuating velocity v about 30 % of the mean, while an integral length scale is roughly half the channel depth, i.e. $l \approx 1$ m. Water at 20 Celsius has a kinematic viscosity of 1×10^{-6} m^2/s. Therefore, the Kolmogorov length and time scales are about 0.2 mm and 0.07 s respectively. Assuming a diffusivity $D_m \approx 1 \times 10^{-9}$ m^2/s for a typical chemical dye tracer, the Batchelor scale is 0.009 mm, or 32 times smaller than the Kolmogorov microscale. Thus, one would expect a much finer structure of the concentration field than the velocity field. Corresponding values for the Brisbane River are $v = 0.3$ m/s and $l = 5$ m yielding Kolmogorov length and time scales of 0.12 mm and 0.014 s, respectively. The relatively small difference in terms of scales between the two estuaries is because the increase in Kolmogorov length and time scales caused by the larger channel depth is countered by the reducing effect of the increase in mean velocity.

Appendix II: Pollution of Brisbane River and Eprapah Creek: 2001 Court Case

R v Hobson, Moore & Universal Abrasives Pty Ltd (2001) District Court Queensland, Forno DCJ, 15 June 2001, 1606/01.

R v Moore, 1 Qd R 205 (QCA, 2001).

R. v Moore – [2003] 1 Qd R 205, Court of Appeal, Williams JA, Jones J, Douglas J [2001] QCA 431 [C.A. 162/2001] 5, 12 October 2001

Queensland Court decision (2000) R. v Hobson & Moore & Universal Abrasives Pty Ltd.

In EPA v Universal Abrasives Pty Ltd and Moore and Hobson (Brisbane District Court, 2001), a company and two directors were charged with offences under the environmental protection (EP) Act in relation to the disposal of spent abrasive blasting product from a ship cleaning business in Brisbane. On 28 September 1998, the company released liquid waste containing high concentrations of heavy metals including lead, zinc, copper, arsenic, chromium, cadmium, selenium and biocide tributyltin (TBT) to a stormwater drain connected to the Brisbane River at Bulimba. The discharge was analysed and found to contain 2,700,000 μg/L TBT: that is, more than a million times the ANZECC limit of 2 μg/L. The company had also stored abrasive blasting material adjacent to the stormwater drain in a manner contravening its licence conditions. In addition, the same abrasive blasting material was stored in a manner that had the potential to cause serious environmental harm to a mangrove estuary at Eprapah Creek, Thornlands. The company failed to carry out environmental protection orders to clean up the affected sites.

The company and directors pleaded not guilty to causing serious environmental harm and to other offences, but were found guilty by a jury. That was the company and directors; second conviction under the EP Act, and the trial judge found that they showed no remorse. The company was fined $375,000. One director was given a suspended sentence of 9 months imprisonment (suspended for 3 years from sentences totalling 3 years to be served concurrently) and was fined $50,000. The other director was sentenced to 18 months actual imprisonment (based on sentences totalling 7.5 years to be served concurrently) and fined $100,000. In EPA v Moore [2001] QCA 431, the Queensland Court of Appeal rejected an appeal against one of the sentences.

References

Allen CR, Gunderson LH (2011) Pathology and failure in the design and implementation of adaptive management. J Environ Manage 92:1379–1384

Antoine P, Giraud A, Meunier M, Van Asch T (1995) Geological and geotechnical properties of the "Terres Noires" in southeastern France: weathering, erosion, solid transport and instability. Eng Geol 40:223–234

Aomame R (2007) The continuous pursuit of sustainable forest management & living in harmony with nature – Sumitomo Forestry Co., Ltd. Toward a sustainable Japan–corporations at work article series no.62, Japan for Sustainability. Available at: http://www.japanfs.org/en_/business/corporations62.html. Accessed 19 Oct 2012

Batchelor GK (1959) Small-scale variations of convected quantities like temperature in turbulent fluid. J Fluid Mech 5:113–133

Bilger RW, Atkinson MJ (1992) Anomalous mass transfer of phosphate on coral reef flats. Limnol Oceanogr 37:261–272

BOM (2009) Average annual rainfall map (1961–1990). Commonwealth of Australia, Bureau of Meteorology. Available: http://www.bom.gov.au/jsp/ncc/climate_averages/rainfall/index.jsp?period=an#maps. Accessed 1 Sept 2012

Bradshaw P (1971) An introduction to turbulence and its measurement, The commonwealth and international library of science and technology engineering and liberal studies, thermodynamics and fluid mechanics division. Pergamon Press, Oxford, 218 pp

Bradshaw P (1976) Turbulence, vol 12, Topics in applied physics. Springer, Berlin, 335 pp

Brown R, Chanson H (2012) Suspended sediment properties and suspended sediment flux estimates in an urban environment during a major flood event. Water Resour Res, AGU, vol. 48, Paper W11523, p.15. doi: 10.1029/2012WR012381

Brugnot G, Cassayre Y (2002) De la politique française de restauration des terrains en montagne à la prévention des risques naturels. (From the French politics of highland restoration to the prevention of natural hazards). In: Proceedings of the workshop Les pouvoirs publics face aux risques naturels dans l'histoire, Grenoble, MSH Alpes Publication, 22–23 Mar 2001, 11 pp (in French)

Catterall CP, Storey R, Kingston M (1996) Assessment and analysis of deforestation patterns in the SEQ 2001 area 1820-1987-1994: final report. Faculty of Environmental Sciences, Griffith University, Brisbane, 65 pp

Chanson H (1999) The hydraulics of open channel flow: an introduction. Edward Arnold, London, 512 pp

Chanson H (2008) Field observations in a small subtropical estuary during and after a rainstorm event. Estuar Coast Shelf Sci 80 (1):114–120. doi:10.1016/j.ecss.2008.07.013

Chanson H (2009) Applied hydrodynamics: an introduction to ideal and real fluid flows. CRC Press/Taylor & Francis Group, Leiden, 478 pp

Chanson H, Trevethan M (2010) Chapter 4: Turbulence, turbulent mixing and diffusion in shallow-water estuaries. In: Lang PR, Lombargo FS (eds) Atmospheric turbulence, meteorological modeling and aerodynamics. Nova Science Publishers, Hauppauge, pp 167–204

Connell D, Miller G (1998) Moreton Bay catchment: water quality of catchment rivers and water storage. In: Tibbetts IR, Hall NJ, Dennison WC (eds) Moreton Bay and catchment. School of Marine Science, The University of Queensland, Brisbane, pp 153–164

Davie P, Stock E, Choy DL (1990) The Brisbane River a source book for the future. The Australian Littoral Society Inc. in association with the Queensland Museum, Brisbane, 427 pp

Davies PL, Eyre BD (1998) Nutrient and suspended sediment input to Moreton Bay – the role of episodic events and estuarine processes. In: Tibbetts IR, Hall NJ, Dennison WC (eds) Moreton Bay and catchment. School of Marine Science, The University of Queensland, Brisbane, pp 545–552

Dennison WC, Abal EG (1999) Moreton Bay study: a scientific basis of the Healthy Waterway Campaign. SE Qld Regional Water Quality Management Strategy, Brisbane, 246 pp

DERM (2010) Land cover change in Queensland 2008–09: a Statewide Landcover and Trees Study (SLATS) report, 2011. Department of Environment and Resource Management (DERM), Brisbane, 100 pp

DERM (2011) South East Queensland event monitoring summary (6th–16th January 2011) preliminary suspended solids loads calculations. South East Queensland event monitoring program summary report, Department of Environment and Resource Management (DERM), Brisbane, 4 pp

Digby MJ, Saenger P, Whelan MB, McConchie D, Eyre B, Holmes N, Bucher D (1998) A physical classification of Australian estuaries. Report prepared for the Urban Water Research Association of Australia by the Centre for Coastal Management, Report no 9, LWRRDC occasional paper 16/99, Southern Cross University, Lismore

Dobson J (1990) Physical/engineering aspects of the estuary. In: Davie P, Stock E, Choy DL (eds) The Brisbane River a source-book for the future. Australian Littoral Society/Queensland Museum, Brisbane, pp 203–211

Doyle M, Drew CA (2008) Large-scale ecosystem restoration: five case studies from the United States. Island Press, Washington, DC, 344 pp

Dyer KR (1973) Estuaries. A physical introduction. Wiley, London, 140 pp

Dyer KR (1997) Estuaries. A physical introduction, 2nd edn. Wiley, New York, 195 pp

Event Monitoring Group (2011) South East Queensland event monitoring summary (6th–16th January 2011). Preliminary suspended sediment solid loads calculations. South East Queensland event monitoring (Water quality and aquatic ecosystem heath), Queensland Department of Environment and Resource Management, Australia, 4 pp

Fischer HB, List EJ, Koh RCY, Imberger J, Brooks NH (1979) Mixing in inland and coastal waters. Academic, New York, 483 pp

Graf WH (1971) Hydraulics of sediment transport. McGraw-Hill, New York, 513 pp

Hall HJ (1990) 20 000 years of human impact on the Brisbane River and environs. In: Davie P, Stock E, Choy DL (eds) The Brisbane River a source-book for the future. Australian Littoral Society/Queensland Museum, Brisbane, pp 175–182

Hinze JO (1975) Turbulence, 2nd edn. McGraw-Hill Publisher, New York, 790 pp
Holland I, Maxwell P, Grice A (2002) Chapter 12: Tidal Brisbane River. In: Abal E, Moore K, Gibbes B, Dennison WC (eds) State of south-east Queensland waterways report 2001. Moreton Bay Waterways and Catchments Partnership, Brisbane, pp 75–82
IFN (2003) Massif des lands de Gasogne 1998-1999-2000 + Résultats après la tempête du 27/12/1999 – Résultats et commentaires. Inventaire Forestier National (IFN), République Française, Ministère de l'agriculture, de l'alimentation, de la pêche, et des affaires rurales, France, 72 pp
Kelly CA, Ayoko GA, Brown RJ, Swaroop CR (2004) Underwater emissions from a two-stroke outboard engine: a comparison between an EAL and an equivalent mineral lubricant. Mater Design 26(7):609–617
Lesieur M (1997) Turbulence in fluids, 3rd edn. Kluwer Academic, Dordrecht, 515 pp
Lewis R (1997) Dispersion in estuaries and coastal waters. Wiley, Chichester, 312 pp
McAllister T, Patterson D (1999) Task hydrodynamics: exchange and mixing (HD). Final report for the South East Queensland Regional Water Quality Management Strategy, WBM Oceanics, Brisbane
McLeod R (1978) A short history of the dredging of the Brisbane River, 1860 to 1910. J R Hist Soc Qld 10(3):137–148
Nakao T (1993) Research and practice of hydraulic engineering in Japan. J Hydrosci Hydraul Eng Jpn, Special issue SI-4 River Engineering
Neil DT (1998) Moreton Bay and its catchment: seascape and landscape, development and degradation. In: Tibbetts IR, Hall NJ, Dennison WC (eds) Moreton Bay and catchment. School of Marine Science, The University of Queensland, Brisbane, pp 3–54
Nezu I, Nakagawa H (1993) Turbulence in open-channel flows, IAHR monograph series. Balkema Publisher, Rotterdam, 281 pp
Nielsen P (1992) Coastal bottom boundary layers and sediment transport, vol 4, Advanced series on ocean engineering. World Scientific, Singapore, 324 pp
Nishimura H (1989) How to conquer air pollution: a Japanese experience. Elsevier, Amsterdam, 301 pp
QDERM (2010) Moreton Bay Marine Park map. Queensland Department of Environment and Heritage Management QDERM, Brisbane. Available at: http://www.nprsr.qld.gov.au/parks/moreton-bay/zoning/pdf/map3-zoningplus.pdf. Accessed 19 Oct 2012
QOESR (2011) Queensland government population projections: local government areas report. Queensland Office of Economic and Statistical Research QOESR, Brisbane. Available at: http://www.oesr.qld.gov.au/products/publications/qld-govt-pop-proj-lga/index.php. Accessed 19 Oct 2012
Redlands Shire Council (undated) Eprapah creek waterway management plan – final. Redlands Shire Council. Available at: http://web01.redland.qld.gov.au/robo/plans/Eprapah_Creek_WMP/Eprapah_Creek_Waterway_Management_Plan.htm. Accessed 19 Aug 2012
Reynolds WC (1990) The potential and limitations of direct and large eddy simulations, in whither turbulence? turbulence at the crossroads. Lect Notes Phys 357:313–343
Roberts PJW, Webster DR (2002) Turbulent diffusion. In: Shen HH, Cheng AHD, Wang KH, Teng MH, Liu CCK (eds) Environmental fluid mechanics: theories and applications. ASCE, Reston, pp 7–45
Savenije HHG (2005) Salinity and tides in alluvial estuaries. Elsevier, Amsterdam, 194 pp
SEQHWP (2007) South East Queensland healthy waterways strategy 2007–2012. South East Queensland Healthy Waterways Partnership, Brisbane. (www.healthywaterways.org). Accessed 12 Oct 2012
Tennekes H, Lumley JL (1972) A first course in turbulence. MIT Press, Cambridge, MA, 300 pp
Tibbetts IR, Hall NJ, Dennison WC (1998) Moreton Bay and catchment. School of Marine Science, The University of Queensland, Brisbane, 645 pp
Trevethan M, Chanson H (2010) Turbulence and turbulent flux events in a small estuary. Environ Fluid Mech 10(3):345–368. doi:10.1007/s10652-009-9134-7
Trevethan M, Chanson H, Takeuchi M (2007) Continuous high-frequency turbulence and sediment concentration measurements in an upper estuary. Estuar Coast Shelf Sci 73(1–2):341–350. doi:10.1016/j.ecss.2007.01.014
Trevethan M, Chanson H, Brown R (2008a) Turbulent measurements in a small subtropical estuary with semi-diurnal tides. J Hydraul Eng ASCE 134(11):1665–1670. doi:10.1061/(ASCE)0733-9429(2008)134:11(1665)
Trevethan M, Chanson H, Brown R (2008b) Turbulence characteristics of a small subtropical estuary during and after some moderate rainfall. Estuar Coast Shelf Sci 79(4):661–670. doi:10.1016/j.ecss.2008.06.006

Hervey Bay and Its Estuaries

Joachim Ribbe

Abstract

Hervey Bay and its estuaries are located along the east coast of Australia just to the south of the Great Barrier Reef Marine Park. The region has long been recognised as one of Australia's most biodiverse marine environments and including the Great Sandy Strait in the south of the Bay, it is referred to as the Great Sandy Biosphere. The United Nations Educational, Scientific and Cultural Organisation (UNESCO) included the area in its list of 580 designated biospheres located in 114 countries.

Although widely recognised for its exceptional biodiversity, little is known about the physical processes and climate characteristics that shape its natural marine environment. Only recently, the Bay has been classified as a large low inflow and predominantly hypersaline system. River runoff and discharge from its many estuaries is very small. It has almost been absent during the Australian *Millennium Drought* lasting the first decade of the twenty-first century. During other times, freshwater inflow is only significant following flooding as a result of tropical/subtropical depressions, which often is amplified during La Nina events. A positive freshwater balance leads to a net loss of water establishing hypersaline conditions. This appears to prevail throughout the year and is re-established shortly after storm events. Hydrodynamic modelling suggests that predominant southeasterly to easterly trade winds establish a cyclonic water renewal pathway, with Hervey Bay water exiting along the western shoreline. Hypersalinity and the characteristics of an inverse estuarine circulation are evident from observations and modelling.

This chapter reviews our understanding of the environmental processes that shape Hervey Bay and its estuaries in the context of its climate. Future changes in the regional freshwater balance indicate a continued trend toward drier and warmer conditions. It leads to an intensification of hypersaline and possible inverse circulation states of the Bay. Insight into the environmental forces shaping Hervey Bay, its estuaries, and a unique and biodiverse environment, informs continued sustainable natural resource management and policy development. It is anticipated that over the next few decades physical processes associated with climatic trends and variability are likely to impact more dramatically upon the natural environment of the region than direct human activities such as fishing, aquaculture, tourism and continued local population and urbanisation trends.

Keywords

Hervey Bay • Great Sandy Strait • UNESCO biosphere • Hypersalinity • Inverse circulation • Australia • Biodiversity • Climatic variability • Freshwater balance

J. Ribbe (✉)
Department of Biological and Physical Sciences, University of Southern Queensland, Toowoomba 4350, QLD, Australia
e-mail: Joachim.Ribbe@usq.edu.au

E. Wolanski (ed.), *Estuaries of Australia in 2050 and Beyond*, Estuaries of the World,
DOI 10.1007/978-94-007-7019-5_11, © Springer Science+Business Media Dordrecht 2014

Box 1

Joachim Ribbe studied Hervey Bay, one of Australia's most biodiverse marine environments and a UNESCO designated biosphere. River runoff and discharge from its many estuaries is very small. It has almost been absent during the Australian *Millennium Drought* lasting the first decade of the twenty-first century. During other times, freshwater inflow is only significant following occasional flooding, which often is amplified during La Nina events. Hypersalinity prevails most of the year and is re-established shortly after storm events. In the context of its climate, the regional freshwater balance suggests a continued trend toward drier and warmer conditions. It leads to an intensification of hypersaline and possible inverse circulation states of the Bay. It is anticipated that over the next few decades physical processes associated with climatic trends and variability are likely to impact more dramatically upon the natural environment of the region than direct human activities such as fishing, aquaculture, tourism and continued local population growth and urbanisation trends.

Introduction

Like many of world's coastal systems, Hervey Bay and its estuaries undergo observed environmental changes due to many anthropogenic and natural factors. These include pollution from agriculture and industry as well as climate change. An improved understanding of the physical processes that shape these natural marine environments is needed to insure future sustainable resource utilisation and management of a unique biodiverse region.

The marine state of any estuary or large coastal embayment is, to a large extent, a result of the local freshwater balance. Many Australian estuaries and large coastal bays such as Hervey Bay can be characterised as low or no inflow estuaries. A surface layer outflow of freshwater is absent, although during short and intermittent rainfall events a classical estuarine circulation could be established (see for a more complete review of low inflow estuaries Largier (2010)).

Hypersalinity as well as a possible inverse circulation develops in many Australian low inflow estuaries and coastal Bays. In these coastal systems, salinity is higher than in the ocean since evaporation often exceeds the supply of freshwater from precipitation and river discharges. Therefore, the net loss of water results in an elevated salinity, the hypersalinity zone, a surface inflow of ocean water with lower salinity, and potentially an outflow of high salinity water near the seafloor establishing an inverse circulation. This applies to Hervey Bay, which only recently has been characterised as a low inflow system with predominately hypersaline conditions (Ribbe 2006; Ribbe et al. 2008) and an established inverse circulation pattern for some periods throughout the year (Gräwe et al. 2010).

Australia's continental wide freshwater balance is dominated by evaporation (Fig. 1) and many coastal systems are potentially characterised by low-runoff for most of the year. It would favour the establishment of hypersaline zones and some of those have been documented in the scientific literature. These include small tropical estuaries (Wolanski 1986; Ridd and Stieglitz 2002), subtropical water ways (Benfer et al. 2007), large coastal bays such as the Gulf of St Vincent and Spencer Gulf in South Australia (e.g. Samarasinghe and Lennon (1987), Nunes Vaz et al. (1990), Petrusevics (1993)), Shark Bay in Western Australia (e.g. Nahas et al. 2005), and large coastal zones such as the Great Barrier Reef coastal waters (Andutta et al. 2011) and the Great Australian Bight (Petrusevics et al. 2009); clearly evidence of an Australia-wide feature. Large hypersaline systems exist in many other parts of the world including the Gulf of California, the Mediterranean Sea and the Red Sea which is the most saline region of the world's ocean (Tomczak and Godfrey 2003). These systems are often also characterised by an inverse circulation and act like salt fountains by ejecting high salinity water into the adjacent ocean.

Climate change and variability lead to changes in the hydrological cycle and impact upon the estuarine characteristics. The impact of persistent drought, climate change and variability has been documented for Hervey Bay (Gräwe et al. 2010) and Port Philip Bay (Lee et al. 2012), with Port Philip Bay having switched to a hypersaline system during the Australian *Millennium Drought* in the first decade of the twenty-first century.

The particular focus of this chapter is on the hydrological cycle that determines the freshwater water balance of Hervey Bay and its estuary and reviews the physical oceanography of the Bay. Climate change and variability are associated with changes in evaporation (E), precipitation (P) and river runoff (R). Any changes in these climate elements are leading to changes in the physical characterises of Hervey Bay. Recent research has significantly improved

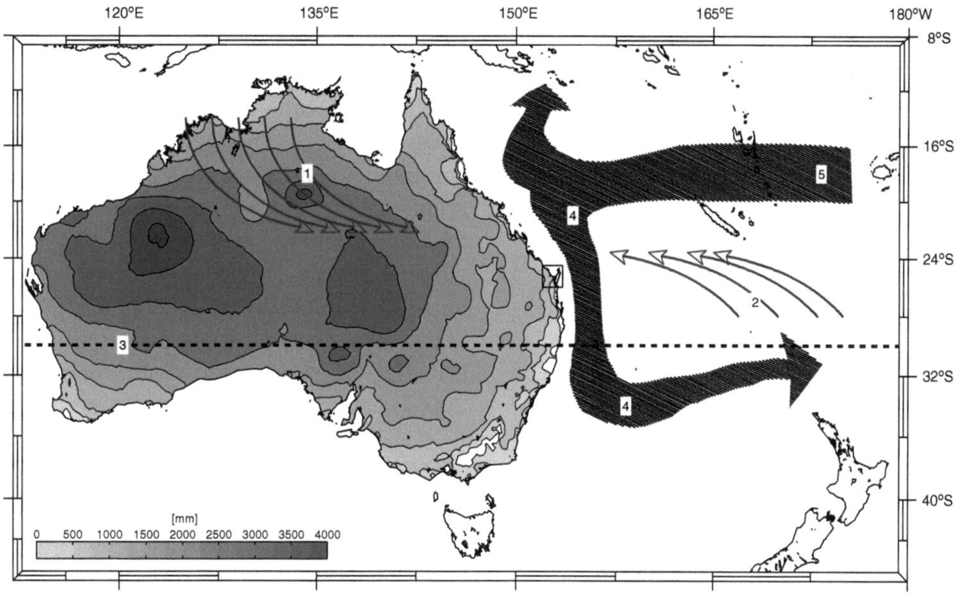

Fig. 1 Map of Australia and the region with Hervey Bay (see Fig. 2) located at about 24.5°S along the subtropical climate influenced mid-east coast of Australia. Dominant and precipitation driving climatological features schematically shown include the *1* Australia-Asian Monsoon during the southern hemisphere summer, *2* the east to southeasterly trade winds, *3* approximate position of the subtropical ridge, *4* the East Australian Current (EAC) and *5* the equatorial Pacific Ocean circulation. The mean annual positive continental water balance (Evaporation – Precipitation) (Data Source: Australian Bureau of Meteorology) is *shaded*. *White* areas indicate access of precipitation, which is limited to small regions along the east Australian coasts, south western Australia and Tasmania

our understanding of all these physical properties and the circulation of Hervey Bay. Both climate and physical ocean features combined largely determine the state of its marine environment and control the exchange of water between its many estuaries and the open ocean.

Physical Setting

The Site

Hervey Bay and its estuaries are located along the eastern subtropical coast of Australia at about 152.5°W and 24.5°S and just to the south of the Great Barrier Marine Park (Fig. 1). The Bay is bowl or U-shaped with a northward facing opening of about 80 km width, narrowing southward, and a north to south extend of about 60 km (Fig. 2). It has a mean depth of about 20 m. Fraser Island, with a length of about 120 km and an approximate width of about 24 km, is the world largest sand island and forms the eastern boarder of Hervey Bay. The most northern point of the Island is Sandy Cape and beyond Sandy Cape, the Break Sea Spit extends below the surface an additional 30 km to the north of Fraser Island (Fig. 2). To the north of Break Sea Spit, the Australian continental margin changes orientation and curves to the west. It is here where marine sands derived from the south and transported northward forming the sand islands of southeast Queensland, are delivered via a coastal transport system into the deep sea (Boyd et al. 2008), a transport system that has been operating for more than 700,000 years (Schröder-Adams et al. 2008). At the narrow southern end of Hervey Bay and Fraser Island, the Great Sandy Strait, which is a very shallow (less than 2 m) and about 80 km long passage, connects Hervey Bay proper with the coastal shelf and open ocean to the south of Fraser Island (Fig. 2).

The marine environment of Hervey Bay is one of the most biodiverse and unique ones found in Australia. It is a prominent resting place for humpback whales on their annual migration route between the cold Antarctic and warm tropical waters (Chaloupka et al. 1999; Vang 2002). Franklin et al. (2011) report sightings of more than 10,000 animals during 770 observing days from 1992 to 2005. Hervey Bay is also home to one of the largest seagrass areas along the east coast of Australia (Sheppard et al. 2007). This supports a dugong population that is regarded as being vulnerable to extinction (Preen and Marsh 1995). The region is home to six of the world's seven marine turtles including endangered loggerhead (*Caretta caretta*) and green turtles (*Chelonia mydas*) (Hazel and Gyuris 2006). Populations of dolphins have been observed (Cagnazzi et al. 2011) and many other unique marine species. This includes reef-building coral communities within the southern regions of Hervey Bay identified as the furthest south located along the east coast of Australia (Zann et al. 2012).

The State Government of Queensland has recognised the value of the natural marine environment of this region by

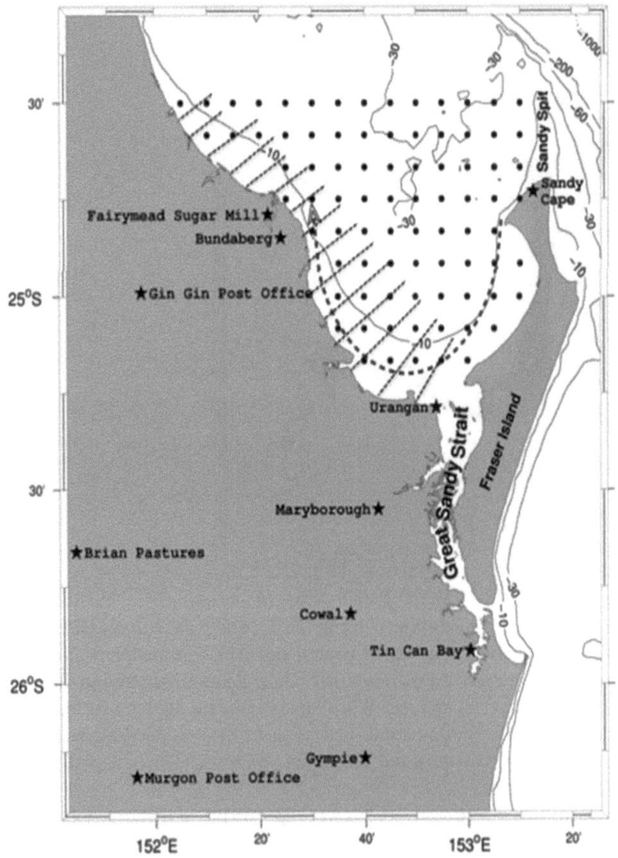

Fig. 2 Map of Hervey Bay and sample locations covered during five hydrographic surveys in September 2004, August and December 2007 and May and June 2008. Land based locations shown have been used to document precipitation and evaporation rates. *Dashed arrow* indicative of mean climatological wind-driven cyclonic circulation and water renewal pathway for Hervey Bay. *Dash lines* indicate location of hypersalinity zone along the west coast of the Hervey Bay region. Bathymetry is shown for 10, 30, 60 and 200 m depth intervals (Data source: GeoscienceAustralia 2009)

establishing the Great Sandy Marine Park. This park extends from Bundaberg in the north to Tin Can Bay in the south including Hervey Bay, the Great Sandy Strait and the coastal ocean to about three nautical miles east of Fraser Island. It is one of only three state marine parks in Queensland, which includes the Great Barrier Reef Coast Marine Park to the north and the Moreton Bay Marine Park to the south of Hervey Bay. Furthermore, the park is listed as World Heritage Natural Icon and the Great Sandy Strait has been noted under the Ramsar Convention on Wetlands recognising the Strait as one of Australia's most biodiverse region. In 2009, UNESCO included the Great Sandy Biosphere as one of a total of 580 designated biosphere located in 114 countries.

Major coastal urban centres in the Hervey Bay region also referred to as the Fraser Coast, include Bundaberg, Gympie and Maryborough with a coastal population of more than 290,000 (Fig. 1). About one million visitors a year are attracted into the region making tourism together with fisheries and aquaculture an important economic driver for regional prosperity. In 2011, the Great Sandy Regional Marine Aquaculture Plan (Queensland So 2011b) was released which is the first plan developed for any marine coastal region off Queensland, underpinning the future economic value of the marine park to the region. The footprint of human activity in the region is evident. Documented adverse impacts include the accidental boating–related death of marine species such as turtles (Hazel and Gyuris 2006), the contamination from herbicide that enter the estuaries of Hervey Bay with runoff (McMahon et al. 2005) and the accumulation of pesticides in seagrasses (e.g. Haynes et al. 2000).

There are several rivers draining coastal catchments into Hervey Bay. These include the Mary River in the south as well as the Burrum, Elliott and Burnett rivers along the west coast. The most important of those are the Mary River in the south and the Burnett River in the north near Bundaberg. The former is about 310 km long and drains a catchment of about 9,595 km^2 at the southern end of Hervey Bay into the Great Sandy Strait. The Burnett River is about 475 km long with a catchment of about 32,500 km^2; it drains the third largest catchment in Queensland into the northern area of Hervey Bay. Freshwater water discharges from these two systems are most important for the large-scale freshwater balance of the Bay and its estuaries. The contribution from other sources, including potential groundwater and freshwater discharge from Fraser Island, is more than a factor 10 smaller.

A signature of river discharges is found within fluvial sediments collected within Hervey Bay during September 2004 (Fig. 3). These fluvial deposits dominate the western Bay and are characterised by coarser sediments (>1 mm, i.e. gravel fraction). This is indicative for the pathway of river plumes being confined to the western region of the Bay, which is support through computational studies of floodwater dispersion into Hervey Bay below (see below). Hydrodynamic fining of the sediments occurs eastward. Fine sediments (<63 μm, i.e. silt and clay fraction) are found in proximity to Fraser Island. Mid-size grains (about 125–500 μm; i.e. sand fraction) are observed in the northeast to southeast interior of the Bay.

Geomorphology, sedimentary environment, and water balance lead to estuarine classifications. In terms of its geomorphology, Hervey Bay is a coastal plain estuary, or drowned river valley formed as a result of global sea level rise following the last deglaciation. This resulted in the drowning of the Mary River valley with the previous river delta during glaciation being located to the north of Fraser Island and its extension, referred to as the Great Sandy Spit. The old north–south

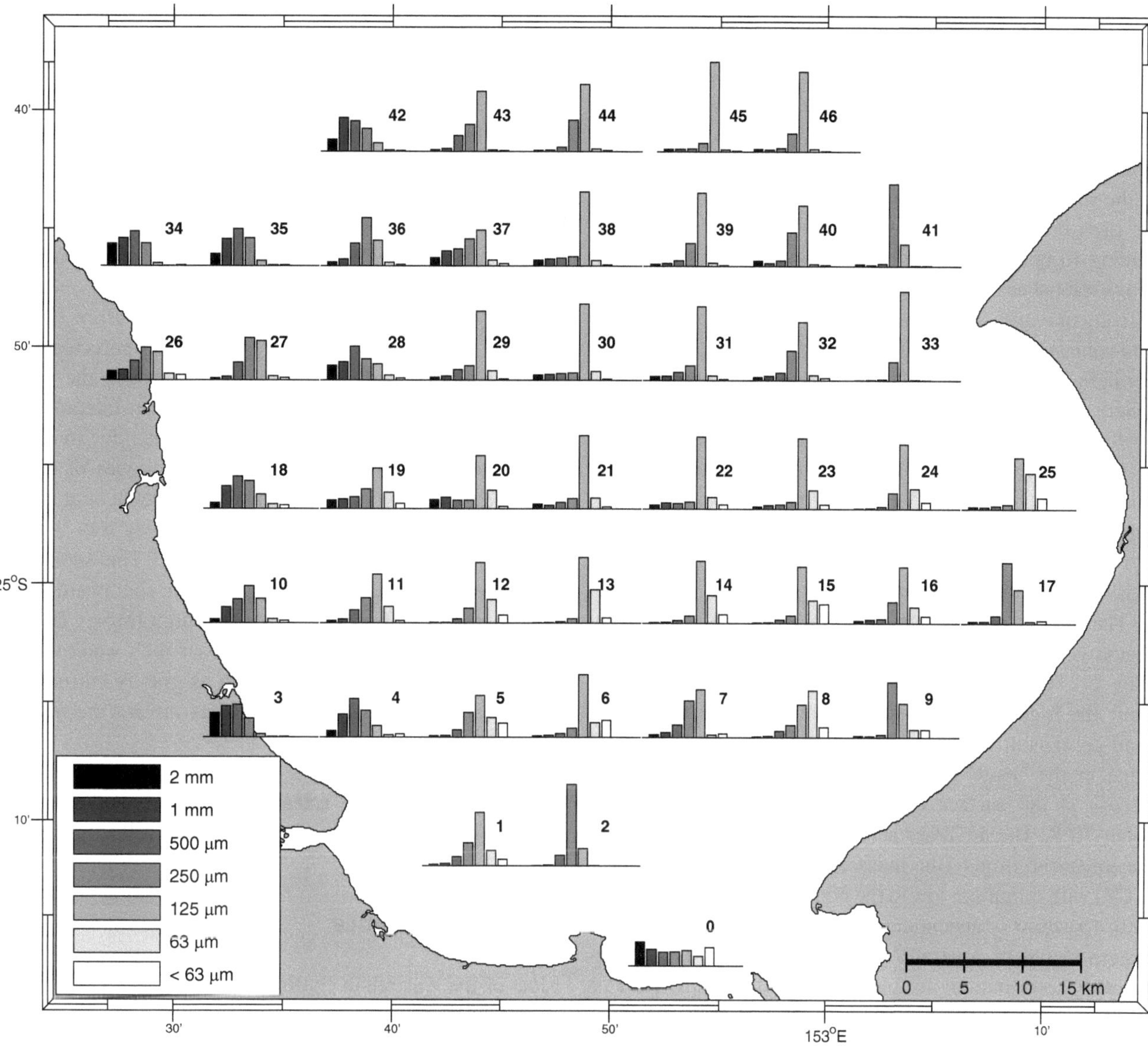

Fig. 3 Grain size distribution observed during September 2004. The western region is characterised by coarser sediments (>1 mm, i.e. gravel fraction) and most of finest sediments (63 μm and less, i.e. silt and clay fraction) are found in proximity to Fraser Island. Mid-size grains (about 125–500 μm; i.e. sand fraction) are observed in the northeast to southeast interior of the Bay

oriented Mary River valley now forms the deeper parts within Hervey Bay. Boyd et al. (2004) applied a classification developed by Roy et al. (2001) for Australian estuaries and defined Hervey Bay as a shoreline divergent estuary based on its sedimentary environment. An alternative estuarine classification is based on the local water balance for Hervey Bay (Ribbe 2006). This leads to it being classified as a low inflow system that is often characterized by a coastal hypersalinity zone and intermittent inverse states. Oceanographic observations and computational modeling leading to this classification are reviewed in sections "Hydrographic observations" and "Hydrodynamic modelling".

Climate and Weather

Hervey Bay and its estuaries are located in a subtropical climate with mean annual maximum and minimum temperature of about 26.2 and 16.7 °C respectively (Stern et al. 2000). Winter in the region is drier than summer with climatological rainfall varying from a minimum in July of about 49 mm to a maximum of about 154 mm in December. Total mean annual rainfall is about 915 mm with about 440 mm delivered during the wetter summer months of December, January, February and about 150 mm during the drier winter months of June, July and August.

Rainfall in southeast Queensland and the Hervey Bay region is influenced by the location of the subtropical ridge (Fig. 1), a zonally oriented band or zone of semi-permanent high pressure that migrates north- and southward with the seasons across the Australian continent. During the southern hemisphere summer, the high-pressure belt is located further to the south at about 40°S. The Australian continent warms; air rises drawing in southward moving moist tropical air leading to potential rainfall penetrating further southward. This establishes the Austral-Asian summer monsoon, which delivers moisture from the Australian tropics and southward into subtropical regions of southeast Queensland during the northern Australian wet season. During the southern hemisphere winter, the high-pressure belt migrates north to about 30°S and mid-latitude rain delivering east coast low pressure systems penetrate further to the north driving winter rainfall. These synoptic scale cyclonic systems are associated with significant wind and often lead to coastal erosion. However, Hervey Bay is largely protected by Fraser Island limiting erosion along its west coast.

The predominant climatological wind in the Hervey Bay region is associated with the east to southeast trade winds (Fig. 1). These winds of the southern hemisphere emanate from the high-pressure systems that establish the belt of semi-permanent high pressure or the so-called subtropical ridge in the South Pacific Ocean. Trade winds are most intense at the surface of the southeast Pacific Ocean at about 30°S. The air flows across the Pacific Ocean toward the equatorial trough (the Intertropical Convergence Zone – ITCZ) with minimum low to the north of Australia. Along its path, moisture is accumulated and delivered all along the eastern slopes of Australia's Great Dividing Range. A significant contribution to climatological rainfall is also associated with sub-synoptic atmospheric features such as tropical cyclones and other tropical depressions such as storm events. All contribute, in addition to the northwest to southeast monsoonal circulation, to the delivery of significant rainfall during the wet season.

Variability in large-scale oceanic and atmospheric circulation patterns impact upon the climatological delivery of rainfall (see Fig. 1). This leads to climate variability in Queensland and the Hervey Bay region. The most prominent features associated with climate variability in the region include the El Nino Southern Oscillation (Murphy and Ribbe 2004), the Pacific Decadal Oscillation (Power et al. 1999), the Madden-Julian Oscillation (Marsh et al. 2005; Donald et al. 2006), variability in the location of the subtropical ridge (Williams and Stone 2009) and longer-termed climate trends (Murphy and Timbal 2008; Timbal and Drosdowsky 2012). Thus, the notion of a climatological or mean rainfall is somehow a conservative view of the climate systems. Only in the last two decades, climate is more fully understood as being characterised by significant variability, and expected rainfall in the region departs from the climatological mean on time scales ranging from weeks to decades. For example, decadal variability associated with the Pacific Decadal Oscillation is likely to have had a positive rain-enhancing impact during the 2010/2011 La Nina event in Southeast Queensland and Hervey Bay (Cai and van Rensch 2012). This climate variability is superimposed on current warming trends associated with long-term shifts in atmospheric and oceanic circulation pattern.

The East Australia Current (EAC) is the most prominent feature of the large-scale ocean circulation in the western South Pacific Ocean (e.g. Tomczak and Godfrey (2003)). The EAC is formed in the Coral Sea and transports Coral Sea Water along the continental slope, and just to the east of Fraser Island, southward. The coastal ocean interacts with the EAC through the exchange of shelf water due to a range of physical processes. For example, at latitudes of Hervey Bay and to the east of Fraser Island, Middleton et al. (1994) argue to have observed a signature of Hervey Bay Water at about 240 m water depths within the EAC. This shelf water entrained within the EAC is then advected southward. Ribbe (2006) concludes from observations within Hervey Bay that the Bay could indeed supply this denser shelf water through marine connectivity processes such as gravity currents and possibly strong residual tidal currents toward the north of Fraser Island.

Climate and precipitation control the regional freshwater balance and determine estuarine conditions within Hervey Bay and its estuaries. This is reviewed in the following section.

Freshwater Balance

Most of the Australian continent is characterised by a positive climatological freshwater balance, i.e. averaged over the period of 1 year, evaporation is significantly larger than precipitation (Fig. 1). This reflects the fact, that Australia is known as the driest continent and largely classified as desert and dry grassland country (Stern et al. 2000). However, there are seasonal variations that lead to a negative balance in the wet tropics during the summer and in the southwest and southeast of Australia during winter, which reflects the rainfall delivery in the north by the summer monsoon and by mid-latitude frontal systems associated with passing low-pressure systems in the south during winter (Fig. 4). Western Tasmania is the only region that has a negative freshwater balance throughout the year, i.e. precipitation is always larger than evaporation. Most of the mid-latitude subtropics of Australia are climatologically characterised by a positive balance, i.e. many of the estuaries and coastal regions are likely to be classed as low inflow estuaries that only assume a classical estuarine structure temporarily during wet seasons or as a result of significant storm and high river discharge events.

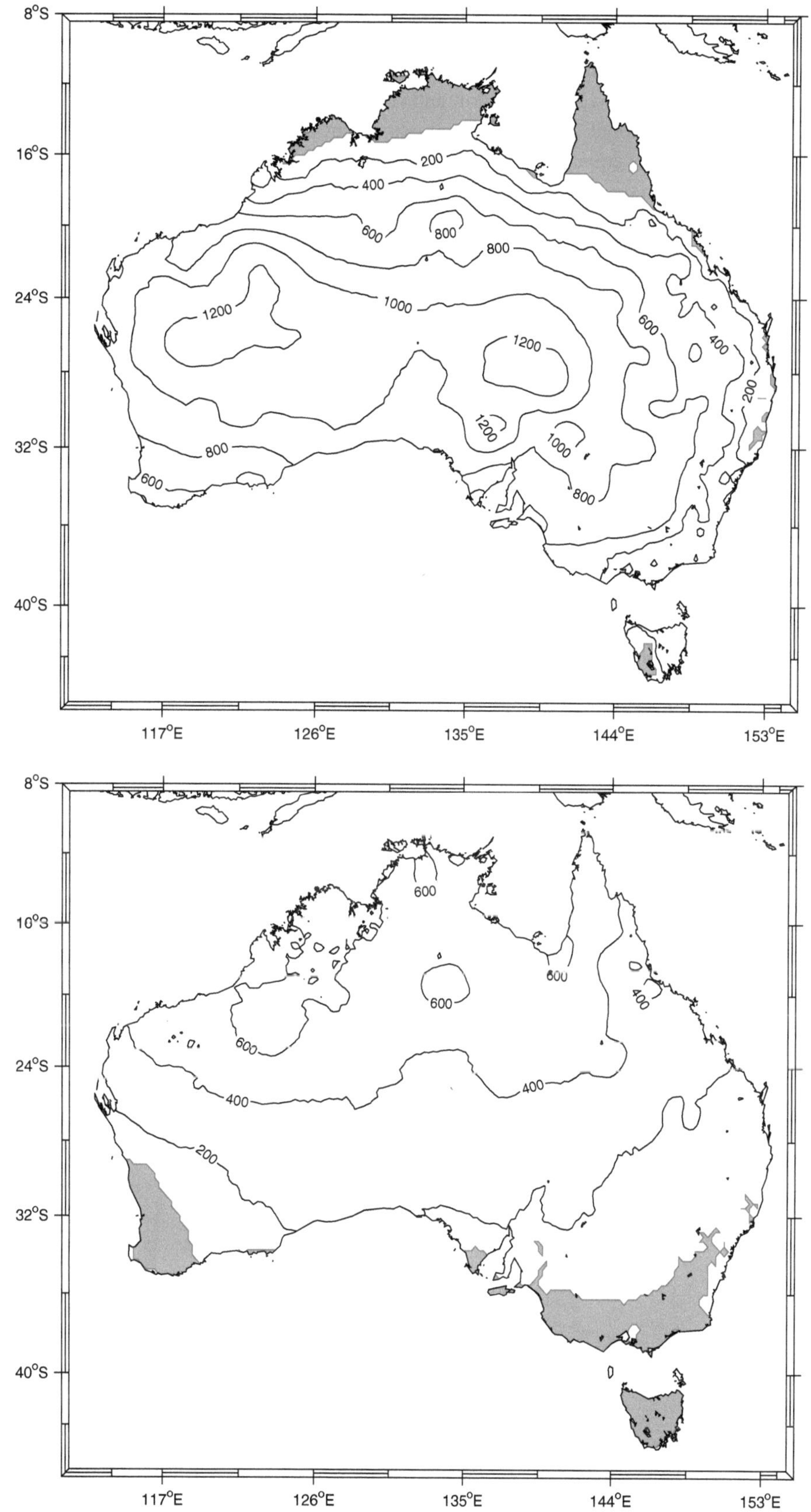

Fig. 4 Climatological continental freshwater balance (E-P) for Australia in [mm]. *Top panel*: southern hemisphere summer, and *bottom panel*: southern hemisphere winter. Contour level is 200 m. *Shading* indicates a negative balance, i.e. precipitation is larger than evaporation. Southeast Queensland including Hervey Bay is characterised by a positive balance

Table 1 summarises the freshwater balance for the Hervey Bay region. Evaporation is more than twice the amount of precipitation as well as the combined amount of freshwater delivered via precipitation and river runoff. The annual supply of about 238 mm rainfall equivalent of river discharge (Table 1) equates to about 1 km^3 of water added to Hervey Bay via river systems per year. This is about 1/80th of the Hervey Bay volume, i.e. about 80 years of average Mary and Burnett River runoff are required to refill the Bay.

Table 1 Climatological freshwater balance

Period	1966–2010 (mm)	2004–2008 (mm)	Difference (mm)
Evaporation	2,078	2,171	93
Precipitation	914	812	−102
River runoff	238	123	−115
E-P	1,164	1,359	195
E-P-R	926	1,236	310

Note: Data Source: Australian High Quality Climate Data (Lavery et al 1997 and update since available http://www.bom.gov.au/climate/change/hqsites/). Closest to the region available evaporation mean comes from locations 039083 (Rockhampton) and 040428 (Brain Pastures). Rainfall mean comes from the following locations: 039037 (Fairymead Sugar Mill), 040428 (Brain Pastures), 040013 (Cowal), 039040 (Gin Gin) and 040152 (Morgan). Annual information on total river flow discharges into Hervey Bay are provided by the State Government of Queensland, data source: http://watermonitoring.derm.qld.gov.au/host.htm. This data has been converted into rainfall equivalent by assuming that the flow is distributed across the surface area of Hervey Bay (4,000 km^2). Thus the river discharge equates for an additional input of rainfall across the region rather than being a point source

From 1997 to 2009 Australia experienced its most severe drought of the twentieth and twenty-first century, which is now referred to as the Millennium Drought. During this period, the first Bay-wide hydrographic observations from Hervey Bay were obtained (2004–2008, see below). This period was drier than the base period 1966–2010 with evaporation increased to 2,171 mm, precipitation and river runoff reduced to 812 and 123 mm respectively; a difference in the freshwater balances for these two periods of about 310 mm (Table 1).

The freshwater balance (E-P-R for the period 1966–2010) results in a net water loss in the Hervey Bay region in the order of about 930 mm/year or 3 mm/day. This yields a volume loss of about 140 m^3/s if applied over the area of the Hervey Bay proper region of 4,000 km^2. The water balance for the Bay requires a supply of water into the Bay mostly across a northern boundary of about 80 km width and 20 m depths. The net water loss results in an evaporative driven inflow velocity of about $8.75*10^{-5}$ m/s or about 8 m/day. With an inflow of 140 m^3/s, it would take about 18 years to refill the Bay volume ($V = 4{,}000\ km^2 * 20\ m$) with inflowing ocean water if all water would be lost due to evaporation. This is equivalent to an evaporation driven residence or water renewal time scale of about 18 years. However, oceanic processes such as the wind-driven circulation, tidal mixing, and residual currents result in much shorter residence times as discussed below.

Similar to many other regions of Australia, near surface mean air temperature in the Hervey Bay region has increased in particular during the second half of the twentieth century (Fig. 5). The annual linear trend is about 0.024 °C or a total of

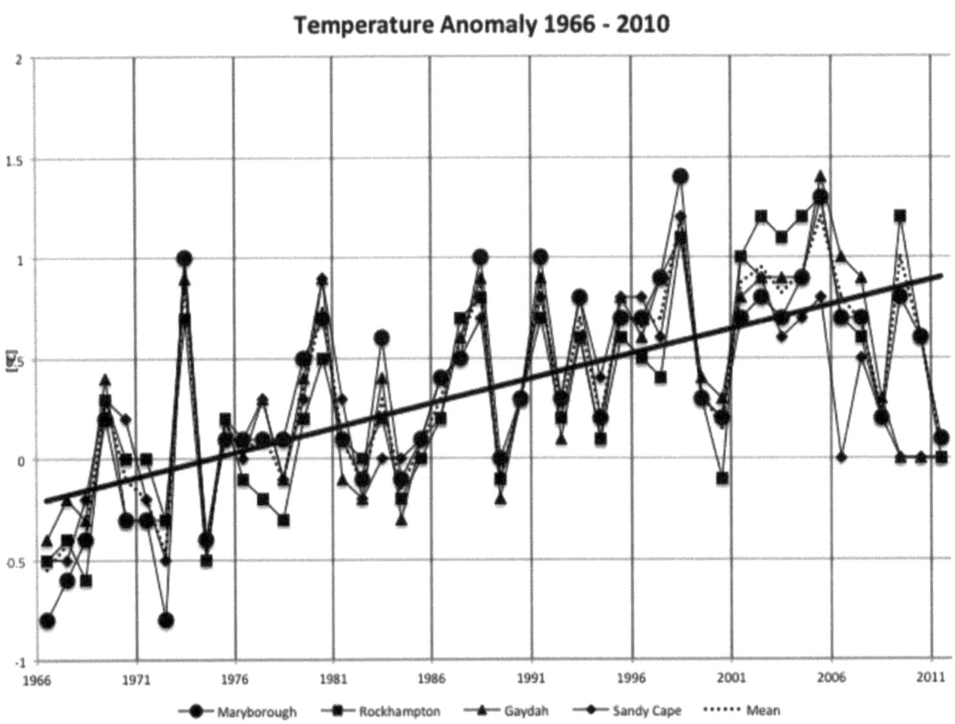

Fig. 5 Temperature anomaly [°C] in the Hervey Bay region from 1966 to 2011. Data is shown for several locations in the region as well as presented as a mean between those and the linear trend for the mean (Data source: Australian High Quality Climate Data Set)

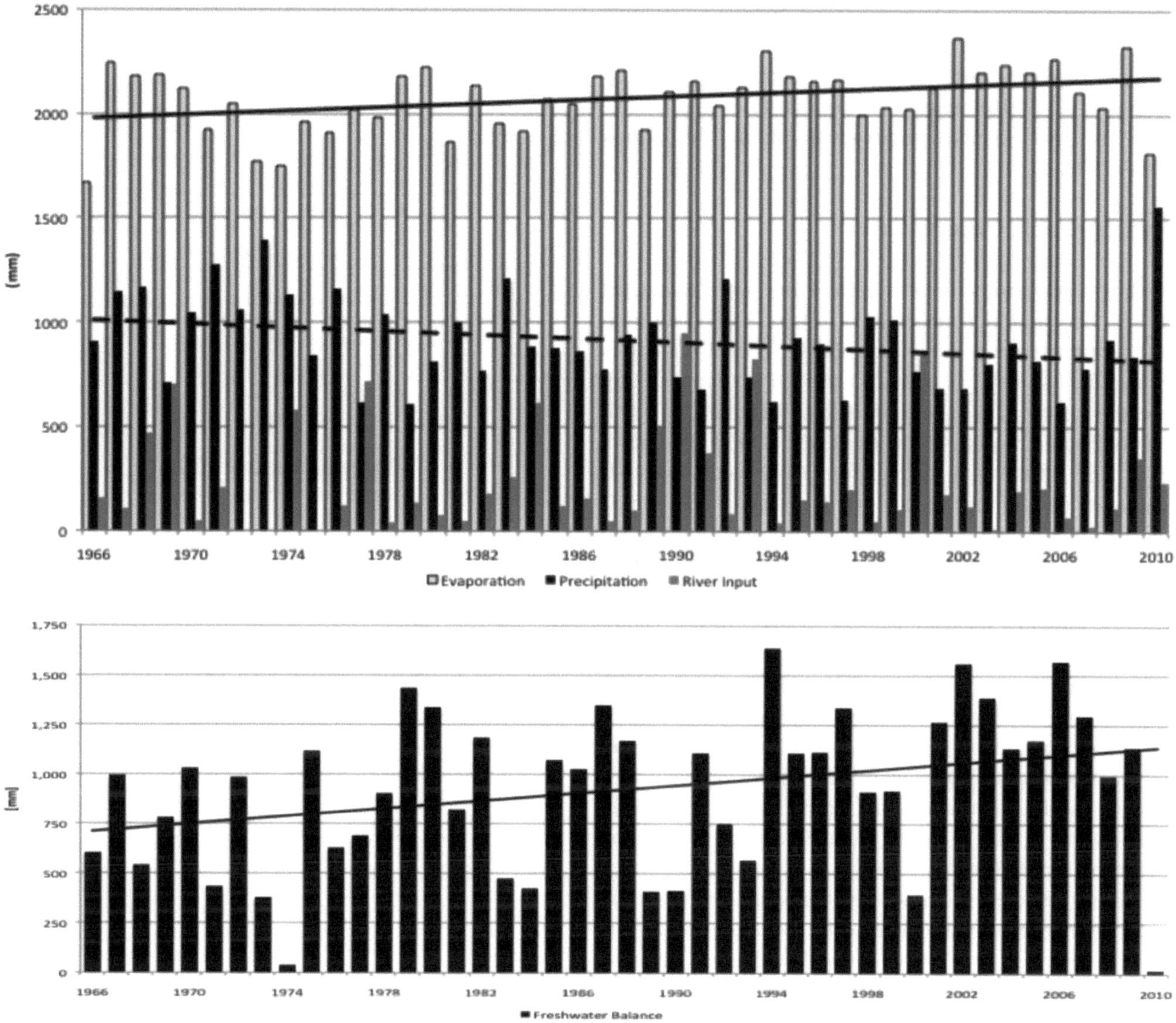

Fig. 6 (*Top panel*) Evaporation [mm], precipitation [mm] and river flows [mm] with linear trends of precipitation (*dashed declining line*) and evaporation (*solid inclining line*); (*lower panel*) total freshwater balance (E-P-R) in [mm] for the Hervey Bay region and positive linear trend of about 9.7 mm/year (Data source: Australian High Quality Climate Data Set)

about 1.08 °C during the period 1966–2010. During the same period, the annual freshwater balance (E-P-R) for Hervey Bay (Fig. 6, lower panel) indicates a linear trend of about +9.7 mm per year or a total decline in freshwater of 437 mm during the 45-year period from 1966 to 2010. The decline is mostly due to increasing evaporation (solid line in Fig. 6, top panel) and decreasing precipitation (dashed line in Fig. 6, top panel). The annual trend for precipitation is a decline of about 4.4 mm per year and evaporation increases by about 4.5 mm per year or about 198 mm (precipitation) and 203 mm (evaporation) in total. The reminder of the trend in the freshwater balance (less than 1 mm per year or about 36 mm for the period under consideration) is due to a decline in river flow discharges primarily from the Mary River into the southern region of Hervey Bay.

This is a significant change in the regional Hervey Bay freshwater balance and indicative of a distinct drying trend. Evaporation dominates the freshwater balance, which is positive every year (Fig. 6, lower panel). Total annual precipitation has decreased by about 20 %, total annual evaporation increased by about 10 %, and in turn, assuming linear trends for the period of interest, the total (positive) freshwater balance changed from about 703 mm to about 1,140 mm. This is an increase of the positive, evaporation-dominated balance by about 63 %. The hydrographic observations reviewed below were obtained from 2004 to 2008, i.e. during the drier part of the documented period.

The freshwater balance underpins the definition of a Hervey Bay and its estuary as a low water inflow system. In fact, the observed changes in the freshwater balance clearly support a long-term trend in the mean estuarine conditions toward increased hypersaline episodes. This is clearly supported through recent modelling studies by Gräwe et al. (2010) reviewed below and following a summary of hydrographic observations during the period 2004–2008.

Hydrographic Observations

Hydrographic observations in Hervey Bay were conducted during the period September 2004 to June 2008. Data from all surveys are reported by Ribbe (2008). A detailed analysis of the 2004 survey has been documented in (Ribbe 2006), and the survey data have been used to benchmark several hydrodynamic modelling studies (Ribbe et al. 2008; Gräwe et al. 2009, 2010). Key findings from those modelling studies are reviewed in the section "Hydrodynamic modelling".

Here, the depth-averaged distribution of salinity along west to east Conductivity-Temperature-Depth (CDT) transects are shown (Fig. 7) in distance [km] away from the first sampling location near the western shallow shoreline of the Bay. Sampling locations are indicated in Fig. 2 with spacing in longitudinal and latitudinal direction of 5 nautical miles. Not every section shown in Fig. 2 was sampled in each of the surveys. The water column was vertically mixed in most cases, although several times a vertical stratification was found in salinity and density as well as the daily temperature stratification. For detailed survey information see the report by Ribbe (2008).

Clearly evident from all hydrographic surveys, and observed in each of the seasons, is the existence of a hypersalinity zone (Fig. 7). This zone is well established along the western shoreline during early spring (September 2004), late winter (August 2007), early summer (December 2007) and late autumn (May 2008). The salinity difference along the west to east transects range from 0.4 to 0.6 between the western near shoreline to the interior and eastern region of the Bay. The maximum salinity gradients with values of about 1.0 and 1.3 are observed from the southwest to the north east of the Bay toward the open ocean in September 2004 and August 2007 respectively. This is indicative of a trapping of high salinity water in the most southwestern region of the Bay and indicative of longer water residence times in this part of the Bay.

Depth-averaged density (not shown) is found to be highest along the western to south western region of the Bay and decreases toward the east to north east of Fraser Island, i.e. the open ocean (see Ribbe (2008)). Maximum horizontal density differences of 0.75, 1.32 and 1.4 kg/m^3 are found during the September 2004, August 2007 and May 2008 surveys respectively, with the highest density in the most western and south western regions of Hervey Bay. During December 2007, no distinct depth-averaged density difference could be established most likely due to strong seasonal surface layer heating opposing the effect of increasing salinity on density.

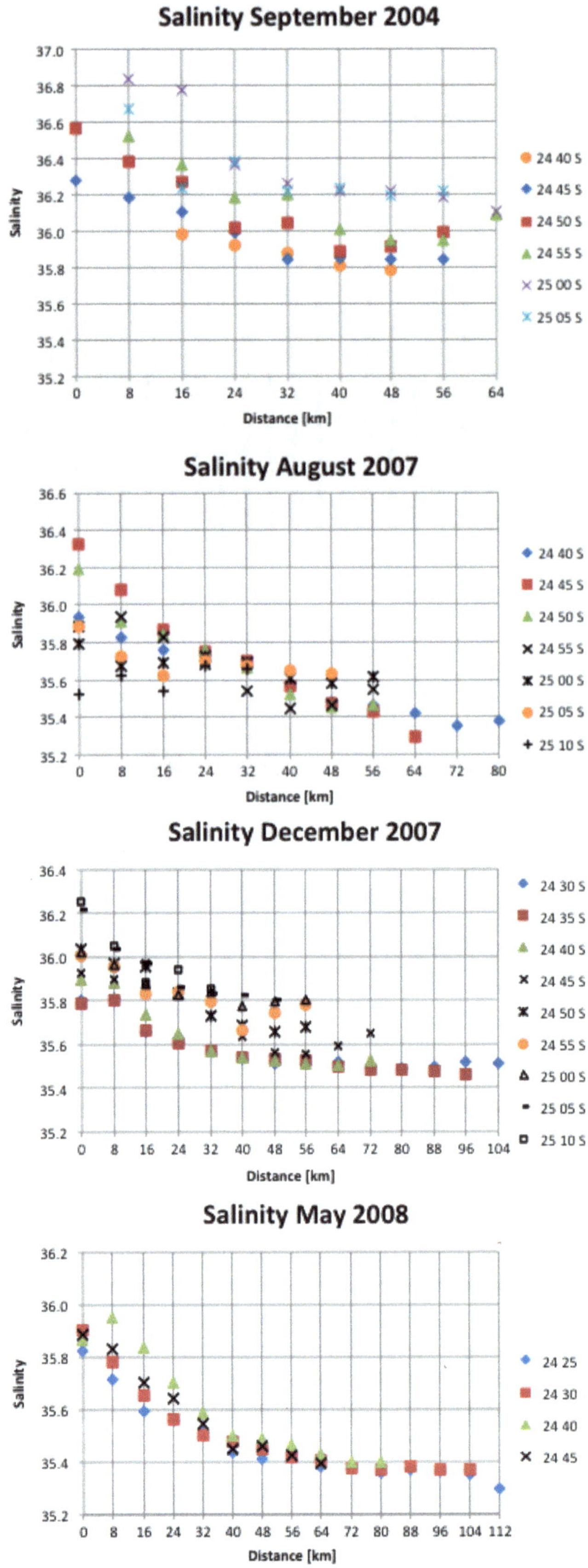

Fig. 7 Depth-averaged salinity observed during four hydrographic surveys into Hervey Bay during the period 2004–2008. (**a**) *Top*: September 2004, *bottom*: August 2008, (**b**) *top*: December 2007, *bottom*: May 2008

The most northern zonal transect (see Fig. 2) was sampled during a survey in May 2008 to the north of Hervey Bay at 24° 25′ S. This particular survey had to be aborted after sampling along latitude 24° 25′ S at the entrance to the Bay due to a significant storm event. A June 2008 survey following the storm event documented an erosion of the coastal salinity zone. This was due to the significant discharge of freshwater from the Mary River and many smaller estuaries along the western coastal boundary of the region in the Hervey Bay.

Nevertheless, the May 2008 survey outside Hervey Bay proper (see Fig. 2) documents the continuation of a hypersalinity zone toward the north of Hervey Bay and along the western coast. It is likely contributing to a hypersaline boundary zone that possibly extends all along most of the eastern coast of Queensland. This zone is a distinct characteristic for most, possible all, of the coastal zone of the Great Barrier Reef (see Andutta et al. (2011)) which is most developed during the low precipitation, low run-off season of the year (winter), and could possibly exist throughout most of the year. Although not documented through sustained and continuous observations along the east coast, the existence of this large-scale hypersaline boundary zone is supported through the continental freshwater balance with evaporation being larger than precipitation throughout the year for most of eastern Australia.

An important tool in coastal and estuarine management is the concept of a flushing time scale. It provides a simple assessment of the time it takes to remove a pollutant or any other property from a confined body of water. The hydrographic data sampled in Hervey Bay provide an opportunity to assess this time scale for the first time by applying a simple zero-dimensional water and salinity budget model. This model then yields information about water renewal or flushing time scales in the region. It assumes that the hypersaline zone is fully developed and in a steady state (see Ribbe (2006) for details). The build-up of salinity is finalised, and additional salt due to ongoing evaporation is to be transported away from the region. Salt-enriched water takes the role of a pollutant in this application, which is flushed away from the Bay.

A simple schematic of this salt/water balance model is represented in Fig. 8. The model defines a balance between outflowing high salinity water and incoming fresher, low salinity water, which is derived from the open ocean to the north and east of Hervey Bay. The balance or conservation equations yield a simple expression for the computation of the water renewal time scale (τ) as a function of the salinity difference (ΔS), the evaporation rate (e), and the geometry of the domain (surface area = A, depth = h), assuming that the system is in balance (Ribbe 2006).

The model was applied previously to the September 2004 data and a flushing time scale of about 89 days was computed by Ribbe (2006). Here the model is applied to all data and a summary of flushing time scales is shown in Table 2. Mean flushing times using observed depth averaged salinity differences along west to east transects and maximum differences from southwest to northeast yield values of between about 59–89 days or 2–3 months. The water renewal process is associated with the discharge of about 8,500 kg/s of salt from the Bay. This water renewal process potentially explains the injection of Hervey Bay Water into the EAC as postulated but not observed directly by Middleton et al. (1994).

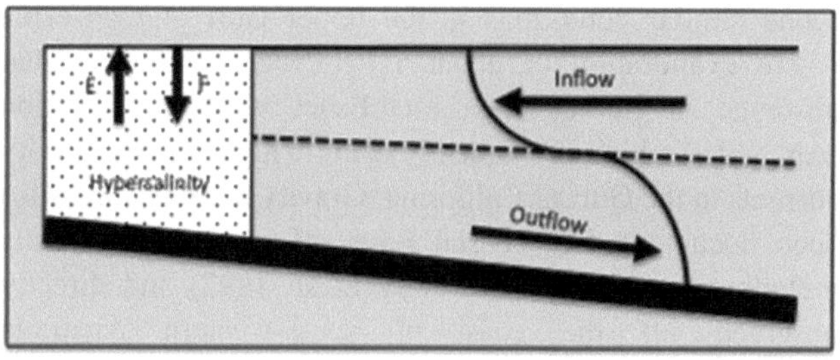

Fig. 8 Simple schematic of an inverse estuarine circulation (Adopted from Ribbe 2006). *Inflow* of low salinity water at the surface is balanced by *outflow* of high salinity water along the seabed. In hypersaline inverse estuaries, evaporation (*E*) is larger than the supply of freshwater (*F*) leading to an import of ocean water

Table 2 Estimates of Hervey Bay flushing times

Date	S_B	S_o	ΔS	τ(days)
Sept 2004	36.5	36.0	0.5	57
August 2007	36.0	35.4	0.6	67
Dec 2007	36.0	35.6	0.4	40
May 2008	35.9	35.4	0.5	59
Mean	36.1	35.6	0.5	59
Date	S_B	S_o	ΔS	τ(days)
Sept 2004	36.8	35.8	1.0	116
August 2007	36.3	35.0	1.3	67
Dec 2007	36.3	35.5	0.8	89
May 2008	36.0	35.3	0.7	75
Mean	36.4	35.4	1.0	89

Note: *Top table*: flushing times using west–east salinity gradient
Lower table: using maximum salinity gradient

In addition to costly field surveys, hydrodynamic modelling yields time dependent and spatial information as an alternatives means to assess the functioning of a marine system. This is reviewed in the following section also confirming the above time scales. Furthermore, this modelling provides insight into some of the actual physical processes that drive water renewal processes and the removal of saline water from Hervey Bay and its estuaries. This includes the establishment of a weak vertical stratification that can lead to the development of gravity currents, in particular during winter at neap tide. Lavin et al. (1998)

found similar conditions in the upper Gulf of California where evaporation of about 1.1 m/year, similar to that observed in Hervey Bay, establishes weak stratification ($\Delta\sigma_t \sim 0.2$) which subsequently leads to high salinity gravity currents in the Gulf of California. Gravity currents have also been documented in coastal Bays off southeast Australia including Jervis Bay (Holloway et al. 1992) and directly observed off the coast of south-western Australia (Pattiaratchi et al. 2011). The existence for Hervey Bay has also been postulated by Gräwe (2009) from modeling studies but direct observational evidence is not available yet.

Hydrodynamic Modelling

Three-dimensional hydrodynamic modelling of the continental shelf region of Hervey Bay is limited to studies by Ribbe (2006), Gräwe (2009) and Gräwe et al. (2009, 2010). Yet, this research provides significant and new insight into important ocean processes that contribute to the environmental health of the Bay and its many estuaries. The modelling informs present and future regional and marine resource management such as natural fisheries and aquaculture. These economic activities add millions of dollars to the regional economy in addition to recreational fisheries tourism that is estimated to deliver an additional $140 million (Robinson 2001).

The modelling studies provide information about flushing time scales, physical ocean transport processes including tidal mixing and mean circulation pattern, flood events and freshwater dispersal pathways, impact of climate variability processes such as the El Nino Southern Oscillation (ENSO), trends in salinity and inverse states of the Bay. Furthermore, yet to be observed phenomena such as gravity currents have been documented. These contribute to ventilating other large hypersaline coastal systems including Spencer Gulf in South Australia (Lennon et al. 1987) and the Great Australian Bight (Petrusevics et al. 2009). A summary of key findings from hydrodynamic modelling is presented in the following sections, all of which yield important natural resource management tools.

Ribbe et al. (2008) investigate mean circulation, water renewal pathways and ventilation times scales in a series of computational experiments utilising tidal, climatological forcing for wind, heat and freshwater fluxes and the COHRENS model (Luyten et al. 1999). The model is able to reproduce the large-scale patterns of the observed temperature and salinity distribution using only climatological forcing (Ribbe 2006). Climatological easterly winds establish a cyclonic, i.e. clockwise water renewal pathway in Hervey Bay with faster flushing time scales in the east and slower toward the south and western shore of the Bay (see Fig. 2). Water renewal times scales vary from several days in the east to several 10s of days in the west and more than 100 days in the interior of the Bay.

Gräwe et al. (2009) build on the previous modelling study and investigate the impact of tides upon mixing as well as hypersalinity within Hervey Bay. Observed monthly forcing data were used to simulate the circulation within Hervey Bay over a period of about 20 years.

Despite a significant and predominantly semidiurnal tidal range of about 3.5 m to sometimes over 4 m, the residual circulation is found to be only small. Tides are discovered to be responsible primarily for vertical mixing only. Instead, wind driven flow of about 5–10 cm/s and baroclinic forcing dominate the circulation within Hervey Bay and the very small residual currents (<1 cm/s) are an intrinsic, climatological feature of the Bay. A wind-driven predominantly cyclonic flow of 5–10 cm/s, and established by easterlies, corresponds to a basin-wide water renewal time scale of about 2 months.

Tidal mixing erodes any stratification for most of the time, however, with the potential for a weak stratification to develop only during neap (weak) tides and cooling (e.g. during winter) of the surface layer. This would then result into the development of outflow below the surface layer in form of gravity currents, i.e. inverse circulation pattern. Gräwe et al. (2010) find that hypersaline conditions prevail on average for about 240 days and that the system is in an inverse state for about 108 days throughout the year. Changes in the hydrological balance impose a trend of 2.7 and 1.8 days per year upon both hypersaline and inverse conditions. This equates to about 49 and 32 days during the period 1990–2007 or a total of 289 days and 140 days in 2007. During the same period, the export of salt or salt flux has increased by about 22 %. The removal of this saline water occurs in a cyclonic fashion with inflow of less saline (fresher) water in the east and outflow of high salinity water in the west. This confirms Ribbe et al. (2008) initial finding of a predominately wind-driven cyclonic water renewal pathway. However, Gräwe et al. (2009) also find that the western regions of the Bay exhibit the characteristics of a west–east inverse circulation, which is superimposed upon the mean circulation.

In addition to the impact of trends in hydrological parameters on the marine environment of Hervey Bay, Gräwe et al. (2010) investigate the response to short-time, episodic/quasi-catastrophic weather events such as storms and cyclones and that of interannual climate phenomena such as ENSO on Hervey Bay. Both short term and interannual variability is superimposed upon the documented long termed trends of drying and warming (Figs. 5 and 6).

Storm events impact dramatically upon the marine environment of Hervey Bay and its estuaries. For example, two major floods and a cyclone during a 3-week period in 1992 resulted in the destruction of most seagrass habitats (about 1,000 km^2) in Hervey Bay and an unprecedented number of dugong deaths. The dugong population declined to about 2.2 % of the pre-flood numbers (Preen and Marsh 1995; Preen et al. 1995). Both floods and cyclone occurred during an ENSO cold event or La Nina, which is often associated with above average rainfall and enhanced cyclone/storm activity in the Australasian region. This enhance rainfall delivery is clearly evident in a significant increase in total annual rainfall from about 677 mm in 1991 to about 1,210 mm in 1992 (Fig. 6).

The 1992 flood events are reproduced in simulations by Gräwe et al. (2010). The modelling suggests that major flooding of Hervey Bay is limited to a narrow coastal strip along the western shoreline. During the period 1990–2008, three major floods occurred, each time the floodwaters establish a narrow freshwater band with most of the Bay not being affected and most of the eastern Bay continues to be characterised by hypersalinity. The floodwaters discharged from the Mary River in the south and the smaller tributaries along the west coast of Hervey Bay establish a frontal structure along its western regions. The recovery time for the system following a significant flood event was established with about 2 months, corresponding to the earlier determined flushing times scales.

Summary and Future Outlook

Oceanographic observations in conjunction with detailed computational modelling of the Hervey Bay region provide significant new insight into a marine environment long been recognised as one of Australia's most biodiverse and unique marine habits. These findings include:

- Discovery of elevated salinity within the Bay and a hypersalinity zone along the western shoreline that extends beyond the Bay and its estuaries northward;
- Water balance modelling leads to the classification of Hervey Bay as a low inflow Bay or large estuary;
- The Bay is mostly vertically mixed due to tidal action and salinity appears mostly trapped within the south/southwest region of the Bay, although some observations as well as modelling provide evidence of an inverse circulation; most likely to occur during neap tide;
- Application of a simple salt balance model gives an estimate for the residence or flushing time scale in the order of about 2–3 months. Numerical studies support this basin-wide estimate, which varies regionally from as little as a few days to over 100 days within the interior of the Bay; the residence time scale derived from the balance equation is also equivalent or even greater than the time scale needed to develop hypersalinity;
- Numerical modelling indicates that the residual tidal driven circulation is at least a factor 10 smaller than the wind-driven circulation;
- The wind-driven circulation is predominately of a cyclonic nature with low salinity water inflow in the east and high salinity water outflow along the western shoreline;
- The pathway for floodwaters is limited to the western shoreline with a northward flow and establishing a strong frontal structure that maintains hypersaline conditions to the east of the fronts, however with salinity decreasing toward the northeast;
- It is possible that gravity currents as observed in other hypersaline coastal systems, contribute to the discharge of saline water to the northeast.

The above findings provide information and tools that aid future estuarine natural resource management with sustained pressure on the marine environment of Hervey Bay and its estuaries due to natural and anthropogenic factors. These include continued sea ranching and aquaculture, population growths, pollution from agriculture within catchments, and projected future climate change. The latter is superimposed with phenomena of climate variability resulting to changes in climate system processes such as ENSO, frequency and/or intensity of tropical cyclones and other disturbances that alter the freshwater balance within the region. A sound understanding of future natural and human impacts is essential to maintain the conservation status and natural beauty of this so unique coastal environment (Fig. 9).

An assessment of climate projections by the Australian Bureau of Meteorology and the Commonwealth Scientific Industrial and Research Organisation (BOM and CSIRO 2007) for the Hervey Bay region (Queensland So 2011a) indicate that current trends in climate elements such as temperature, precipitation and evaporation are to continue (see Figs. 4, 5 and 6). Against the historical annual mean of about 20.5 °C, the projected increases range from an additional warming of about +0.9 °C in 2030 to about +1.5 to +2.9 °C in 2070. Annual precipitation is likely to decline by about −3 % in 2030 and by about −5 to −10 % in 2070, although precipitation projections show significant larger variations between different climate models than corresponding projections for temperature. Projected trends in evaporation are positive and evaporation is to increase by about +3 % in 2030 and to about +6 to +11 % in 2070.

The direction of projected climatic changes are likely to lead to continued drying in the Hervey Bay region and a positive freshwater balance that already controls physical conditions in Hervey Bay. It results in hypersaline conditions associated with possible inverse circulation patterns. The drying trends have been shown by Gräwe et al. (2010) to drive an increase in the number of hypersaline/inverse states of the systems during the last 20 years.

Fig. 9 *Top left*: View of the world largest sandy but densely vegetated Fraser Island toward the south. *Top right*: Sandy Cape Lighthouse at the northern end of Fraser Island where the island transitions into the submerged 30 km long Break Sea Spit. *Bottom left*: Urangan Pier at the southern end of Hervey Bay; 1,100 m long when constructed in 1910 of which 880 m remain. *Bottom right*: A typical beach long the south western shore line of Hervey Bay where hypersaline water is formed in shallow water

These are likely to increase in the future shifting the system further to potentially more persistent quasi-permanent hypersaline/inverse conditions. The past and possible continued future drying trend is associated with changes in climate system processes such as the widening of the tropics/subtropics due to a southward shift of the subtropical ridge. This is the likely cause of reduced winter rainfall along the eastern coast of Australia (Cai et al. 2012) .

The projected future regional warming is also likely to be compounded by large-scale ocean warming. This is already increasing coastal temperature along the east Australian coast. Trends of about 0.4 °C per decade have been attributed by Lima and Wethey (2012) to increased transport of warm water via the EAC southward (Cai et al. 2005). Similarly, local salinity increases are likely to be moderated by observed global changes in the hydrological cycle. Changes have already led to a global intensification of the global hydrological cycle with increased evaporation and changes in ocean salinity (Durack et al. 2012). The so-called rich get richer scenario sees dry regions becoming drier and wet region to be wetter in a warming world. Projections of future large scale ocean temperature and salinity changes show a warming and freshening of the ocean off most of eastern Australia (Gupta and McNeil 2012). However, global projections yield no or little real information about likely changes in the coastal ocean and its estuaries where local exchange processes between the ocean and coastal region as well as local weather and climate pattern are important. Our knowledge of regional coastal ocean changes is yet very limited and uncertain and regional assessments are needed.

Increases in ocean temperature and changes in salinity within the region are likely to impact on marine species. Current climate variability and trends have shown to impact on species such as corals, turtles, seagrasses, mangroves and many fish species that have been recognised as iconic features of Hervey Bay, the Great Sandy Strait and many of its estuaries. Rasheed and Unsworth (2011) find that elevated temperature and reduced freshwater supply is correlated with lower seagrass biomass. In the past, this was associated with a reduction in sea grass meadows linked to reduced populations size of marine species such as dugongs (Preen and Marsh 1995).

Marine turtles, having some ability to adapt to climatic changes, are sensitive to temperature increases and other changes in climatic conditions, affecting their breeding and migration pattern (Poloczanska et al. 2009). Adverse impacts of ocean warming on coral reef building communities have been widely documented (e.g. Hoegh-Guldberg (1999). Zann et al. (2012)) find that vulnerability due to climatic changes increases in particular in the Hervey Bay region, where reef building corals are already impacted on by human activity due to their close proximity to coastal communities.

River discharges and freshwater inputs have been linked as contributing to primary productivity in estuaries (Gillanders et al. 2011). Loneragan and Bunn (1999) document that high river flows have a strong positive effect on commercial and recreational fisheries in southeast Queensland. Meynecke et al. (2006) and Meynecke and Lee (2011) explore this further for all of coastal Queensland finding that 30 % of all total fish catch variations can be explained by rainfall and sea surface temperature variability. For southeast Queensland including Hervey Bay, a strong positive link is found with high/low rainfall during ENSO warm/cold events. Continued future climatic changes such as a decline in precipitation and reduced river flows, or changes in the climate and rainfall drivers such as ENSO, could compound human influences. This includes the construction of river flow regulating dams such as the standing proposal for a Mary River dam, which would impact on the regional economy and impact on marine health and biodiversity. Runoff is more likely to decrease than increase in most catchments (Whitfield et al. 2010).

Ward et al. (2003) propose that several southern temperate fish species migrate during autumn and winter to subtropical waters of southeast Queensland including Hervey Bay to spawn. Sea temperature conditions are favourably low and similar to the southern summer sea temperatures during this period. Future warming may impact on these migration patterns. Other physical conditions that impact on Hervey Bay and its estuaries in the future are likely to change. These include sea level rise, wave climate and storm surges leading to increase threats of coastal inundation.

According to an assessment of future changes for Queensland's coastal zone (Whitfield et al. 2010) sea level is to rise by 0.8 m. Changes in storm surges are found to be largest in areas such as Hervey Bay where the water is funnelled into the Bay. Significant amplification of storm and tidal surge is found for the shallow region of Hervey Bay and storm surge modelling indicates a 19 % increase in storm tide levels (Queensland So 2004).

In addition to impacts upon Hervey Bay and its estuaries from natural processes such climate variability and change, and potential changes in the hydrology of catchments due to e.g. the aforementioned construction of dams, other direct human impacts are likely to arise from population growth, urbanisation, associated fragmentation of landscapes, tourism, pollution, coal mining, aquaculture, coastal constructions and expansions of harbours.

Hervey Bay and the Fraser Coast are located in the north of Southeast Queensland where population increases are largest along the fringes of the region. In 2006, Hervey Bay City had the highest population growths (5.3 %) out of a total of 70 Australian cities. The population is projected to increase from about 105,000 in 2011 to about 165,000 in 2031 (Queensland So 2012a). This will necessitate the future construction of about 19,400 new homes. It contributes to the continued loss of vegetation cover due to urbanisation, agriculture and grazing. This has been estimated with 56 % since settlement of Southeast Queensland (Queensland So 2012b), thus potentially further threatening the environmental integrity of the region. The transient population in the region due tourism is likely to increase by about 57 % from 2009 to 2021. Although tourism is estimated to contribute over $440 million in the Fraser Coast regions economy (Queensland So 2011c), it adds additional pressures on local resource. Other regional economic activities include the potential expansion of coal mining near Maryborough (Queensland So 2011c). If assessed as viable, this may prompt expansion of existing coastal infrastructure such as the Port of Bundaberg, construction of loading terminals and jetties, and increased shipping. The potential for marine environmental pollution is likely to accompany all these human activities. It includes contamination from herbicides (McMahon et al. 2005) or accidents such as the escape of four million litres of untreated black water from wastewater treatment plants into local creeks (Chronicle TFC 2012). The longer flushing times scales in a low inflow and intermittent inverse coastal systems such as Hervey Bay, are likely to prolong the potentially adverse impacts from marine pollution. However, a comprehensive assessment of all human and natural impacts along the Fraser Coast and for Hervey Bay is not available.

In conclusion, it remains a major challenge for policy makers and natural resource managers to address two priorities. Firstly, continue supporting activities that improve our understanding of the physical world and basc future natural resource management decision on a sound scientific basis, and secondly, negotiate and mediate between the many stakeholders that depend on the region to continue providing sustainable ecosystem services into the future and enhance the future economic prosperity of the Fraser Coast.

References

Andutta FP, Ridd PV, Wolanski E (2011) Dynamics of hypersaline coastal waters in the Great Barrier Reef. Estuar Coast Shelf Sci 94 (4):299–305

Benfer NP, King BA, Lemckert CJ (2007) Salinity observations in a subtropical estuarine system on the Gold Coast, Australia. In: The International Coastal Symposium (ICS). Gold Coast, pp 646–651

Boyd R, Ruming K, Davies S, Payenberg T, Lang S (2004) Fraser Island and Hervey Bay – a classic modern sedimentary environment. In: Boult PJ, Johns DR, Lang SC (eds) Eastern Australian basins symposium II. Petroleum Exploration Society of Australia, Special Publication, Adelaide, South Australia, pp 511–521

Boyd R, Ruming K, Goodwin I, Sandstrom M, Schröder-Adams C (2008) Highstand transport of coastal sand to the deep ocean: a case study from Fraser Island, southeast Australia. Geology 36(1):15–18

BOM and CSIRO (2007) Climate change in Australia. Commonwealth of Australia, Canberra
Cagnazzi DDB, Harrison PL, Ross GJB, Lynch P (2011) Abundance and site fidelity of Indo-Pacific Humpback dolphins in the Great Sandy Strait, Queensland, Australia. Mar Mamm Sci 27(2):255–281
Cai W, van Rensch P (2012) The 2011 southeast Queensland extreme summer rainfall: a confirmation of a negative Pacific Decadal Oscillation phase? Geophys Res Lett 39(8):L08702. doi:10.1029/2011gl050820
Cai W, Shi G, Cowan T, Bi D, Ribbe J (2005) The response of the Southern Annular Mode, the East Australian Current, and the southern mid-latitude ocean circulation to global warming. Geophys Res Lett 32(23):L23706. doi:10.1029/2005gl024701
Cai W, Cowan T, Thatcher M (2012) Rainfall reductions over Southern Hemisphere semi-arid regions: the role of subtropical dry zone expansion. Sci Rep 2:702
Chaloupka M, Osmond M, Kaufman G (1999) Estimating seasonal abundance trends and survival probabilities of humpback whales in Hervey Bay (east coast Australia). Mar Ecol Prog Ser 184:291–301. doi:10.3354/meps184291
Chronicle TFC (2012) Sewage process scrutiny. The Fraser Coast Chronicle, 10 February 2012
Donald A, Meinke H, Power B, Maia AHN, Wheeler MC, White N, Stone RC, Ribbe J (2006) Near-global impact of the Madden-Julian Oscillation on rainfall. Geophys Res Lett 33(9):L09704. doi:10.1029/2005gl025155
Durack PJ, Wijffels SE, Matear RJ (2012) Ocean salinities reveal strong global water cycle intensification during 1950 to 2000. Science 336(6080):455–458. doi:10.1126/science.1212222
Franklin T, Franklin W, Brooks L, Harrison P, Baverstock P, Clapham P (2011) Seasonal changes in pod characteristics of eastern Australian humpback whales (Megaptera novaeangliae), Hervey Bay 1992–2005. Mar Mamm Sci 27(3):E134–E152
Geoscience Australia (2009) Australian bathymetry and topography grid. Geoscience Australia, Canberra
Gillanders BM, Elsdon TS, Halliday IA, Jenkins GP, Robins JB, Valesini FJ (2011) Potential effects of climate change on Australian estuaries and fish utilising estuaries: a review. Mar Freshw Res 62 (9):1115–1131
Gräwe U (2009) Hervey Bay – insights from numerical modelling into the hydrodynamics of an Australian subtropical Bay. Carl von Ossietzky University of Oldenburg, Oldenburg
Gräwe U, Wolff JO, Ribbe J (2009) Mixing, hypersalinity and gradients in Hervey Bay, Australia. Ocean Dyn 59(5):643–658
Gräwe U, Wolff JO, Ribbe J (2010) Impact of climate variability on an east Australian bay. Estuar Coast Shelf Sci 86(2):247–257
Gupta AS, McNeil B (2012) Variability and change in the ocean. In: Henderson-Sellers A, McGuffie K (eds) The future of the world's climate. Elsevier, Amsterdam, pp 141–165
Haynes D, Muller J, Carter S (2000) Pesticide and herbicide residues in sediments and seagrasses from the Great Barrier Reef world heritage area and Queensland coast. Mar Pollut Bull 41(7–12):279–287. doi:10.1016/s0025-326x(00)00097-7
Hazel J, Gyuris E (2006) Vessel-related mortality of sea turtles in Queensland, Australia. Wildl Res 33(2):149–154. doi:10.1071/wr04097
Hoegh-Guldberg O (1999) Climate change, coral bleaching and the future of the world's coral reefs. Mar Freshw Res 50(8):839–866. http://dx.doi.org/10.1071/MF99078
Holloway PE, Symonds G, Nunes VR (1992) Observations of circulation and exchange processes in Jervis Bay, New South Wales. Mar Freshw Res 43(6):1487–1515
Largier J (2010) Low-inflow estuaries: hypersaline, inverse and thermal scenarios. In: Valle-Levinson A (ed) Contemporary issues in estuarine physics. Cambridge University Press, Cambridge, p 326
Lavery B, Joung G, Nicholls N (1997) An extended high-quality historical rainfall dataset for Australia. Aust Meteorol Mag 46:27–38
Lavin MF, Godinez VM, Alvarez LG (1998) Inverse-estuarine features of the upper Gulf of California. Estuar Coast Shelf Sci 47 (6):769–795. doi:10.1006/ecss.1998.0387
Lee RS, Black KP, Bosserel C, Greer D (2012) Present and future prolonged drought impacts on a larget temperate embayment: Port Phillip Bay, Australia. Ocean Dyn 62(6):907–922
Lennon GW, Bowers DG, Nunes RA, Scott BD, Ali M, Boylet J, Wenju C, Herzfeld M, Johannsson G, Nield S, Petrusevics P, Stephenson P, Suskin AA, Wijffels SEA (1987) Gravity Currents and the release of salt from an inverse estuary. Nature 327:695–697
Lima FP, Wethey DS (2012) Three decades of high-resolution coastal sea surface temperatures reveal more than warming. Nat Commun 3:704. http://www.nature.com/ncomms/journal/v3/n2/suppinfo/ncomms1713_S1.html
Loneragan NR, Bunn SE (1999) River flows and estuarine ecosystems: implications for coastal fisheries from a review and a case study of the Logan River, southeast Queensland. Aust J Ecol 24(4):431–440
Luyten PJ, Jones JE, Proctor R, Tabor A, Tett P, Wild-Allen K (1999) COHERENS – a Coupled Hydrodynamical-Ecological Model for Regional and Shelf Seas: user documentation, Brussels
Marsh H, De'Ath G, Gribble N, Lane B (2005) Historical marine population estimates: triggers or targets for conservation? The dugong case study. Ecol Appl 15(2):481–492. doi:10.1890/04-0673
McMahon K, Nash SB, Eaglesham G, Muller JF, Duke NC, Winderlich S (2005) Herbicide contamination and the potential impact to seagrass meadows in Hervey Bay, Queensland, Australia. Mar Pollut Bull 51(1–4):325–334. doi:10.1016/j.marpolbul.2004.10.045
Meynecke J-O, Lee SY (2011) Climate-coastal fisheries relationships and their spatial variation in Queensland, Australia. Fish Res 110 (2):365–376. doi:10.1016/j.fishres.2011.05.004
Meynecke J-O, Lee SY, Duke NC, Warnken J (2006) Effect of rainfall as a component of climate change on estuarine fish production in Queensland, Australia. Estuar Coast Shelf Sci 69:491–504
Middleton JH, Coutis P, Griffin DA, Macks A, McTaggart A, Merrifield MA, Nippard GJ (1994) Circulation and water mass characteristics of the southern Great Barrier Reef. Mar Freshw Res 45(1):1–18
Murphy BF, Ribbe J (2004) Variability of southeastern Queensland rainfall and climate indices. Int J Climatol 24(6):703–721. doi:10.1002/joc.1018
Murphy BF, Timbal B (2008) A review of recent climate variability and climate change in southeastern Australia. Int J Climatol 28 (7):859–879. doi:10.1002/joc.1627
Nahas EL, Pattiaratchi CB, Ivey GN (2005) Processes controlling the position of frontal systems in Shark Bay, Western Australia. Estuar Coast Shelf Sci 65(3):463–474
Nunes Vaz RA, Lennon GW, Bowers DG (1990) Physical behaviour of a large, negative or inverse estuary. Cont Shelf Res 10(3):277–304. doi:10.1016/0278-4343(90)90023-f
Pattiaratchi C, Hollings B, Woo M, Welhena T (2011) Dense shelf water formation along the south-west Australian inner shelf. Geophys Res Lett 38(10):L10609. doi:10.1029/2011gl046816
Petrusevics PM (1993) SST fronts in inverse estuaries, South Australia – indicators of reduced gulf-shelf exchange. Aust J Mar Freshw Res 44:305–323
Petrusevics P, Bye JAT, Fahlbusch V, Hammat J, Tippins DR, van Wijk E (2009) High salinity winter outflow from a mega inverse-estuary—the Great Australian Bight. Cont Shelf Res 29 (2):371–380. doi:10.1016/j.csr.2008.10.003
Poloczanska ES, Limpus CJ, Hays GC (2009) Vulnerability of marine turtles to climate change. In: Sims DW (ed) Advances in marine biology. Academic, New York, pp 151–211
Power S, Tseitkin F, Mehta V, Lavery B, Torok S, Holbrook N (1999) Decadal climate variability in Australia during the twentieth century. Int J Climatol 19:169–184

Preen A, Marsh H (1995) Response of dugongs to large-scale loss of seagrass from Hervey Bay, Queensland, Australia. Wildl Res 22(4):507–519

Preen AR, Long WJ, Coles RG (1995) Flood and cyclone related loss, and partial recovery, of more than 1000 km^2 of seagrass in Hervey Bay, Queensland, Australia. Aquat Bot 52(1–2):3–17

Queensland So (2004) Queensland climate change and community vulnerability to tropical cyclones, project synthesis report. Queensland Government, Brisbane

Queensland So (2011a) Climate change in the Wide Bay Burnett Region. Queensland Government, Brisbane

Queensland So (2011b) Great sandy regional marine aquaculture plan. Queensland Government, Brisbane

Queensland So (2011c) Wide Bay Burnett regional plan. Queensland Government, Brisbane

Queensland So (2012a) Population and dwelling profile, Fraser coast regional council. Queensland Government, Brisbane

Queensland So (2012b) State of the environment report 2011. Queensland Government, Brisbane

Rasheed MA, Unsworth RKF (2011) Long-term climate-associated dynamics of a tropical seagrass meadow: implications for the future. Mar Ecol Prog Ser 422:93–103. doi:10.3354/meps08925

Ribbe J (2006) A study into the export of saline water from Hervey Bay, Australia. Estuar Coast Shelf Sci 66(3–4):550–558. doi:10.1016/j.ecss.2005.10.012

Ribbe J (2008) Monitoring and assessing salinity and temperature variations in Hervey Bay. Final report prepared for the Burnett Mary Regional Group, Bundaberg

Ribbe J, Wolff J-O, Staneva J, Gräwe U (2008) Assessing water renewal time scales for marine environments from three-dimensional modelling: a case study for Hervey Bay, Australia. Environ Model Software 23(10–11):1217–1228. doi:10.1016/j.envsoft.2008.02.007

Ridd PV, Stieglitz T (2002) Dry season salinity changes in arid estuaries fringed by mangroves and saltflats. Estuar Coast Shelf Sci 54(6):1039–1049. doi:10.1006/ecss.2001.0876

Robinson J (2001) The economic value of Australia's estuaries: a scoping study. CRC for Coastal Zone Estuary and Watery Management, Brisbane

Roy PS, Williams RJ, Jones AR, Yassini I, Gibbs PJ, Coates B, West RJ, Scanes PR, Hydson JP, Nichol S (2001) Structure and function of south-east Australian estuaries. Estuar Coast Shelf Sci 53:351–384

Samarasinghe JRS, Lennon GW (1987) Hypersalinity, flushing and transient salt-wedges in a tidal gulf – an inverse estuary. Estuar Coast Shelf Sci 24:483–498

Schröder-Adams CJ, Boyd R, Ruming K, Sandstrom M (2008) Influence of sediment transport dynamics and ocean floor morphology on benthic foraminifera, offshore Fraser Island, Australia. Mar Geol 254(1–2):47–61. doi:10.1016/j.margeo.2008.05.002

Sheppard JK, Lawler IR, Marsh H (2007) Seagrass as pasture for seacows: landscape-level dugong habitat evaluation. Estuar Coast Shelf Sci 71(1–2):117–132. doi:10.1016/j.ecss.2006.07.006

Stern H, de Hoedt G, Ernst J (2000) Objective classification of Australian climates. Aust Meteorol Mag 49(2):87–96

Timbal B, Drosdowsky W (2012) The relationship between the decline of Southeastern Australian rainfall and the strengthening of the subtropical ridge. Int J Climatol 33:1021–1034

Tomczak M, Godfrey S (2003) Regional oceanography: an introduction, 2nd edn. Daya Publishing House/Elsevier, Delhi/Amsterdam

Vang L (2002) Distribution, abundance and biology of Group V humpback whales megaptera novaeangliae: a review. Queensland State Government, Brisbane

Ward TM, Staunton-Smith J, Hoyle S, Halliday IA (2003) Spawning patterns of four species of predominantly temperate pelagic fishes in the subtropical waters of southern Queensland. Estuar Coast Shelf Sci 56:1125–1140

Whitfield L, Oude-Egberink K, Wecker B, Cravigan L, Dalla Pozza R, Hernaman V, Scott J, Chidzambwa S (2010) Climate change in Queensland: what the science is telling us. State of Queensland, Brisbane

Williams AAJ, Stone RC (2009) An assessment of relationships between the Australian subtropical ridge, rainfall variability, and high-latitude circulation patterns. Int J Climatol 29(5):691–709

Wolanski E (1986) An evaporation-driven salinity maximum zone in Australian tropical estuaries. Estuar Coast Shelf Sci 22:415–424

Zann M, Phinn S, Done T (2012) Towards marine spatial planning for Hervey Bay's coral reefs. Paper presented at the 12th international coral reef symposium, Cairns, 9–13 July 2012

Moreton Bay and Its Estuaries: A Sub-tropical System Under Pressure from Rapid Population Growth

Badin Gibbes, Alistair Grinham, David Neil, Andrew Olds, Paul Maxwell, Rod Connolly, Tony Weber, Nicola Udy, and James Udy

Abstract

Moreton Bay and its associated estuaries are an example of a complex aquatic system that is under increasing pressure from rapid population growth and urbanisation. Although the extent of decline in ecosystem health within Moreton Bay and its associated estuaries is significant and well documented, a range of innovative management responses have been implemented to reverse current declines. An overview of the development of Moreton Bay is provided, highlighting the dynamic and resilient nature of the system over geological time. The ecological responses that occur at decadal timeframes are presented along with a summary of the current state of the Bay's ecology. The future challenges that are posed by predicted population increases, urbanisation and changes to the region's climate are also discussed. The highly variable nature of the system over relatively short timeframes (i.e. flood vs non-flood conditions) as well as the ability of the system to adapt to long term changes (i.e. past morphological and ecosystem shifts) suggests that Moreton Bay and its associated estuaries have significant capacity to adapt to change. Whether the current rate of anthropogenically induced change is too rapid for the system to adapt, or whether such adaptions will be undesirable, is unable to be ascertained in any detail at this stage. Notwithstanding the above, the combination of a science-based management framework and the collaborative decision making processes that have been implemented to halt the decline of Moreton Bay have shown remarkable progress in a relatively short period of time.

Keywords

Moreton Bay • Sub-tropical estuary • Water quality • Ecosystem health • Ecosystem resilience • Marine park • Environmental management

B. Gibbes (✉) • A. Grinham
School of Civil Engineering, The University of Queensland, Brisbane, QLD 4072, Australia
e-mail: b.gibbes@uq.edu.au

D. Neil
School of Geography Planning and Environmental Management, The University of Queensland, Brisbane 4072, Australia

A. Olds • P. Maxwell • R. Connolly
Australian Rivers Institute – Coast and Estuaries and School of Environment, Griffith University, Gold Coast 4222, Australia

T. Weber
Catchments and Receiving Environments, BMT WBM, Brisbane 4000, Australia

N. Udy
Marine Resource Management, Queensland Department of National Parks, Recreation, Sport and Racing, Brisbane 4002, Australia

J. Udy
Science and Innovation Program, Healthy Waterways Pty Ltd., Brisbane, QLD 4003, Australia

E. Wolanski (ed.), *Estuaries of Australia in 2050 and Beyond*, Estuaries of the World,
DOI 10.1007/978-94-007-7019-5_12, © Springer Science+Business Media Dordrecht 2014

Box 1

Badin Gibbes and colleagues studied Moreton Bay and its associated estuaries that are under increasing pressure from rapid population growth and urbanisation. They document the decline in ecosystem health as well as a range of innovative management responses to reverse current declines. The ecological responses occur both at decadal timeframes as well as over relatively short timeframes (i.e. flood vs. non-flood conditions). Whether the current rate of anthropogenically induced change is too rapid for the system to adapt, or whether such adaptions will be undesirable, is unknown at this stage. Nevertheless the science-based management framework that has been implemented to halt the decline of Moreton Bay has shown remarkable progress in a relatively short period of time.

Introduction

Moreton Bay and its associated estuaries provide an example of a complex aquatic system that is under increasing pressure from rapid population growth and urbanisation. The region has a human population of more than three million people and is one of the fastest growing regions in Australia with an expected population of over four million people by 2026 (QOESR 2011). The region's history of rapid population growth has also shaped the current condition and function of its waterways. The region has experienced significant land use change, dominated by the removal of native vegetation since European settlement approximately 200 years ago. Although the extent of decline in ecosystem health within Moreton Bay and its associated estuaries is significant and well documented, the characteristics and values of the region have also inspired the development and application of a range of innovative management responses in an attempt to reverse current declines. In this regard it provides an instructive case study for management of estuarine systems that are undergoing rapid transformations. This chapter provides an overview of the development of Moreton Bay and highlights the dynamic and resilient nature of the system over geological time, as well as investigating the ecological responses that occur at decadal timeframes. A summary of the current state of the Bay's ecology is also provided before discussing the future challenges that are posed by predicted population increases, urbanisation and changes to the region's climate.

Site Geomorphological and Hydrological Settings

Moreton Bay is a semi-enclosed subtropical embayment of considerable geomorphic, ecological and economic significance, and an important recreational and aesthetic resource for the people of southeast Queensland. The Bay lies between 27° and 28° south latitude, approximately 110 km north to south, and has its major opening to the north (Fig. 1). It is roughly triangular in shape with a 15.5 km wide north entrance opening (Skirmish Point to Comboyuro Point) tapering to the mouth of the Nerang River in the south. The seabed in Moreton Bay slopes from west to east with a gradual slope near the western shore that transitions to a steep slope on the eastern shoreline. The deepest waters of the Bay are at 20–29 m along the west coast of Moreton Island and the northwest margin of the South Passage flood tide delta.

The Bay is defined on the east by the large dune island barriers of Moreton Island (198 km^2; 37 km long), and North Stradbroke Island (285 km^2; 36 km long), and the barrier island South Stradbroke Island (26 km^2; 20 km long), and the Southport Spit. Formation of the eastern margin of the Bay was by aeolian dune building, onshore sand transport and northward longshore spit formation during the sea level oscillations of the late Quaternary, with the modern shoreline a product of the last stages of the post-last glacial sea level rise and the late Holocene sea level high stand. Although these sand islands have formed over several hundred thousand years (Ward 2006), continual changes to their

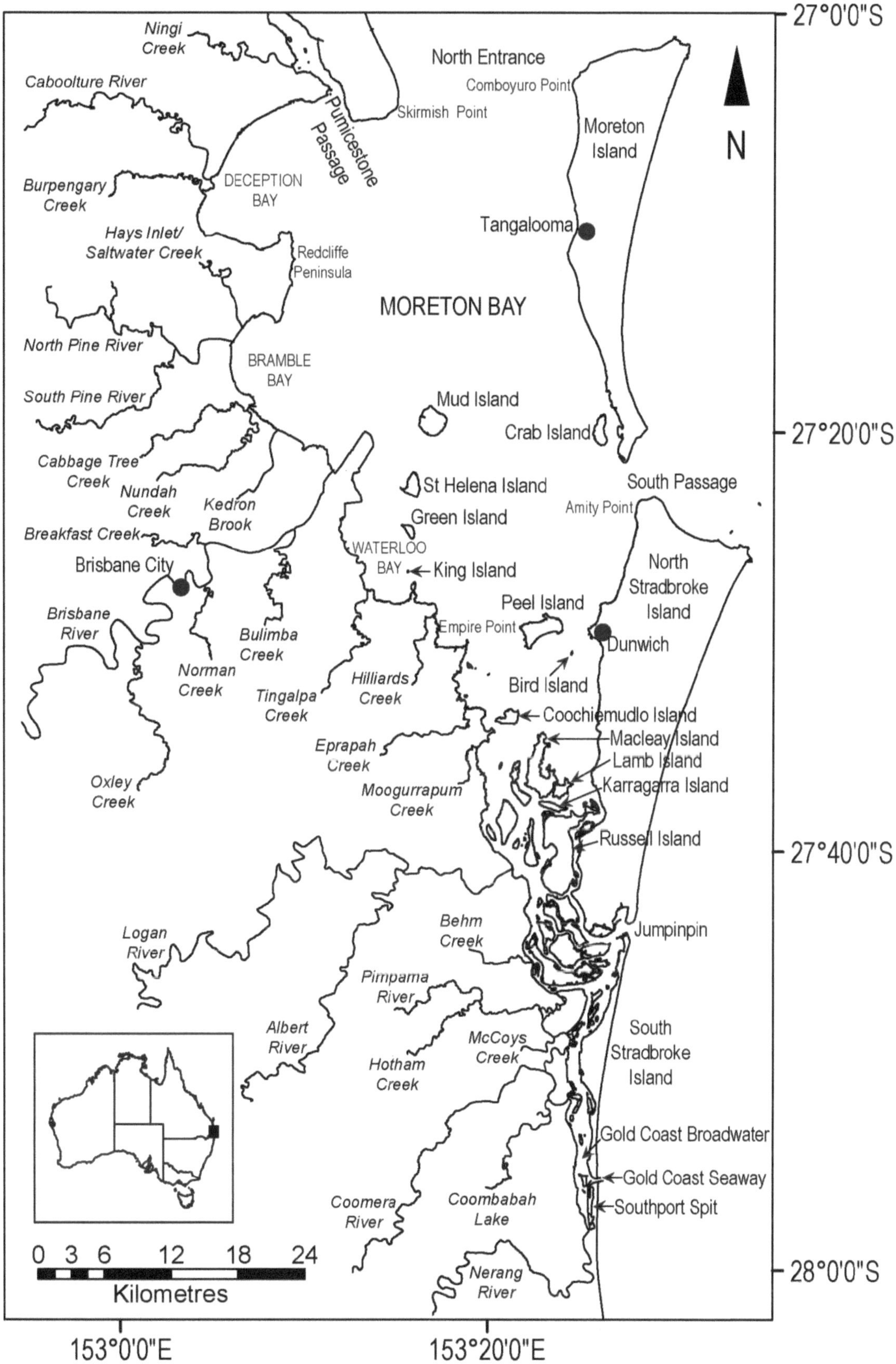

Fig. 1 The Moreton Bay estuary illustrating the major river systems draining to the Bay and the major tidal entrances that connect the Bay to the adjacent Coral Sea

Fig. 2 Photographs of the Moreton Bay estuary illustrating (**a**) the relatively undeveloped shoreline of the eastern Bay at Tangalooma, (**b**) marina and new residential canal estate development (*foreground*) adjacent to a relatively undeveloped island (*middle* to *background*) in the southern Bay, (**c**) urbanisation of the lower Nerang River, and (**d**) the Luggage Point sewage treatment plant (*foreground*) and Port of Brisbane (*background*) at the mouth of the Brisbane River

shorelines during the course of the last several thousand years have in turn resulted in changes to the Bay's bathymetry, tidal exchange, flushing and the existence and characteristics of islands and ecosystems within the Bay. The eastern shoreline of the Bay is largely in a natural state with very limited areas of human settlements, infrastructure and disturbance (Fig. 2a).

Four entrances provide oceanic exchange by tidal flushing of the Bay. The North Entrance is the largest and the most significant contributor to oceanic exchange. It is $\approx$15.5 km wide and the subtidal banks penetrate up to 18 km into the Bay. South Passage is $\approx$3.7 km wide with associated flood tide delta deposits (intertidal and subtidal) that extend 13.5 km, more than half way across the Bay. The Jumpinpin Entrance between North and South Stradbroke Islands is $\approx$0.8 km wide. Both the South Passage and Jumpinpin entrances are dynamic in response to changing wave and sediment transport conditions. The Southport Entrance between South Stradbroke Island and the mainland Southport Spit was replaced by an artificial, rockwall stabilised entrance (the Gold Coast Seaway) opened in 1986, which has increased tidal flushing of the southern Bay and Broadwater areas. Moreton Bay has semi-diurnal tides and, in the eastern Bay (at Amity Point), the mean tidal range is 1.48 m (springs) and 0.84 m (neaps). The tidal range is about 15 % higher in the western Bay (Brisbane Bar). The coastal ocean to the west of Moreton Bay is dominated by the East Australian Current (EAC) which allows the tidal exchange of warm tropical water (and associated biota) with the Bay through its four entrances. The EAC adjacent to Moreton Bay is relatively consistent in both flow rate and water temperature, and has a low frequency of upwelling events (i.e. there is little evidence to suggest that oceanic upwelling is a source of nutrients for the Bay, however this aspect still needs further investigation).

The west (mainland) coast of the Bay is characterised by a number of relatively erosion resistant headlands of Tertiary volcanics with extensive deposits of Quaternary alluvium in intervening embayments and adjacent to river and creek mouths. From the Pumicestone Passage in the North to the Gold Coast Broadwater in the south there are 20 estuaries (as well as many smaller ephemeral creeks) that connect to the western shoreline of the Bay which has seen significant alteration and disturbance by human activities. Habitat disturbance by both commercial and recreational activities is widespread (e.g. Fig. 2b–d). Land 'reclamation' and seawall construction is ubiquitous between Redcliffe Peninsula and Redland Bay, including works for the Brisbane Airport and Port of Brisbane (both located adjacent to the mouth of the Brisbane River – see Fig. 2d). Urban areas dominate the immediate hinterland of much of the western shoreline. Greater than 10 % of the catchment area of the Bay is urban and less than 25 % is remnant natural vegetation with the dominant land use agricultural (grazing ≈65 %; cropping ≈5 %) (Capelin et al. 1998).

There are no islands in the northern 30 km of the Bay. Between Mud Island (7 km northeast of the Brisbane River mouth) and the Logan River mouth there are reefal islands (comprised largely of biogenic marine sediments; Mud, Green, King, Bird), high (continental; bedrock) islands (topographic prominences associated with mainland terrains; eg. St Helena, Peel, Coochiemudlo, Macleay, Lamb, Karragarra and Russel Islands) and tidal delta islands. South from the Logan River mouth to the Nerang River estuary the southern Bay is 'choked' with numerous tidal delta islands. The high islands have been largely subdivided for residential and other development. The reefal islands, although not settled per se, have been subject to significant disturbance, including clearing of littoral rainforest and noxious weed invasion (Green Island; Neil 2000) and dredging of the adjacent reef flat resulting in erosion and mangrove mortality (Mud Island; Allingham and Neil 1995). Both Mud and Green Islands are of considerable geomorphic significance as high latitude occurrences of reef island types which otherwise only occur on the northern Great Barrier Reef (Allingham and Neil 1995; Neil 2000).

Moreton Bay is characterised as a modified wave dominated estuary with semi-diurnal tides (Digby et al. 1998). Key physical characteristics of the Moreton Bay estuary (Table 1) include the large catchment area to water area ratio (≈15:1) from which it can be inferred that, while the Bay is wave dominated for the majority of the time, the large catchment area is able to produce significant inflows that are capable of transforming the hydrodynamics and ecosystem processes in the system. This is particularly true of the two large river systems (Brisbane and Logan-Albert Rivers, catchment area >1,000 km^2) which exert a significant influence on the sediment and water quality characteristics of Moreton Bay during large rainfall events (Davies and Eyre 1998; DERM 2011).

Table 1 Summary of physical characteristics of Moreton Bay (After Digby et al. 1998)

Parameter	Units	Value
Back barrier	km^2	436.6
Central basin	km^2	1057.0
Fluvial Bay head	km^2	35.9
Flood ebb delta	km^2	149.0
Intertidal flats	km^2	75.7
Mangrove	km^2	80.3
Saltmarsh	km^2	22.8
Tidal sand banks	km^2	422.8
Seagrass meadow	km^2	189.0[a]
Rocky reef	km^2	0.5
Coral	km^2	13.5[a]
Channel	km^2	77.1
Floodplain	km^2	25.4
Bedrock perimeter	km	8.1
Perimeter	km	297.6
Catchment area	km^2	22,807.0
Water area	km^2	1,493.7
Maximum length	km	78.0
Maximum width	km	33.8
Entrance width	km	21.6
Mean wave height	M	0.7
Mean wave period	S	5.9
Maximum wave height	M	1.8
Maximum wave period	S	9.8
Tidal range	M	1.6

Note: [a]Indicates data not included in Digby et al. (1998)

Geomorphology

To explore the current and potential future states of Moreton Bay and its estuaries it is instructive to understand the range of different states that this system has existed in over geologic timeframes. In particular the ability of the system to transition between these states (i.e. from a non-marine river valley to a more oceanic embayment) provide a guide as to the possible range of future states that the system could operate within when subject to human population increases, land modification and changes in climate. Neil (1998) provides a comprehensive overview of the geomorphology of Moreton Bay and this section is largely based on this work and references therein with some more recent sources.

Orbitally-forced global-scale oscillations in temperature have resulted in significant variation in the accumulation of ice caps which, in turn, have resulted in variations in sea level of amplitude >100 m and period of >100 ka. Over the last several million years (>20) such oscillations have occurred. During warm (interglacial) periods sea levels may remain relatively constant for several thousand years. Reviewing the record of such fluctuations, it can be inferred that, over the last several million years, Moreton Bay, in something like its present form, has existed for <10 % of the time, and then each time of occurrence of the Bay differs according to maximum sea level height and duration and the antecedent conditions. The present area of Moreton Bay is a non-marine, broad river valley for about 50 % of the time. At low sea levels, most of the catchment drains east and north across the present Bay and flows north along the west coast of Moreton Island (Lang et al. 1998).

The Bay filled from about 11 ka as sea levels rose following the last glacial maximum. Present sea level was reached about 7.8 ka and maximum sea level, about 1.5 m higher than present, was reached by about 7.4 ka (Sloss et al. 2007), establishing the general form of the current Moreton Bay, but with very different characteristics. With sea level about 1.5 m higher the mainland coast was, in places (e.g. Deception Bay, Brisbane River mouth, Coomera River mouth) up to 9 km west of its present location. Most of the fluvially-derived sediment in the western Bay was not yet present. Although wave energy would have been higher, deeper water and the relative absence of muddy sediments are likely to have resulted in lower rates of wave resuspension of bottom sediments. Thus the Bay was wider, deeper, more open to seaward, better flushed and more oceanic in character with higher wave energy but lower sediment resuspension and turbidity than is the case today.

These morphological characteristics were accompanied by warmer air temperatures, warmer sea surface temperatures, higher rainfall and reduced rainfall intensity and variability – a so called 'climatic optimum' and a pre El Nino–Southern Oscillation (ENSO) climate. A mid-Holocene climatic optimum has been widely reported from locations around the world including eastern Australia, although direct evidence from the Moreton Bay region is lacking. At the mid-Holocene sea level high stand, sea surface temperature (SST) in Moreton Bay would have been both warmer (similar to Hervey Bay or Shoalwater Bay today) and less variable, due to deeper water and more oceanic exchange. Catchment characteristics under climatic optimum, very weak ENSO conditions are likely to have been of higher rainfall, higher vegetative cover and higher but less variable runoff carrying low sediment and nutrient concentrations. Thus catchment impacts on the water quality and ecosystems of the Bay would have been much lower than are currently observed.

From the mid-Holocene optimum to the time of European settlement conditions in both the Bay and its catchment deteriorated. Drier and more variable conditions associated with the onset of stronger ENSO forcing (about 3 ka; Donders et al. 2008) resulted in decreased vegetative cover, increased soil erosion and sediment yield, and lower but more variable runoff (more flood events). In the Bay, a decrease in water depth associated with a marked fall in sea level (about 2 ka; Sloss et al. 2007; Lewis et al. 2008), and coastal progradation on both the mainland coast and on the seaward Bay margins resulted in reduced volume, reduced tidal flushing and declining water quality. These changes in the physical environment of the Bay altered habitats and species composition of the Bay. Indications of these changes can be seen in the transition from Acroporid corals to mangrove communities at Empire Point on the mainland coast (Flood 1978), from coral fragments to shell fragments in a beach ridge sequence on Green Island (Neil 2000) and from Acroporid to Favid corals at Peel Island (first reported by Stutchbury 1854 (Saville-Kent 1893)). Additional conclusions of Lybolt et al. (2011) relevant to this discussion were that the depth of coral growth (controlled for the fall in sea level) had decreased by 2 m during the late Holocene as a result of decreased volume, decreased flushing, increased thermal stress and increased flood impacts in the Bay (all attributed to sea level fall) and coral growth in the Bay was episodic, not continuous. Such 'switch on – switch off' patterns of coral growth have been reported from marginal reef environments elsewhere (e.g. Smithers et al. 2006). Five phase shifts occurred over the 7,000 year age range of the coral death assemblage, demonstrating the feasibility of reversible phase shifts in coral communities in the Moreton Bay environment, at least under pre-European settlement conditions.

The physical changes in the Bay and its catchment thus led to a change from oceanic to estuarine, decreased sea SST and increased SST variability, a decline in water quality and declining habitat quality for some taxa (e.g. corals), improved for others (e.g. saltmarsh, mangrove, seagrass) and a mixed outcome for yet others (e.g. dugongs). Costs/benefits of the changes in the Bay are also likely to have varied spatially for many taxa. Using reconstructions of aeolian sediment transport rates from swamp sediments on North Stradbroke Island, McGowan et al. (2008) report significant climate instability in the Moreton Bay region during this period. Thus it seems likely that the late Holocene transition in the geomorphology and ecology of the Bay was a complex combination of gradual change, both synchronous and asynchronous between forcing factors, punctuated by episodic events of changing recurrence intervals and infrequent episodic phase shifts.

It has been suggested (Walters 1989; Hall 1990) that aboriginal land use ("fire stick farming" (Jones 1969)) was responsible for "changing the ecosystem to one more suitable to their needs" where increased erosion led to the "formation of large areas of mud and sand flats covered with shallow turbid waters and seagrass beds, permitting the evolution of fish stocks on a scale which today form the basis of large contemporary commercial fisheries". Based on a number of admittedly untested assumptions, Neil (1998) argued that the likely contribution of aboriginal burning to sediment yield was probably <10 %. Sediment yield occurs naturally, sediment yield was increasing due to climate shifts (e.g. onset of ENSO) and the aboriginal population was small (c. 5,000 (Hall 1990)). Given that sediment yield was just one of many factors driving geomorphic and ecological change in the Bay, it seems likely that the role of aboriginal people in this transition was negligible.

Hydrology

The sub-tropical climate of Moreton Bay is dominated by a distinct seasonal rainfall pattern that is characterised by high rainfall during summer months that can lead to large runoff events and occasional floods. Large scale floods are typically caused by degraded tropical cyclone or east coast lows that can persist over the region for several days and produce large volumes of rainfall in short periods of time. This seasonal, event-driven hydrology can result in rapid shifts between two distinct hydrological modes: (1) Wind, wave and tidally dominated; and (2) Freshwater inflow dominated. It has been hypothesised that the rapid and highly variable nature of the shifts between these two modes has influenced the ecosystems capacity to alter its processing pathways to adapt (and therefore become more resilient) to these episodic inputs of freshwater, sediment and nutrients.

The region also experiences a significant east–west rainfall gradient with average annual rainfall on the eastern edges of the Bay (e.g. Cape Moreton Lighthouse) exceeding that of the western edge by approximately 26 % (i.e. Cape Moreton Lighthouse cf. Brisbane Aero stations; Table 2). A similar comparison of annual average rainfall at the western edge of the Bay compared to the western edge of the larger river catchments that drain to the Bay shows a 35 % decrease in rainfall near the inland boundaries of these catchments (e.g. The University of Queensland Gatton cf. Brisbane Aero stations; Table 2). This gradient results in more rainfall in the smaller coastal catchments compared to the larger catchments which have the majority of their area to the west of the Bay. This in turn causes the smaller estuaries to be characterised by more frequent, short-duration, episodic inflows of freshwater compared to the larger catchments. During these event flows the smaller estuaries can be flushed before developing significant vertical stratification for short periods (days to weeks).

In the absence of significant rainfall events the catchments are characterised by very little or no flow (i.e. baseflow is minimal during the winter dry season). During these dry periods a residual clockwise circulation pattern is established within the Bay due to the asymmetry of the flood and ebb tide flows through the passages (i.e. North Passage, South Passage, Jumpinpin and the Gold Coast Seaway – refer to Fig. 1) that allow water exchange with the adjacent Coral Sea (Dennison and Abal 1999). This residual circulation creates a pattern of northward water movement on the western edge of the Bay and southern movement on the eastern edge. These circulation patterns combine with seasonal wind patterns and the complex morphology in the southern parts of the Bay to establish a strong gradient in water residence time in different parts of the Bay (refer to Table 3). It is noted that, while the residence times of the embayments on the western edge of the Bay can be significant (50–60 days) the residual northerly movement of water provides an important pollutant removal mechanism for these areas which receive large pollutant (predominantly sediment and nutrients) loads from the adjacent river estuaries.

Table 2 Selected long-term meteorological statistics

		Station		
Meteorological statistic	Units	Cape Moreton Lighthouse	Brisbane Aero	University of Queensland Gatton
Station reference number	–	40043	40223	40082
Period	–	1869–2012[a]	1950–2000	1897–2012[a]
Mean annual rainfall	mm	1,494.1	1,186.2	771
Mean annual days of rain	d	142	122.7	90.6
Highest recorded daily rainfall	mm	339.9	307.4	199.4
Mean 9 am air temperature	°C	20.8	20.8	20.4
Mean 9 am relative humidity	%	75	66	67
Mean 9 am wind speed	km h^{-1}	25.2	11.2	10.4

Source: Australian Bureau of Meteorology
[a]The period used for individual statistics varies within this range with maximum available periods used for all statistics

Table 3 Summary of average water residence times in different parts of Moreton Bay and is estuaries (After Dennison and Abal 1999)

Site	Residence time [d]
Ocean boundaries	3–5
Central Bay	50–55
Mouth of Brisbane River	63–68
Lower Brisbane River	110–120
Middle Brisbane River	154–162
Bremer/Brisbane junction	187–189
Bramble Bay	59–62
Deception Bay	54–57
Pumicestone Passage	43–53
Pine River	55–62
Caboolture River	53–57
Logan River	66–75

Ecology: Turbidity, Bio-sedimentary Aspects, Living Communities and Processes

Bio-sedimentary Processes

Under current conditions terrigenous sediment and nutrient input to Moreton Bay is delivered in a significantly variable (over both space and time) manner. This is primarily due to the concentration of catchment development along its southern and western shores and regional weather patterns described above. Over 30 major sewage plants and industrial wastewater treatment plants discharge directly into the Bay (e.g. Fig. 2d) and its associated waterways and these are the largest source of nutrients during average years (Eyre and McKee 2002). Episodic flows associated with high rainfall events deliver the majority of sediment inputs through highly turbid inflows into Moreton Bay consisting primarily of suspended silts and clays. These small particles are highly charged and carry a relatively large nutrient (2.8 % organic carbon; 0.3 % nitrogen and 0.1 % phosphorus) and metal (4.8 % iron and 0.3 % manganese) load (Grinham et al. 2012). This results in western and southern areas having the highest sediment mud content, the highest nutrient availability (Heggie et al. 1999) and the lowest water clarity (Longstaff 2003) relative to northern and eastern areas of Moreton Bay (Fig. 3). In the western Bay, particularly north of the Brisbane River, muds and silts extend to 10–20 km offshore (Jones and Stephens 1981). Sediments in the eastern Bay are predominantly tidal delta sands which extend from the North Entrance banks along the west coast of Moreton Island to the southern extent of the South passage flood tide delta south of Dunwich (Fig. 3). In the southern Bay, adjacent to South Stradbroke Island, sediments of the eastern Bay are predominantly tidal delta sands with fluvial sands and muds adjacent to the mainland coast (Lockhart et al. 1998).

The impact of this nutrient and sediment loading on autotrophic primary productivity suggest the system is undergoing a typical response of shallow-water ecosystems where benthic primary production decreases in favour of pelagic primary productivity (Meyer-Reil and Koster 2000). Pelagic primary productivity is estimated to have increased tenfold since European settlement of the area (McEwan et al. 1998). Large declines in benthic microalgal productivity have occurred particularly in subtidal habitats of western and southern Moreton Bay, this has resulted in an estimated 50 % reduction in baywide primary productivity compared with pre-European settlement (Grinham 2007). This decline in benthic primary productivity as well as a decline in clean sand facies within Moreton Bay is of concern as further declines could result in periodic anoxia of the sediment surface particularly in western and southern areas. This would directly couple sediment microbial nutrient remineralisation to the water column, further stimulating pelagic productivity and allowing these degraded conditions to persist (Meyer-Reil and Koster 2000).

Seagrass and Ecosystem Functioning: Processes That Promote Resistance to Impact and Restrict Recovery

Moreton Bay supports 189 km^2 of seagrass (Figs. 3 and 4) comprised of seven different species (Roelfsema et al. 2009). Seagrasses in coastal and nearshore environments, like Moreton Bay, perform a range of services, which are lost as seagrasses decline. Large-scale loss of seagrass meadows to unvegetated substrate has occurred particularly in western embayments of Bramble Bay (the receiving body for the Brisbane River) and southern Deception Bay (Fig. 4) (Dennison and Abal 1999). This loss of seagrass to unvegetated substrate is not easily reversed, and considerable resources have been invested to aid ecosystem recovery in the Bay and worldwide, but with little success (de Jonge et al. 2000; van der Heide et al. 2007). This section describes the processes that promote resistance of seagrass to impacts in Moreton Bay and those that prevent its recovery once lost.

Seagrasses are particularly susceptible to increases in nutrients and sediments, which in Moreton Bay are attributed to riverine inflows to the western and southern embayments (Dennison and Abal 1999). These can reduce the cover and extent of seagrass through smothering, by reducing light availability and promoting overgrowth of epiphytic algae (Neckles et al. 1993; Abal and Dennison 1996). Seagrass meadows do, however still exist in areas of the Bay that regularly experience poor water quality, which provides evidence of their ability to resist these impacts. Seagrasses resist impact through three processes: the uptake of nutrients from the water column, which

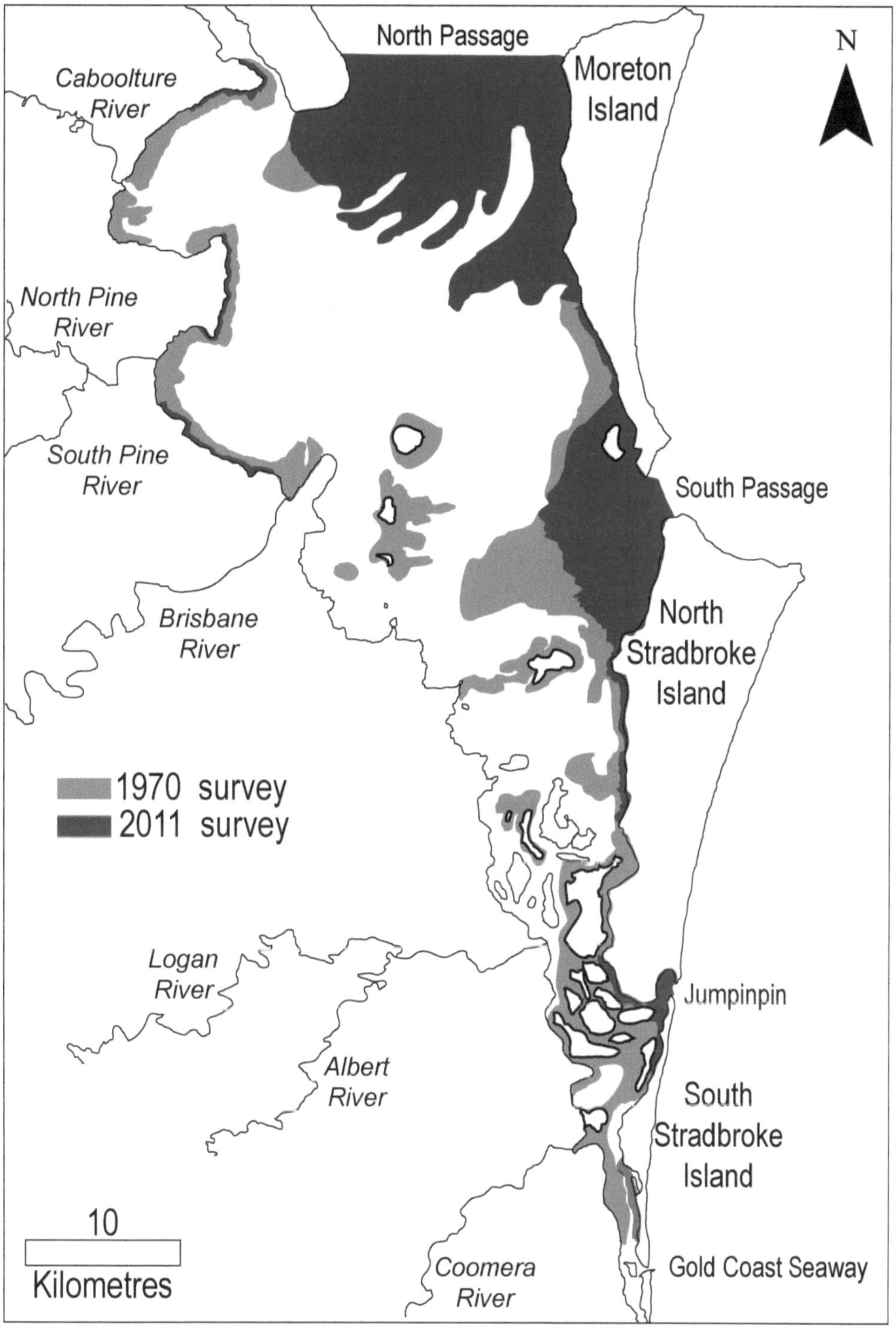

Fig. 3 Summary of sediment survey results showing clean sand substrate distribution in Moreton Bay. Results indicate a 20 % (260 km^2) decline of clean sand facies in Moreton Bay over a 30 year period from 1970 to 2011 (Maxwell 1970; O'Brien et al. 2012)

reduces the amount available for algal growth, the trapping of sediments from the water column, which improves water clarity, and the harbouring of vertebrate and invertebrate grazers that minimize the growth of epiphytic algae (Cornelisen and Thomas 2004; Heck and Valentine 2007; Carr et al. 2010) (Fig. 6b). In areas where seagrass has been lost, sediments are more easily resuspended, nutrients are released into the water column making them available for algal growth and grazing rates of algae are reduced, which limits the potential for seagrass recovery.

In January 2011, the Brisbane River flooded and reduced water clarity throughout most of Moreton Bay. Despite the

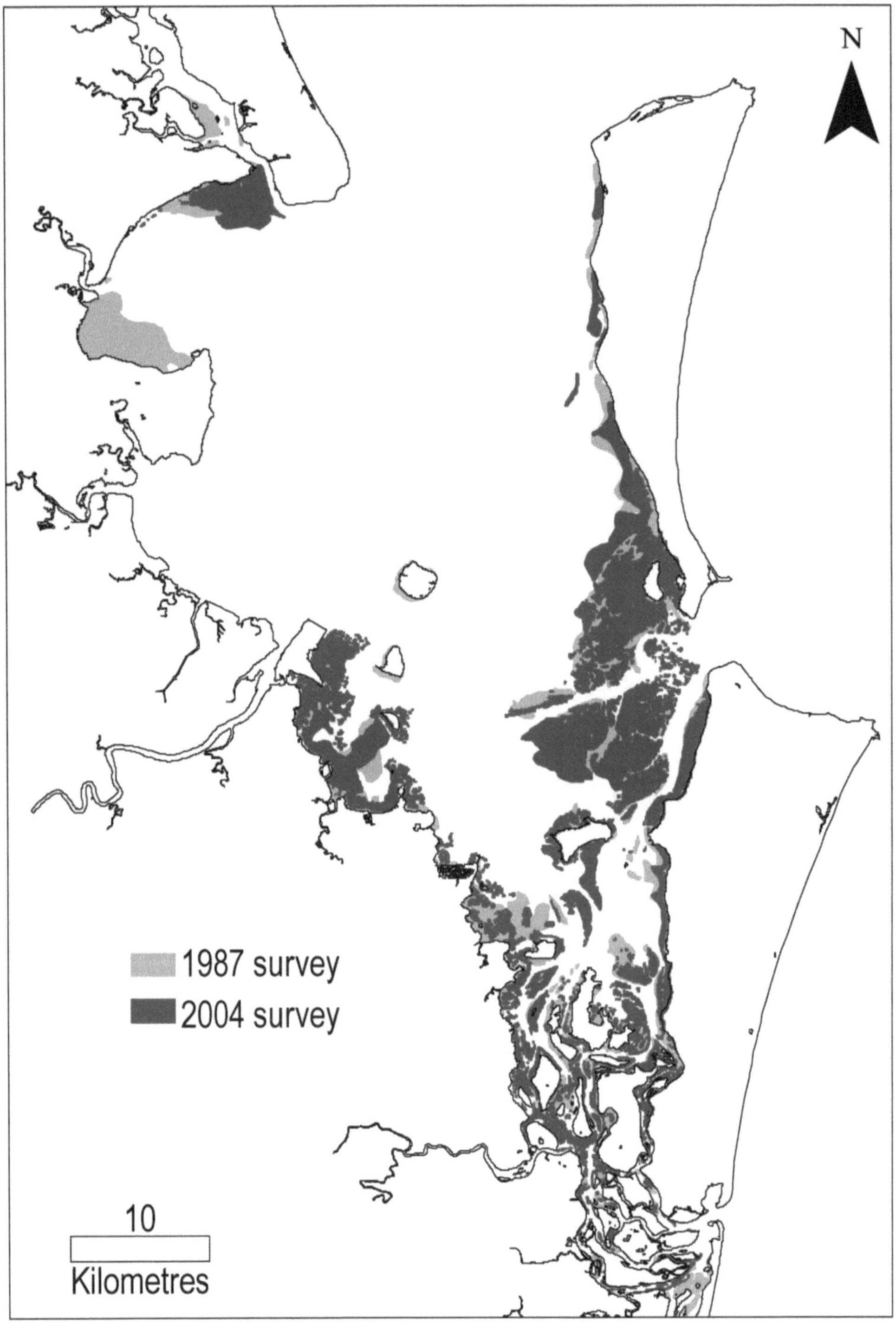

Fig. 4 Summary of seagrass survey results showing a 20 % (49 km^2) loss in seagrass coverage over a 20 year period from 1987 to 2004 (Hyland et al. 1989; Roelfsema et al. 2009)

impact of the flood plume, subtidal seagrass (*Zostera muelleri*) meadows closest to the river mouth had higher rates of nutrient uptake and algal grazing and lower sediment resuspension than meadows less impacted by the flood plume. As a result, seagrass biomass remained constant throughout the year at meadows close to the impact, but declined at meadows further away. The differing response to the flood impact was correlated with morphological differences between the two meadow types. Meadows closest to the river had longer and wider leaves with greater concentrations of chlorophyll-a than those further away, an adaptation which allows greater light capture and sediment baffling (Abal et al. 1994). In contrast, nutrient uptake and grazing rates were consistently lower, and sediment resuspension rates higher, in unvegetated areas following the flood than rates recorded in seagrass meadows. This resulted in sub-optimal light availability for seagrass colonisation and recovery for much of the year, and highlights the importance of understanding both the processes that promote resistance to impacts and those that restrict recovery following seagrass loss. Given the dynamic nature of the system at geological timescales, it is likely to have the capacity to shift between states, although further knowledge of local ecology is required to better understand whether the current rate of anthropogenically induced change is too rapid for the system to adapt.

Current management actions have focussed on reducing nutrient point source discharge into the system and appear to have shown ecosystem health recovery. Recent surveys in areas of historical complete seagrass loss have shown strong recovery of coverage (Fig. 5b). However, there is a need to reduce sediment loads during flood events as these both increase nutrient loading and allow persistence of turbid conditions following sediment resuspension events. This is potentially a crucial step in recovery or maintenance of current levels of ecosystem health as the likelihood of extreme rainfall events and associated inflows is projected to increase in this region (Cai and van Rensch 2012). Recovery of seagrass beds from areas of previous reported complete loss, highlights the need for better understanding as to why relatively minor flood events can result in complete loss whilst recovery occurs after major events in some areas. The extensive and obvious coverage of seagrass suggests current monitoring methods might need improvement.

Fish and Ecosystem Functioning: Effects of Connectivity, Coastal Development and Marine Reserves

Moreton Bay supports a high abundance and diversity of finfish and crustaceans, and is an important nursery for harvested species (Tibbetts and Connolly 1998). The embayment is a heterogeneous seascape containing a mosaic of estuarine habitats (e.g. saltmarsh, seagrass, mangroves and mudflats), shallow fringing coral reefs (interspersed with mangroves and seagrass) and deeper soft sediments (e.g. Stevens and Connolly 2005) (Fig. 5a). The level of connectivity among, and spatial arrangement of, these habitats affects both the distribution of fish and crustaceans (Skilleter et al. 2005; Olds et al. 2012b), and the productivity of dependent fisheries (Manson et al. 2005; Meynecke et al. 2008). The region has, however, been impacted by coastal development, terrestrial runoff and fishing, and supports modified habitats and altered fish assemblages (e.g. Lybolt et al. 2011; Waltham et al. 2011). The fish assemblages of Moreton Bay have been reviewed elsewhere (Tibbetts and Connolly 1998), and so this section describes the roles of fish in maintaining key ecological processes in the Bay.

The distribution and abundance of fish in central Moreton Bay is affected by the degree of connectivity among reefs, mangroves and seagrass (Olds et al. 2012b). Connectivity between reefs and adjacent mangroves is particularly important and enhances the ability of local marine reserves to promote the abundance of harvested and herbivorous fish species (Olds et al. 2012a). The synergistic effects of connectivity and marine reserves increase both the biomass and species richness of herbivorous fish, and thereby promote herbivory, which reduces algae cover and enhances coral recruitment (Olds et al. 2012c) (Fig. 5c). These effects on coral-algae recruitment dynamics and benthic succession serve to increase reef resilience (i.e. the capacity to absorb disturbance and regenerate without degrading, or changing state), and suggest that targeted seascape conservation may improve the resilience of other similarly degraded seascapes.

Extensive networks of canals and lakes have been constructed for residential purposes on the estuaries of Moreton Bay (Waltham and Connolly 2011). The Bay has the largest cluster of artificial estuarine waterways (about 250 km in length) outside of the USA (and almost ten times the extent of Venice). As fish habitat, canal estates have lower productivity and diversity than shallow vegetated habitats in the Bay (Morton 1989). Where the artificial waterways have been constructed in terrestrial habitat, however, they nevertheless provide a major new estuarine habitat. Fish aggregate at the canal edges (Waltham and Connolly 2007) and around jetties and pontoons (Moreau et al. 2008). Fisheries species have shown remarkable plasticity in diet to adapt to systems lacking conspicuous vegetated habitat. It has been demonstrated using stable isotope (Connolly 2003) and stomach content analyses (Waltham and Connolly 2006), for example, that snub-nosed garfish (Family: Hemiramphidae) consume algae (energy source) and insects (protein source) in canals, whereas in natural wetlands the species utilises seagrass and crustaceans.

Ecosystem Functioning and Resilience

Moreton Bay is a diverse and productive ecosystem, it supports a range of subtropical and temperate species, is socially and culturally important as a focus for recreation and fisheries, but is also heavily impacted by development, runoff and habitat modification. The ecological resilience of Moreton Bay relies upon its ability to resist or adapt to change without changing its structure and function. Components of the broader ecosystem that confer this ecological resilience

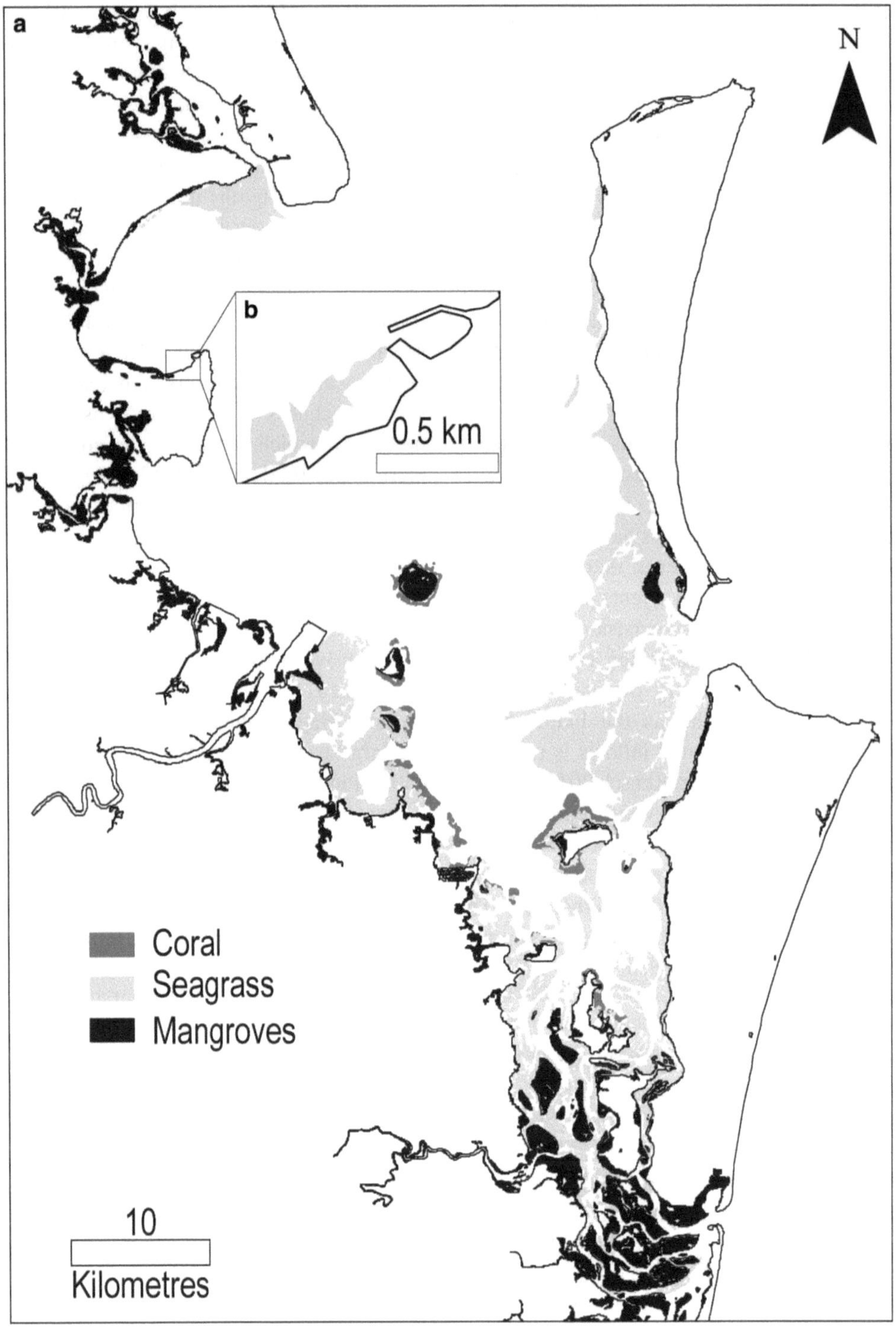

Fig. 5 Substrate distribution in Moreton Bay including: (**a**) Current best available estimated seagrass, mangrove and coral habitat distribution (Olds et al. 2012a) and (**b**) Recent survey (October, 2012) from historical seagrass loss area showing extensive recovery of seagrass beds

include: habitat-scale ecological processes that promote resistance and adaptation (e.g. seagrass feedback mechanisms and morphological flexibility), connectivity (e.g. links between mangroves and coral reefs), food web plasticity (e.g. fish in canals) and functional redundancy (e.g. herbivore diversity) (Fig. 6d). An increased understanding of how these components interact to drive resilience, and their incorporation into management decision-making, can underpin the long-term maintenance of ecosystem functioning in Moreton Bay.

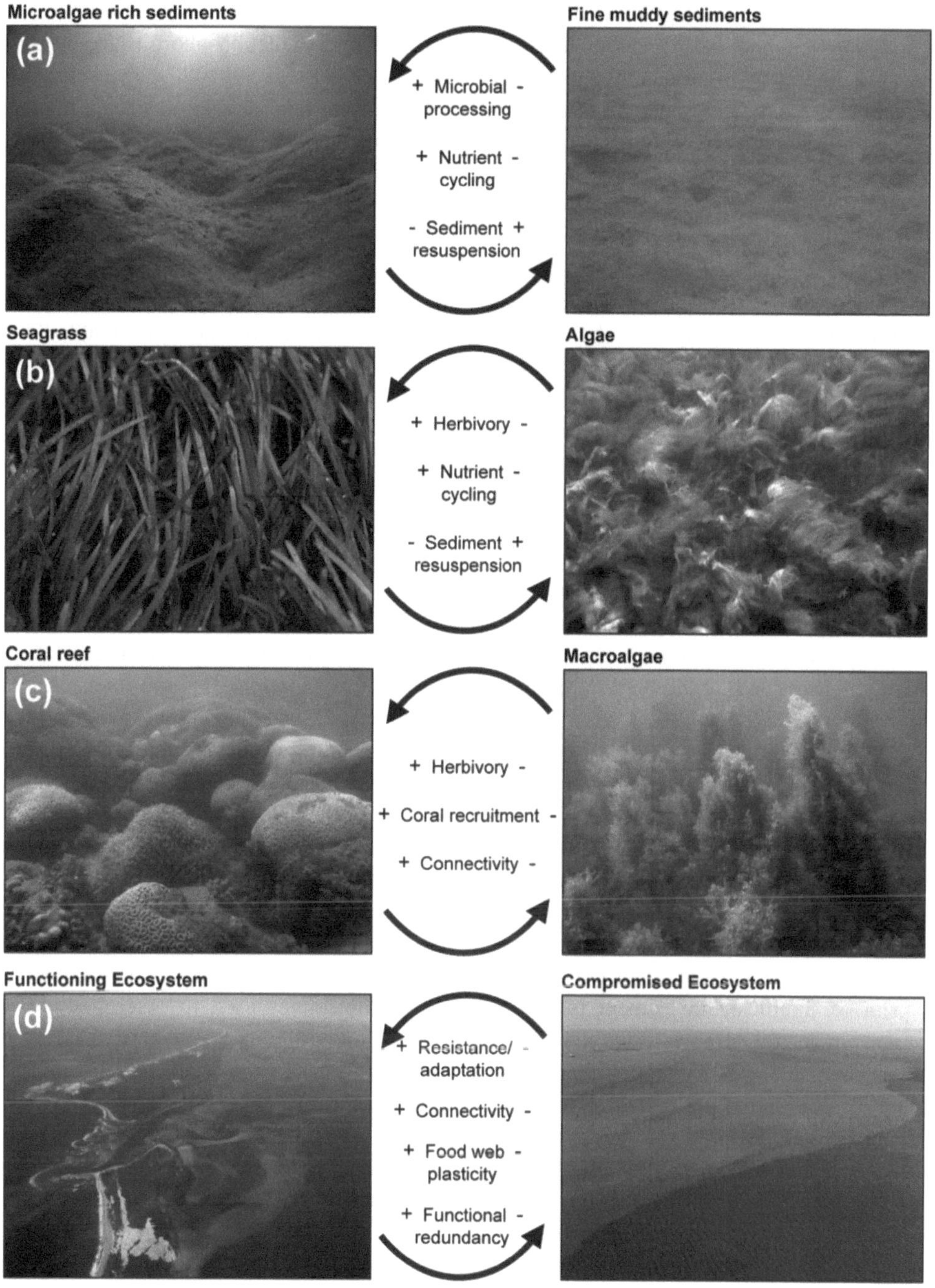

Fig. 6 Ecological processes that maintain or erode the resilience of (**a**) deep microalgae rich sediments, (**b**) seagrass meadows, and (**c**) coral reef seascapes in Moreton Bay, and (**d**) underpin the functioning of the ecosystem at the Bay scale (+/– symbols depict direction of effect)

Anthropological Influences: Resources, Pressures, Impacts and Remediation

The condition of the Moreton Bay estuary has been classified as 'modified' or 'extensively modified' with the major modifiers including: sewage treatment plant discharges, dams and weirs, wetland loss, urbanisation, dredging and entrance modification (Digby et al. 1998). Furthermore sediment and nutrient loads to Moreton Bay from its adjacent catchment systems have increased as a result of significant land use transformation from naturally vegetated catchments to a condition of extensive native vegetation loss as a result of agricultural and urban development within 200 years (e.g. Fig. 2b). Moreton Bay receives most of the development pressure on its western shore with most pollution being discharged into Moreton Bay from its four large estuaries (Logan, Brisbane, Pine, Caboolture – see Fig. 1). Both point source discharges from sewage treatment plants and industry as well as diffuse pollution from both urban and rural land uses have a chronic negative impact on water quality and aquatic ecosystem health in the western and southern sections of Moreton Bay (Deception Bay, Bramble Bay, Waterloo Bay and the Southern Bay), while the northern and eastern sections of Moreton Bay still have high water quality and ecosystem health, showing limited impacts of anthropogenic pollution (EHMP 2007). During periods of extreme weather, such as the 1974 and 2011 floods, the amount of sediment and nutrients delivered to the Bay increase dramatically, with the sediment delivered by the 2011 flood being estimated at 10–20 million tonnes. This equates to approximately 20–50 years of average annual sediment delivery in a single event (i.e. average sediment delivery in years without major floods). During the past 10 years significant investments have been made to upgrade sewage treatment plants and reduce the point source nutrient discharges into Moreton Bay. This has reduced the proportion of available nutrients, especially nitrogen, that are contributed from sewage treatment effluent and resulted in a decline in the occurrence of phytoplankton blooms in the western embayments, particularly Bramble Bay (SEQHWP 2007).

Sediment and Nutrient Loads

Sediments and nutrients have been identified as the major 'pollutants' of concern for Moreton Bay and the reduction of these pollutant loads have been the focus of recent management interventions (SEQHWP 2007). Initial management interventions aimed to reduce point source nutrient loads to the Bay which, prior to significant investment in sewage treatment plant and industrial process upgrades, were estimated to be approximately 3,383 t of nitrogen and 1,182 t of phosphorous per year (Eyre and McKee 2002). By 2011 these loads had reduced to approximately 995 $\pm$ 134 t of nitrogen and 536 $\pm$ 52 t of phosphorous per year (QDSITI 2011) as a result of major infrastructure investments and despite an increasing human population. However, as the human population transitions from three million to over four million people by 2026 (QOESR 2011) it is estimated that point source nitrogen and phosphorous loads will steadily increase over the long term if current management practices are maintained.

In the absence of suitable monitoring data, catchment-derived diffuse source sediment and nutrient (total nitrogen and total phosphorous) inputs to Moreton Bay have been estimated using a range of catchment modelling approaches (Chiew et al. 2002; WBM 2005; Stewart 2009). Recent estimates, using data from the Queensland Land Use Mapping Project (QLUMP) (Witte et al. 2006), suggest that annual total suspended sediment loads (in the absence of significant flood events such as the 1974 and 2011 events) are in the order of 345,000–390,000 t while annual total nitrogen and total phosphorous loads are in the order of 4,000–4,500 t and 500–580 t respectively (Stewart 2009). If the current land development and management practices are maintained it is estimated that these sediment loads will increase by approximately 17 % with diffuse source total nitrogen and total phosphorous loads projected to increase by 14 % and 21 % respectively by 2026 (SEQHWP 2007).

Sediment and nutrient load estimates for different population growth, land use pattern and climate scenarios have formed the basis of the 'sustainable load' management approach that has been applied to reverse the recent decline in ecosystem health of Moreton Bay and its associated estuaries. A sustainable load in this context is the target load which can be readily assimilated by the receiving waters to maintain a sustainable ecosystem health outcome. In the majority of cases predictive models have been used to identify the maximum load that can be delivered to a given system whilst still achieving the resource condition targets (RCTs – a combination of environmental values and associated water quality objectives). Application of this approach to Moreton Bay and its estuaries has shown that the target sustainable load requirements are often difficult to achieve using current approaches as often the sustainable load target is already being exceeded significantly, and even with all of the proposed management actions implemented, the target sustainable load would still be exceeded (Weber and Ramilo 2012). While this result might prompt the use of the sustainable load approach in determining the management action to be questioned, experience in Moreton Bay suggests that the central issues arises from the original water quality objectives used to determine the sustainable load (Weber and Ramilo 2012). These issues are explored in more detail in section "Management approach and Resource Condition Targets (RCTs)" below.

Signs of Successful Management Interventions

Moreton Bay's natural values have long been recognised as being worthy of protection. From as early as 1975 various studies were being undertaken to protect the Bay from the variety of pressures exerted by rapid urbanisation of the catchment and increasing use of the Bay itself. These eventually morphed into a Bay-wide approach to management in 1993 when the Queensland government adopted the Moreton Bay Strategic Plan (DEH 1993) and 3,400 km^2 of the Bay and adjoining Queensland coastal waters were gazetted as a multiple use marine park. Additionally, a significant portion of the Bay was listed as a wetland of international significance under the RAMSAR convention in the same year. The Moreton Bay Strategic Plan guided the management of the marine park until the first zoning plan

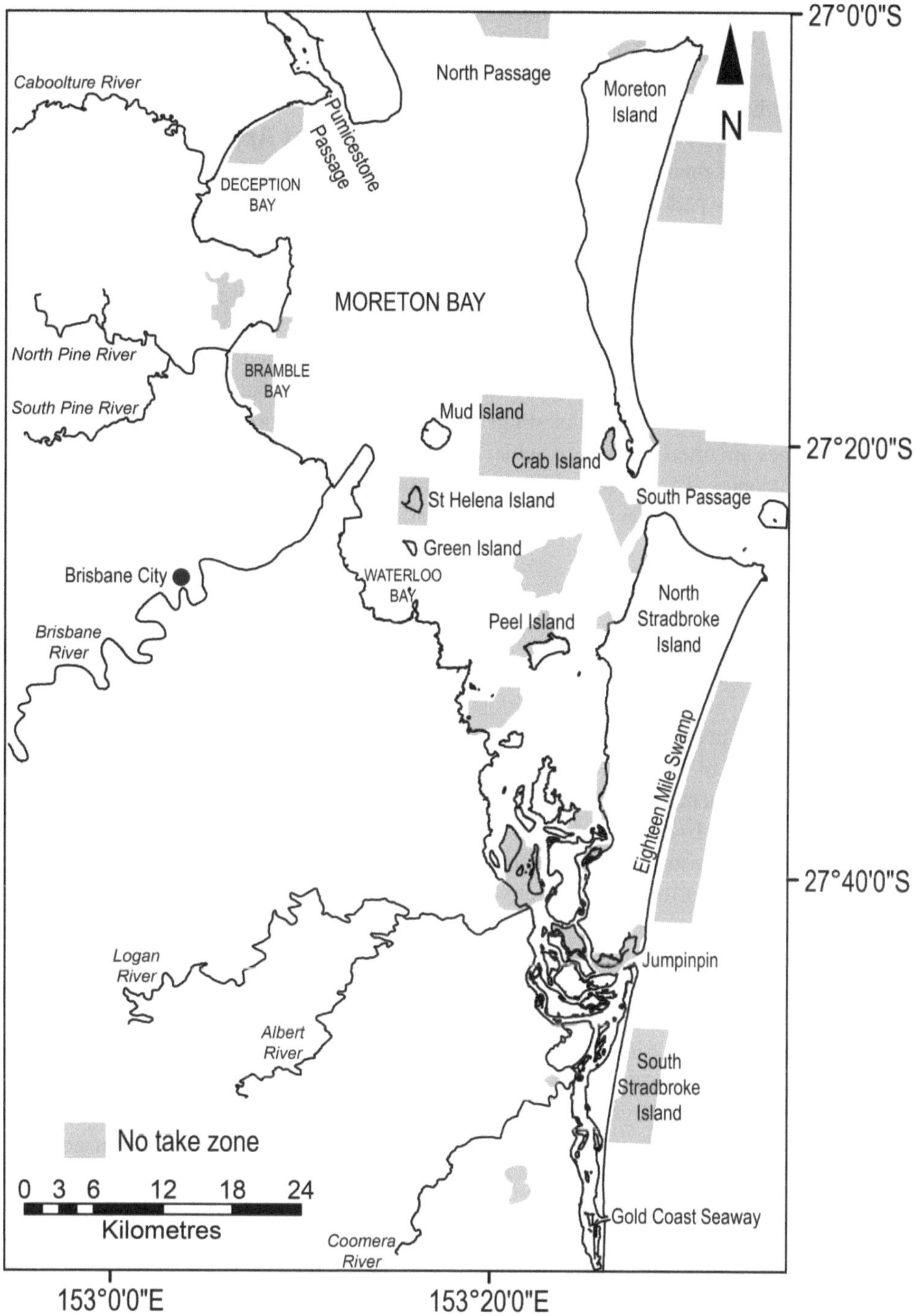

Fig. 7 Extent of "no take" zones in the Moreton Bay Marine Park

was implemented in 1997. At that stage, six areas, comprising 0.5 % of the marine park were protected from all extractive activities (e.g. fishing). These areas protected small examples of the iconic or well known habitats – mangrove forests, seagrass beds and inshore and offshore coral reefs. While several of these areas were delivering some conservation benefits (Pilans 2006), by the mid 2000s it was clear that the management of Moreton Bay as a whole was lagging behind world standards for marine park management and it was timely then that a legislative requirement provided the opportunity for the zoning plan to be thoroughly reviewed. This was undertaken based on the CAR (comprehensive, adequate, representative) principle to address biophysical needs, with the revised zoning plan taking effect in 2009 with considerable changes to the size and type of habitats protected (refer to Fig. 7). It is noted that, while several of the river estuaries draining to Moreton Bay are included in the marine park, few river estuaries have designated 'no take' zones (i.e. the level of conservation/protection in estuaries is lower than for other areas that are designated as habitat protection, conservation or marine national park zones). Socio-economic impacts of zoning were also considered resulting in 16 % of the marine park currently protected from all extractive activities and early monitoring of these areas suggests that the revised zoning is achieving good conservation outcomes such as increased abundance of fish and crab species (DERM 2010, 2012; Olds et al 2012a). The formation of the Moreton Bay Marine Park represents one of the earliest large-scale, planned conservation activities undertaken in the system and these recent results suggest that relatively long timeframes are often required to achieve the desired conservation outcomes (i.e. improvements not clearly evident until many years after management intervention).

Summary and Discussion: Moreton Bay and Its Estuaries from 2050 and Beyond

A significant feature of the recent management interventions targeted at halting the decline of ecosystem health in Moreton Bay has been the prominence of a science-based management framework. A science-based management was deemed necessary as past experience and knowledge of the system was limited and also was not likely to effectively predict the response of the system under different future conditions (i.e. different population growth, land use and climate scenarios). The science-based approach, combined with a significant investment in collaborative decision making processes that include community, industry and government stakeholders, has seen significant investment in actions to improve the state of waterways in the region. For example in excess of AUD$300 M was invested in wastewater treatment plant upgrades from 1998 to 2006 resulting in a 44 % reduction in point source nitrogen load entering Moreton Bay in 2006 compared to pre-2001 loads, with early indications of a positive ecosystem response (SEQHWP 2007). While the priority and cost-effectiveness of such management action can be argued there is general consensus that a science-based management framework has been successful and should be retained as a key feature when shaping the future of Moreton Bay and its estuaries from 2050 and beyond. Similarly there is a need to maintain the collaborative decision making and management framework to allow the future of Moreton Bay in 2050 and beyond to be determined by local communities, industry and government stakeholders. With these features (i.e. science-based management framework and a collaborative decision making and management framework) taken as a basis for the future the authors offer the following ideas to promote discussion on what sort of estuaries do we want to see in 2050 and beyond for Moreton Bay.

Management Approach and Resource Condition Targets (RCTs)

An adaptive management framework has been implemented by the various organisations involved in the management of Moreton Bay (SEQHWP 2007) however the long-term effectiveness of this approach remains uncertain. As noted by Allen and Gunderson (2011) since its initial introduction, adaptive management has been hailed as a solution to endless trial and error approaches to complex natural resource management challenges. However, its implementation has failed more often than not. It does not produce easy answers, and it is appropriate in only a subset of natural resource management problems. Furthermore Allen and Gunderson (2011) highlight that adaptive management functions best when both uncertainty and controllability are high as there is high potential for learning and the system can be manipulated. In the case of Moreton Bay and its estuaries, uncertainty is high however controllability of the system is low which suggests scenarios are a more appropriate approach and allow for the exploration of potential future outcomes of present actions (Baron et al. 2009). The management of Moreton Bay and its estuaries is shifting towards scenario planning based management (i.e. the SEQHWP's Science Program has extensively used predictions of aquatic ecosystem response to a well-defined set of potential future land use and climate scenarios to inform the development of management actions) although most scenarios under consideration have short timescales in the order of 15–50 years (e.g. current Healthy Waterways vision has a 2026 target date for system improvements). These timeframes reflect the current transition in management approach from a mainly short-term economic based option to a more long-term economic-ecological approach.

A strong theme of past commentary on management approaches for Moreton Bay has been the idea of "management for millennia" (Davie et al. 1990; Tibbetts et al. 1998). While there are very practical reasons for using the current shorter timeframes for planning and management of Moreton Bay it might be timely to also incorporate longer term views (e.g. 1,000 year planning horizons). While it may be argued that there is little practical advantage in attempting to set specific management targets (and associated actions) for such extended periods the process of framing and examining such long-term questions is likely to allow the relatively short term (i.e. 2026) targets to be more effectively contextualised in relation to the broader dynamics of the system and the level of uncertainty associated with our current understanding of the processes that drive them.

A central element to the current approach for defining the desired future for Moreton Bay and its estuaries is a series of defined resource condition targets (RCTs). These resource condition targets use a combination of environmental values and specific water quality objectives to classify the current and future state of the region's waterways. Examples from the South East Queensland Healthy Waterways Strategy 2007–2012 include: (1) In 2026, 100 % of SEQ waterways classified in 2007 as having high ecological values retain this classification; (2) In 2026, 100 % of SEQ waterways classified in 2007 as meeting their water quality objectives retain this achievement; (3) By 2026, waterways classified as disturbed and/or degraded in 2007 have their ecosystem health and ecological processes reinstated. These RCTs represent the outcome of a pragmatic approach, based on the Queensland Environmental Protection (Water) Policy 2009, with the baselines for both the RCTs and Community Targets referenced to the 2007 levels established by accepted monitoring or benchmarking programs. The current set of targets are clearly designed to halt the current decline in water quality and ecosystem health. This is an important first step in any natural resource management process. When considering the desired future state(s) of Moreton Bay over the extended timeframes discussed above (i.e., >100 years) it is useful to note that the 2007 reference conditions used to set the short-term (i.e. 2026) RCTs are likely to be substantially different from a pristine condition or reference condition for biological integrity (as discussed in more detail by Stoddard et al. (2006)).

When considering the future state of Moreton Bay from 2050 and beyond it might be useful to consider whether the current goals of 'halting the decline' are appropriate in the longer term. Without resource limitations it might be argued that a more worthy condition target would be to move beyond some 'historical condition' or 'least/minimally disturbed condition' to the 'best attainable condition'. The 'best attainable condition' for Moreton Bay from 2050 onwards might be one in which ecosystem integrity is less than that currently observed (i.e. current degradation has resulted in an ecosystem shift that is not able to be reversed). Alternatively the 'best attainable condition' might be one that represents an enhancement over some reference condition for biological integrity. The current gaps in scientific understanding of the ecosystems of Moreton Bay and their function largely prevent us from defining the 'best attainable condition' with any certainty. This does not diminish the substantial progress that has been made in understanding Moreton Bay and its function. Rather, it highlights that additional information is needed to better explore what futures are achievable for this system. The 'best attainable condition' will also be determined by socio-economic factors (an aspect explored in more detail in the following section). Regardless of its feasibility the concept of managing Moreton Bay to achieve net positive gains (i.e. a condition that represents an enhancement over a pristine condition or reference condition for biological integrity) is worth investigating. Such thinking appears to have been deferred in past management planning processes due to the focus on the urgent need to halt the decline in ecosystem health but is likely to be useful to consider when framing the next revision of RCTs and associated management actions.

Cost-Benefit Framework to Attract and Guide Investment

As suggested above the 'best attainable condition' for Moreton Bay in 2050 and beyond will involve more than just a bio-physical assessment of the system. Moreton Bay and its estuaries show signs of significant impact and little assimilative capacity for increased pollutant loads. The catchments for these estuaries are also areas where population growth has been identified and as such balancing economic growth of an area while at the same time reducing pollution loads into an impacted estuary is extremely challenging. As highlighted by Weber and Ramilo (2012) environmental constraints in such systems are often viewed as sacrosanct, either in legislation or by the community or both, but at the same time that same community wishes to live in a region that has job provision, a range of services (schools, shops, health care) and future prospects for their community. As such socio-economic factors play a significant role in determining what the 'best attainable condition' might be with regard to the region's economic and cultural constraints. An effective cost-benefit framework will be required to balance the achievement of suitable environmental outcomes with competing sources and uses of water in the face of strong development pressure.

The current approach in the Moreton Bay region is to attempt to balance the desire for economic growth while attempting to reduce discharges to waterways from sewage, stormwater and agricultural land uses so that the community's environmental values are maintained. This is often done in the absence of information on the cost and effectiveness of different and often competing management actions. Therefore the current method of cost-effectiveness evaluation is largely a process that identifies a solution that delivers the required objectives or standards of service for the least overall cost and represents a simplistic view of what true least cost planning approaches should deliver (Weber and Ramilo 2012). If the approach does not adequately address all externalities, then some costs may not be properly accounted for (see Lane et al. 2010). In addition, if the benefits, both tangible and intangible, of such approaches are not accounted for, true cost-effectiveness may not be able to be delivered. Also, it may be that the cost burden for

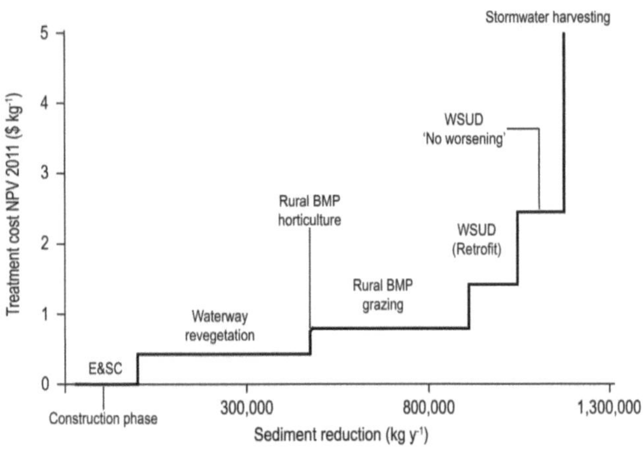

Fig. 8 Example of a cost-treatment curve (After Weber and Ramilo 2012). These cost curves can also show a specific target in terms of a pollutant load reduction or potable supply demand reduction. Sometimes those targets are beyond that which can be adequately achieved with current technology or within current resources availability. *BMP* Best Management Practice, *WSUD* Water Sensitive Urban Design

proposed management actions may be too high for the community to bear in order to achieve the required RCTs or environmental objectives. To address these issues approaches that include extended cost-effectiveness analysis (Hall 2011) can be usefully applied in decision making processes to demonstrate to decision makers how far towards achieving objectives they may get for a certain level of investment (refer to example in Fig. 8). Experience from the application of such an approach in the Moreton Bay region has been well received by political decision makers especially where benefits may not be adequately quantified (Weber and Ramilo 2012).

Concluding Remarks

Experience with the current management approach for the Moreton Bay estuary has highlighted the need to adopt a coordinated, multi-stakeholder approach to the management of the system. For example, while the existing marine park legislation provides an effective tool to address and manage direct impacts on the Bay's resources and values, it has no ability to ensure that activities in the catchment are managed in the best interests of the marine park. Furthermore it has been demonstrated that while the conservation approach taken within the Bay (via the marine park zoning) is beginning to show positive effects there is less value in protecting discreet areas without ensuring that the overall health of the Bay is maintained. Connectivity between habitats is important as well as the connectivity provided by the water itself.

Placed within the broad backdrop of global eutrophication and the predicted acceleration in the rate of land modification and urbanisation it is unclear as to whether the RCTs for Moreton Bay and its estuaries can be met by 2026. More fundamentally it is unclear as to whether the current RCTs are an appropriate long term (i.e. >100 years) target given that investment decisions that are being made now are resulting in the development of infrastructure with a design life much greater than the 2026 targets that have been adopted. Furthermore other pressures such as population growth, climate variability, sea level fluctuations and changes in ocean chemistry will also act over longer timeframes than those considered by the current management framework. It is anticipated that these issues will form part of the discussion to inform the process of updating the RCTs for the regional natural resource management planning process. Consideration of longer timeframes in conjunction with the development of locally relevant tools and methods that are able to effectively represent a range of relatively sophisticated management scenarios and their associated socio-economic cost-benefit profile appear to be some of the key components that are needed in the next evolution in our attempts to manage this complex system.

Overall the current gaps in our understanding of the ecosystems of Moreton Bay and their function largely prevents us from defining the best attainable future condition with any certainty. The highly variable nature of the system over relatively short timeframes (i.e. flood vs non-flood conditions) as well as the capacity of the system to adapt to long term changes (i.e. past morphological and ecosystem shifts) suggests that Moreton Bay and its associated estuaries have significant capacity to adapt to change. Whether the current rate of anthropogenically induced change is too rapid for the system to adapt (or whether such adaptions will be undesirable) is a key question when considering how the system may function from 2050 and beyond – and one that we current do not have the capacity to answer in any detail. Notwithstanding the above it can be argued that the combination of a science-based management framework and the collaborative decision making processes that have been implemented to halt the decline of Moreton Bay have shown remarkable progress in a relatively short period of time. This suggests we can be cautiously optimistic about our future capacity to manage the system well.

Acknowledgements Resilience research on corals, seagrass and fish by Griffith University was supported financially by the Australian Research Council and the Queensland Government. Sections of this chapter have been based on or contain data kindly provided by the State of Queensland (Department of Science, Information Technology, Innovation and the Arts).

References

Abal E, Dennison W (1996) Seagrass depth range and water quality in southern Moreton Bay, Queensland, Australia. Mar Freshw Res 47(6):763–771

Abal E, Loneragan N, Bowen P, Perry C, Udy J, Dennison W (1994) Physiological and morphological responses of the seagrass Zostera capricorni Aschers, to light intensity. J Exp Mar Biol Ecol 178(1):113–129

Allen CR, Gunderson LH (2011) Pathology and failure in the design and implementation of adaptive management. J Environ Manage 92:1379–1384

Allingham DP, Neil DT (1995) The supratidal deposits and effects of coral dredging on Mud Island, Moreton Bay, southeast Queensland. Z Geomorphol 39:273–292

Baron J, Gunderson L, Allen CD, Fleishman E, McKenzie DH, Meyerson LA, Oropeza J, Stephenson N (2009) Adapting to climate change: options for national parks and reserves. Environ Manage 44(6):1033–1042

Cai W, van Rensch P (2012) The 2011 southeast Queensland extreme summer rainfall: a confirmation of a negative Pacific Decadal Oscillation phase? Geophys Res Lett 39:L08702. doi:10.1029/2011GL050820

Capelin M, Kohn P, Hoffenberg P (1998) Land use, land cover and land degradation in the catchment of Moreton Bay. In: Tibbetts IR, Hall NJ, Dennison WC (eds) Moreton Bay and catchment. School of Marine Science, The University of Queensland, Brisbane, pp 55–66

Carr J, D'Odorico P, McGlathery K, Wiberg P (2010) Stability and bistability of seagrass ecosystems in shallow coastal lagoons: role of feedbacks with sediment resuspension and light attenuation. J Geophys Res 115(G3):G03011

Chiew F, Scanlon P, Vertessy R, Watson F (2002) Catchment scale modelling of runoff, sediment and nutrient loads for the south-east Queensland EMSS, Cooperative research centre for catchment hydrology report no. 02/1. Monash University, Melbourne

Connolly RM (2003) Differences in trophodynamics of commercially important fish between artificial waterways and natural coastal wetlands. Estuar Coast Shelf Sci 58:929–936

Cornelisen CD, Thomas FIM (2004) Ammonium and nitrate uptake by leaves of the seagrass Thalassia testudinum: impact of hydrodynamic regime and epiphyte cover on uptake rates. J Mar Syst 49(1–4):177–194

Davie P, Stock E, Low Choy D (1990) The Brisbane River a source book for the future. The Australian Littoral Society Inc in association with the Queensland Museum, Brisbane, 427 pp

Davies PL, Eyre BD (1998) Nutrient and suspended sediment input to Moreton Bay – the role of episodic events and estuarine processes. In: Tibbetts IR, Hall NJ, Dennison WC (eds) Moreton Bay and catchment. School of Marine Science, The University of Queensland, Brisbane, pp 545 552

de Jonge VN, de Jong DJ, van Katwijk MM (2000) Policy plans and management measures to restore eelgrass (Zostera marina L.) in the Dutch Wadden Sea. Helgoland Mar Res 54(2):151–158

DEH (1993) Moreton Bay strategic plan. Queensland Government (Department of Environment and Heritage), Brisbane

Dennison WC, Abal EG (1999) Moreton Bay study. A scientific basis for healthy waterways campaign. South East Queensland Regional Water Quality Management Strategy, Healthy Waterways, Brisbane, 245 pp

DERM (2010) Moreton Bay Marine Park monitoring program interim report September 2010. Queensland Government (Department of Environment and Resource Management), Brisbane

DERM (2011) South East Queensland event monitoring summary (6th–16th January 2011) preliminary suspended solids loads calculations. South East Queensland Event Monitoring Program summary report. Department of Environment and Resource Management (DERM), Brisbane, 4 pp

DERM (2012) Moreton Bay Marine Park Monitoring Program February 2012. Queensland Government (Department of Environment and Resource Management), Brisbane

Digby MJ, Saenger P, Whelan MB, McConchie D, Eyre B, Holmes N, Bucher D (1998) A physical classification of Australian Estuaries. Report prepared for the Urban Water Research Association of Australia by the Centre for Coastal Management, Report No 9, LWRRDC occasional paper 16/99. Southern Cross University, Lismore

Donders TH, Wagner-Cremer F, Visscher H (2008) Integration of proxy data and model scenarios for the mid-Holocene onset of modern ENSO variability. Quat Sci Rev 27(5–6):571–579

EHMP (2007) Ecosystem health monitoring program 2005–06 annual technical report. South East Queensland Healthy Waterways Partnership, Brisbane

Eyre BD, McKee LJ (2002) Carbon, nitrogen, and phophorus budgets for a shallow subtropical coastal embayment (Moreton Bay, Australia). Limnol Oceanogr 47(4):1043–1055

Flood PG (1978) The significance of two contrasting sedimentary environments (the fringing coral reef and the tidal mud flat) presently in juxtaposition along the southwestern shore of Moreton Bay, Queensland. Pap Dept Geol Univ Qld 8:44–65

Grinham AR (2007) Downstream effects of land use on shallow-water benthic microalgal communities in Moreton Bay, Australia and Marovo Lagoon, Solomon Islands. PhD thesis, The University of Queensland

Grinham A, Gibbes B, Gale D, Watkinson A, Bartkow M (2012) Extreme rainfall and drinking water quality: a regional perspective. In: Brebbia CA (ed) Water pollution XI: 11th international conference on modelling, monitoring and management of water pollution, New Forest, 10–12 July 2012, pp 183–194

Hall J (1990) 20 000 years of human impact on the Brisbane River and environs. In: Davie P, Stock E, Choy DL (eds) The Brisbane River: a source-book for the future, Brisbane. Australian Littoral Society and the Queensland Museum, Moorooka, pp 175–182

Hall MR (2011) Prudent and efficient: the use of extended cost-effectiveness analysis for total water cycle management plans. In: Proceedings of the science forum, Urban Water Security Research Alliance, Brisbane, September 2011

Heck K, Valentine J (2007) The primacy of top-down effects in shallow benthic ecosystems. Estuar Coast 30(3):371–381

Heggie D, Holdway D, Tindall C, Fredericks D, Fellows M, Berelson W, Longmore A, Cowdell R, Nicholson G, Lowering M, Udy J, Logan D, Prange J, Watkinson A, Schmidt A, Capone D, Burns J (1999) Task Sediment Nutrient Toxicant Dynamics (SNTD) Phase 2 final report South East Queensland Regional Water Quality Management Strategy, Brisbane, 222 pp

Hyland SJ, Courtney AJ, Butler CT (1989) Distribution of seagrass in the Moreton region from Coolangatta to Noosa, Queensland Department of Primary Industries report Q189010. Queensland Government, Brisbane

Jones R (1969) Fire-stick farming. Aust Nat Hist 16:224–228

Jones M, Stephens AW (1981) Quaternary geological framework and resource potential of Moreton Bay. In: Hofmann GW (ed) 1981 field conference, Brisbane-Ipswich Area. Geological Society of Australia, Queensland Division, Brisbane, pp 17–23

Lane J, de Haas D, Lant P (2010) Life cycle impacts of the Gold Coast urban water cycle. Paper presented at OzWater 2010, Brisbane, March 2010

Lang SC, McClure ST, Grosser M, Lawless M, Herdy T (1998) Sedimentation and coastal evolution, northern Moreton Bay. In: Tibbetts IR, Hall NJ, Dennison WC (eds) Moreton Bay and catchment. University of Queensland, Brisbane, pp 81–92

Lewis SE, Wüst RAJ, Webster JM, Shields GA (2008) Mid–late Holocene sea-level variability in eastern Australia. Terra Nova 20:74–81

Lockhart DA, Lang SC, Allen GP (1998) Sedimentation and coastal evolution of southern Moreton Bay. In: Tibbetts IR, Hall NJ, Dennison W (eds) Moreton Bay and catchment. University of Queensland, Brisbane, pp 93–106

Longstaff BJ (2003) Investigations into the light requirements of seagrasses in Northeast Australia. PhD thesis. The University of Queensland

Lybolt M, Neil D, Zhao J, Feng Y, Yu K, Pandolfi J (2011) Instability in a marginal coral reef: the shift from natural variability to a human-dominated seascape. Front Ecol Environ 9(3):154–160

Manson FJ, Loneragan NR, Harch BD, Skilleter GA, Williams L (2005) A broad-scale analysis of links between coastal fisheries production and mangrove extent: a case-study for northeastern Australia. Fish Res 74:69–85

Maxwell WJH (1970) The sedimentary framework of Moreton Bay, Queensland. Aust J Mar Freshw Res 21:71–88

McEwan J, Gabric AJ, Bell PRF (1998) Water quality and phytoplankton dynamics in Moreton Bay, south-eastern Queensland. II. Mathematical modelling. Mar Freshw Res 49:227–239

McGowan HA, Petherick LM, Kamber BS (2008) Aeolian sedimentation and climate variability during the late Quaternary in southeast Queensland, Australia. Palaeogeogr Palaeoclimatol Palaeoecol 265:171–181

Meyer-Reil LA, Koster M (2000) Eutrophication of marine waters: effects on benthic microbial communities. Mar Pollut Bull 41:255–263

Meynecke J-O, Lee SY, Duke NC (2008) Linking spatial metrics and fish catch reveals the importance of coastal wetland connectivity to inshore fisheries in Queensland, Australia. Biol Conserv 141:981–996

Moreau S, Peron C, Pitt KA, Connolly RM, Lee SY, Meziane T (2008) Opportunistic predation by small fishes on epibiota of jetty pilings in urban waterways. J Fish Biol 72:205–217

Morton RM (1989) Hydrology and fish fauna of canal developments in an intensively modified Australian estuary. Estuar Coast Shelf Sci 28(1):43–58

Neckles HA, Wetzel RL, Orth RJ (1993) Relative effects of nutrient enrichment and grazing on epiphyte-macrophyte (*Zostera marina* L.) dynamics. Oecologia 93(2):285–295

Neil DT (1998) Moreton Bay and its catchment: landscape and seascape, development and degradation. In: Tibbetts IR, Hall NJ, Dennison WC (eds) Moreton Bay and catchment. University of Queensland, Brisbane, pp 3–54

Neil DT (2000) Characteristics and significance of a sub-tropical 'low wooded island': Green Island, Moreton Bay, Australia. J Coast Res 16(2):287–294

O'Brien K, Tuazon D, Grinham A, Callaghan D (2012) Impact of mud deposited by 2011 flood on marine and estuarine habitats in Moreton Bay. Final report to Healthy Waterways Ltd., Brisbane

Olds AD, Connolly RM, Pitt KA, Maxwell PS (2012a) Habitat connectivity improves reserve performance. Conserv Lett 5:56–63

Olds AD, Connolly RM, Pitt KA, Maxwell PS (2012b) Primacy of seascape connectivity effects in structuring reef fish assemblages. Mar Ecol Prog Ser 462:191–203

Olds AD, Pitt KA, Maxwell PS, Connolly RM (2012c) Synergistic effects of reserves and connectivity on ecological resilience. J Appl Ecol 49:1195–1203. doi:10.1111/1365-2664.12002

Pilans S (2006) Effectiveness of no take marine reserves in Moreton Bay, subtropical Australia. PhD thesis, The University of Queensland

QDSITI (2011) Point source database 2010/11, data provided by the State of Queensland's Department of Science, Information Technology, Innovation (QDSITI). Brisbane, Accessed Sept 2012

QOESR (2011) Queensland Government population projections: local government area report. Queensland Office of Economic and Statistical Research (QOESR), Brisbane. Available at http://www.oesr.qld.gov.au/products/publications/qld-govt-pop-proj-lga/index.php

Roelfsema C, Phinn S, Udy N, Maxwell P (2009) An integrated field and remote sensing approach for mapping seagrass cover, Moreton Bay, Australia. J Spat Sci 54(1):45–62

Saville-Kent W (1893) The Great Barrier Reef of Australia: its products and potentialities. W.H Allen & Co Ltd., London, 387 pp

SEQHWP (2007) South East Queensland healthy waterways strategy 2007–2012, South East Queensland Healthy Waterways Partnership, Brisbane. www.healthywaterways.org

Skilleter GA, Olds A, Loneragan NR, Zharikov Y (2005) The value of patches of intertidal seagrass to prawns depends on their proximity to mangroves. Mar Biol 147:353–365

Sloss CR, Murray-Wallace CV, Jones BG (2007) Holocene sea level change on the southeast coast of Australia: a review. Holocene 17:999–1014

Smithers SG, Hopley D, Parnell KE (2006) Fringing and nearshore coral reefs of the Great Barrier Reef: episodic Holocene development and future prospects. J Coast Res 22:175–187

Stevens T, Connolly RM (2005) Local-scale mapping of benthic habitats to assess representation in a marine protected area. Mar Freshw Res 56:111–123

Stewart J (2009) The South East Queensland WaterCAST model: development of flow and load time series inputs to the CSIRO RWQM V3. Report prepared by BMT WBM for the South East Queensland Healthy Waterways Partnership, Brisbane, 33 pp

Stoddard JL, Larsen DP, Hawkins CP, Johnson RK, Norris RH (2006) Setting expectations for the ecological condition of streams: the concept of reference condition. Ecol Appl 16(4):1267–1276

Stutchbury S (1854) Twelfth tri-monthly report on the geological and mineralogical survey of New South Wales. Votes and Proceedings Legislative Council NSW, Sydney

Tibbetts IR, Connolly RM (1998) The nekton of Moreton Bay. In: Tibbetts IR, Hall NJ, Dennison WC (eds) Moreton Bay and catchment. University of Queensland, Brisbane, pp 395–420

Tibbetts IR, Hall NJ, Dennison WC (eds) (1998) Moreton Bay and catchment. School of Marine Science, The University of Queensland, Brisbane, 645 pp

van der Heide T, van Nes EH, Geerling GW, Smolders AJP, Bouma TJ, Katwijk MM (2007) Positive feedbacks in seagrass ecosystems: implications for success in conservation and restoration. Ecosystems 10(8):1311–1322

Walters IN (1989) Intensified fishery production at Moreton Bay, southeast Queensland, in the late Holocene. Antiquity 63:215–224

Waltham NJ, Connolly RM (2006) Trophic strategies of garfish, Arrhamphus sclerolepis, in natural coastal wetlands and artificial urban waterways. Mar Biol 148:1135–1141

Waltham NJ, Connolly RM (2007) Artificial waterway design affects fish assemblages in urban estuaries. J Fish Biol 71:1613–1629

Waltham NJ, Connolly RM (2011) Global extent and distribution of artificial, residential waterways in estuaries. Estuar Coast Shelf Sci 94:192–197

Waltham NJ, Teasdale PR, Connolly RM (2011) Contaminants in water, sediment and fish biomonitor species from natural and artificial estuarine habitats along the urbanized Gold Coast, Queensland. J Environ Monit 13:3409–3419

Ward WT (2006) Coastal dunes and strandplains in southeast Queensland: sequence and chronology. Aust J Earth Sci 53:363–373

WBM (2005) Moreton Bay water quality improvement plan – receiving water quality modelling, final calibration report. Report prepared by BMT WBM Pty Ltd for the Moreton Bay Waterways and Catchments Partnership, WBM Ref: R.B15545.003.01, vol 1/2, Brisbane

Weber T, Ramilo NJ (2012) Integrated water cycle management – dealing with dilemmas. Presentation to the 7th international conference on water sensitive urban design, Melbourne Cricket Ground, Melbourne, 21–23 Feb 2012

Witte C, van den Berg D, Rowland T, O'Donnell T, Denham R, Pitt G, Simpson J (2006) Mapping land use in Queensland – technical report on the 1999 land use map for Queensland, Department of Natural Resources, Mines and Water, Brisbane. Available at: http://www.derm.qld.gov.au/science/lump/pdf/witte_1999_mapping_land_use_technical_report.pdf

Water Resource Development and High Value Coastal Wetlands on the Lower Burdekin Floodplain, Australia

Aaron M. Davis, Stephen E. Lewis, Dominique S. O'Brien, Zoë T. Bainbridge, Christie Bentley, Jochen F. Mueller, and Jon E. Brodie

Abstract

The lower Burdekin floodplain in north Queensland houses the combination of northern Australia's largest and most intensively developed agricultural floodplain with one of the largest concentrations of high value freshwater, estuarine and marine wetlands in Australia. The area has a long history of supporting one of Australia's most economically important sugarcane growing districts, most of which is located upstream of this complex of internationally and nationally significant wetland environments. A unique management feature of agriculture in the region is the total reliance on supplemental flood irrigation to meet crop water demands. Agricultural developments in the catchment area, particularly the establishment of water resource schemes to support this extensive irrigated agriculture, pose significant threats to the integrity of the downstream receiving wetlands. Cumulative (and ongoing) changes to water regimes and the chemistry of both surface and subsurface waters now pose major threats to both the long-term viability of wetlands and large sections of the sugar industry itself. Substantial shifts in societal perceptions and expectations regarding the value of wetlands and water resources at national and global levels are reflected in the lower Burdekin region. The legacy of earlier perceptions and associated policy decision-making are, however, going to provide some of the most enduring management challenges for lower Burdekin coastal wetlands, and ultimately the viability of irrigation areas themselves.

Keywords

Irrigated agriculture • Water quality • Hydrology • RAMSAR • Wet-dry tropics

A.M. Davis (✉) • S.E. Lewis • D.S. O'Brien • Z.T. Bainbridge
J.E. Brodie
Catchment to Reef Research Group, Centre for Tropical Water and Aquatic Ecosystem Research (TropWATER), Australian Tropical Science and Innovation Precinct, James Cook University, Building 145, Townsville Q 4811, Australia
e-mail: aaron.davis@jcu.edu.au

C. Bentley • J.F. Mueller
National Research Centre for Environmental Toxicology (Entox), University of Queensland, 39 Kessels Rd, Coopers Plains, QLD 4108, Australia

E. Wolanski (ed.), *Estuaries of Australia in 2050 and Beyond*, Estuaries of the World,
DOI 10.1007/978-94-007-7019-5_13, © Springer Science+Business Media Dordrecht 2014

Box 1

Aaron Davis and colleagues studied the lower Burdekin floodplain, which houses both intensively developed agricultural floodplain and high value freshwater, estuarine and marine wetlands. Irrigated agricultural developments pose significant threats to the integrity of the downstream receiving wetlands. The legacy of earlier decisions provides enduring management challenges for lower Burdekin coastal wetlands, and ultimately the viability of irrigation areas themselves.

Introduction

Many of the floodplains along the coastal margins of tropical north-eastern Australia have been substantially modified by agricultural development, with increasing deterioration of aquatic habitat condition within these catchments being linked to agricultural activities (Arthington et al. 1997; Tait and Perna 2001; Rayment 2002). Within these intensively developed floodplains in this region approximately 80 % of the naturally occurring wetlands have been cleared, predominantly for agricultural development (Johnson et al. 2000; Kemp et al. 2007). Given these dramatic losses of wetland habitat, effective conservation and management of remaining high quality wetland environments takes on added significance for governments and communities. While northern Australia's tropical rivers as a whole discharge approximately 70 % of the continent's surface runoff (Kennard et al. 2010), most rivers in the region remain largely undeveloped in terms of large-scale intensive agricultural and water resource initiatives (NLWRA 2002). With Australia being the world's driest continent, there is understandably continual interest in exploring and developing agriculture in northern Australia (Bristow et al. 2004). While northern Australia is receiving increasing development attention from irrigated agriculture, there is also increasing community appreciation of the iconic ecological and heritage status of the natural environments of tropical Australia that warrant conservation (Zander et al. 2010). There have accordingly been marked shifts in government and public opinion on the values and appropriate management of water resources. The residual effects of some past perceptions and policies have in several cases, however, left profound and ongoing impacts on natural environments that should serve as a lesson for future development (Petheram et al. 2008a).

The lower Burdekin floodplain in north Queensland's dry tropics presents the juxtaposition of northern Australia's largest and most intensively developed agricultural floodplain (Petheram et al. 2008b) being located directly adjacent to one of the largest concentrations of freshwater and marine wetlands in Australia. The Burdekin River catchment in its entirety is Queensland's second largest east coast river basin after the Fitzroy River Basin. Including the coastal plains between Giru and Bowen, the catchment covers 136,000 km^2, or 8 % of the area of Queensland. The Burdekin River Delta drains one of Australia's largest floodplain delta environments (ca. 1,250 km^2), with the lower areas of the four coastal subcatchments comprising a single floodplain which is considered to contain the single most important and productive wetland system along the Queensland coast (Dight 2009). It includes many large, permanent freshwater wetlands, long lengths of perennially-flowing creeks, and estuarine wetlands with 47,274 ha of this wetland constituting one of just five internationally-recognised RAMSAR wetlands in Queensland (including one of only two RAMSAR – listed internationally significant wetlands in the Great Barrier Reef (GBR) World Heritage Area) as well as several contiguous wetlands listed in the National Directory of Important Wetlands (Dight 2009). The lower Burdekin also supports northern Australia's largest and longest established irrigation areas, with over 120,000 ha of irrigated crops, dominated (~100,000 ha) by sugarcane (Fig. 1). The Burdekin region as a whole is one of Australia's largest sugar producing regions, accounting for approximately 30 % of Australia's total sugar production (Thorburn et al. 2011), and providing some of the highest quality and quantity crop yields (average productivity *ca.* 120 tonnes/ha) in Australia, if not the world (Anon 2010). The perceived value of the wetland environments of the area has evolved from that of a convenient mechanisms for transporting and draining irrigation schemes, to an appreciation of ecosystem services provided by wetlands (Queensland Department of the Premier and Cabinet 2003), to environments worth conserving for their own intrinsic values (Kelly and Lee Long 2012). There is also an increasing realisation that these systems have undergone (and likely continue to undergo) a major phase shift in their function (Burrows and Butler 2007). Regional Natural Resource Management (NRM) planning instruments now implicitly accommodate these realisations, with an emphasis on maintaining the ecosystem health, self-sustainability and functionality of these coastal wetland ecosystems, rather than any apparent expectation of returning these significantly

Fig. 1 Map of lower Burdekin waterways depicting major land uses and catchment water quality sampling locations. Note the area of internationally (RAMSAR) and nationally recognised wetlands. Land use categories are based on 2004 Queensland Land Use Mapping (*QLUMP*) shape files obtained from the Queensland Department of Natural Resources and Water (*DNRW*) for the Burdekin catchment. Boundaries of the major floodplain irrigation scheme areas are indicated by red boundaries; *BHWSS* Burdekin Haighton Water Supply Scheme; *NBWB* North Burdekin Water Board; *SBWB* South Burdekin Water Board

modified systems to pristine states (see Burdekin Dry Tropics Board 2005).

Site Setting

The history of human settlement and agricultural development on the lower Burdekin floodplain

The lower Burdekin floodplain region has a substantial indigenous Australian history predating European settlement (c. 1860 A.D.), with several tribes living for at least part of the year across the entire region (Bruinsma et al. 1999; Scheltinga and Heydon 2005). Traditional owner peoples lived and subsisted throughout the region, including Magnetic Island, Cape Bowling Green and Cape Cleveland where archaeological sites such as burial grounds, shell middens and other sites of cultural and historical importance can still be found (King 2003). Tribes that historically lived for at least part of the year in the region include the Manbarra Tribe of the Palm Island Group, Bindal tribe of the Townsville region, Wulgurukaba tribe of Magnetic Island and Cape Cleveland, Juru tribe of the Ayr region, the Wothan clan and the Bumbarra group of the Gia tribe of the Cape Upstart region (GBRMPA 2004; Bruinsma et al. 1999; Scheltinga and Heydon 2005). Cape Cleveland and Cape Bowling Green still provide important ongoing cultural, traditional hunting and subsistence activities for traditional owners (Kelly and Lee Long 2012). Traditional owner groups in the lower Burdekin also have an ongoing interest and desired role in natural resource management on the floodplain (Tait and Veitch 2007).

The lower Burdekin was first settled by Europeans in 1861 (Kerr 1994). The region is also northern Australia's most established cane growing system, with the first sugarcane grown in the Burdekin Delta area in the 1860s where it has remained the principal crop ever since. Irrigation commenced almost immediately in the lower Burdekin on the highly transmissive and fertile alluvial soils of the Burdekin River Delta, when surface water from floodplain lagoons of the Delta was coupled with groundwater extraction from the underlying Burdekin River aquifer to irrigate cane (Credlin 1979; Kerr 1994; Petheram et al. 2008b). The invention of the spear system for extracting groundwater was introduced to the lower Burdekin in 1887 (Kerr 1994), allowing growers to circumvent the limited surfacewater supply available for irrigation. Close to 2,000 large production bores currently extract around 330,000–550,000 megalitres/year of groundwater from the Burdekin Delta aquifer for sugarcane production (Scheltinga and Heydon 2005). Due to 'excessive' extraction within the Delta it was reported that the Burdekin Delta aquifer was overdrawn in the 1930s following expansion of the sugarcane industry and a long period of drought. Lowered aquifer water levels resulted in an advance of saline/fresh water groundwater interface inland (Hopley 1978; Credlin 1979). As a management response, two autonomous water management boards, the North Burdekin Water Board (NBWB) and the South Burdekin Water Board (SBWB) (Bristow et al. 2001) were established in the mid 1960s. Independently funded by sugar growers and millers in their respective regions, these boards have the charter to maintain groundwater levels through the use of large artificial groundwater recharge pits and to maintain associated infrastructure (e.g. sand dams, channel networks and a series of earthen barrages that restrict the upstream migration of sea water in tidal inlets) (Petheram et al. 2008a). The potential for saltwater intrusion has since become more concerning under scenarios of sea-level rise (Kelly and Lee Long 2012).

Today, the lower Burdekin region is predominately used for sugarcane production, with some other areas under mango, citrus, tropical fruit and vegetable farms. The gross value of production from sugar and horticulture in the lower Burdekin has been estimated at $450 million (Beare et al. 2003). Tourist accommodation ($9 m) and commercial ($20 m) and recreational ($40 m) fishing are other significant industries in the immediate area (Beare et al. 2003) together with the aquaculture industry. While contributions from tourism in the lower Burdekin area itself are relatively small, tourism in the entire Great Barrier Reef region is valued at more than $6 billion per annum (Access Economics 2007).

The lower Burdekin's region's wet-dry monsoonal climate (average rainfall ca. 1,000 mm/year) makes full supplemental irrigation, predominantly via furrow irrigation, an important, and distinguishing feature of local sugarcane production (Thorburn et al. 2011). The Burdekin is the only region in Queensland where cane is grown under full irrigation with approximately 95 % of the sugarcane in the district being furrow (flood) irrigated (Charlesworth et al. 2002). The average use of irrigation water on Burdekin farms ranks as the highest in Queensland (i.e. 8–15 megalitres/ha), and while extensive water use efficiency research conducted in the district has significantly improved irrigation practices, application efficiency remains comparatively low at around 45–65 % (Holden et al. 1998; Tilley and Chapman 1997; Qureshi et al. 2001). Average district cane yields are higher than other northern Queensland cane districts due to higher rates of solar radiation and a readily available water supply. This combination of factors including high productivity, high crop inputs and essentially total reliance on supplemental flood irrigation mark the Burdekin as unique in terms of the risks as well as management considerations required to mitigate environmental impacts (see Thorburn et al. 2011; Davis et al. in press).

Until the 1980s, cropping activities in the region were predominantly confined to the better alluvial and lighter textured soils of the Burdekin Delta (Fig. 1) and the riparian lands of the lower Burdekin and lower Haughton River (Christian et al. 1953; Fleming et al. 1981). A substantial expansion of the Burdekin cane industry, and one that has had major repercussions for functioning of local coastal wetlands, was facilitated by the increased security of surface water supplies afforded by the completion of the Burdekin Falls Dam in the late 1980s, taking the form of the Burdekin River Irrigation Area (BRIA). Prior to the development of the dam, the surface water resources of the Burdekin River system were subject to only minor regulation (Fleming et al. 1981). The BRIA was originally conceptualised and designed in the 1950s, with the principal objective of the Burdekin Falls Dam being to provide water supplies for the irrigation of new sugarcane, rice and other crops in the fertile, mainly heavier clay soils, of the floodplain adjacent to Bowling Green Bay, extending from the north bank of the Burdekin River to the Haughton River (Petheram et al. 2008b). Completed in 1987, the Burdekin Falls Dam is the largest water reservoir in Queensland (1,860,000 megalitres), located ~80 km from the river mouth. The subsequent agricultural expansion on the lower Burdekin floodplain, which commenced in 1988, was developed by SunWater, a government owned corporation, and mainly occurred in the area of what is now termed the Burdekin-Haughton Water Supply Scheme (BHWSS).

The BHWSS was at the time the largest land and water conservation initiative undertaken in Queensland (Kerr 1994).

The lower Burdekin and lower Haughton Rivers now allocate large amounts of water for irrigated agriculture from the Burdekin Falls Dam through the BHWSS, with a current allocation of 150,000 megalitres per annum. Irrigation Burdekin Falls Dam water is diverted from the Burdekin River at Clare Weir (downstream of Burdekin Falls Dam) into balancing ponds and then into an extensive irrigation distribution system which consists of several distinct networks of channels, pipelines, balancing storages, unlined irrigation channels and creek and river systems including the Haughton River, Barratta and Sheep Station Creek (Petheram et al. 2008a). Previous to this agricultural expansion, much of the eventual BHWSS area, which constitutes a large proportion of the RAMSAR wetland watershed in the Barratta Creek and Haughton River catchments, had been devoted to beef production on rainfed pastures. With development of the new farming area in the late 1980s and early 1990s the area under irrigation on the Burdekin floodplain essentially doubled, and the scheme currently supplies water to approximately 43,000 ha of irrigable land (DNR 1999), which is predominantly used for the irrigation of sugarcane. The BHWSS was designed with the expectation of relying on furrow irrigation for the foreseeable future, with furrow irrigation efficiency accordingly an important issue examined through better design and management of irrigated cane fields across the range of BHWSS soil types (Raine and Bakker 1996; Raine and Shannon 1996). As a whole the lower Burdekin is a conjunctive use scheme (i.e. water is sourced from both surface and groundwater), though the Delta region uses mainly groundwater (80–90 %) and the BHWSS uses mainly surface water (80 %). SunWater is responsible for the maintenance of the open, earth-lined channel and drainage infrastructure and the supply of bulk water to growers in the BHWSS, with water supplies being metered and landholders charged a fee based on consumption.

Geomorphology

The coastal zone of the lower Burdekin consists of nearly flat alluvial lowlands with scattered mountains that rise from sea level to over 1,200 m. The central southern zone consists of maturely dissected rugged hills and mountains and extensive undulating country with highland areas reaching over 1,000 m above sea level, and the western zone ranges in altitude from 90 to ~600 m (Christian et al. 1953). The Burdekin-Haughton-Barratta alluvial plains form a gently sloping, slightly depressed plain approximately 900 km^2 that is underlain by fluvial sediments, forming a complex pattern of channels, levees and floodplains of various ages (Fleming et al. 1981; Fielding et al. 2006). It is bounded by the Burdekin River and delta on the east, by the Haughton River on the west and by the swamps and saline flats of the littoral zone in the north (Fleming et al. 1981). The lower Burdekin Delta and alluvial floodplain consist of a complex layering of unconsolidated Quaternary sediments of limited lateral continuity overlying a basement of igneous origin (Hopley 1970; Fielding et al. 2006). Quaternary sediments greater than 100 m thick presently overlie the granitic basement rock that cradles the lower Burdekin basin (Hopely 1970). Basement contours constructed by Hopely (1970) and revised by McMahon (2004) indicate that bedrock underlying the lower Burdekin alluvium slopes away from the basin margins at 2–20 m/km. Sedimentation of the lower Burdekin was strongly influenced by Quaternary sea-level fluctuations. The most substantial successions of sediments were deposited over the crystalline basement rocks during the Pleistocene epoch (1.8–0.03 Ma) (Fielding et al. 2006). During this period terrigenous and marine sediments as old as 30 ka were deposited within in the upper delta plain (Fig. 1). The deposits of the Burdekin Delta area tend to be comprised of coarser material and have fewer clay and fine sediments than the upper alluvial deposits on the left and right banks of the Burdekin River upstream of Mount Kelly (Petheram et al. 2008a). The finer material in the Barratta Creek catchment and BHWSS is likely to have resulted from the deposition of clays and fine silts during overbank flow events. The upper clay layer in the BHWSS has a thickness of between 2.5 and 20 m, and directly overlies either bedrock or more coarse sand and gravel deposits.

Fielding et al. (2006) concluded that the sea had transgressed considerable distances inland during the Pleistocene, inundating the geographic region presently used for irrigated agriculture. It is generally accepted that during the last glacial cycle sea levels in the western Pacific were up to 130 m below present (Lewis et al. in press). During this time the Burdekin River incised the alluvial sediments to the west of Mount Kelly (Fig. 1), draining northward toward the Haughton River (Hopley 1970; Fielding et al. 2006). Numerous palaeochannels and associated river-mouth deposits have been identified (Hopley 1970; Fielding et al. 2006), and it is estimated that the Burdekin River changed its course some 13 times over the last 8–10 ka (Fielding et al. 2005). The present-day surface morphology of the lower Burdekin is dominated by low-relief (<5 m) undulations formed by active and abandoned channels and floodplains (Fielding et al. 2006). The lower Burdekin floodplain is drained by the minor rivers and creeks of the Burdekin Region (Haughton River, Barratta Creek and Sheep Station Creek) that run into Bowling Green Bay (Lewis et al. 2006). The main contributing catchments to Bowling Green Bay include the Haughton River, Barratta Creek, Barramundi Creek, Sheep Station Creek and many other small creeks. Land of the lower Burdekin region has been built up from soils deposited by the Burdekin and Haughton River systems. The lower floodplain soils of the Burdekin-Haughton region are derived from marine sediments ranging in permeability from highly permeable sands on levees to heavy uniform cracking clays of low permeability. The soil pattern is complex, consisting of sands and clay, and frequent changes to river courses and sea level have assisted in the formation of the lower Burdekin flood plain (Dalla Pozza 2005). The Burdekin Delta alluviums are 10–80 m thick and rest on eroded granite (Hopley 1978). The main water bearing material is coarse sand that grades into gravel in places or fine sands (Hopley 1978). This material is separated by relatively impermeable silts, silty clays, and clays. The aquifers range from 30 to 50 % of the total thickness of alluvium and are unconfined over the majority of the area (Hopley 1978).

Prior to the Burdekin River breaking through the 'rocks' region near Mt Kelly (Mt Kelly is a bedrock high approximately 60 m above the floodplain) in the late Holocene, the northern part of the Barratta-Haughton floodplain (the current BHWSS) may have been a deltaic area of the Burdekin and Haughton Rivers (Hopley 1970), and is referred to by Fielding et al. (2006) as the 'inactive' part of the Burdekin Delta. The Barratta Creek floodplain is a complex of distributary channels and drainage depressions that appears to have been superimposed on this former delta of the Burdekin (Fielding et al. 2006). Landform pattern consists of a flood plain with elements of backplain, bank, bar, drainage depression, levee, plain, stream bed, stream channel, and swamp. This complexity of wetland types and aerial extent in the coastal zone provides a regionally important role in nutrient assimilation and sediment stabilization to adjacent and downstream areas in the Great Barrier Reef lagoon (Bruinsma 2001). The Bowling Green Bay wetland complex is also considered a unique system in north-eastern Australia because of the large scale flows/discharge, sand/sediment inputs and floodplain habitat mosaic associated with the Burdekin River catchment and delta floodplain system.

The Haughton River gently descends into Bowling Green Bay with the average width of the channel estimated at approximately 50 m, widening to 2 km at the river mouth (Dalla Pozza 2005). Sand bars and islands are scattered throughout the channel, indicative of large quantities of sand originating from Hervey's Range and Mount Elliot Complex (Dalla Pozza 2005). Analysis of aerial photographs indicates shoreline retreat of approximately 100 m on the western side of the estuary since 1942 (Dalla Pozza 2005). In this study it was considered that the Haughton River estuary is a sink for marine sediments. The Burdekin and Haughton Rivers supply large quantities of sand and clay sediments during periods of high flow. Prevailing south-east winds and longshore drift transport plumes to the north and west, and highly terrigenous sediments are deposited in the nearshore zone. Sand is generally deposited on beaches and spits by wave action and the fine sediments gradually settle as muds in Bowling Green Bay and other inlets (Fleming et al. 1981). A unique geomorphological feature of the region is the Cape Bowling Green sand spit, which results from sand transported from the nearby Burdekin River via longshore sediment transport, and prevailing wind and tide-driven coastal currents (Kelly and Lee Long 2012). Approximately 80–90 % of the fine sediment particles exported from the Burdekin River are trapped in Bowling Green Bay, with sediment accumulation rates of 18.0 kg/m^2/year adjacent to the tip of Cape Bowling Green sand spit and 4 kg/m^2/year at the eastern inshore area of the bay (Orpin et al. 2004).

Bowling Green Bay sand spit has grown over the last 50 years with long term growth suggested to be 20 m/year between 1960 and 1970 (Hopley 1970) with the main supplier of sediment considered to be the Burdekin River. It was considered that the Burdekin Falls Dam would cut off the supply of sediment to the sand spit, although this is yet to be quantified. The integrity of the Cape Bowling Green sand spit is considered critical for the maintenance of present conditions in Bowling Green Bay (Fleming et al. 1981). Cape Bowling Green sand spit is a critical component for high tide roosts for migratory shorebirds and nesting habitat for marine turtles and critical indicators for oceanographic processes and coastal geomorphology (Kelly and Lee Long 2012).

Hydrology and Oceanography

The large complex of coastal wetlands that fringe Bowling Green Bay and the eastern portion of Cleveland Bay form a large part of the lower floodplain of the Burdekin River catchment. The site occurs in the "wet-dry tropics" of Queensland, a region characterised by two broad seasons including a cool/warm dry season and a warm/humid wet season. Annual rainfall averages 1,000–1,200 mm/year on the coastal Burdekin floodplain (Bureau of Meteorology 2009), but is highly variable from year-to-year. Streamflow also exhibits high inter-annual variability and is also highly seasonal, with approximately 80 % of annual flows generally occurring between December and April (Lough 2007). Annual potential evaporation is 2,080 mm/year and, on a monthly basis exceeds rainfall except in February (Petheram et al. 2008a). Hence, it is not uncommon for rivers within the Burdekin catchment (including the floodplain) to cease flowing between May and November (Roth et al. 2002).

The coastal environment of Bowling Green Bay is shaped by semi-diurnal tides, with the highest astronomical tides of 2.25 m above mean sea level and mean high water springs of 1.15 m above mean sea level. Tidal range is between 3–4 m. Marine and estuarine influences can reach more than 10 km inland and hydrological interactions between tidal and freshwater systems (surface and groundwater) manifest across a very broad coastal zone. Wetlands in this area may therefore be subject to extreme periods of tidal and/or freshwater inundation at multiple times during each year, providing a marine-freshwater mosaic of great complexity. Freshwater and oceanographic hydrological patterns are critical to maintaining this extensive and diverse mosaic of coastal wetlands and wetland dependant biota, with the complex and dynamic interplay of freshwater and saline systems over such a large wetland system uncommon in north-eastern Australia (Kelly and Lee Long 2012).

The lower Burdekin catchment is drained seaward into Bowling Green Bay by the Haughton River, three larger (Barramundi, Barratta and Sheep Station Creeks) and many small creeks. Two large creeks (Cocoa Ck and Alligator Ck) and a number of minor creeks discharge into Cleveland Bay at the north western end of the site; a few small creeks drain into Upstart Bay at its southern end. The mountainous area to the south of Cape Cleveland and Feltham Cone are drained by seasonally active creeks into Cleveland and Bowling Green Bays. McKenzie and Emmett Creeks drain seasonal runnoff from the steep northeastern slopes of Mt Elliott into extensive sedge swamps situated in a shallow basin whose eastern edge is fringed by The Cone, Storth Hill and Feltham Cone. However, it is considered that in periods of high flow, the entire lower Burdekin can become a floodplain with little distinct demarcation of catchments (Kelly and Lee Long 2012).

With a catchment length of approximately 70 km, Barratta Creek's source lies in the Leichhardt Range and for the early part of its course it is a small stream flowing from the coastal ranges onto the coastal plain. In the latter part of its course (i.e. to the north of Woodhouse Mountain) it spreads out and becomes a functioning part of the Burdekin-Haughton floodplain (Environment Australia 2001). Barratta Creek is largely a floodplain drainage which owes some of its geomorphic origin to being a historic channel of the lower Burdekin River, for which it still acts as a distributary system in large flood events. During large floods of the Burdekin River, floodwaters overtop the left bank levee in the vicinity of Clare and flow into the Barratta system. It has been estimated that this occurs when discharge reaches 28,000 m^3/s. This discharge has been exceeded at least five times between 1919 and 1999, approximating 1:20 year flood events (Tait and Veitch 2007). A notable hydrologic feature of the area is that due to local topography (high riverbank levees), the majority of the overland drainage between the Burdekin River and the Haughton River to the north (including the BHWSS) flows into the Barratta Creek system prior to discharging into adjacent estuarine-marine environments (see Fig. 1).

Groundwater on the lower Burdekin floodplain is stored in two main aquifers. Older, deeper aquifers extend from the slopes of Mount Elliot across the Haughton River to the Burdekin Delta, while younger and shallower aquifers occur around the Haughton River bed (Lenahan and Bristow 2010). The coastal floodplain groundwater system, which underlies a geographic area of ~1,600 km^2, is largely unconfined with aquifer sediments comprised of siliclastic non-carbonate materials (McMahon et al. 2000; Fass et al. 2007). Holocene sediments comprise the shallowest portions of the unconfined aquifer system which is in direct hydraulic connection with the Burdekin River (McMahon 2004; Petheram et al. 2008a). Deeper groundwater resides in Pleistocene sediments and is hydraulically connected to the shallow Holocene units. The main sources of groundwater are unconsolidated sediments associated with the Burdekin Delta and Haughton River valley area (Hopley 1978). In general, groundwater flows north toward the coast, however large fluctuations in groundwater levels due to extraction of large volumes of groundwater and enhanced recharge of irrigation water make it difficult to construct past and present potentiometric surfaces that are meaningful in terms of predicting groundwater flow paths (Lenahan and Bristow 2010). Natural groundwater variability in the lower Burdekin is not well understood. In some systems groundwater may mimic surface water variability. The hydrological regime of the site is complex and currently not well understood, and information is needed particularly on the hydrological connectivity of various aquifers and interactions with surface and ground water along the Cape Bowling Green coastal margin and adjacent areas of the Burdekin River Delta and Haughton River catchments (Cook et al. 2011; Kelly and Lee Long 2012).

Coastal and oceanographic processes including tide and wind driven currents, water levels, waves and mixing are largely responsible for transport, mixing and the deposition of sediments in the bay and alongshore. The oceanographic and tidal hydrodynamic processes also facilitate breeding migrations, larval transport and recruitment of finfish and crustaceans into nursery areas. A key feature of the coastal and oceanographic processes here is that they help to sustain the Cape Bowling Green sand spit, other beaches and spits, intertidal sand and mudflats, and sub-tidal profiles within Bowling Green Bay (Kelly and Lee Long 2012). Conversely, the resultant coastal geomorphology (Cape Bowling Green sand spit in particular) also has a strong influence on the behaviour of tide and wind driven forces within Bowling Green Bay.

The marine environment of Bowling Green Bay comprises seasonal estuaries, large coastal wetlands (mud flats, mangroves and some salt marshes, tidal freshwater wetlands, patches of seagrass) and shallow coastal waters. In Bowling Green Bay the substrate composition varies from muddy substrate on the east side to sandy to the west

(Fleming et al. 1981). Despite being the largest sediment contributor to the central Great Barrier Reef, the great majority of the Burdekin output is trapped in the eastern portion of Bowling Green Bay (Neil et al. 2002; Orpin et al. 2004). Modelling of the dry season dynamics of the coastal and estuarine waters in Bowling Green Bay has suggested that tidal dynamics in these shallow waters generates significant coastal trapping, resulting in waters that are very poorly flushed in the dry season, mixing minimally with offshore waters (Wolanski and Ridd 1990). The combined effects of evaporation greatly exceeding rainfall, numerous small coastal bays with minimal flushing produce hypersaline coastal waters (salinity ~37) which mix minimally with offshore water, although some longitudinal water transport between bays does occur (Andutta et al. 2011). The wetland-fringed estuaries (tidal creeks) are flushed even slower, with a residence time in the order of several weeks, because of the lack of freshwater runoff in the dry season and a landward oceanic water mass flux to compensate for the evaporative loss of water (Wolanski 2007). The slow flushing makes Bowling Green Bay a pollutant trap during the dry season, impacting on its functioning as an ecosystem and the ecosystem services that it provides.

Ecological Processes and Communities

The lower Burdekin region (Fig. 1) supports a very large (~188,000 ha) and mostly intact mosaic of coastal wetland complexes with a wide zone of mixing between estuarine, marine and freshwater habitats in the dry tropics region of Queensland, north-eastern Australia (Fig. 2b, d). The site is particularly distinctive for providing a large contiguous area of coastal wetland vegetation and constitutes one of the most expansive coastal wetland complexes along the entire east coast of Australia (Environment Australia 2001). The extensive low-lying coastal geomorphology of the lower Burdekin supports a particularly expansive and diverse dry-tropics mangrove community (approximately 17 species) compared to more southern areas and this diversity in turn provides habitat for a range of species listed as vulnerable or rare. In terms of size and complexity the only comparable areas of coastal wetland vegetation remaining on the east-Australian coast are Hinchinbrook Channel (ca. 200 km N; 30,682 ha) and Broadsound (212,042 ha), St Lawrence (Bruinsma 2001; Kelly and Lee Long 2012). Each location exhibits different mangrove species assemblages – attributed to rainfall. Bowling Green Bay is characterised by low-profile mangrove

Fig. 2 (**a**) Riparian *Livistona* spp. palm forest in freshwater reaches of Barratta Creek; (**b**) Bowling Green Bay freshwater wetland mosaic (Photo provided by NQ Dry Tropics); (**c**) Brolga (*Grus rubicunda*) dancing on Bowling Green Bay saltpans (Photo by Dominique O'Brien); (**d**) Bowling Green Bay estuarine mangrove-saltpan wetland mosaic (Photo provided by NQ Dry Tropics)

communities and extensive areas of saltpan (typical of a dry tropics location) whereas Hinchinbrook Channel displays (typically wet-tropics) taller mangrove communities and less areas of saltpan that are devoid of vegetation (Bruinsma 2001). Similarly, Broadsound has a markedly different species composition and lacks the freshwater habitats found at Bowling Green Bay (Kelly and Lee Long 2012). The wetlands of Bowling Green Bay are, therefore, considered rare in the "Northeast Coast" drainage division and the "Northeast Province" bioregion for their particularly extensive mosaic of freshwater, marine and estuarine habitat in such a near-natural state in the Northeast Drainage Division. The littoral communities in Bowling Green Bay occur along intricate patterns of tidal inlets, where mangrove communities may extend up to 9 km upstream (Fleming et al. 1981). Thin fringes of mangroves in the intertidal zone are backed by complex patterns of saltmarsh and saltwater couch grasslands, with such an extensive area of 'dry tropics' mangrove communities uncommon in north-eastern Australia (Kelly and Lee Long 2012).

On the 22nd of October 1993, 47,274 ha of the Bowling Green Bay wetland was listed as a Ramsar wetland of international importance under the Ramsar Convention (Ramsar Website: http://www.ramsar.org/pdf/sitelist_order.pdf) in recognition of the extent and diversity of its coastal wetland systems, and is therefore protected under the Australian Environmental Protection and Biodiversity (EPBC) Act, 1999. At the date of listing, the site met eight of the current Ramsar criteria. Bowling Green Bay is also listed in the Australian Directory of Important Wetlands (Environment Australia 2001). It is also listed as a "special unique site" of the Great Barrier Reef World Heritage Area and forms part of the Commonwealth Great Barrier Reef Marine Park under the Australian Government's Great Barrier Reef Marine Park Act 1975. Bowling Green Bay and Cleveland Bay are additionally protected as declared Fish Habitat Areas under the Queensland Fisheries Act 1994 as they are considered critical nursery grounds for fish and crustacean species (Danaher 1995). Further protection of values within the site occurs under the Queensland Fisheries Act 1994 as a Dugong Protection Area with Cleveland Bay declared under protection level "A" and Bowling Green Bay under protection level "B". These levels of protection regulate commercial netting activities that are permitted within designated areas.

The lower Burdekin's extensive mosaic of diverse wetland types plays a pivotal role in regulation of water flows and quality entering the coastal zone and Great Barrier Reef lagoon (Kelly and Lee Long 2012). This same extent and diversity of wetland types is integral to supporting a significant diversity and abundance of wetland flora and fauna, including threatened and migratory species, regionally significant baitfish aggregations and fisheries stocks. The Bowling Green Bay Ramsar site is crucial to large populations of waterbird (Fig. 2c), marine reptile and marine mammal species for foraging, protection and breeding purposes. The site is also of significant cultural, spiritual and provisioning value to local Indigenous Australians and is important in the region and Australia for nature-based science and recreation. The critical components and processes that underpin these ecosystem services are numerous and complex, reflecting the diversity and extent of wetland types that occur across the transition from freshwater to brackish and coastal marine systems. The critical wetland habitats contributing to the sites' ecological character include a mixture of palustrine (freshwater and brackish swamps dominated by the bulkuru sedge, *Eleocharis* sp.), coastal dune and swale systems, mangrove and saltmarsh tidal wetlands, intertidal seagrass meadows, intertidal sand and mudflats, freshwater and estuarine streams, creeks and riverine habitats. The particularly large sand spit at Cape Bowling Green, plus several smaller coastal spits, provide additional critical habitat for several wetland fauna species that migrate, roost and/or nest here. Within the area, the ecological communities that are listed as 'threatened ecological communities' under the Environment Protection and Biodiversity Conservation (EPBC) Act 1999 include microphyll/notophyl vine forest to semi-deciduous vine thicket on coastal dunes, riverine *Melaleuca leucadendra* open forest and bulkuru sedge habitat. The vegetation communities listed as "Of Concern" under the Queensland Vegetation Management Act (1999) include the dune swale beach ridge habitat, the bulkuru sedge habitat and a small area of riverine melaleuca habitat.

The lower Burdekin floodplain's biological diversity includes a great number of freshwater brackish and marine faunal species. However, there is a select group of faunal species that highlight the region's unique ecological character (Kelly and Lee Long 2012). The area provides feeding habitats for at least four species of marine turtle, which occur regularly and in substantial numbers; the loggerhead (*Caretta caretta*, listed as Endangered under the EPBC Act); and the green turtle (*Chelonia mydas*), hawksbill turtle (*Eremochelys imbricata*) and flatback turtle (*Natator depressus*) are all listed as Vulnerable under the EPBC Act. Marine turtles are recognised internationally as species of conservation concern and all of the species found at the site are listed in the 2000 International Union for Conservation of Nature Red List of Threatened Animals (International Union for Conservation of Nature 2009) and the Convention on International Trade in Endangered Species of Wild Fauna and Flora (CITES) list. All resident turtle species are similarly considered a priority for conservation under the Convention on the Conservation of Migratory Species of Wild Animals (the Bonn Convention). Threatened species of marine mammals utilising the area include dugong (*Dugong dugon*), the Australian snubfin (*Orcaella heinsohni*) and Indo-Pacific humpback (*Sousa chinensis*) dolphins are listed as Rare under the Queensland's Nature Conservation Act 1992 and are classified as Data Deficient by the International Union for Conservation of Nature (2009). The intertidal seagrass meadows found in Cleveland Bay, and within the Bowling Green Bay Ramsar site hold intrinsic value to dugong and turtle conservation, and more significantly, are one of four areas along the Queensland east coast assigned a 'high' conservation value to dugong.

The snubfin dolphin was recently described as a new species and is the only cetacean endemic to Australian waters (Beasley et al. 2005). Genetic studies on Indo-Pacific humpback dolphins indicate that the Australian populations may also represent a different species that is only found in Australian waters (Frère et al. 2008). Estuarine crocodiles, 4 species of threatened waterbirds, migratory shorebirds and seabirds, and waterbirds that feed and/or breed at the site (brolgas, magpie geese and colonial waterbirds) are characteristic features of the site, The large stocks of fisheries species (particularly barramundi, mangrove jack and mud crab) and baitfish produced within the site are considered particularly important in the Northeast Province bioregion (IMCRA Version 4.0).

As a contracting party to the Ramsar Convention, Australia is committed to managing all Australian wetlands listed under the Ramsar Convention, with the objective of maintaining the ecological character of the wetland. Information for its Ramsar sites (documented in the Ramsar Information Sheet 1999) is to be updated every 6 years or when significant changes occur to the ecological character of the site. Under the Convention, a notification of change is required if the ecological character of a site has changed, is changing, or is likely to change as a result of technological developments, pollution or other human interference. To assist in fulfilling these obligations, the Australian government has recently implemented the preparation of Ecological Character Descriptions (ECDs) for all of Australia's Ramsar-listed wetlands (Department of Environment, Water, Heritage and the Arts 2008). ECDs aim to supplement the description of the ecological character contained in the Ramsar Information Sheet and form an official record of the ecological character by more clearly identifying and defining the critical ecosystem components and ecosystem processes that support the wetland. They also provide qualification of the natural variation and limits of acceptable change where possible. ECDs thereby assist in informing at the earliest possible time if the ecological character of any wetland in its territory, and included in the Ramsar List, is under threat from anthropogenic interference. The Ramsar Information Sheet update emerging from the Bowling Green Bay ECD re-interpreted the old (pre-1999) criteria against the new Ramsar criteria. On the basis of current data, information is only sufficient to confirm that the site meets seven of the nine new criteria (Kelly and Lee Long 2012). Data are insufficient to confirm whether criterion 8 (the site regularly supporting more than 20,000 waterfowl) continues to be met. Data are similarly insufficient to confirm whether the site regularly supports 1 % of the individuals in a population of one species or subspecies of wetland dependent non-avian animal species (Ramsar Criterion 9).

Barratta Creek and Jerona Aggregation

In addition to the Bowling Green Bay Ramsar wetland itself, there are several contiguous wetland areas of considerable conservation value on the lower Burdekin floodplain, some of which play pivotal roles in the current and future functional integrity of the Bowling Green Bay Ramsar site. Barratta Creek, one such system, is a small subcatchment of the lower Burdekin floodplain where the major land use is grazing on natural pastures with approximately 31 % of the land area also used for irrigated sugar production (Dight 2009). The catchment also constitutes one of the major upstream catchment and drainage lines in the watershed of the Bowling Green Bay Ramsar wetland. The site was largely excluded from agricultural development during the most recent major expansion of the BHWSS occurring in the late 1980s. During development of the BHWSS, a wildlife corridor extending ca. 1 km laterally from the system was retained along much of the length of Barratta Creek. The retention of such significant areas of remnant floodplain and wetland habitat within an interconnected upper catchment to coastal-marine corridor network was unprecedented for a Government funded irrigation scheme development in Australia (particularly at the time). Preservation of this vegetated riparian corridor was, however, largely aimed at mitigating the groundwater table rises predicted to occur under irrigated agriculture, rather than biodiversity management. Regardless of the intent behind it's conservation, the value of such alluvial landform floodplain and wetland habitat remnants in the context of the relatively intensively developed east coast of Queensland, where most equivalent habitat types have been lost to development, is very high (Tait and Veitch 2007). This Barratta Creek vegetation corridor has suffered several negative influences from the surrounding irrigation area, inappropriate fire regimes and invasive weeds, and appears to be slowly degrading (Dight 2009). Barratta Creek does, however, currently retain considerable riparian integrity (Tait and Perna 2001; Burrows and Butler 2007), certainly in comparison to most other lower Burdekin floodplain creek systems in irrigated areas which have suffered extensively from riparian clearing and diffuse pollution (Perna and Burrows 2005). Analysis of riparian habitat along Barratta Creek indicates that this subcatchment has declined from 'excellent condition' in the 1970s to 'relatively good' condition by the mid 2000s, due largely to floodplain clearing, an increase of bare soil on the floodplain and increased scalding. Despite these changes, the area continues to provide a wide range of riparian habitats (Dight 2009), with diverse plant assemblages, high levels of bank structure, fallen timber and riparian shading to the stream (Fig. 2a). With high levels of vegetative recruitment and regeneration, this system is likely to maintain good riparian condition into the future, particularly relative to other floodplain riparian wetlands. The site has been excluded from cane expansion in recognition of these values; with land use comprising light cattle grazing, or no direct use. Barratta Creek is thus considered a valuable remnant wetland (Environment Australia 2001).

The Barratta Creek complex provides a very good representative example of a flood distributary system on a large tropical flood plain and yields valuable insights into the development of floodplains (Environment Australia 2001). Barratta Creek represents some of the most important, healthy and productive creek systems in the Burdekin region and includes many large, permanent wetlands and long lengths of perennially-flowing creek system (Dight 2009). No significant fish passage barriers are present and migratory catadromous fish species including commercially and recreationally important species (i.e. Barramundi, Mangrove Jack, Mullet) range throughout the site including to inland lagoon systems (Tait and Veitch 2007). Indeed, Burrows and Butler (2007) note that Barratta Creek is the only major watercourse in the entire coastal region between the Proserpine River, ~150 km S and the Herbert River ~150 km N, that has no fish passage barriers. With increased availability of good condition habitat, and no passage barriers, Barratta Creek provides a major regional freshwater nursery for desirable species such as barramundi and mangrove jack that compensate for the extensive losses of their habitats elsewhere in the region (Burrows and Butler 2007). It has been estimated that the majority of small mammals remaining on the lower Burdekin floodplain are reliant on riparian vegetation associated with the Barratta Creek vegetation corridor.

The Barratta Creek riparian corridor likely provides critical habitat for the bare rumped sheathtail bat *Saccolaimus saccolaimus nudicluniatus* (Nce, Sr), listed as critically endangered in the Commonwealth Environmental Protection and Biodiversity Conservation Act (1999). It is restricted to coastal woodland and is dependent on hollows in old poplar gum (*Eucalyptus platyphylla*) trees for roosting and breeding. One of very few records of this bat comes from Jerona, immediately to the north, and more recent echolocation call data suggest it still resides in the area. The remnant riparian corridor associated with the Barratta Creek- Jerona aggregation probably represent the only suitable remaining habitat for it on the lower Burdekin floodplain (Environment Australia 2001). In recognition of its natural values the Barratta Creek Complex has been inscribed on the Register of the National Estate (2002) as the 'Barratta Channel Aggregation', and also inscribed on the Directory of Important Wetlands in Australia (Environment Australia 2001), meeting 4 of the 6 criteria for inclusion; it is a good example of a wetland type occurring within a biogeographic region in Australia; it is a wetland which plays an important ecological or hydrological role in the natural functioning of a major wetland system/complex; it is a wetland which is important as the habitat for animal taxa at a vulnerable stage in their life cycles, or provides a refuge when adverse conditions such as drought prevail; and the wetland supports native plant or animal taxa or communities which are considered endangered or vulnerable at the national level.

Anthropological Influences: Resources, Pressures, Impacts, and Remediation

Threats to the ecological character of lower Burdekin wetlands are multiple, complex and are occurring both within the site and external to the site. Issues such as inappropriate fire regimes and exotic weeds pose threats within the site (Environment Australia 2001; Tait and Veitch 2007). Agricultural developments in the catchment area external to these wetlands, however, pose the most significant threats to the integrity of the site, particularly the lowland areas (Environment Australia 2001; Kelly and Lee Long 2012). The potential greatest impact posed by these threats is through cumulative changes to water regimes and the chemistry of both surface and subsurface waters, and it is these issues which will be elaborated on below. As already evident within the coastal region, such influences initiate rapid changes in biological communities and degrade the natural functioning of wetlands within the landscape and their value as habitat (Ramsar Information Sheet 1999).

Changes to Wetland Hydrology

The long history of water resource development across the lower Burdekin region has considerably altered the underlying function of most wetland environments across the floodplain including river flow regulation, increased nutrient and pesticide loads, altered sedimentation patterns, losses of lateral and longitudinal wetland connectivity, vegetation clearing and invasions by introduced plants and animals (Tait and Veitch 2007). The expansion of cropping in the catchment, including establishment of the Burdekin Haughton Water Supply Scheme in particular has allowed for significant expansion in sugarcane cropping and increasing diversity of irrigated horticultural and tree crop areas immediately upstream of coastal zone wetland complexes. The close hydrological linkages between coastal and near-coastal wetlands with the irrigation area exist through groundwater and surface water supplies. Both groundwater extraction and irrigation immediately upstream of the Ramsar site have had significant effects on the hydrological integrity of these coastal wetlands. The hydrology of Barratta Creek and the Haughton River, the two main floodplain catchments draining into the Ramsar site, have in particular, been substantially modified by the BHWSS.

With Bowling Green Bay listed as a Ramsar wetland in 1993, the hydrological impacts to wetlands caused by the development of the BHWSS are thought to have begun manifesting soon after (mid 1990s) (Kelly and Lee Long 2012). Hydrological changes to the flow regime of Barratta Creek were, however, already emerging several years prior to Ramsar listing. The BHWSS was specifically designed to be irrigated by an 'open-ended system', where water goes in at one end and a lesser amount comes out at the other, a process necessary to make the system work (Environment Australia 2001; Petheram et al. 2008b). Similar large-scale irrigation scheme designs also occurred in the Ord River Irrigation Area, a surface water irrigation development established at a similar time to the BHWSS in north-western Australia (Petheram et al. 2008b). Unlike other floodplain subcatchments such as the Haughton River and Sheep Station Creek which are used as irrigation water transfer conduits, the Barratta Creek complex has no direct role in water transfer in the BHWSS. However, the BWHSS was designed for excess irrigation tail water to be discharged into natural drainage lines and large areas of the BWHSS are drained predominantly by Barratta Creek. Considerable resources were invested in the design and construction of tail water drainage infrastructure within the BHWSS, but were concerned primarily with managing tail water from a quantity perspective, and that the effects of tail water quantity, quality and timing releases on receiving wetland ecosystems were given relatively little consideration (Petheram et al. 2008b).

Subdivision and the development of new farms within the BHWSS during the late 1980s onwards were staged, and it took some time for irrigated canefarming on the flood plain between the Burdekin and the Haughton to become the dominant floodplain land-use. Up until 1986, Barratta Creek was an intermittent seasonal floodplain creek system that rarely carried flowing water for more than 4 months during the wet season. Since late 1987 creek flow in many sections of the Barratta complex has been continuous. Even in the very early stages of development (between 1989 and 1991) rainfall increased 47 % but the discharge of the Barrattas increased by 323 % (Environment Australia 2001). While direct inputs of farm irrigation tailwater run-off into Barratta Creek are a large contributor to perennial flow maintenance, other factors such as irrigation channel overflow, BHWSS infrastructure leakage and possible interception of groundwater rise to the Barratta Creek system and wetlands have also been implicated in the major changes evident in the hydrological regime of the system since the time of Ramsar listing (Petheram et al. 2008b; Bennet 2012).

With the introduction of surface water for irrigation following completion of the Burdekin Falls Dam in 1987, watertable levels across the BHWSS also exhibited marked changes. Impacts to groundwater and surface water flows from the BHWSS have resulted in a reported rise in the water table from 10 to 2 m below ground level upstream from the site since the inception of the surface water scheme in 1987 (Roth et al. 2002). Bristow et al. (2001) report that ground water salinity in the Burdekin River Irrigation Area also increased from less than 1,000 μS/cm (microSiemens per centimetre – a measure of the electrical conductivity of water) over the same period to around 2,500 μS/cm by the year 2000. Moreover, watertable levels are expected to continue to rise in most districts of the BHWSS under future wet periods (Petheram et al. 2008b). At present, it is estimated that the BHWSS aquifer is receiving an average of 67,576 ML/year more than it would have under pre-development conditions and this is the source of rising groundwater in the study area. The average rate of salt importation, from surface water alone, was estimated at 32,027 tonnes of salt per year (Bennet 2012). These dramatic rises in groundwater are due to sub-optimal water efficiency and inadequate deep drainage management (i.e. subsurface drainage was not installed in the BHWSS) (Petheram et al. 2008b). With limited connectivity between the groundwater system and the adjacent Burdekin River, the BHWSS lacks the presence of a deeply incised river to provide an effective discharge mechanism. Because the unsaturated zone underlying much of the BHWSS has decreased in depth, its capacity to buffer against recharge events is considerably reduced. Further, as the specific yield of the heavy textured upper soil layers that make up much of the BHWSS is smaller than the coarse, deep soil layers, future rises in watertable level are likely to be of greater magnitude than in the past.

Deep drainage following an irrigation event is often inevitable and in many cases a desirable outcome as a small amount of deep drainage is needed to leach salts from the root zone. However, without appropriate management these accessions can cause a water imbalance whereby groundwater levels rise until the groundwater discharge equals the aquifer recharge. The long-term experiences of irrigation practices in many different climates around the world have highlighted that irrigation schemes almost invariably require sub-surface drainage management in order to be sustainable (Petheram et al. 2008b).

Petheram et al. (2008b) noted that although hydrogeological investigations were conducted prior to development of the BHWSS, the number of reports published on the 'suitability' of the underlying hydrogeology for irrigation were out-numbered by the number of reports on the soil and land capability (e.g. Thompson 1977; Gardner and Coughlan 1982) and surface water (e.g. DNR 1976; Fleming et al. 1981). Numerous workers expressed concerns about the implication of irrigation on salinity and hydrology of the BHWSS area, with several studies concluding that problems with rising water tables and salinity were inevitable in the region (e.g. Gunn 1981; Gardner and Coughlin 1982; Shaw et al. 1983, 1984; Shaw 1986, 1988). Why more substantial measures were not introduced at the inception of the BHWSS schemes or during the development of more recent districts within the scheme, even after problems were evident, is unclear. Early management efforts to compensate for likely groundwater level rise consisted of a conjunctive water use policy introduced in 1989, where farmers were allowed to extract 1 ML of groundwater (at extraction cost only) for every 8 ML of surface water applied. However, this policy has not been effective at preventing groundwater level rise in the BHWSS.

As well as the hydrological changes emerging in the BHWSS and downstream area, there is also concern that the large carry over storages in the Burdekin River have stopped the downstream supply of coarse sediments, which is changing the morphology of the lower river reaches and surrounding coastal areas (Roth et al. 2002). Reduction in coarse sediments may compromise the longer term integrity of the Cape Bowling Green Spit, which is fundamental to maintaining the local environment of Bowling Green Bay (Fleming et al. 1981; Kelly and Lee Long 2012). The recent Ecological Character Description of the Bowling Green Bay Ramsar site noted that the hydrological elements of ecological character (flow, level and salinity regimes of surface and groundwater) have likely trended towards their limits of acceptable change (Kelly and Lee Long 2012). In the absence of statistical information, however, formal Limits of Acceptable Change were not set, making determination of any exceedences of these limits impossible on the basis of current data.

Changes to Wetland Water Quality

Riverine and wetland ecosystems are not just dependent upon the quantity and timing of water, but also water quality. Significant deterioration of water quality has been observed in the streams entering Barratta Creek and the Bowling Green Bay Ramsar site since the establishment of the BHWSS. This includes increases in pesticides (particularly herbicides) and nutrients, as well as fine sediment from agricultural irrigation tailing-waters. At the time of Ramsar listing (1993), the water quality within the site was still considered to have suffered relatively low impact from agricultural land-use and water management, although changes resulting from upstream and adjacent agricultural practices were underway. Pre-development reviews of the BHWSS identified the potential for such changes but stated that adverse effects of the scheme on the environment would not be significant (Petheram et al. 2008b). Water quality research focused on the adjacent Great Barrier Reef marine environment (i.e., Bennell 1979; Bell and Gabric 1991) had, even before this time, highlighted the risks to downstream ecosystems from diffuse-source, catchment-derived pollutants. This issue, as demonstrated below, clearly received minimal attention when considering the likely landscape-scale effects of major irrigation area development on a high ecological value floodplain.

Since irrigation in the BHWSS area began, the water in lower Burdekin floodplain systems such as the Haughton River and Barratta and Sheep Station Creeks has remained permanently turbid, due to the input of turbid water from the Burdekin Falls Dam. An unforseen consequence of the Burdekin Falls Dam has been higher turbidity levels at low flows (Griffiths and Faithful 1996; Burrows and Faithful 2003). The Burdekin Falls Dam now traps a large volume of turbid water (which would otherwise have passed downstream), with minimal colloidal settlement, with the impoundment now persistently turbid all year round. Water is released daily from the Burdekin Falls Dam for use by downstream irrigators, thus the river length below that dam is also persistently rather than episodically turbid. This turbid water from the river is also pumped into the extensive creek and wetland system on the lower Burdekin floodplain for irrigation water delivery, also increasing the turbidity persistence of many of these formerly clear waterbodies (Tait and Perna 2001; Perna and Burrows 2005). Thus, both the hydrological regime and the water clarity of systems such as the Haughton River, Barratta Creek and Sheep Station Creek has been greatly altered because of supplemental water flows. Another unforseen consequence of the higher turbidity levels in the Burdekin River following dam construction has been the effect of fine sediment adversely affecting the infiltration characteristics of the aquifer recharge pits operated by the north and south Burdekin water boards. In recent years the Water Boards have encouraged farmers to source their irrigation water directly from supply channels rather than groundwater, in order to minimise recharge pit maintenance (Petheram et al. 2008b).

The effect of chronic turbidity on floodplain ecosystems such as aquatic plants is uncertain. Beds of aquatic plants throughout the system now tend to be scattered, sparse and dominated by emergents; probably as a response to perennial turbidity (Environment Australia 2001). Important habitat provided by fringing water lilies and submerged aquatic

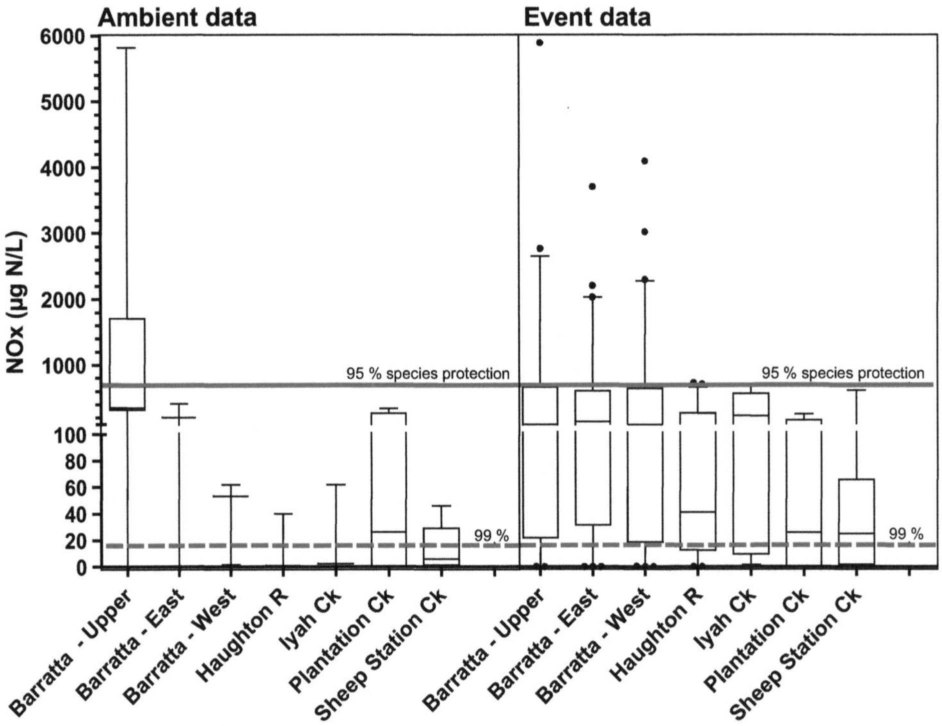

Fig. 3 Boxplots summarising in-stream nitrate concentration during ambient (low flow) and wet-season flood events in lower Burdekin floodplain waterways. Summary boxplots present the median value (line in box), interquartile range (containing 50 % of values), extreme values (values more than 3 box-lengths from the 75th percentile) and whiskers extending to highest and lowest values (excluding outliers). The *solid* and *red dotted red lines* represent the ANZECC and ARMCANZ (2000) 95 and 99 % guidelines for freshwater ecosystem protection

plants is now largely absent from the lower East Barratta Creek system. This turbid water is also compounded with the addition of considerable inorganic nutrient loading from tailwater runoff and groundwater leachate inputs sourced from local cane farms. Nitrate concentrations in many surfacewater systems such as Barratta Creek are now consistently well above Australian water quality guidelines (ANZECC and ARMCANZ 2000) for ecosystem protection (Fig. 3). These sustained inflows of turbid, nutrient rich water, and a lack of dry season draw-down have also impacted floodplain and lower catchment wetlands by favouring dominance by exotic pasture species Olive Hymenachne (*Hymenachne amplexicaulis*) and Para grass (*Urochloa mutica*) which are often not accessible to cattle grazing due to sustained higher water levels, as well as floating exotic weeds such as salvinia (*Salvinia molesta*) and water hyacinth (*Eichhornia crassipes*) (Perna and Burrows 2005; Tait and Veitch 2007). The development of impenetrable surface growth of Para and Hymenachne grass is also contributing to further water quality deterioration via organic loading and dissolved oxygen deprivation (Perna and Burrows 2005). Elevated water levels are also implicated in the waterlogging and associated die back of riparian trees on some lagoon systems (Tait and Veitch 2007). Rich submerged and floating aquatic macrophyte beds are best developed in off-channel waterholes isolated from the main Barratta Creek complex. This appears to be largely because waterholes connected to the main channels receive turbid tail water with resultant chronic turbidity issues, whereas the main supply for the isolated holes may be comparatively clear groundwater (Environment Australia 2001).

As the area of global cropland increases (Scanlon et al. 2007), particularly land converted to irrigated agriculture, the environmental consequences of off-site pesticide movement on aquatic ecosystem health have also emerged as a prominent natural resource management issue (Clark et al. 1999; Graymore et al. 2001). A substantial quantity of local research is available from the lower Burdekin cane growing region addressing pesticide dynamics at a range of scales extending from paddock-scale and 'end of farm', through sporadic sampling of floodplain irrigation and drainage systems, as well as limited insights into pesticide-groundwater interactions (Bauld et al. 1996; Hunter et al. 1998; Müller et al. 2000). The lower Burdekin catchment pesticide surfacewater quality monitoring dataset (Davis et al. 2008, 2012, in press; Smith et al. 2012) is probably the most extensive for any river system in northern Queensland. The Australian sugar industry is heavily reliant on herbicides in pest management strategies (Johnson and Ebert 2000), particularly for the control of weeds in ratoon crops, a reliance that is reflected in water quality monitoring in the surrounding floodplain wetlands. A range of herbicide residues, predominantly associated with the sugarcane industry, have been detected in many lower Burdekin waterways during both low flow and flood event conditions (Davis et al. 2008, 2012, in press; Smith et al. 2012). While over 20 different herbicides have been detected, detections are dominated by diuron, atrazine, hexazinone and ametryn, four of the five herbicides whose usage was recently regulated under the Great Barrier Reef Protection Amendment Bill (2009). Sampling of wet season flood plumes from local catchments in 2007 detected levels of two herbicides, atrazine and diuron, being transported to offshore marine environments, although concentrations ($\leq$0.08 μg/L) were below current water quality ecosystem protection guidelines (Davis et al. 2012). Annual loads of measured herbicides being discharged into Bowling Green Bay via the Haughton River and Barratta Creek complex are consistently in the order of 250+ kilograms (Davis et al. 2008, 2012). Atrazine, it's degradate desethylatrazine, and diuron contribute approximately 90 % of the annual 'end-of-catchment' herbicide load, with early 'first-flush' events accounting for the majority of herbicide loads leaving the catchment. Diuron has been the only herbicide residue detected in Bowling Green Bay estuarine or inshore marine benthic sediments, being found at two-thirds of sample sites (Davis et al. 2012). Detectable sediment diuron concentrations ranged between 0.12 μg/kg in Bowling Green Bay up to 1.62 μg/kg in the Barratta Creek estuary.

One of the more significant outcomes emerging from sub-catchment monitoring results are the considerable and persistent pesticide levels found in some lower Burdekin waterways, such as the Barratta Creek system during the low flow (dry-season) conditions that prevail throughout much of the year. While several lower Burdekin waterways exhibited frequent low concentrations of pesticides throughout much of the year under dry season conditions, the freshwater reaches of the Barratta Creek system stood out with relatively elevated pesticide levels documented over the long term during the dry season. These concentrations were often consistently above available ecosystem protection guidelines for several herbicides (Davis et al. 2008, in press). Dry season pesticide concentrations throughout Barratta Creek are often more than an order of magnitude higher than concentrations documented for similar caneland dominated systems in the GBR (Davis et al. 2008).

The highest in-stream herbicide concentrations in Barratta Creek are typically detected under dry-season conditions (when crops in the BHWSS are being irrigated) and the system receives large volumes of tailwater from farms during this period. The soils of the Barratta Creek catchment are predominantly sodic with very low permeability (Nelson 2001), so tailwater run-off is quite substantial on many farms in this area (Dight 2009; Thorburn et al. 2011). As a result, the dry-season flow regime of the middle and lower reaches of the Barratta Creek drainage complex, a previously intermittent system, is now dominated by local irrigation tailwater input. The resultant diffuse input of undiluted farm tailwater into Barratta Creek appears to produce a continual background 'irrigation tailwater fingerprint' for pesticides during low flow conditions (Davis et al. in press).

Recent pesticide monitoring was extended from freshwater reaches in Barratta Creek (i.e., Davis et al. 2008, 2012) through to estuarine reaches to assess pesticide dynamics within the Bowling Green Bay Ramsar site. Initial results from polar passive samplers in the form of Chemcatchers® highlight a progressive attenuation of pesticide concentrations from the upper catchment monitoring sites in the upper BHWSS through to downstream estuarine reaches in the Bowling Green Bay Ramsar site (Fig. 4, Site 4). Passive pesticide samplers also expanded on the range of pesticides detected compared to traditional discrete 'grab' sampling, with 34 different pesticides (again predominantly herbicides) detected across the monitored reaches.

This reduction through the catchment in pesticide concentration is likely due to several factors such as increased groundwater inputs in middle reaches and additional dilution by tidal flushing in the most downstream reaches of the Barratta estuary. While considerable downstream dilution is apparent, concentrations of several herbicides such as diuron and atrazine still consistently exceed available ANZECC and ARMCANZ (2000) ecosystem health guidelines for long periods of time during the year. Currently a number of challenges are faced in the meaningful interpretation of these sort of monitoring program results including poor knowledge of the possible impacts of sub-lethal pesticide effects over chronic duration, and the cumulative effects of ecosystem exposure to the complex 'cocktail' of compounds typically present at any one time in receiving environments (Eggen et al. 2004; Davis et al. 2008). Results do, however, highlight Barratta Creek, one of the key watersheds draining into an internationally listed Ramsar wetland, as one of the creek systems most polluted by agrichemicals within Australia. Results also indicate that freshwater and to some extent estuarine, ecosystems (often with their own high ecological values) in close proximity to agricultural land, rather than marine habitats, face the greatest herbicide-associated risks of any Great Barrier Reef receiving environment.

These broader 'downstream effects' of irrigation area development when natural drainage lines are used to receive tailwater are not limited to the lower Burdekin in northern Australia. The Ord River Irrigation Area (located in Western Australia) was designed as a similar flow through system, where water extracted from the Ord River Diversion Dam flows through the irrigated area via a series of canals and furrows, is used on farm, before draining back into the Ord River below the irrigation area (Petheram et al. 2008b). This tailwater drainage has very similar water quality composition to that evident in the lower Burdekin, and has been implicated in a range of deleterious effects on receiving aquatic ecosystems such as increased eutrophication, altered

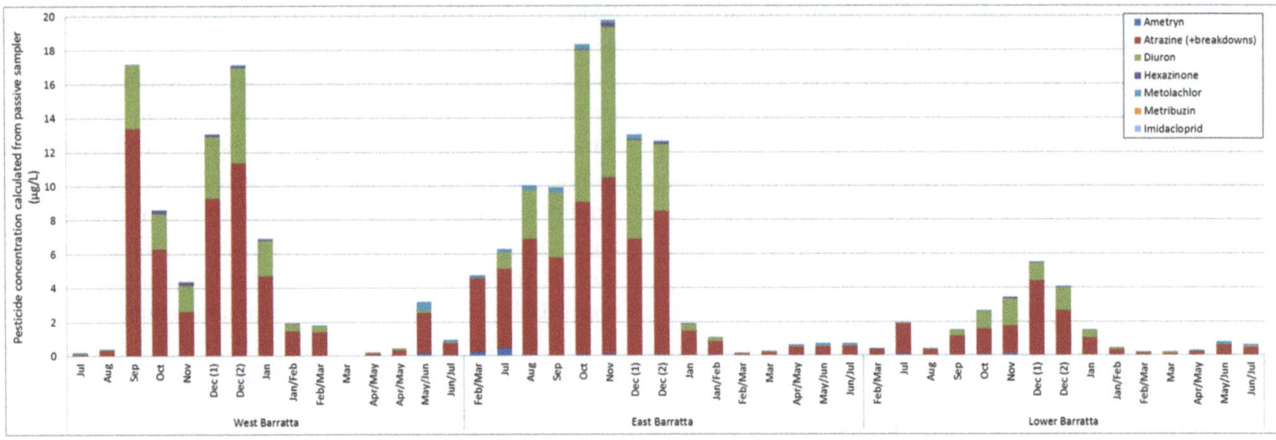

Fig. 4 Average monthly pesticide concentrations (µg/L) derived from passive pesticide samplers along the Barratta Creek catchment for the period July 2011–June 2012

community composition and fish kills (Oliver and Kookana 2006a, b; Petheram et al. 2008b).

Remediation and Management

Legislation

There is a range of relevant Commonwealth legislation pertaining to the conservation and protection of the Bowling Green Bay Ramsar site and Barratta Creek; including the EPBC Act 1999 (Commonwealth); the Great Barrier Reef Marine Park Act 1975 (Commonwealth); Coastal Protection and Management Act 1995; and Environmental Protection Act 1994. Most of these instruments relate to management of the site itself (or future developments). Bowling Green Bay in many ways, however, epitomises the global challenge of management of 'upstream effects' due to hydrology on the integrity of high value conservation areas, particularly deltaic environments in the lower sections of watersheds (see Pringle 2001). Few of these legislative acts have the capacity to exert direct management change on the more diffuse hydrological and water quality issues that have been identified as posing some of the greatest threats to the integrity of the site. A number of planning documents are directed more at the effects of diffuse agriculture, at least in their underlying intent.

The Ramsar site forms part of the Burdekin Dry Tropics region, which is supported by the North Queensland Dry Tropics board (NQDT), a natural resource management group established by the Australian Government tasked with developing and implementing an accredited regional Burdekin Dry Tropics Natural Resource Management Plan (Burdekin Dry Tropics Board 2005). This regional natural resource management (NRM) plan aims to sustainably manage regional natural resources in conjunction with the Queensland and Australian Governments, the community, Traditional Owner groups, sub-regional groups and advisory committees of the region on behalf of the Burdekin catchment community. The NQDT NRM plan has a number of longer-term aspirational targets for the BHWSS area such as; by 2050 ensuring a sustainable landscape integrating conservation, primary production and community aspirations; cooperative management and protection of wetland systems of high environmental value and importance to the community; all lower Burdekin water bodies having ambient water quality that maximises environmental productivity, diversity and ecological processes. Short term NRM goals include improvement in water quality (suspended sediments, nutrients, pesticides) at a sub-catchment and catchment level based on 2005 levels and establishment and implementation of environmental flows by 2015.

A major vehicle for delivery of water quality improvements under the NRM plan, however, has been the development of a Water Quality Improvement Plan (WQIP) under the Coastal Catchments Initiative (CCI). The CCI is the Australian Government's primary mechanism for meeting its commitments under the United Nations Environment Programme Global Action Program for the Protection of the Marine Environment from Land-based Activities. The CCI is conceptually similar to the United State's Total Maximum Daily Load Program. The CCI in a Queensland context was mainstreamed in its explicit application to protect water quality of the Great Barrier Reef, under the overarching direction of the Reef Water Quality Protection Plan or 'Reef Plan' (Queensland Department of the Premier and Cabinet 2003). As a result the WQIP has similar visions for management of freshwater and estuarine ecosystems as the local NRM plan; freshwater, estuarine and marine ecosystems are ecologically healthy, productive, resilient, enjoyed and valued; terrestrial ecosystems sustainably managed for good water quality; surface and ground water sustainably managed for good water quality; less pollutants in surface and ground water (sediment, nutrients and pesticides); hydrological conditions support functional aquatic systems, appropriate water levels and productive uses; and rivers and wetlands are in good ecological condition and function. All of the explicit 'resource condition targets' of the WQIP outlining requisite percentage reductions in nutrients and herbicides, however, relate solely to end-of-catchment water quality entering the Great Barrier Reef (GBR) (Dight 2009).

The Australian and Queensland Governments released the Reef Plan in 2003 (Queensland Department of the Premier and Cabinet 2003) with the aim of halting and reversing the decline in water quality entering the GBR within 10 years, and strictly focuses on diffuse pollution from agriculture. In 2009, Reef Plan 2003 was revised and updated (Queensland Department of the Premier and Cabinet 2009), with better defined targets and actions. Some of the targets and actions directly relate to catchment wetlands, but the language highlights some of the enduring perceptions of the perceived value of wetlands from a catchment water quality perspective, particularly when concerning the GBR. If there can be any criticisms of the GBR Reef Plan, it is the implied value of coastal wetlands as a filter for poor quality water entering the GBR lagoon (Brodie et al. 2012). Reef Plan and associated supporting statements certainly recognise that the health of freshwater ecosystems in the GBR catchment area is impaired by agricultural land use, hydrological change, riparian degradation and weed infestation (Queensland Department of the Premier and Cabinet 2009). The Reef Plan has two key objectives. The first aims to reduce the amount of pollutants entering the waterways and the Reef, while the second promotes protection and improvement of 'natural filters that capture these pollutants prior to entering the Reef' and to 'rehabilitate and conserve areas of the Reef catchment that have a role in removing waterborne pollutants'. Reef Plan ascribes the value of these coastal floodplain wetlands as primarily important for improving the quality of water leaving the land and reaching waters of the GBR, rather than constituting high value ecosystems in their own right, that play key roles in many ecosystem processes (beyond water quality buffering) that underpin the functional integrity of the GBR.

Despite this somewhat debatable appreciation for the value of coastal wetlands, several of the key programs underpinning the roll-out of Reef Plan will almost certainly have indirect benefits to freshwater and estuarine systems like the lower Burdekin floodplain. Reef Plan catalysed the GBR Coastal Wetlands Protection Programme (GBRCWPP), announced by the Australian Government in 2003 with the aim to develop and implement measures for the long term conservation and management of priority wetlands in the Great Barrier Reef Catchment. Outputs include management investment strategies for wetlands such as Barratta Creek (Tait and Veitch 2007). Program outputs have had positive impacts on many localised issues such as weed, vegetation management and fish passage constraints (Australian Government 2007), but have limited capacity to address the complex, broad-scale and multijurisdictional challenges of water resource management on the lower Burdekin floodplain.

The Federal Government also implemented the Reef Rescue initiative, an AU $200 million investment for on-ground works, monitoring, research and partnerships over 5 years (Australian Government 2007). This voluntary program's objective is to improve the water quality of the GBR lagoon by increasing the adoption of land management practices that reduce the run-off of nutrients, pesticides and sediments from agricultural land. Reef Rescue was developed following an initiative by the Reef Water Quality Partnership, including agricultural industries, regional NRM bodies and natural resource and environment managers. It built on key activities conducted under Reef Plan, including the development of local and regional Water Quality Improvement Plans (e.g. Dight 2009; Drewry et al. 2008; Kroon 2009 and references therein), and the Nutrient Management Zones process (Brodie 2007). Since 2008, Reef Rescue funded many on-ground land management projects across the GBR catchment area, mainly in the sugarcane and grazing industries, but also in dairy farming and horticulture. Projects include the introduction of new farming practices; fencing along streams for cattle management with off-stream watering points; machinery modifications including harvesters, fertiliser and pesticide application gear; and cultivation and tillage equipment and practices (Brodie et al. 2012).

While much WQ research in Queensland has focussed on the GBR marine environment, recent research has highlighted that freshwater and estuarine ecosystems likely face the highest risks from WQ pollution (Davis et al. 2008, in press). Focus on the health of GBR ecosystems has in many respects over-ridden focus on the often more pressing NRM issues facing many northern Queensland coastal wetlands. That said, with attenuation of water quality concentrations with progression through the freshwater to marine continuum, the freshwater-estuarine habitats will very likely derive the greatest water quality benefits that hopefully emerge from Reef Plan.

Future Directions for Lower Burdekin Floodplain Management

Regardless of the range of different programs and plans relevant to conservation of the lower Burdekin floodplain, there is almost certainly a need for a carefully considered reappraisal of the ultimate goals of floodplain management and restoration. Even ignoring the lack of baseline data, serious doubts have to be raised that pre-development conditions, functions, and processes on the lower Burdekin floodplain are even remotely reversible, a scenario facing many similar habitats around the world (Zweig and Kitchens 2010). The recent calls for more progressive paradigms, such as enhancing ecosystem services and increasing resilience to future change, are exciting new directions that hold considerable potential for profoundly altered environments like the lower Burdekin floodplain (Zweig and Kitchens 2010; Suding 2011). The term "novel ecosystem" has recently emerged in the field of restoration and management (Chapin and Starfield 1997) in recognition of the response of an ecosystem to current and future climatic events (Seastedt et al. 2008). A novel ecosystem is simply an assemblage of species and environmental conditions that have never before existed in a landscape, caused by climate, human activities, or stochastic events (Zweig and Kitchens 2010). Acknowledging the novelty of a system helps redefine restoration targets in terms of realistic and attainable goals. The value of simplicity in restoration targets cannot be overstated, particularly in novel systems where landscape responses to change are unpredictable (Zweig and Kitchens 2010).

Like many coastal wetlands, there is increasing recognition that the current lower Burdekin floodplain is vastly different from the pre-drainage system (Burrows and Butler 2007). The aquatic ecosystems of Barratta Creek and the Bowling Green Bay wetland complex are considered to be highly disturbed, notwithstanding their significant ecological value. While there may be community interest in 'returning' these wetlands and riparian vegetation to their original state, this new condition may be irreversible. In some cases, such as Barratta Creek, the new ecological communities may now actually be dependent upon the perennial tail water flows. Although this change from a seasonal to a perennial stream is a very significant departure from its natural condition, Barratta Creek is considered to be one of the most important, healthy and productive creek systems in the Burdekin region and includes many large, permanent wetlands and long lengths of perennially-flowing creek where there are no major fish passage barriers. The idea that although it has departed significantly from its natural condition, but retains (and has probably enhanced) some functional values, is further discussed in Burrows and Butler (2007). Appropriate management questions should not focus on how to return the system back to its pre-development flow regime, but what is the minimum level of water and rate of flow required to prevent the system suffering the ill effects of stagnation and other water quality problems that would reduce its health and productivity. Currently, the tailwater coming off the farms, although not perfect in its quality, is essentially provided as a free environmental flow. Burrows and Butler (2007) suggest even better result would be obtained by replacing tailwater with environmental allocations direct from the Burdekin Falls Dam as this water contains less contaminants, especially farm contaminants. Barratta Creek was largely neglected in the recent Burdekin basin Water Resource Plan (Brizga et al. 2005). The Burdekin River Resource Operations plan (which outlines the practical management of the Water Resource Plan) only briefly mentioned Barratta Creek with the desired aim to 'ensure that there are no further impacts on the natural creek flows in the Barratta Creek system'. All Queensland water resource plans have the pre-determined generic goal of mimicking natural flow regimes, not allowing for situations where non-natural flow regimes would provide the greatest environmental benefit. Similarly, there is an overt focus in the current WRP process on managing water quantity, with minimal consideration of associated water quality. Management of major issues such as invasive weeds and inappropriate fire regimes may also demand innovative or counter-intuitive management approaches. For example, introduction of strategic and controlled pastoral grazing in riparian and wetland areas, traditionally regarded as an undesirable land management practice, may offer some of the most pragmatic and cost-effective management options for invasive weeds and associated fire risks in lower Burdekin remnant wetlands (Tait 2011).

Shifts in perception regarding appropriate management of water resources in Australia

To fully appreciate the contemporary nature of environmental and natural resource management processes, recognition of antecedent societal and policy drivers provides much of the required context. There is little doubt that the current objectives of tropical river management and those of water resource management in Australia generally, have undergone a major paradigm shift from a historically narrow resource development focus to include consideration of a broader range of social, ecological and economic values and perspectives (Hussey and Dovers 2006; Jackson et al. 2008). The concept of the BHWSS for example, first emerged in the 1950s, an era of pre-occupation with harnessing Australia's 'wasted' water resources, which saw Australian states and territories, aided by Commonwealth funding, promote national development and agricultural settlement through public provision of irrigation infrastructure and settlement schemes, culminating in massive dam-building exercises through the 1950–1970s. In remote areas, in particular, water management was predominantly a matter of expanding supply, with demand management and environmental considerations lagging far behind (see Hussey and Dovers 2006; Petheram et al. 2008b).

Australia was then prominent in the global trend toward integration of water with land and vegetation management in the form of 'integrated' or 'total catchment management', with emerging salinity issues a key catalyst of this NRM shift in the 1980s (Hussey and Dovers 2006). This catchment scale focus has subsequently evolved to a point where major resource management programs are delivered largely through regional, often catchment-defined, organizations (exemplified in the lower Burdekin by the NQ Dry Tropics regional body). The ecologically sustainable development era from the early 1990s in Australia grappled with the emerging challenge of integrating environmental, social and economic dimensions in natural resource management. The growing societal awareness of the environmental impacts of irrigation developments and contribution made by healthy rivers to human well-being and cultural identity has been particularly influential in this change (Jackson et al. 2008; Petheram et al. 2008b). Interestingly, however, rather than resource management experience or environmental concern, economic policy has also been identified as providing the biggest policy shifts of all (Hussey and Dovers 2006). Recent water reform frameworks such as the Council of Australian Governments (comprising the heads of Australian State, Territory and Commonwealth governments) and the National Water Initiative (Council of Australian Governments 2004), include provision for environmental flows (allowing for quality and seasonal variability), as well as volume, the establishment and trading of water rights, intergovernmental coordination and institutional development, and regional and catchment scale planning water trading, environmental restoration, indigenous interests, and regional development in water policy (Connell et al. 2005). While tensions inevitably arise between the policy agenda of different stakeholders, particularly in heavily pressured or over-allocated water systems, environmental and long-term economic viability of new water resource developments clearly assume much greater influence in the contemporary Australian water policy forum (Hussey and Dovers 2006).

These shifting trajectories in broader scale environmental policy are also reflected in a shift in the perception of the coastal wetlands of Bowling Green Bay at societal and government levels. The perceived value of the area has evolved from a convenient mechanism for transporting and draining water in irrigation schemes, to an appreciation of ecosystem services provided by wetlands (Queensland Department of the Premier and Cabinet 2003), to environments worth conserving for their own intrinsic values (Kelly and Lee Long 2012). There is also an increasing realisation that these systems have undergone (and likely continue to undergo) a major phase shift in their function (Burrows and Butler 2007). Regional NRM Plans now apparently accommodate these realisations, with management targets for the lower Burdekin emphasising aspirations such as ensuring system resilience and functional integrity, rather than aspiring to restore pristine conditions (Burdekin Dry Tropics Board 2005).

Retrospective appraisals of the viability of large scale irrigation schemes (i.e. Petheram et al. 2008b) highlight the fundamental need to understand the hydrogeology of the whole system and emphasize that water quantity should not be the only parameter of interest in irrigation scheme planning. The timing of release and water quality must also be properly considered, especially in areas where water quality plays a large role in ecosystem functioning. The impacts of diffuse pollutants on receiving ecosystems is one that has garnered scientific and policy attention in many northern hemisphere regions for more than 50 years (Diaz and Rosenberg 2008). There is little doubt, however, that the issue of 'downstream effects' of diffuse agricultural pollution in the Australian environment received minimal attention until relatively recently. Even with regard to the GBR, while concerns were raised as early as the 1970s (Bennell 1979), the issue of catchment-derived pollution, a threat that ultimately produced Reef Plan, only received major scientific scrutiny in the early-mid 1990s (Yellowlees 1991; Bell and Gabric 1991; Hunter et al. 1996), sometime after the development-centric national ethos that prompted agricultural expansion in earlier decades. The much greater research investment in the ecohydrological repercussions of prospective irrigation expansion in northern Australia of late (i.e. Erskine et al. 2003; Jackson et al. 2008; Petheram et al. 2008c; Chan et al. 2012), suggests policy makers are currently assessing both costs and benefits of development and wetland conservation in determining the long-term consequences of tropical river developments.

Regardless of the considerable change in community attitudes towards maintenance of environmental values since the inception of established schemes such as the BHWSS, the enduring legacy of earlier perceptions are going to provide ongoing management issues for lower Burdekin coastal wetlands, as well as the viability of irrigation areas themselves. Farmers on the floodplain, most of whom have invested heavily in purchasing and developing farms, face uncertain futures for their enterprises, as well as negative public perceptions stemming from the environmental impacts of farming operations, all resulting from what now appears a less than optimally designed irrigation scheme. The key challenges now lie in how to sustainably manage such pressured environments in light of significant current and future pressures and appropriately aligning management goals and expectations. Continuing neglect is probably the greatest future threat to the region. It must be recognised that these wetland systems, when basically utilised as drainage outlets for upstream agriculture, require careful management to continue to function in a beneficial manner.

Acknowledgements The Australian and Queensland Governments and NQ Dry Tropics are acknowledged for funding of sub-catchment and catchment monitoring, under programs including The Coastal Catchments Initiative, Water Quality Improvement Plans and Reef Plan and support from the Reef Rescue Initiative. Data for the upper Barratta Creek site were supplied from the Queensland State Government's Great Barrier Reef Loads Monitoring program. This research was also supported by the Marine and Tropical Sciences Research Facility, implemented in North Queensland by the Reef and Rainforest Research Centre Ltd.

References

Access Economics (2007) Measuring the economic and financial value of the Great Barrier Reef. Report, Access Economics PTY Ltd. for Great Barrier Reef Marine Park Authority, 2005–2006. Online at: http://www.gbrmpa.gov.au/corp_site/about_us/documents/economic_values_report.pdf

Andutta FP, Ridd PV, Wolanski E (2011) Dynamics of hypersaline coastal waters in the Great Barrier Reef. Estuar Coast Shelf Sci 94:299–305

Anon (2010) CANEGROWERS annual report 2009–2010. Queensland Cane Growers Organisation Limited, Brisbane

ANZECC, ARMCANZ (2000) Australian and New Zealand guidelines for fresh and marine water quality. Australian and New Zealand Environmental and Conservation Council, Agriculture and Resource Management Council of Australia and New Zealand, Canberra

Arthington AH, Marshall J, Rayment G, Hunter H, Bunn S (1997) Potential impacts of sugarcane production on the riparian and freshwater environment. In: Keating BA, Wilson JR (eds) Intensive sugar cane production: meeting the challenges beyond 2000. CAB International, Wallingford, pp 403–421

Australian Government (2007) Labour's reef rescue plan. Election 2007 policydocument. http://www.alp.org.au/download/071028___labors_reef_rescue_plan.pdf

Bauld J, Leach LL, Sandstrom MW (1996) Abstract: impact of land use on groundwater quality in the Burdekin River delta and the Burdekin River Irrigation Area. In: Hunter HM, Eyles AG, Rayment GE (eds) Downstream effects of land use. Department of Natural Resources, Brisbane, pp 195–196

Beare S, Bell R, Blias A, Gooday P, Heaney A, Hooper S, Langenkamp D, Love G, Levantis C, Mues C, Qureshi E, Riley C (2003) Natural resource management in the Burdekin River catchment. Integrated assessment of resource management at the catchment scale. A case study. Australian Bureau of Agricultural and Resource Economics. ABARE report 03.18

Beasley I, Roberston KM, Arnold P (2005) Description of a new dolphin, the Australian snubfin dolphin *Orcaella heinsohni* sp. n. (Cetacea, Delphinidae). Mar Mamm Sci 21:365–400

Bell P, Gabric AJ (1991) Must GBR pollution become chronic before management reacts. Search 22:117–119

Bennell N (1979) Coastal land development – a shadow over the Great Barrier Reef. Aust Parks Recreat November 1979: 41–45

Bennet B (2012) Rising water tables in the Burdekin Groundwater Management Area Part A – an estimate of the impacts of irrigation and water distribution activities on groundwater levels, 2002–2010. State of Queensland (Department of Natural Resources and Mines), Brisbane

Bristow KL, Charlesworth PB, Lowis B, Laidlow G, Gilbey P (2001) The lower Burdekin initiative: an industry/science partnership to facilitate improved water management. In: Proceedings of the ANCID conference, Bunbury, Western Australia, 29th July – 1 August 2001

Bristow KL, Kellett B, Charlesworth PB (2004) Northern Australia irrigation futures. In: Proceedings of the Australian national committee on irrigation and drainage, national conference and exhibition, Tanunda, Barossa Valley, South Australia, 10–13 October 2004

Brizga SO, Kapitzke R, Brodie J, Burrows D, Butler B, Cappo M, Dowe J, Lait R, Pearson RG, Pusey B, Werren GL (2005) Burdekin basin draft water resource plan. Department of Natural Resources, Mines and Water, Brisbane

Brodie J (2007) Nutrient management zones in the Great Barrier Reef catchment: a decision system for zone selection Nutrient management zones (NMZ) technical report. ACTFR report no. 06–07

Brodie JE, Kroon FJ, Schaffelke B, Wolanski EC, Lewis SE, Devlin MJ, Bohnet I, Bainbridge ZT, Waterhouse J, Davis AM (2012) Terrestrial pollutant runoff to the Great Barrier Reef: An update of issues, priorities and management responses. Mar Pollut Bull 65:81–100

Bruinsma C (2001) Queensland coastal wetland resources: cape tribulation to bowling green bay, information series QI01064. Department of Primary Industries, Brisbane

Bruinsma C, Danaher K, Treloar P, Sheppard R (1999) Coastal wetland resources of the Bowen region: Cape Upstart to Gloucester Island. Department of Primary Industries, Brisbane

Burdekin Dry Tropics Board (2005) Burdekin dry tropics natural resource management plan (2005–2010). http://www.burdekin-drytropics.org.au/

Bureau of Meteorology (2009) www.bom.gov.au. Accessed Sept 2012

Burrows DW, Butler B (2007) Determining end-point goals and effective strategies for rehabilitation of coastal wetlands: examples from the Burdekin River, North Queensland. In: Wilson AL, Dehaan RL, Watts RJ, Page KJ, Bowmer KH, Curtis A (eds) Proceedings of the 5th Australian stream management conference, Albury, pp 49–54

Burrows DW, Faithful JF (2003) From blue to brown: persistently elevated turbidity resulting from damming of the tropical Burdekin River. In: Proceedings of the ninth international river regulation conference, Albury July 2003. Abstract

Chan TU, Hart BT, Kennard MJ, Pusey BJ, Shenton W, Douglas MM et al (2012) Bayesian network models for environmental flow decision making in the Daly River, Northern Territory, Australia. River Res Appl 28(3):283–301

Chapin FS, Starfield AM (1997) Time lags and novel ecosystems in response to transient climatic change in arctic Alaska. Clim Chang 35:449–461

Charlesworth PB, Bristow KL, Chinn C, Ham G (2002) Healthy crop and healthy groundwater – sugarcane in the Burdekin delta – contributed paper to the IAA conference, 20 to 24 May 2002, Sydney

Christian CS, Paterson SJ, Perry RA, Slatyer RO, Stewart GA, Traves DM (1953) Survey of the Townsville – Bowen Region, North Queensland, 1950, Land research series no. 2, Commonwealth Scientific and Industrial Research Organization, Australia, Melbourne

Clark GM, Goolsby DA, Battaglin WA (1999) Seasonal and annual load of herbicides from the Mississippi River Basin to the Gulf of Mexico. Environ Sci Technol 33:981–986

Connell D, Dovers S, Grafton Q (2005) A critical analysis of the national water initiative. Australas J Nat Resour Law Policy 10:81–107

Cook PG, Lamontagne S, Stieglitz T, Cranswick R, Hancock G (2011) A re-evaluation of groundwater discharge from the Burdekin Floodplain aquifer using geochemical tracers. NCGRT report, 105 p

Council of Australian Governments (2004) Intergovernmental agreement on a national water initiative. Council of Australian Governments, Canberra. Retrieved from http://www.coag.gov.au/meetings/250604/iga_national_water_initiative.pdf

Credlin BL (1979) Water resources of the Burdekin Basin and their development. Queensland Water Resources Commission, Brisbane

Dalla Pozza R (2005) A Holocene sand budget for the seasonally wet tropics region of north Queensland. PhD thesis, James Cook University

Danaher K (1995) Coastal wetlands resources investigation of the Burdekin Delta for declaration as fisheries reserves: report to ocean rescue 2000. Department of Primary Industries, Brisbane

Davis A, Lewis S, Bainbridge Z, Brodie J, Shannon E (2008) Pesticide residues in waterways of the lower Burdekin region: challenges in ecotoxicological interpretation of monitoring data. Australas J Ecotoxicol 14:89–108

Davis AM, Lewis SE, Bainbridge ZT, Glendenning L, Turner RDR, Brodie JE (2012) Dynamics of herbicide transport and partitioning under event flow conditions in the lower Burdekin region, Australia. Mar Pollut Bull 65:182–193

Davis AM, Thorburn PJ, Lewis SE, Bainbridge ZT, Attard SJ, Milla R, Brodie J (in press) Herbicide run-off dynamics of furrow irrigated sugarcane farms and associated drainage systems on the Burdekin River Floodplain, north-eastern Australia. Agric Ecosyst Environ: dx.doi.org/10.1016/j.agee.2011.06.019

Department of Environment, Water, Heritage and the Arts (2008) National framework and guidance for describing the ecological character of Australia's Ramsar Wetlands. Module 2 of the national guidelines for Ramsar Wetlands – implementing the Ramsar convention in Australia. Australian Government Department of the Environment, Water, Heritage and the Arts, Canberra

Department of Natural Resources (1999) Annual water statistics. Department of Natural Resources, Scientific Publishing, Brisbane

Diaz RJ, Rosenberg R (2008) Spreading dead zones and consequences for marine ecosystems. Science 321:926–929

Dight I (2009) Burdekin water quality improvement plan. NQ Dry Tropics, Townsville

DNR (1976) Ord irrigation project. An outline of its history, resources and progress. Western Australia. Department of National Resources/ Australian Government Publishing Service, Canberra

Drewry J, Higham W, Mitchell C, Rohde K, Masters B, Galea L (2008) Water quality improvement plan. Turning environmental values into water quality objectives and targets. Mackay Whitsunday Natural Resource Management Group, Mackay

Eggen RIL, Behra R, Escher BI, Schweigert N, Burkhart-Holm P (2004) Challenges in ecotoxicology. Environ Sci Technol 38:58A–64A

Environment Australia (2001) A directory of important wetlands in Australia, 3rd edn. Environment Australia, Canberra

Erskine WE, Begg GW, Jolly P, Georges A, O'Grady A, Eamus D, Rea N, Dostine P, Townsend S, Padovan A (2003) Recommended environmental water requirements for the Daly River, Northern Territory, based on ecological, hydrological and biological principles. Supervising scientist report 175 (National River health Program, Environmental Flow Initiative, Technical Report 4). Supervising Scientist, Darwin, Northern Territory

Fass T, Cook PG, Stieglitz T, Herczeg AL (2007) Development of saline groundwater through transpiration of seawater. Ground Water 45:703–710

Fielding CR, Trueman JD, Alexander J (2005) Sedimentology of the modern and Holocene Burdekin River Delta of North Queensland, Australia – controlled by river output, not by waves and tides. In: Giosan L, Bhattacharya JP (eds) River deltas – concepts, models, and examples, vol 83, SEPM. Special publication. SEPM, Tulsa, pp 467–496

Fielding CR, Trueman JD, Alexander J (2006) Holocene depositional history of the Burdekin River Delta of Northeastern Australia: a model for a low accumulation, Highstand Delta. J Sediment Res 76:411–428

Fleming PM, Gunn RH, Reece AM, McAlpine JR (1981) Burdekin project ecological study. August 1980. CSIRO and Department of National Development and Energy, Canberra

Frère CH, Hale PT, Porter L, Cockcroft VG, Dalebout ML (2008) Phylogenetic analysis of mtDNA sequences suggests revision of humpback dolphin (*Sousa* spp.) taxonomy is needed. Mar Freshw Res 59:259–268

Gardner EA, Coughlin KJ (1982) Physical factors determining soil suitability for irrigated crop production in the Burdekin-Elliot River area. Queensland Department of Primary Industries, Agricultural Chemistry Branch, Technical report no. 20. Queensland Department of Primary Industries, Brisbane

GBRMPA (2004) GBR traditional owners. http://www.reefed.edu.au/explorer/traditional_owners/index.html

Graymore M, Stagnitti F, Allinson G (2001) Impacts of atrazine in aquatic ecosystems. Environ Int 26:483–495

Griffiths DJ, Faithful JW (1996) Effects of the sediment load of a tropical north-Australian river on water column characteristics in the receiving impoundment. Arch Hydrobiol 113(Supp):147–157

Gunn RH (1981) Soils. In: Fleming PM (ed) Burdekin project ecological study. AGPS, Canberra, pp 137–142

Holden J, Hussey B, McDougall A, Mallon K, Shannon E (1998) Water check-improving irrigation efficiencies in the Queensland Sugar Industry. Proc Aust Soc Sugar Cane Technol 20:126–132

Hopley D (1970) The geomorphology of the Burdekin Delta, North Queensland. James Cook University of North Queensland, Department of Geography Monograph No:1

Hopley D (1978) Geographical studies of the Townsville area. Monograph series occasional paper no. II. Department of Geography, James Cook University, Townsville

Hunter HM, Eyles AG, Rayment GE (1996) Downstream effects of land use. Department of Natural Resources, Brisbane

Hunter HM, Hargreaves PA, Rayment GE (1998) Appendix II: pesticide residues in surface waters of the Burdekin River irrigation area. In: Hunter HM, Lukacs G (eds) Nutrient control in irrigation drainage systems using artificial wetlands, milestone report no. 4, LWRRDC Project QPI 26, 30th June 1998. Department of Natural Resources and James Cook University, Townsville

Hussey K, Dovers S (2006) Trajectories in Australian water policy. J Contemp Water Res Educ 135:36–50

International Union for Conservation of Nature (2009) Red list of threatened species – international union for conservation of nature. Website http://www.iucnredlist.org

Jackson S, Stoeckl N, Straton A, Stanley O (2008) The changing value of Australian tropical rivers. Geogr Res 46:275–290

Johnson AKL, Ebert SP (2000) Quantifying inputs of pesticides to the Great Barrier Reef Marine Park – a case study in the Herbert River catchment of north-east Queensland. Mar Pollut Bull 41:302–309

Johnson AKL, Ebert AP, Murray AE (2000) Land cover change and its environmental significance in the Herbert river catchment, North-east Queensland. Aust Geogr 31:75–86

Kelly KE, Lee Long WJ (2012) Ecological character description for the Bowling Green Bay Ramsar site. Report for the Department of Sustainability, Environment, Water, Population and Communities

Kemp JE, Lovatt RJ, Bahr JC et al (2007) Pre-clearing vegetation of the coastal lowlands of the Wet Tropics Bioregion, North Queensland. Cunninghamia 10:285–329

Kennard MJ, Pusey BJ, Olden JD, Mackay SJ, Stein JL, Marsh N (2010) Classification of natural flow regimes in Australia to support environmental flow management. Freshw Biol 55:171–193

Kerr JD (1994) Black snow and liquid gold: a history of the Burdekin Shire. Burdekin Shire Council, Ayr

King G (2003) Social atlas, 5th edn. Community and Cultural Services Department, Townsville City Council, Townsville

Kroon FJ (2009) Draft tully water quality improvement plan. Final report for Terrain NRM. CSIRO: Water for a Healthy Country National Research Flagship. ISSN: 1835-095X. ISBN: 978 0 643 09822 0

Lenahan M, Bristow K (2010) Understanding subsurface solute distributions and salinization mechanisms in a tropical coastal floodplain groundwater system. J Hydrol 390:131–142

Lewis SE, Davis A, Brodie J, Ledee E, Alewijnse M (2006) The spatial extent of delivery of terrestrial materials from the Burdekin Region in the Great Barrier Reef Lagoon, ACTFR Report Number 06/02. Australian Centre for Tropical Freshwater Research, James Cook University

Lewis SE, Sloss CR, Murray-Wallace CV, Woodroffe CD, Smithers SG (in press) Post-glacial sea-level changes around the Australian margin: a review. Quatern Sci Rev

Lough JM (2007) Tropical river flow and rainfall reconstructions from coral luminescence: Great Barrier Reef, Australia. Paleoceanography 22:PA2218

McMahon GA (2004) An integrated hydrogeological/hydrogeochemical approach to characterising groundwater zonations within a quaternary coastal deltaic aquifer: the Burdekin River Delta, North Queensland. PhD thesis, Queensland University of Technology, Brisbane

McMahon GA, Arunakumaren NJ, Bajracharya K (2000) Hydrogeological conceptualisation of the Burdekin River Delta. In: Proceedings of the 3rd international hydrology symposium of the institution of engineers, pp 208–213

Müller JF, Duquesne S, Ng J, Shaw GR, Krrishnamohan K, Manonmanii K, Hodge M, Eaglesham GK (2000) Pesticides in sediments from Queensland irrigation channels and drains. Mar Pollut Bull 41:294–301

Neil DT, Orpin AR, Ridd PV, Bofu Y (2002) Sediment yield and impacts from river catchment to the Great Barrier Reef lagoon. Mar Freshw Res 53:733–752

Nelson P (2001) Diagnosis and management of sodic soil under sugar cane. Cooperative Centre for Sustainable Sugar Production, Townsville

NLWRA (National Land & Water Resources Audit) (2002) Australians and natural resource management. Natural Heritage Trust, Canberra

Oliver DP, Kookana RS (2006a) Minimising off-site movement of contaminants in furrow irrigation using polyacrylamide (PAM). I. Pesticides. Aust J Soil Res 44:551–560

Oliver DP, Kookana RS (2006b) Minimising off-site movement of contaminants in furrow irrigation using polyacrylamide (PAM). II. Phosphorus, nitrogen, carbon, and sediment. Aust J Soil Res 44:561–567

Orpin AR, Brunskill GJ, Zagorskis I, Woolfe KJ (2004) Patterns of mixed siliciclastic–carbonate sedimentation adjacent to a large dry-tropics river on the central Great Barrier Reef shelf, Australia. Aust J Earth Sci 51:665–683

Perna C, Burrows D (2005) Improved dissolved oxygen status following removal of exotic weed mats in important fish habitat lagoons of the tropical Burdekin River floodplain, Australia. Mar Pollut Bull 51:138–148

Petheram C, Bristow KL, Nelson PN (2008a) Understanding and managing groundwater and salinity in a tropical conjunctive water use irrigation district. Agric Water Manag 95:1167–1179

Petheram C, Tickell S, O'Gara F, Bristow KL, Smith A, Jolly P (2008b) Analysis of the lower Burdekin, Ord and Katherine-Douglas-Daly irrigation areas: implications to future design and management of tropical irrigation. CRC for irrigation futures; CSIRO Land and Water Science, Report No. 05/08; No. 19/08, Townsville

Petheram C, McMahon TA, Peel MC (2008c) Flow characteristics of rivers in northern Australia: implications for development. J Hydrol 357:93–111

Pringle CM (2001) Hydrological connectivity and the management of biological reserves: a global perspective. Ecol Appl 11:981–998

Queensland Department of the Premier and Cabinet (2003) Reef water quality protection plan; for catchments adjacent to the Great Barrier Reef World Heritage Area. Queensland Department of Premier and Cabinet, Brisbane. http://www.reefplan.qld.gov.au/about/assets/reefplan-2003.pdf

Queensland Department of the Premier and Cabinet (2009) Reef water quality protection plan 2009: for the Great Barrier Reef World Heritage Area and adjacent catchments, Queensland Department of Premier and Cabinet, Brisbane. http://www.reefplan.qld.gov.au/resources/assets/reef-plan-2009.pdf

Qureshi ME, Wegener MK, Harrison SR, Bristow KL (2001) Economic evaluation of alternative irrigation systems for sugarcane in the Burdekin delta in north Queensland, Australia. In: Brebbia CA, Anagnostopoulos P, Katsifarakis K, Cheng AHD (eds) Water resources management. WIT Press, Boston, pp 47–57

Raine SR, Bakker D (1996) Increased furrow irrigation efficiency through better design and management of cane fields. Proc 18th Conf Aust Soc Sugar Cane Technol 119–125

Raine SR, Shannon EL (1996) Improving the efficiency and profitability of furrow irrigation for sugarcane production. Sugarcane: research towards efficient and sustainable production – sugar 2000 symposium, pp 211–212

Ramsar Information Sheet (1999) Information sheet on Ramsar Wetlands – Bowling Green Bay

Rayment GE (2002) Land use and surface water quality in sugar catchments. CRC for Sustainable Sugar Production, Indooroopilly

Roth CH, Lawson G, Cavanagh D (2002) Overview of key natural resource management issues in the Burdekin Catchment, with particular reference to water quality and salinity: Burdekin Catchment Condition Study Phase I. Burdekin Dry Tropics Board, Townsville

Scanlon BR, Jolly I, Sophocleous M, Zhang L (2007) Global impacts of conversions from natural to agricultural ecosystems on water resources: quantity versus quality. Water Resour Res 43:W03437. doi:10.1029/2006WR005486

Scheltinga DM, Heydon L (2005) Report on the condition of estuarine, coastal and marine resources of the Burdekin dry tropics region. Commissioned by the Burdekin Dry Tropics Board. Cooperative Research Centre for Coastal Zone, Estuary and Waterway Management

Seastedt TR, Hobbs RJ, Suding KN (2008) Management of novel ecosystems: are novel approaches required? Front Ecol Environ 6:547–553

Shaw RJ (1986) Predicted hydrology, salinity and sodicity changes under irrigation development in the lower Burdekin right bank, pages 138–1 to 138–16, in landscape, soil and water salinity. In: Proceedings of the Burdekin regional salinity Workshop, Ayr, April 1986. Queensland Department of Primary Industries, conference and workshop series QC86003

Shaw RJ (1988) Predicted deep drainage loss under dryland and irrigation managements for Burdekin soils. In: Rayment GE, Eldershaw VJ (eds) Soils of the Burdekin River Irrigation Area, workshop proceedings Ayr, August 1988

Shaw RJ, Thorburn PJ, McShane TJ, Maltby JE, Robson Ck (1983) The effectiveness of drainage in a region of variable aquifer hydraulic conductivity in the Lower Burdekin region, North Queensland. In: Smith RJ, Rixon AJ (eds) Proceedings of symposium, rural drainage in Northern Australia, Darling Downs Institute of Advanced Education, Toowoomba

Shaw RJ, Eldershaw VJ, Thompson WP, Smith GD (1984) Hydrology and salinity changes under irrigation – Lower Burdekin right bank (Fort site). Queensland Department of Primary Industries Bulletin QB 84004

Smith R, Middlebrook R, Turner R, Huggins R, Vardy S, Warne M (2012) Large-scale pesticide monitoring across Great Barrier Reef catchments – paddock to reef integrated monitoring, modelling and reporting program. Mar Pollut Bull 65:117–127

Suding KN (2011) Toward an era of restoration in ecology: successes, failures and opportunities ahead. Annu Rev Ecol Evol Syst 42:465–487

Tait J (2011) Guidelines for the use of grazing in the management of exotic pasture weeds in wetland and riparian habitats. Wetland Care Australia, Ballina

Tait J, Perna C (2001) Fish habitat management challenges on an intensively developed tropical floodplain. RipRap 19:14–21

Tait J, Veitch V (2007) Freshwater wetlands of the Barratta Creek: Management Investment Strategy. Report No. 07/34. Australian Centre for Tropical Freshwater Research, James Cook University, Townsville

Thompson WP (1977) Soils of the lower Burdekin river – Elliot River Area, North Queensland. Agricultural Chemistry Branch, Department of Primary Industries. Technical Report no 10

Thorburn PJ, Biggs JS, Attard SJ, Kemei J (2011) Environmental impacts of irrigated sugarcane production: nitrogen lost through run-off and leaching. Agric Ecosyst Environ 144:1–12

Tilley L, Chapman L (1997) Benchmarking crop water index for the Queensland sugar industry. Bureau of Sugar Experiment Stations, Brisbane

Wolanski E (2007) Estuarine ecohydrology. Elsevier, Amsterdam

Wolanski E, Ridd P (1990) Coastal trapping and mixing in tropical Australia. In: Cheng RT (ed) Long-term currents and residual circulation in estuaries and coastal seas. Springer, New York, pp 165–183

Yellowlees D (1991) Land use patterns and nutrient loading of the Great Barrier Reef region: proceedings of the workshop held at the James Cook University of North Queensland, 17–18 November, 1990. Sir George Fisher Centre for Tropical Marines Studies, James Cook University of North Queensland, Townsville

Zander KK, Garnett ST, Straton A (2010) Trade-offs between development, culture and conservation – willingness to pay for tropical river management among urban Australians. J Environ Manage 91:2519–2528

Zweig CL, Kitchens WM (2010) The semi-glades: the collision of restoration, social values and the ecosystem concept. Restor Ecol 18:138–142

The Hawkesbury Estuary from 1950 to 2050

Peter Collis

Abstract

This chapter describes the eutrophication of the Hawkesbury-Nepean River estuary that drains Sydney western suburbs. The project started in the 1950s by spot measurements of nutrient concentration and in the 1970s with a hydrodynamic and water quality study of the estuary. This was to gauge the effect of sewage treatment plants already proposed in the 1970s to deal with the growth of Sydney and their projected effects on the flow, chemistry and biology of the system by the year 2000. The study, using a mathematical model of proposed effluents, predicted an increase of nitrogen and phosphorus nutrients that was large enough to significantly degrade the water quality of the estuary. In particular it predicted a large increase in plankton, mainly blue/green algae, if planners did not make careful decisions concerning land use, urbanization and catchment development. A recent detailed study of water quality by Sydney Water has shown that the river system has indeed been degraded with the occasional occurrence of outbreaks of floating flowering plants (macrophytes) in the upper Nepean and blue/green algae in the saline/freshwater interface and "red tides" of toxic diatoms near the mouth of the estuary. This is despite upgrading of sewage treatment plants. More upgrades are needed especially in the area of South Creek. The biggest problem to address is ongoing urbanization and the resulting wet-weather inflow of degraded stormwater, sediment, nutrients and many other contaminants that reach the estuary. A public information program should be started and wet weather runoff should be treated preferably at the source of the runoff, somewhat mimicking the planned new developments at the Gold Coast Broadwater. An estuary, once urbanized will always be degraded, the extent of which however can be managed.

Keywords

Eutrophication • Nutrients • Nitrogen • Phosphorus • Chlorophyll a • Blue green algae • Wet weather runoff • Permeability • Ecology • Algal assemblage

P. Collis (✉)
School of Biological Science, Sydney University, Now at 223
Tinney Rd, Upper Caboolture, QLD 4510, Australia
e-mail: yarrabee@iprimus.com.au

E. Wolanski (ed.), *Estuaries of Australia in 2050 and Beyond*, Estuaries of the World,
DOI 10.1007/978-94-007-7019-5_14, © Springer Science+Business Media Dordrecht 2014

Box 1

Peter Collis studied the Hawkesbury River that drains Sydney growing western suburbs. He pioneered studies in the 1970s that suggested that that the sewage treatment plants already proposed in the 1970s to deal with the growth of Sydney, would increase nitrogen and phosphorus sufficiently to levels sufficient to significantly degrade the water quality of the estuary. In particular it predicted a large increase in plankton, mainly blue/green algae, if planners did not make careful decisions concerning land use, urbanization and catchment development. This warning was largely ignored and the river system is now degraded with the occasional occurrence of outbreaks of floating flowering plants (macrophytes) in the upper Nepean and blue/green algae in the saline/freshwater interface and "red tides" of toxic diatoms near the mouth of the estuary. This is despite upgrading of sewage treatment plants; more upgrades are clearly needed.

The biggest problem to address, and presently largely ignored, is ongoing urbanization, with the population predicted to further double in the next 20–30 years, and increased stormwater runoff. A public information program is needed. Urban stormwater runoff should be treated preferably at the source of the runoff, somewhat mimicking the planned new developments at the Gold Coast Broadwater. There is need to revisit the option first proposed in the 1970s, of diverting some of the sewage to Sydney ocean outfalls.

Introduction

Geographic Setting

The Hawkesbury-Nepean River (Fig. 1) forms a semi-circle around Sydney. Its catchment area is 21,400 km^2 and it is 270 km long in the north–south direction and 145 km long in the east–west direction. It is a drowned river valley, what geomorphologists call a coastal plain type estuary (Wolanski and Collis 1976).

Pre-European History

The catchment area was first inhabited by the Guringal, the Darkinyung and Dharuk people. It appears that the area was settled some 5,000 years ago. They left an unaltered pristine environment prior to white colonization when they lost their lands. The aboriginal living space was determined by the wealth of food resources. They left behind beautiful cave painting, rock art and work places. Their engravings in Hawkesbury sandstone should still be regarded as important historical, if not sacred, sites by modern Australians.

European History

The river drains Sydney's western suburbs. All along its reaches are complex ecological habits and people use it heavily for recreational purposes (Fig. 2). It attracts ten million tourists per year. Sydney, as the first Australian city, is an icon of Australia's most populous cities (Sydney Catchment Authority 2009). It may be argued that the Hawkesbury-Nepean River estuary is the most important river system in N.S.W., supplying drinking water for 97 % of Sydney's growing 4.6 million people and ground waters for the bottled water industry. The river helps generate 23 % of N.S.W state's electrical power. It produces over $1 billion dollars of agricultural produce (including fresh vegetables, flower, fruit, 89 % of the state's eggs, and 30 % of poultry meat. It was the early Australian colony's market garden and still supplies much of Sydney's fresh produce (Fig. 2). It is the state's 2nd largest oyster and prawn production, horse breeding and turf production. It is a source of gravel and sand extraction (80 % of State needs) and supports other intensive manufacturing and processing industries. Coal mining (19 mines in operation) is still extensive in the western and southern sections of the basin. In the saline zone are large oyster beds and a significant commercial fishery is in operation. It is also a valuable cultural asset as the site of the nation's early pioneers and it has large tracts of rugged bush land along the northern shore of the estuary, especially along the Colo River. Nearly one million people live in the catchment area and the river is also used to convey practically all the wastewater from the human population in the river catchment, which is at odds with its other uses in general. The catchment is being rapidly urbanized (Fig. 2).

Despite its importance, the river and its estuary were largely neglected by scientists for nearly 30 years after the 1970s when Eric Wolanski and the author undertook hydraulic, chemical and biological studies (Wolanski and Collis 1976). This makes some of the chapter somewhat

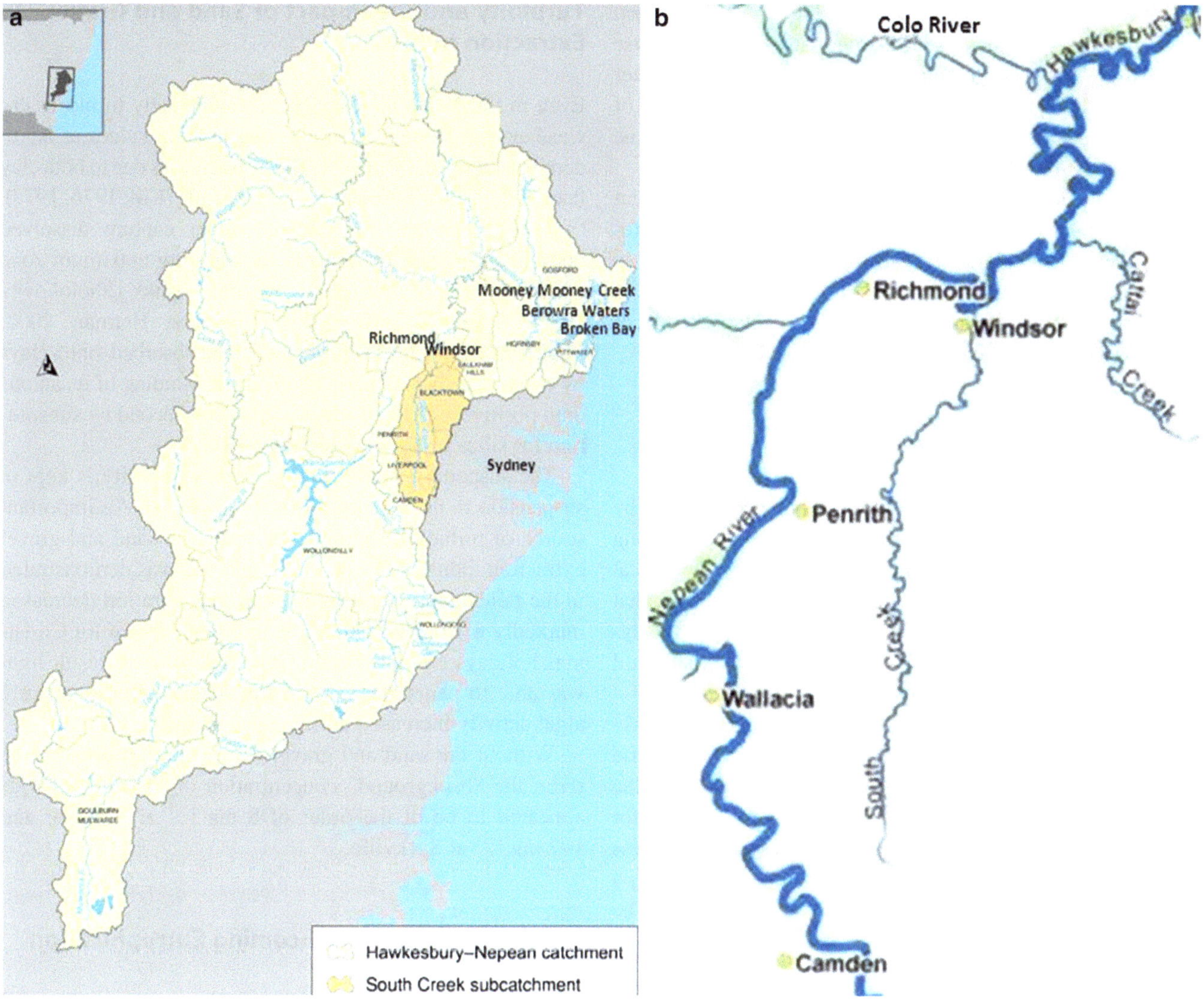

Fig. 1 (**a**) A location map of the Hawkesbury-Nepean River catchment. The tidal limit is located at about 140 km from the river mouth, near Windsor. (**b**) South Creek and Cattai Creek drain to the Hawkesbury Estuary the wastewater and wet weather runoff from Sydney western suburbs

historical, thus explaining the use of older references and also laboratory analytical methods have changed during this period. Nevertheless these data can be compared with more recent data to show historical trends. Surprisingly the modern scientific era for the Hawkesbury starts only 30 years later with the recent intensive surveys of water quality by Sydney Water in 2000–2010 (Sydney Catchment Authority 2009).

Consequence of Urbanization

Sewage Treatment Plants

Sydney water runs 15 plants discharging wastewater in the Hawkesbury-Nepean catchment, and thus ultimately in the Hawkesbury Estuary (Sydney Water 2012; M. Blackmore, personal communication, 2013). Their location is shown in Fig. 3. Sewage treatment plants are called by a number of names but basically do the same operation, namely biologically breaking down waste organic matter in sewage, which is good clean drinking water prior to various uses including mainly flushing toilets. Sydney Water has published the types of works and their operation (w.w.w. sydneywater.com.au). Sydney Water discharges wastewater at a rate of 114.47 Ml day^{-1} to the estuary, and this water contains 652.5 kg day^{-1} of total nitrogen and 18 kg day^{-1} of total phosphorus. At present 18 % of average flow in the Hawkesbury is due to discharge from sewage treatment plants (Bickford and Smith 2006).

The largest plants are at Penrith, West Camden, Quakers Hill, St. Marys and West Hornsby. The Nepean section of the river takes the most effluent but the saline intrusion also receives a large wastewater discharge; ultimately all the wastewater ends up in the estuary. To its credit Sydney

Water treats all the effluents to tertiary level using nitrogen removal type systems and at St. Marys it even uses reverse osmosis. This can produce water of drinking water standards. It also uses some of the effluent for irrigation of golf courses and some crops. In some plants it removes some of the phosphorus by chemical techniques.

It is considered that phosphorus is the prime element controlling the Hawkesbury estuarine eutrophication process, and this explains the efforts of Sydney Water to upgrade sewage treatment plants. However diffuse sources of nutrients (i.e. wet weather runoff and especially that from urbanized areas) are at present not addressed.

Flushing of the Estuary

During dry weather conditions, the turbulence induced by tidal motions is not enough to create complete vertical mixing in the lower estuary (Wolanski and Collis 1976). A vertical salinity gradient results that generates density currents that cause landward advection of ocean water. The tidally-averaged currents are seaward near the surface and landward near the bottom.

During dry weather the freshwater discharge is fairly steady as it is regulated by five storage dams. Under intense rainfall a large amount of freshwater is injected into the estuary; the estuary becomes highly stratified in salinity for a few days to a few weeks; a buoyant freshwater lens forms near the surface, it reaches the mouth, and it is separated by a sharp halocline from the saline waters at the bottom. The characteristic time of residence of wet weather flow was about 2 weeks.

Long periods (~ several months) of dry weather are common during which wet weather conditions do not occur. For these dry weather conditions a hydrodynamic model of the system (Wolanski and Collis 1976) divided the river into three regions, namely section 1 of unidirectional flow upstream of the tide head near Richmond, section 2 which is seaward of the tide head but affected by the tides and still freshwater about 140 km from the river mouth, and section 3, the saline intrusion zone of the estuary from Windsor to the mouth at Broken Bay.

In dry weather (i.e. for freshwater discharges of less than 10 $m^3\ s^{-1}$) tidal motions are the main agent responsible for dispersion and flushing contaminants in the estuary. The tide curve starts as a sinusoidal shape at the mouth at Broken Bay to a non sinusoidal shape at Windsor where a falling tide lasts 2 h longer than the tide at the mouth of the river and where the tide is 180° out-of-phase with the tide at the mouth. The residence time of contaminants discharged at the head of the estuary was estimated to be 14 days to the middle of the estuary.

Turbidity and the Impact of Sand and Gravel Extraction Industries

Even in the 1970s the estuary was moderately turbid in dry weather with suspended solids concentration values at 0.5 m depth of about 20–30 mg l^{-1}. The turbidity was due to both clay flocs and 'Yellow Substance' or 'Gelbstoff' (Kirk 1976, 1977).

These flocs are important to either capture dissolved nutrients or release nutrients in the turbidity maximum zone of the estuary or onto mangrove mud banks (Sholokovitz 1976; Wolanski 2007; Middelburg and Herman 2007; Fitzimons et al. 2011). Phosphates are absorbed onto clays by two mechanisms, firstly by chemical bonding of its anions onto positively charged edges of clay and second by substitution for silica on the clay (Scharpf 1973).

The suspended matter that generates turbidity is kept in suspension in the estuary by tidal turbulence. An important source of turbidity was believed to be the sand and gravel extraction industries in the river, and this was demonstrated in the field as the suspended solids concentration decreased markedly when these extraction industries closed for Christmas holidays (Wolanski and Collis 1976). As a result light was able to penetrate more in the water and consequently algal activity increased during these periods.

Without the sand and gravel extraction industries in the river, the "background" concentration of suspended solids appeared to be of the order of 8 mg l^{-1} at Windsor and 1–2 mg l^{-1} at Sackville.

Early Warnings of Forthcoming Eutrophication

The biochemical oxygen demand (BOD) was also calculated in the 1970s for the then projected sewage plants (Wolanski and Collis 1976). Assuming best practice, the studies indicated that in the Nepean River that the dilution would only be of 1:1, resulting in BOD in the range of 2.3–4.4 mg l^{-1} in the Nepean between Camden and Penrith.

On the effluents standards proposed during the 1970s, the model predicted an approximate 10-fold increase in the nutrient values in the upper estuary at that time. It predicted an increase in phosphorus to the order of 2 mg l^{-1}. This compared with historical phosphorus concentration values of 0.02–.03 mg l^{-1} in the 1950s (Rochford 1951). The Nepean River itself was already eutrophic in the 1970s. In the 1970s, phosphorus values were up to 0.11 mg l^{-1} (total) and 0.02 mg l^{-1} (dissolved) at Windsor and the two were highly correlated ($r = 0.66$). A Sackville the values were up to 0.06 mg l^{-1} (total; highly correlated with the dissolved phosphorus with $r = 0.97$). At Brooklyn, these values were up to 0.02 mg l^{-1} (total) and 0.08 mg l^{-1} (soluble) (also highly correlated with $r = 0.53$).

Fig. 2 Photographs (**a–b**) of the upper Hawkesbury Estuary showing is beauty and use for recreation, (**c**) of a farm irrigated by Hawkesbury River water, (**d**) of rapid urbanization in the southwest catchment of the Hawkesbury River (Photo courtesy of Martin Halliday)

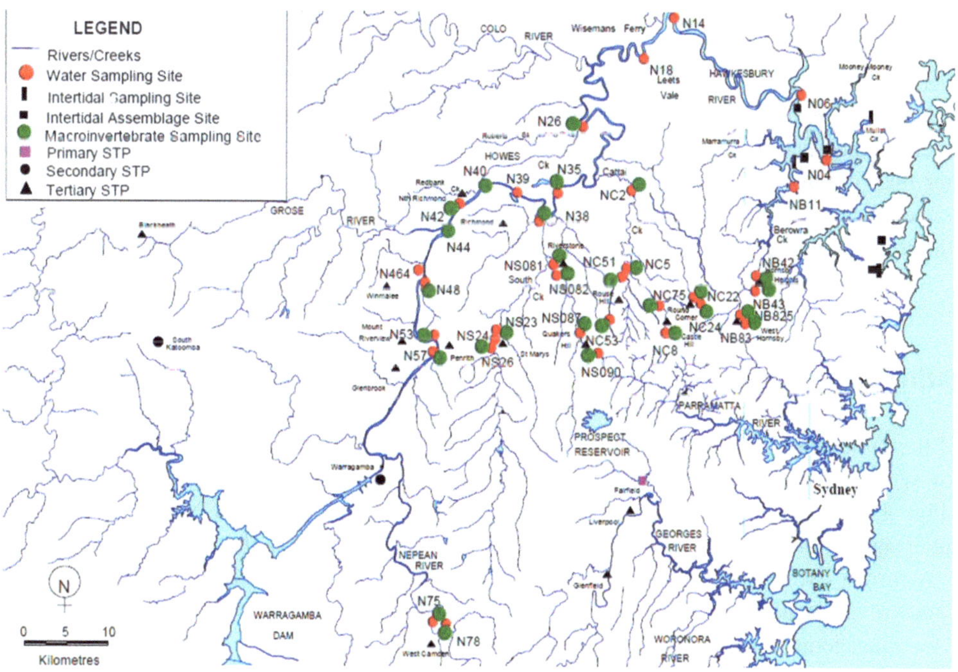

Fig. 3 Location of the major sewage treatment plants and water quality sampling sites in the upper Hawkesbury catchment (Adapted from Bickford and Smith 2006); there are also a few additional treatment plants (not shown) discharging treated sewage in poorly-flushed embayments and tidal creeks in the lower Hawkesbury Estuary principally near Hornsby and in Berowra Waters (see a location map Fig. 1)

Algal bioassays were used to determine the "fertilising power" of wastewater effluent. The effluent was collected at Berowra Waters's treatment plant prior to chlorination and final discharge (Collis 1973). The sample volumes were mixed with effluent and Hawkesbury water and inoculated with *Nitzshia* and *Chlorell*a algae. The effluent was serially diluted to give a total phosphorus of 0.02 mg l^{-1}. Both filtered and unfiltered river water was used. Both types of algae grew prolifically indicating that the river had all the trace elements needed for growth and that algae could obtain phosphorus from suspended matter.

Historical Pre-eutrophication Data

At Windsor the results in the 1970s indicated the rate of primary production was up to 10.2 g C m^{-2} day^{-1}, a very high value.

Little was known in the 1970s about the role of phosphorus in the eutrophication of flowing waters (Carpenter et al. 1970; Weibel 1970; Hutchinson 1973). It was well known however that blue-green algae were able to use most sources of nutrients (Fogg 1973; Fogg et al. 1973). Vollenweider (1968) surveyed North American and European rivers and concluded that flowing waters, that were relatively unpolluted contained less than 0.033 mg l^{-1} and in many cases less than 0.016 mg l^{-1}. However wastewater discharges could cause this to rise to values of order mg l^{-1}. A study of the Potomac Estuary in the USA (Carpenter et al. 1970) showed it to be similar to the Hawkesbury Estuary in many ways including catchment activity. In the saline zone of the estuary 32 km downstream of the city of Washington, DC, which discharged wastewater in the Potomac Estuary, phosphorus concentrations were in the range 0.07 mg l^{-1}. Chlorophyll a levels in this section of the Potomac Estuary were much higher than in the Hawkesbury Estuary. Based that overseas experience, Wolanski and Collis (1976) concluded that the upper Hawkesbury Estuary was already moderately eutrophicated in the 1970s.

Signs of Estuarine Eutrophication

A comparison of the 1950s nutrient data of Rochford, with the 1970s data of Wolanski and Collis and with the recent data shown in Fig. 4, it is apparent that there are large increases in total P and in turbidity all along the estuary since the 1970s. A further extremely large increase in P is apparent near Camden and St Marys – which the low-pass filter, favoured by management authorities, conveniently removes from Fig. 4! However such high values are not aberrations but real data. This leads to accusations that Sydney Water uses the Hawkesbury as a toilet (Sydney Morning Herald 2011). The contaminated water is diluted by freshwater released from Warragamba Dam. Algal growth and macrophytes, such as *Elodea* was and is infesting the river. Further downstream there is also a decrease in P due to inflows from the Colo River. This is still the only natural inflow of natural clean water that comes from an undeveloped catchment. It provides the main flushing of the estuary and it should be vigilantly protected by the NSW Government.

Downstream of the Colo confluence total P levels decrease and this coincides with an area of extensive mangrove/ mudflat ecosystem. Algal mats were observed by the author in 1976 and 2007 growing on the mudflats and mangroves.

From the monitoring study of the estuary (e.g. Frederickson 2010), it appears that the condition of the estuary has degraded since the 1970s. In the upper reaches near Camden there have been large outbreaks of *Elodea densa* and *Alternatha* sp. and an invasion of *Egeria* (Roberts et al. 1999). Indeed the outbreaks covered 347 ha over an 88 km stretch over the river at one point.

Based on such data the Hawkesbury River at Windsor is now permanently moderately eutrophicated with total phosphorus at 0.04 mg l^{-1}, at Sackville the mean phosphorus was 0.024 mg l^{-1} and at Brooklyn it was 0.012 mg l^{-1} (Sydney Water 2012).

Algal blooms have occurred all along the river. Red tides of toxic dinoflaggelates have been recently reported in Berowra Waters and Broken Bay during the warmer months (Peter Coad, Hornsby Council, personal communication, 2013). These red tides mean that oysters from the river may not be eaten by humans due to their potential toxicity. Oysters are now also prone to a new parasite known as Q-X disease, and this is lowering the productivity of the oyster industry (HNCAP 2013). The POM virus has destroyed all Pacific oyster leases worth $2.4 million (NSW Dept. Primary Industries 2013). The cyanobacteria population has changed from predominately *Anacystis* sp. and *Anabaena* sp. (Collis, unpublished data) to new assemblages including blue-green algae.

Urbanisation and Wet Weather Runoff

Wet weather runoff remains largely untreated, and, as shown later, this is a major threat to the estuary because the ultimate sink for all wastes, including sewage and wet weather runoff, is the estuary. The relationship between estuaries and urbanization is poorly understood (Walsh et al. 2012). The term "urban stream syndrome" has been coined to describe the "sick" status of urban streams around the world. The ecology of such streams tends to be dominated by filamentous algae, diatom and flowering macrophytes. They also demonstrate low biological diversity indices and are populated by worms, midge larvae and snails- all tolerant of pollution. The streams that are urbanized tend to suffer from toxic spills, high B.O.D. loads, nutrients and toxicants and sewer overflows. Unfortunately urban streams in the upper Hawkesbury Estuary are suffering in this manner.

Nutrients in wet weather run off vary with land resource properties such as permeability, run-off rate and channel shape. Phosphorus enters surface waters through rainwater, groundwater and waste water discharges. No data were available at the time for groundwater. Rainfall, through wash out of atmospheric pollutants, was already known in the 1970s to contribute phosphorus (P) to surface waters (Weibel 1970). Hutchinson (1973) reported P concentrations from trace levels to 0.049 mg l^{-1}, originating from air pollutants. Some initial studies by the author indeed showed phosphorous in rainfall over Sydney (Collis, unpublished data). This source of pollution will increase with growing urbanization.

Land runoff is a very significant source of P, up to 0.36 mg l^{-1} (Weibel 1970). Wolanski (1977) and Lee and Birch (this book) also observed heavy contamination of runoff in the Parramatta River in the center of Sydney.

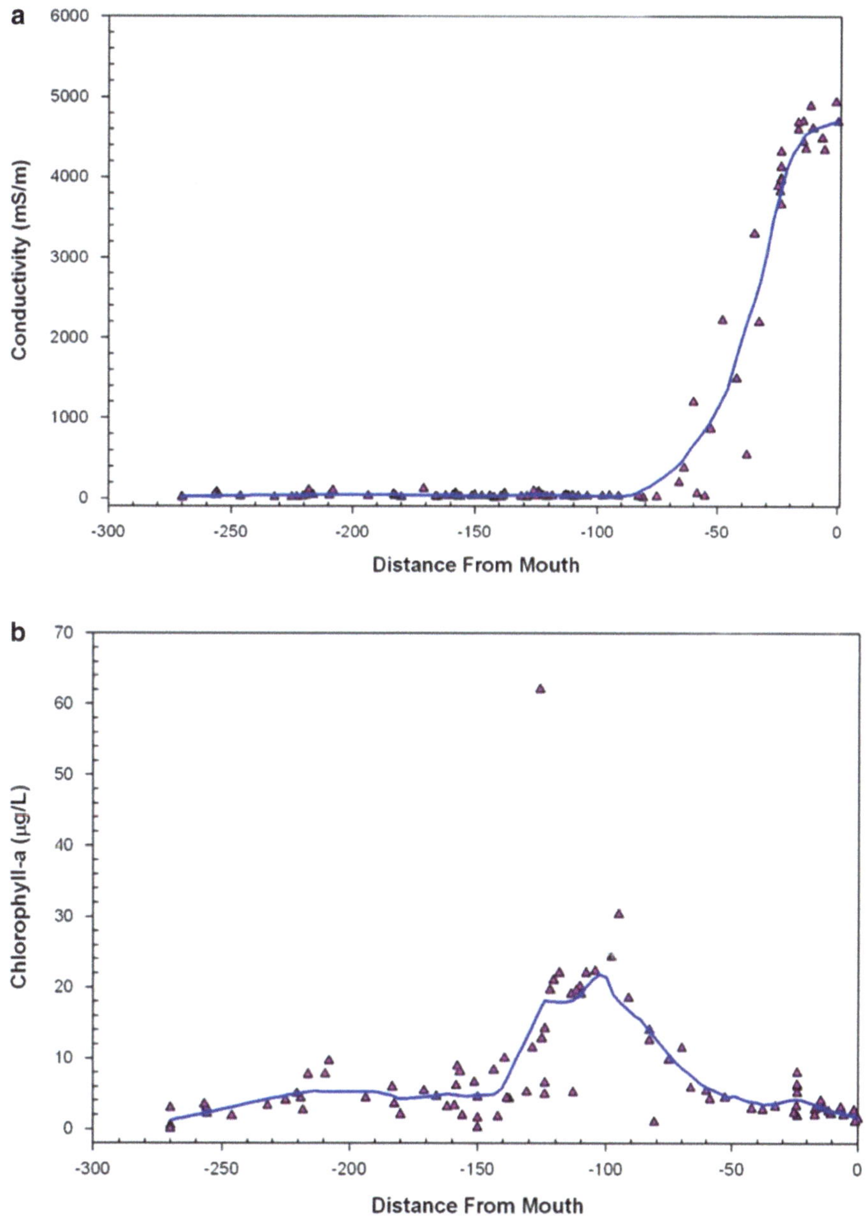

Fig. 4 Along-river concentration profiles in 2010 of near-surface (**a**) conductivity (a parameter for salinity), (**b**) Chlorophyll-a, (**c**) total phosphorus, (**d**) total nitrogen, and (**e**) turbidity. The tidal limit is located about 140 km from the river mouth. As a 1st order approximation, turbidity in NTU can be converted to suspended solid concentration (in mg/L) by multiplying by 2.6. Distance is in km. Figure courtesy of Martin Krogh, NSW EPA; data from Sydney Water Hawkesbury-Nepean Monitoring program. Both the raw data (Δ) and the Loess low-pass filtered data (*line*) are shown. The Loess low-pass filter, chosen by the management authority, excessively smoothens out the high values. These high values however are not spurious values as they are repetitive in sequential sampling and they are indeed signs of severe local eutrophication

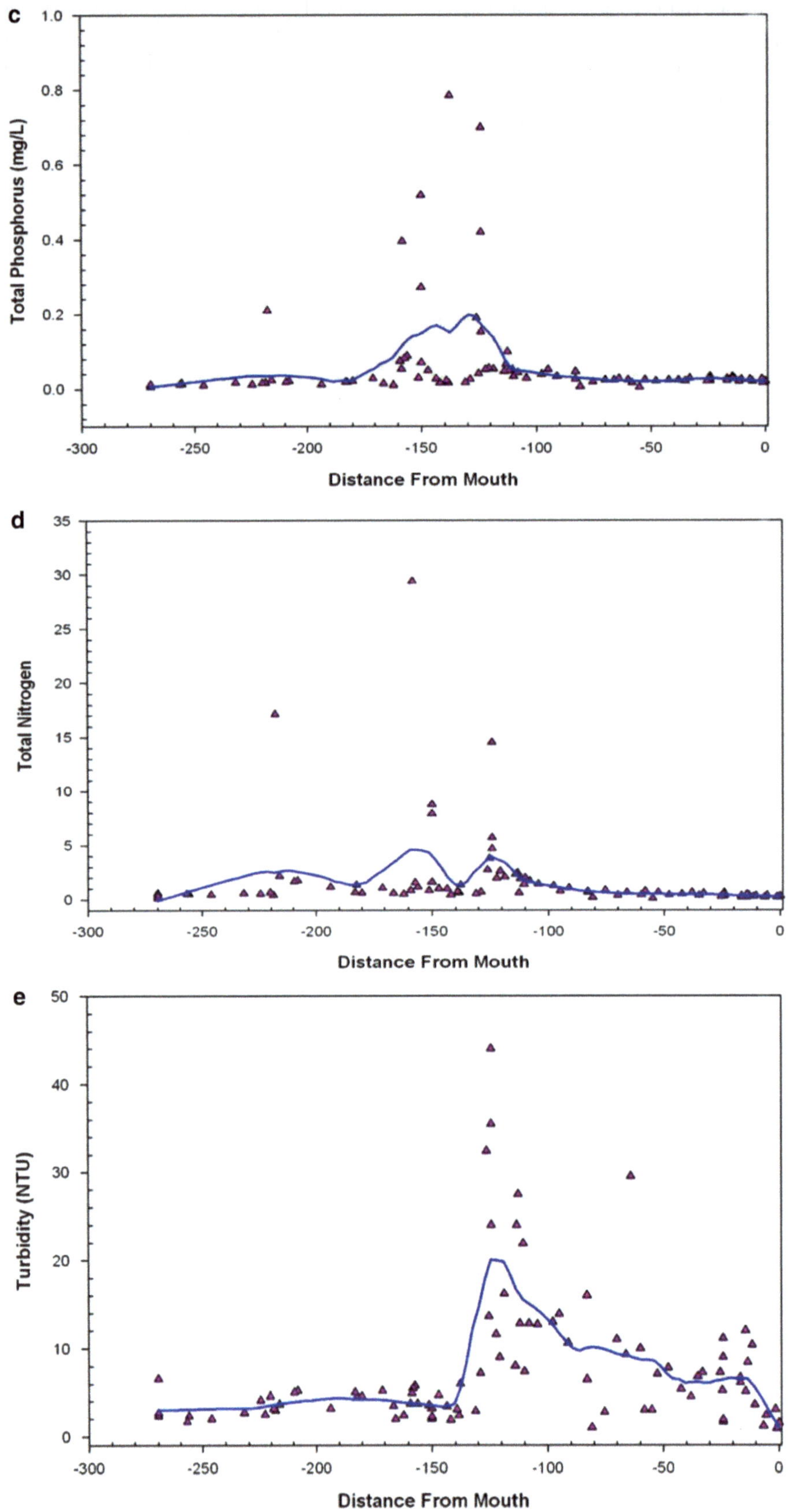

Fig. 4 (continued)

Scott (1978) also observed high nutrient loads from urban areas in Port Hacking, just south of Sydney. Runoff was thus already known as a major source of nutrients in the Hawkesbury estuary in the 1970s. Point sources such as sewage treatment plants, intensive feedlots, and abattoirs (these are particularly centered on the tributaries South Creek and Cattai Creek that drain in the Hawkesbury River) are thus not the only sources of P in the Hawkesbury.

The author also observed on many occasions that creeks (e.g. in Hornsby Shire) in dry weather were running bank-full under very low rainfall conditions. This was probably due to urban people overusing water on gardens, washing driveways and houses in a wasteful manner. This problem is not addressed in management plans of the Hawkesbury River estuary and will only worsen as urbanization increases.

South Creek and Berowra Creek have been studied in detail and the results demonstrate the impact of urbanization on stream water quality (Rae 2007). South Creek enters the Hawkesbury Estuary just below Windsor; it is now considered the most degraded waterway in NSW.

South Creek within 6 years of European settlement in 1789 was changed more than in the 40,000 preceding years of the original occupiers. Timber was cleared, hard hoofed animals were introduced and towns established. In 1825 the creek was used as irrigation water. Blacktown was built (now it is a major urban center) in the 1870s and it now occupies 60 % of the creek's 260 km^2 catchment area. The creek became a site for the disposal of agricultural and sewage wastes. The creek is affected by wet weather and dry weather wastewater releases from five sewage treatment plants and 20 other licensed discharges from abattoirs, dairies, golf clubs and farms. There are also licensed water extractions. High pollutant loads make rehabilitation tasks very difficult and the discharge of South Creek has a high impact on the Hawkesbury estuary. The author saw thick blue/green algal blooms in which daytime dissolved oxygen was so high that the water bubbled in the early 1970s. The South Creek catchment presently has a population of 392,000 people. Urbanisation on the creek is proceeding rapidly with 300,000 home lots planned in the catchment. 10 % of the catchment will become impervious and liable to extreme degradation from storm runoff.

Berowra Waters is a creek that discharges to the Hawkesbury near the mouth. Hornsby Council has a monitoring system in place (Frederickson 2010; Sydney Water 2012). It has experienced a number of red tide events involving toxic diatoms that are indicative of eutrophication. Total nitrogen has reached 1.6 mg l^{-1} and wet weather runoff has a major effect on water quality, increasing suspended solids and nutrients. E. coli levels have been high in wet weather because urban sewerage systems have sewer overflows which operate in wet weather to relieve back pressure into homes but discharge raw effluent into the nearest waterway. Large areas of impervious surfaces connect to the creek. The drains leak or allow infiltration of contaminants. This is common, and the problem is confounded by illegal connections to sewers from backyards, right throughout the Sydney region. In Berowra creek total nitrogen and phosphorus have exceeded the NSW EPA guidelines on many occasions. There are three sewage treatment plants, operated by Sydney Water, discharging to Berowra Creek. Dissolved oxygen at the mouth of the creek during the monitoring period was below 80 % of saturation. Hornsby Council is making a concerted effort to remedy these problems.

Cattai Creek and O'heras Creek (a small tributary of South Creek) drain areas that are undergoing intense urbanization and both creeks drain to the Hawkesbury River (Sydney Water 2012). The headwaters of Cattai Creek are in near pristine condition in bush land. However further downstream the waters are seriously degraded by nutrients and other contaminants draining from impermeable surfaces. Water quality is poor. O'heras Creek was once a chain of ponds but these are now connected by a series of cement pipes, which have degraded the original flow regime. Other large creeks such as Mangrove Creek (near Gosford) and Mooney Mooney Creek have had their headwaters changed to provide significant amounts of water to the NSW urbanized central coast.

Biodiversity

The wide range of aquatic and riparian habitats in the Hawkesbury Estuary supports a diverse assemblage of species, including over 50 finfish species, nine of which were introduced and outcompete the native fish (NSW DPI 2013). Seven species of fish are listed as threatened, including Macquarie perch and the Australian grayling. Three threatened species' silver perch, Murray cod and trout cod have been stocked in the catchment but have problems with competition from exotic species. The region also supports aquatic macro invertebrates including insects, prawns, crayfish and freshwater mussels. These communities are moderate to severely impaired, due to changes in stream flow, water extraction and runoff contamination. Over 60 species of frogs occur in the basin including the threatened giant burrowing frog, the green and golden bell frog, the giant barred frog, the red crowned toad let and others that are all threatened by pollution. The Hawkesbury-Nepean has been listed as an endangered ecological community (NSW DPI 2013).

HNCMA(2008) has published a list of threatened species in the Hawkesbury catchment, and this list includes many woodlands and forests, frogs, birds, sharks, snails, seals, possums, bats , glider possums, koalas, lizards, snakes, goannas, wattle species, orchids etc. It is worthwhile looking at this publication as an indication of the natural resources that could be lost is urbanization continues unchecked.

Conclusions

The Hawkesbury-Nepean Estuary is well on the way to being a "sick urban estuary". The major causes are sewage treatment effluent as direct ongoing discharges and wet weather runoff from urbanized areas. There may also be the ongoing problem of contaminated sediments retaining and recycling phosphorus to the water column.

Remedial measures are difficult to implement given that there are so many bodies involved with river management and that resources are limited. Clearly Sydney Water must continue to upgrade sewage treatment plants particularly in the Berowra Waters and South creek areas.

There is clearly no reason why some of the sewage could not be redirected for discharge at sea through Sydney ocean outfalls as the carrying capacity of the Hawkesbury Estuary and its tributaries seems to be exceeded at present during dry weather conditions.

The wet-weather runoff problem from urban areas must be addressed even if it is by far the most difficult problem to overcome. Sewerage pipes must be checked for leakage and the overflow from sewers in wet weather addressed. Dunn et al. (this book) describes a policy to treat wet-weather runoff from newly urbanized areas on the Gold Coast to avoid exceeding the carrying capacity of the Gold Coast Broadwater; a similar policy needs to be implemented in the Hawkesbury catchment.

Public awareness needs to be made clearer in that people must be discouraged to overuse valuable drinking water for cleaning driveways and cars as well as over-fertilising their gardens.

If sewage treatment plants were further upgraded, basic nutrients would be better removed, and a reduction in algal growth would occur (Simmonds 1973; Monbet 1992; Smith et al. 1999; Smyth et al. 2013). Eutrophication of the Hawkesbury may thus be reversible but there will be a significant time lag because of phosphates bound to sediments.

However there is no long-lasting solution for the Hawkesbury as long as the wet weather runoff from the growing suburbs of Sydney is not intercepted and treated.

References

Bickford GP, Smith SV (2006) Hawkesbury-Nepean river system budgets. http://data.ecology.su.se/MNODE/Australia/HawkNep/hawknepbud.htm

Carpenter JH, Pritchard BW, Whaley RL (1970) Observations of eutrophication and nutrient cycles in some coastal estuaries. In: Eutrophication: causes, consequences, correctives. Proceeding symposium National Academy of Sciences, Washington, DC, pp 210–224

Collis P (1973) School of Biological Science, Sydney University. M.Sc thesis, 1979, Fisher Library, Sydney University

Fitzsimons MF, Lohan MC, Tappin AD, Millward GE (2011) The role of suspended particles in estuarine and coastal biogeochemistry, vol 4, Treatise on estuarine and coastal science. Elsevier, Amsterdam, pp 71–114

Fogg GE (1973) Physiology and ecology of marine blue-green algae. In: Carr NG, Whitton BA (eds) The biology of blue-green algae. Blackwell, Oxford, pp 360–378

Fogg GE, Stewart WD, Fay P, Alsby AE (1973) The blue-green algae. Academic, London/New York

Frederickson P (2010) Water quality monitoring program. http://www.hornsby.nsw.gov.au/media/documents/environment-and-waste/water-catchments/water-quality/Water-Quality-Annual-Report-2009-2010

HNCAP (Hawkesbury-Nepean Catchment Action Plan 2007–2016) (2013) Published by the Hawkesbury-Nepean Catchment Management Authority, Goulburn. http://www.hn.cma.nsw.gov.au/multiversions/6478/FileName/HNCMA%2011-12%20Annual%20Report_LR NoSpreads_4.pdf

HNCMA (Hawkesbury-Nepean Catchment Management Authority) (2008) http://www.hn.cma.nsw.gov.au/http://www.hn.cma.nsw.gov.au/

Hutchinson GE (1973) Eutrophication. The scientific background of a contemporary practical problem. Am Sci 61:269–279

Kirk JTO (1976) Yellow stuff (Gelbstoff) and its contribution to the attenuation of photosynthetic radiation in some inland waters and coastal rivers in S.E. Australia. Aust J Mar Freshw Res 27:61–71

Kirk JTO (1977) Use of a quantum meter to measure attenuation and underwater reflectance in some inland waters and coastal river of S. E. Australian waters. Aust J Mar Freshw Res 28:9–21

Middleburg JJ, Herman PMJ (2007) Organic matter processing in tidal estuaries. Mar Chem 106:127–147

Monbet Y (1992) Control of phytoplankton biomass in estuaries: a comparative analysis of microtidal and macrotidal estuaries. Estuaries 16:563–571

NSW DPI (NSW Dept. Primary Industries) (2013) Fisheries and Agriculture, Hawkesbury River Update http://www.dpi.nsw.gov.au/fisheries/aquaculture/info/poms/hawkesbury-river-update

Rae DJ (2007) Water management in South Creek catchment: current state, issues and challenges. Technical report no. 12/07. CRC for irrigation futures, Richmond

Roberts DE, Church AG, Cummins SP (1999) Invasion of *Egeria* into the Hawkesbury-Nepean river. Aust J Aquat Plant Manag 37:31–34

Rochford DJ (1951) Studies in Australian estuarine hydrology. I. Introductory and comparative features. Aust J Mar Freshw Res 2:1–116

Scharpf IG (1973) Transformation of naturally occurring organophosphorus compounds in the environment. In: Griffiths BJ, Beeton AM, Spencer JM, Mitchell DI (eds) Environmental phosphorus handbook. Wiley, New York, pp 393–412

Scott RB (1978) Phytoplankton and light attenuation in Port Hacking Estuary. J Mar Freshw Res 29:34–44

Sholokovitz ER (1976) Flocculation of dissolved organic matter during mixing of river and seawater. Geochimica Comoschimica Acta 40:831–845

Simmons MA (1973) Experience with algal blooms and the removal of phosphorus from sewage. Water Res 7:255–264

Smith VA, Tilman GD, Nekola JC (1999) Eutrophication: impacts of excess nutrient inputs on freshwater, marine, and terrestrial ecosystems. Environ Pollut 100:179–196

Smyth AR, Thompson SP, Siporin KN, Gardner WS, McCarthy MJ, Piehler MF (2013) Assessing nitrogen dynamics throughout the estuarine landscape. Estuar Coast 36:44–55

Sydney Catchment Authority (2009) Hawkesbury-Nepean river environmental program. http://www.environment.nsw.gov.au/resources/water/09112hnrempfintechrpt.pdf

Sydney Morning Herald (2011) Sydney water using Hawkesbury as a toilet. http://www.smh.com.au/environment/water-issues/sydney-water-using-hawkesbury-as-a-toilet-20111217-1oztv.html

Sydney Water (2012) The Hawkesbury-Nepean river recovery program. http://www.sydneywater.com.au/water4life/ILEP/Hawkesbury/index.cfm

Vollenweider RA (1968) Scientific fundamentals of the eutrophication of lakes and flowing waters with particular reference to nitrogen and phosphorus. O.E.C.D, Paris

Walsh CJ, Leonard AW, Ladson AR, Fletcher TD (2012) Urban stormwater and the ecology of streams. CRCFE/Monash University, Canberra, p 44

Weibel SR (1970) Urban drainage as a factor in eutrophication. In: Eutrophication: causes, consequences, correctives. Proceedings symposium National Academy of Sciences. National Academy of Sciences, Washington, DC, pp 383–403

Wolanski E (1977) The fate of stormwater in Parramatta estuary Sydney. Aust J Mar Freshw Res 29:67–76

Wolanski E (2007) Estuarine ecohydrology. Elsevier, Amsterdam, 157 pp

Wolanski E, Collis P (1976) Aspects of aquatic ecology of the Hawkesbury estuary. Aust J Mar Freshw Res 20:69–76

Part III

Estuaries that Are Still Relatively Pristine

Deluge Inlet, a Pristine Small Tropical Estuary in North-Eastern Australia

Marcus Sheaves, Kátya G. Abrantes, and Ross Johnston

Abstract

Deluge Inlet is a small, tide-dominated estuary on Australia's north-east tropical coast, located in the central part of the Hinchinbrook Island National Park, Australia's largest island National Park. It is situated in Australia's humid tropical zone, and experiences an intense summer wet season and regular impacts of tropical cyclones. Protection by National Parks, World Heritage and Wild Rivers legislation means it remains in near pristine condition. Deluge Inlet sports substantial biodiversity in the form of extensive mangrove forests, seagrass beds, and complex marine mammal, reptile, fish and invertebrate assemblages, all supported by a mosaic of highly interconnected habitat types. The mix of habitats and rich biodiversity makes Deluge Inlet an important nursery for many species, and supports complex food webs. Current threats are from increasing fishing and boating pressure, and effective governance will be needed to ensure Deluge Inlet remains in near-pristine condition into the future.

Keywords

Estuarine assemblages • Fish distribution • Invertebrate distribution • Food webs • Biodiversity • Human impact

M. Sheaves (✉) • K.G. Abrantes • R. Johnston
Centre for Tropical Water and Aquatic Ecosystem Research, School of Marine and Tropical Biology, James Cook University, Angus Smith Drive, Townsville, QLD 4811, Australia
e-mail: marcus.sheaves@jcu.edu.au

E. Wolanski (ed.), *Estuaries of Australia in 2050 and Beyond*, Estuaries of the World,
DOI 10.1007/978-94-007-7019-5_15,

Box 1

Marcus Sheaves and colleagues studied Deluge Inlet, a small, tide-dominated, mangrove-fringed tropical estuary. Protection by National Parks, World Heritage and Wild Rivers legislation means it remains in near pristine condition. They describe the substantial biodiversity in the form of extensive mangrove forests, seagrass beds, and complex marine mammal, reptile, fish and invertebrate assemblages, all supported by a mosaic of highly interconnected habitat types. The mix of habitats and rich biodiversity makes Deluge Inlet an important nursery for many species, and supports complex food webs.

Current threats are from increasing fishing and boating pressure, and benign neglect by the government.

Setting

Deluge Inlet is a small, tide-dominated, near pristine estuary on Australia's north-east tropical coast. It forms the estuary of Boyd's Creek on the landward side of Hinchinbrook Island, the largest island on Australia's eastern tropical coast (Fig. 1), some 37 km long and 10 km wide. The estuary is about 9 km in length, with a maximum width of about 400 m at low tide and a maximum depth of about 4 m. Deluge Inlet flows into the Hinchinbrook Channel, which runs between Hinchinbrook Island and the Australian mainland. The largest source of freshwater to the channel is the 340 km long Herbert River, which enters into the southern end (Fig. 1), delivering sediment and organic matter from its ~10,000 km^2 catchment. Deluge Inlet is surrounded by extensive mangrove forests that comprise part of the Hinchinbrook Channel mangrove complex (Wolanski et al. 1990), one of Australia's largest mangrove swamps (~164 km^2). The location of Deluge Inlet in the central part of Hinchinbrook Island National Park, Australia's largest island National Park (39.3 km^2), its inclusion in the Great Barrier Reef Coast Marine Park (Marine Parks 2009) and Great Barrier Reef World Heritage Area, and the protection of Boyd's Creek (Fig. 2) and surrounding wetlands under a Wild River designation (Hinchinbrook Wild River Declaration 2007), mean it remains close to pristine. The area around Deluge Inlet has a long history of use by traditional people and has been part of the territory of the Bandjin people for at least 2,000 years (Campbell 1982).

Climatically, Hinchinbrook Island is part of Australia's humid tropical zone, with mean annual rainfall greater than 2,000 mm but almost 90 % of rain falling during the wet season (November to May) (Fig. 3) resulting from northern Australia's summer monsoon (Robertson et al. 2006). In the wet season average temperatures range between 27 and 32 °C during the day, and between 17 and 23 °C at night (BoM 2012). Deluge Inlet is in Australia's tropical cyclone belt so is periodically affected by extreme weather events, with the most recent Severe Tropical Cyclone Yasi in 2011. It experiences a mixed semi-diurnal tidal cycle, with a range of over 3.7 m.

Hinchinbrook Island is mountainous, with a central core composed mainly of Paleozoic granitic rocks, and only a narrow terrestrial coastal plain (Stephenson 1990). Boyd's Creek flows down the steep slopes of the island's second highest mountain, Mount Diamantina (955 m), with Deluge Inlet beginning where the mountain slopes abut one of the island's largest areas of, mainly intertidal, coastal sediments (Fig. 2). Intense chemical weathering of the island's granites produces a covering of decomposed rock material on its mountain slopes, so Hinchinbrook Island's substantial annual run-off carries large quantities of sand, as well as fine-grained silts and clays (Bird and Hopley 1969). Boyd's Creek is the largest freshwater stream on Hinchinbrook Island, with continual flow throughout the year. Consequently, large quantities of sand moving downstream result in sandy substrates dominating Deluge Inlet's sub-tidal areas, with large sand banks almost blocking the mouth of the estuary at low tide (Fig. 2). The waters of Deluge Inlet are consistently the clearest of any steam flowing in to Hinchinbrook Channel, probably because the continual inflow of clear fresh water from Boyd's Creek flushes much of the suspended sediment, usually found in mangrove estuaries (Furukawa and Wolanski 1996), out of the system.

Although most of the lower parts of Deluge Inlet are surrounded by intertidal mangrove forests, terrestrial forests abut the estuary on the south-east, where it flows close to low hills, and in the upper parts where it flows through the foothills of Mount Bowen (Figs. 1 and 4). Terrestrial forests are a mix of closed canopy lowland forests and rainforests that are the home to regionally endemic birds (Shoo et al. 2005), arboreal mammals (Trenerry and Werren 1993; Kanowski et al. 2001) and microhylid frogs (Shoo and Williams 2004).

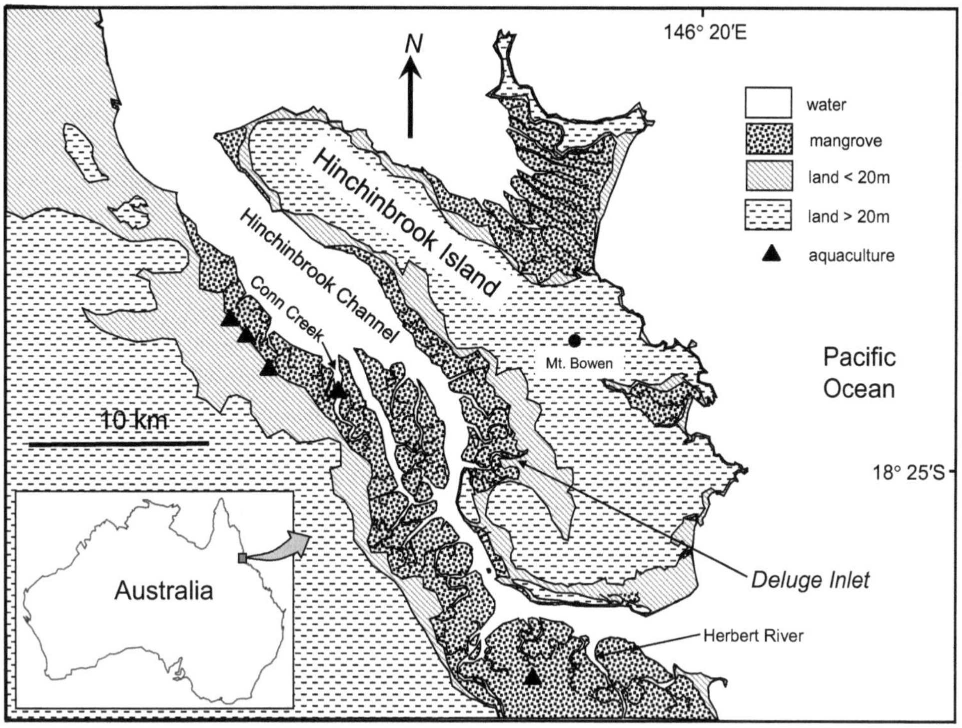

Fig. 1 Location map of the Hinchinbrook region and Deluge Inlet

Fig. 2 Detailed map of Deluge Inlet

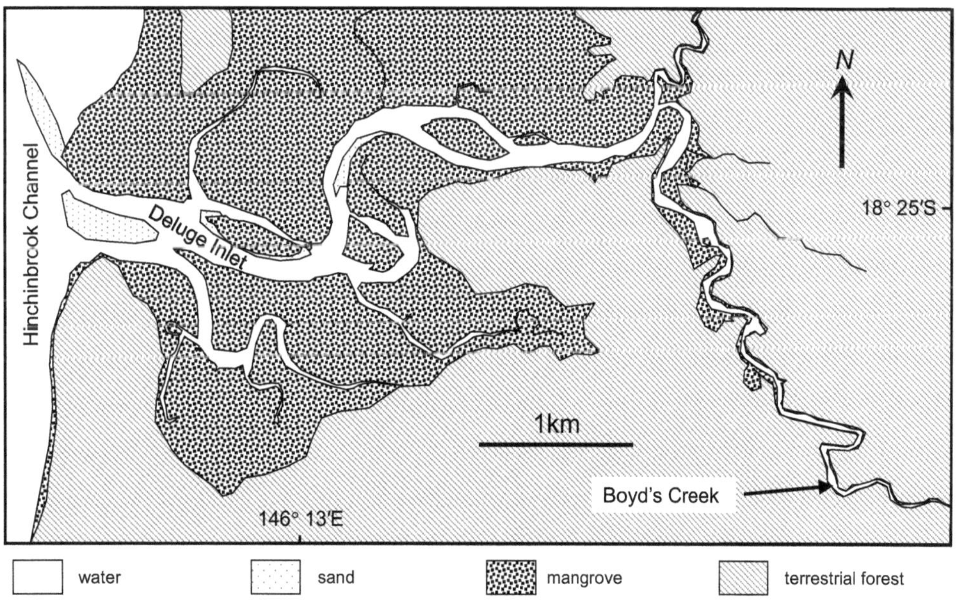

Ecological Capital

Biodiversity

The position of Deluge Inlet in the lee of a high island (Bird and Hopley 1969), and the extensive inputs of sediments from Boyd's Creek, provides the perfect environment for development of extensive mangrove forests (about 5 km^2), which dominate the Inlet's marine ecosystem (Fig. 2). The mangrove forest is rich in species, with the genera *Avicennia, Aegiceras, Acrostichum, Bruguiera, Excoecaria, Heritiera, Lumnitzera, Osbornia, Rhizophora, Sonneratia, Xylocarpus* all common (Bunt and Bunt 1999). In fact Deluge Inlet has been the site of a great deal of mangrove research (e.g. Duke and Bunt 1979; Bunt et al. 1982; Matsui

1998; Bunt and Bunt 1999). Mangroves perform many vital roles in tropical estuaries in trapping fine sediments, converting nutrients to plant biomass that are then available for other organisms, and providing habitat for fish and crustaceans (Bridgewater and Cresswell 1999; Wolanski 2007). Fine sediments are washed into mangrove forests during flooding tides. As the tide slows the suspended sediments settle out assisted by complex above ground mangrove root structures that impede tidal flow. Ebb tide currents are generally too small to resuspend the sediments leading to their retention in the mangrove forest and the formation of mud banks (Furukawa and Wolanski 1996). However, much of the fine sediment is exported from the estuary because of net outflow of water due to freshwater entering at the head of the estuary and the dynamics of water flow within the estuary (Wolanski et al. 1999).

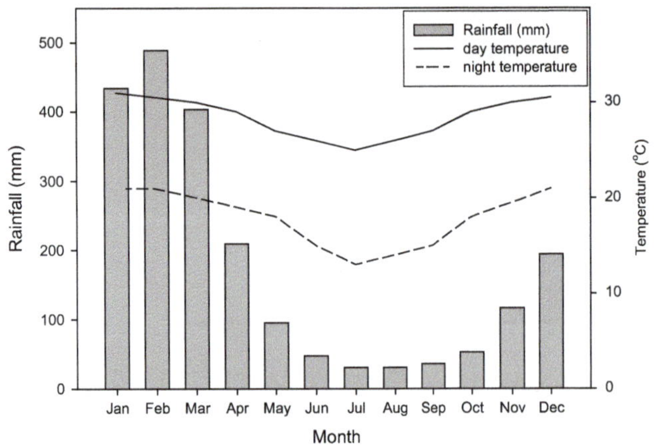

Fig. 3 Mean monthly rainfall, and mean maximum day and night temperatures for Hinchinbrook Island, Australia

Deluge Inlet also has substantial seagrass meadows, principally composed of *Halodule uninervis* and *Halophila ovalis* (Lee Long et al. 1998), and considerable areas of filamentous green algae growing both subtidally and intertidally. Rather than the tall, dense seagrass meadows found in many parts of the world, most of Deluge Inlet's seagrass meadows are short growing forms (only centimetres long) mainly occurring in sparse meadows with individual leaves centimetres apart. In some areas of the upper estuary *Melaleuca* swamp forests occur on higher estuary banks and between the mangrove forests and true terrestrial forests. These provide short-term habitats for marine organisms during seasonal flooding.

Estuarine crocodiles, *Crocodylus porosus*, are permanent residents of Deluge Inlet, with nesting sites common in estuaries on the western side of Hinchinbrook Island (Hamann et al. 2007). Other marine reptiles found in the Inlet include sea snakes, such as Dubois' sea snake, *Aipysurus duboisii* (Wetland Info 2012), and a number of species of large sea turtles, such as Green, *Chelonia mydas*, Loggerhead, *Caretta caretta*, and Hawksbill, *Eretmochelys imbricata*, (Hamman et al. 2007). Dolphins, including Indo-Pacific humpback, *Sousa chinensis*, bottlenose, *Tursiops*

Fig. 4 Deluge Inlet; (**a**) *Rhizophora* mangroves backed by Hinchinbrook Island's central mountain range, (**b**) extensive sand flats in the downstream reaches, (**c**) relatively narrow intertidal areas of mangrove (lighter vegetation) backed by terrestrial forest in mid-estuary reaches, (**d**) little intertidal area with few mangroves mixed with terrestrial vegetation in upstream reaches

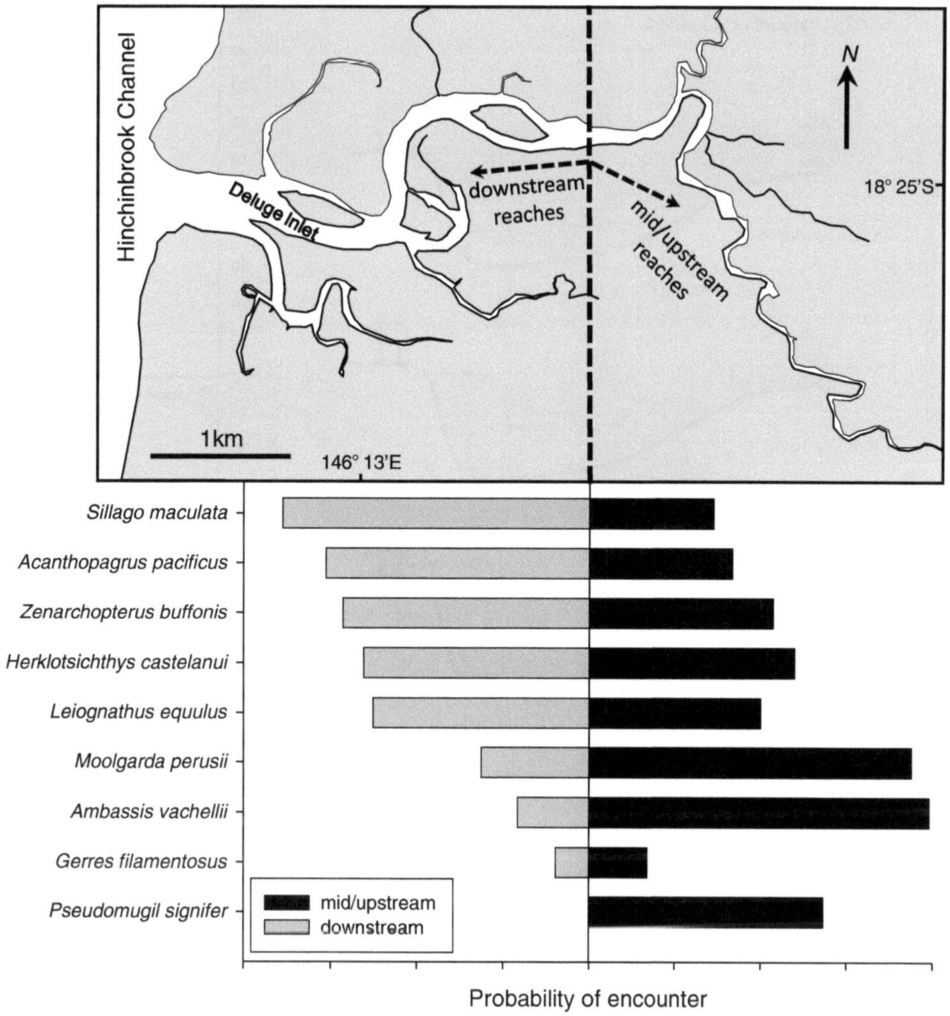

Fig. 5 Longitudinal distribution of common fish species in Deluge Inlet (data are from cast net sampling of downstream, mid- and upstream reaches of the estuary. Data are proportion of nets in which each species occurred for each estuary reach. There was no difference in composition between mid- and upstream reaches so these are presented as a single reach)

spp., and Australian snubfin, *Orcaella heinsohni* (Preen 2000; Lawler et al. 2007; Parra 2007), enter the Inlet where they feed on large schooling fish such as mullet and herring. Seagrass feeding dugongs, *Dugong dugon*, (Marsh and Lawler 2000) are also regular visitors.

Fish assemblages in Deluge Inlet vary depending on habitat type, proximity to channel edges and distance upstream. The assemblages in upstream and mid-estuary reaches differ from those in the lower reaches closer to the junction with Hinchinbrook Channel (Sheaves and Johnston 2009). For example, whiting, *Sillago maculata*, bream, *Acanthopagrus pacificus*, garfish, *Zenarchopterus buffonis*, herring, *Herklotsichthys castelnaui*, and ponyfish, *Leiognathus equulus*, are encountered more frequently in downstream reaches although they all occur throughout the estuary (Fig. 5). Species such as the mullet, *Moolgarda perusii*, glass perchlet, *Ambassis vachellii*, silverbiddy, *Gerres filamentosus*, and blue-eye, *Pseudomugil signifer*, are more likely to be found in mid-estuary and upstream reaches. Shifts in the assemblage composition between downstream reaches and mid/upstream are affected by salinity differences, however, because salinities throughout Deluge Inlet are usually lower than seawater and because most species do occur throughout the estuary (Fig. 5), other factors are probably also involved. Among these factors are differences in substrate type and the extent of mangrove forests between downstream and mid/upstream reaches (Fig. 4). Downstream there are extensive mangrove forests and substantial areas of intertidal muddy substrate, whereas upstream sandy substrates dominate and the extent of intertidal habitat is greatly reduced by the proximity of surrounding hills.

At a finer scale most small fish (<200 mm FL) in Deluge Inlet show strong associations with channel edges and are not often found in deeper mid-channel habitats (Johnston and Sheaves 2008). However, patterns of distribution vary from

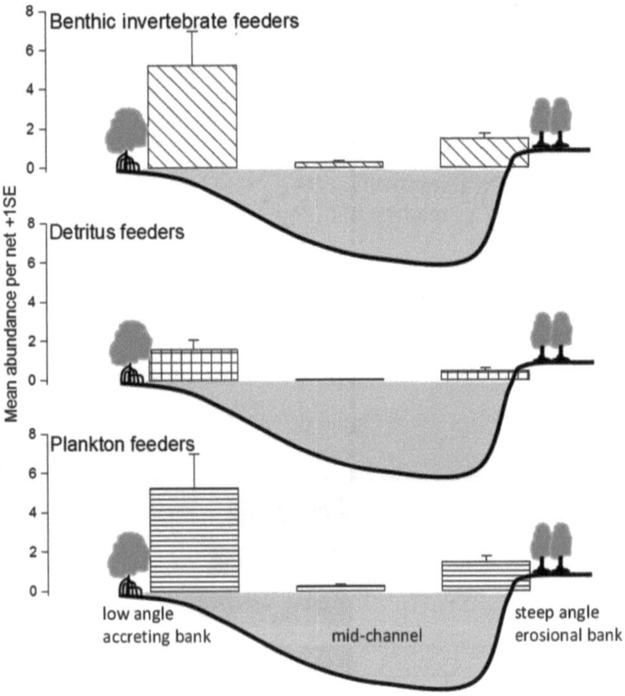

Fig. 6 Cross-channel distribution of fish functional groups in Deluge Inlet

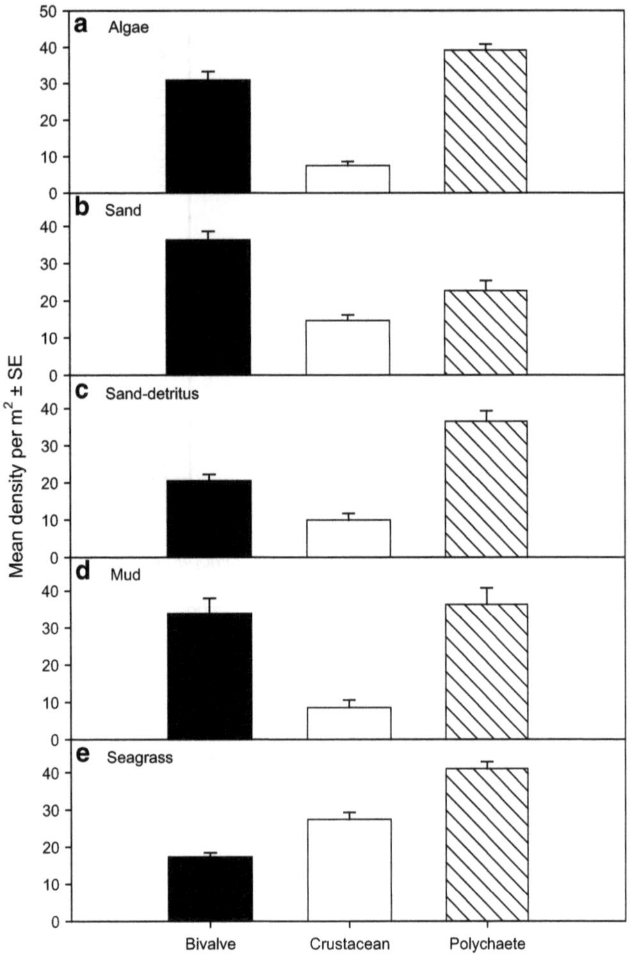

Fig. 7 Average density of three major taxa of benthic invertebrates in five habitat types in Deluge Inlet

species to species, with different functional groups tending to have different distributions. Plankton feeders, benthic invertebrate feeders and detritus feeders are encountered more frequently near the edges of the estuary (Fig. 6), particularly along low angled accreting banks, but have periods of more dispersed distribution when they occur in high abundances in the mid-channel. For example, this pattern of edge-association breaks down on very low tides, with fish distributed right across the channel when depths are less than about 75 cm.

Fish also show preferences for particular micro-habitat types when the water is out of the mangrove forests (Johnston and Sheaves 2007). Muddy substrates are preferred over sandy substrates by a majority of fish species, except some specialist shallow water benthic feeders such as whiting (*Sillago* spp.). Besides these shallow water benthic feeders, most species occur in low abundances in bare sand and mud habitats that lack complex structure, but in high abundances where there are complex structures such as drains returning tidal water from mangrove forests (Johnston and Sheaves 2007) or submerged woody snags and collapsed mud banks (Sheaves 1992). In effect, most fish show a positive response to small-scale habitats that provide particular food resources or greater access to complex habitat likely to provide refuge from predators.

Invertebrates also show strong patterns of spatial distribution, with different habitat types harbouring different invertebrate assemblages and strong changes in distribution across the intertidal gradient. Bivalves are present in all habitat types, with lowest densities in seagrass and sand-detritus habitats and highest densities in unvegetated sand and mud habitats (Fig. 7). Crustacean densities are highest in seagrass habitats with comparatively low densities in other habitats whereas polychaete worms occur in high densities across all habitats except sand. Densities of polychaete worms and bivalve molluscs peak in the low intertidal zone and are lowest at the mangrove edge and in sub-tidal areas, while gastropod mollusc densities peak slightly further up the intertidal gradient (Fig. 8). In contrast, amphipods tend to have broadly similar densities across lower intertidal and sub-tidal zones but are less frequently encountered higher in the intertidal. Highest densities of most taxa in lower and intermediate intertidal zones appear to be a balance between avoidance of unfavourable environmental conditions and reduced predation. The upper part of the intertidal zone dries out most frequently, limiting the occurrence of marine invertebrates, while invertebrates in low intertidal and subtidal zones are vulnerable to fish predation throughout the tide (Quammen 1984; Peterson 1991).

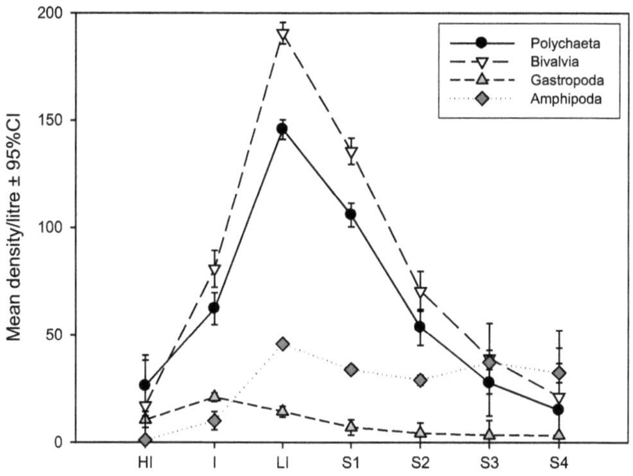

Fig. 8 Average density of four major taxa of benthic invertebrates across the intertidal gradient in Deluge Inlet. *HI* high intertidal, *I* mid intertidal, *LI* low intertidal, *S1 – S4* a gradient of sub-tidal zones of increasing depth

The distribution of invertebrates meshes neatly with the cross-channel distribution of benthic feeding fish. The largest area of intermediate to low-angled intertidal bank, the habitat preferred by most invertebrates (Fig. 8), is associated with accreting banks in Deluge Inlet. Benthic feeding fish also demonstrate a preference for low-angle accreting banks (Fig. 6), thus providing a clear link between invertebrate food sources and their consumers, benthic feeding fish.

Eco-Biophysical Processes

The highly dynamic Deluge estuary encompasses a complex mosaic of interconnected habitats including mangroves, seagrass beds, tidal and sub-tidal habitats and sandy banks, with complex pathways of energy and nutrient flows among habitats. These nutrient flows are mediated by the movement of water as well as the movements of animals like mangrove jack, *Lutjanus argentimaculatus*, and bream, *A. pacificus*, that move into mangrove forests on high tides for feeding forays (Sheaves and Molony 2000).

The ability of mangroves to capture and store large amounts of atmospheric CO_2 makes them among the world's most productive ecosystems (Jennerjahn and Ittekkot 2002). Both the mangrove trees and the upper layers of Hinchinbrook Channel mangrove sediments store large quantities of organic carbon (Clough 1998; Matsui 1998), with the values many times higher than those for other tropical forests (Chmura et al. 2003; Donato et al. 2011). This makes extensive mangrove forests, like those around Deluge Inlet, important natural sinks for atmospheric CO_2 (Bouillon 2011) and important environments for moderating the effects of global climate change (Langley et al. 2009).

Because of the high ratio of mangrove area to open area, mangroves are one of the main sources of organic carbon to the Hinchinbrook Channel, contributing ~56 % of the total carbon input (Alongi et al. 1998). High quantities of nutrients, dissolved and particulate organic matter (Ayukai et al. 1998; Mueller and Ayukai 1998) and up to 1.25 g C m^{-2} of mangrove litter such as leaves and propagules are produced by Hinchinbrook's mangrove forests each day (Bunt 1979; Duke et al. 1981) and can enter foodwebs via the action of a range of organisms. Wooden parts are broken down by woodboring insects and shipworms, while leaves are decomposed by bacteria, fungi and other microorganisms into more easily assimilated detritus that supports species such as sesarmid crabs and mangrove snails (Fratini et al. 2000; Lopez and Levinton 1987; Micheli 1993). Mangrove crabs are important recyclers of nutrients, carrying up to 50 % of the leaf litter back into their burrows (Schories et al. 2003), and so limiting the export of nutrients in these leaves into the estuary. In the canopy, tree climbing crabs feed on plant material (Fratini et al. 2005) and a diverse insect community including ants, moths, crickets and beetles also relies on mangrove productivity (Burrows 2003). These insects are actively preyed upon by archerfish, *Toxotes chatareus*, and can be transported into the water by the wind and rain where they are consumed by other carnivorous species, further strengthening the link between mangrove productivity and the aquatic food web. Exudation of organic compounds from living trees is another pathway for mangrove carbon to enter aquatic food webs. Once in the aquatic environment mangrove-derived dissolved organic matter is decomposed and recycled by highly productive and efficient bacterial populations, entering the microbial loop (Alongi et al. 1989; Boto et al. 1989).

However, although the estimated contribution of mangrove carbon is high (~56 %) it is of poor nutritional quality (Alongi et al. 1989) so few species rely on mangroves as their main source of nutrition, meaning food webs in Deluge Inlet are mostly based on a combination of more easily assimilated aquatic sources (Abrantes and Sheaves 2009; Bouillon et al. 2008; Oakes et al. 2010). Perhaps the most important of these are the tiny unicellular algae, termed microphytobenthos (MPB), which form a thin layer on intertidal mud and sand banks (MacIntyre et al. 1996). MPBs are among the most important of marine primary producers, providing the basis for coastal food webs around the world (MacIntyre et al. 1996; Miller et al. 1996). The large areas of shallow water in Deluge Inlet means the area available for MPBs productivity is great, and a green film of these benthic microalgae can often be observed on sand and mud banks at low tides. Even though minute their productivity is impressive, as they are grazed by invertebrates and fish on every tidal cycle but regrow so rapidly they are continually replaced. While MPBs, seagrass, filamentous algae and

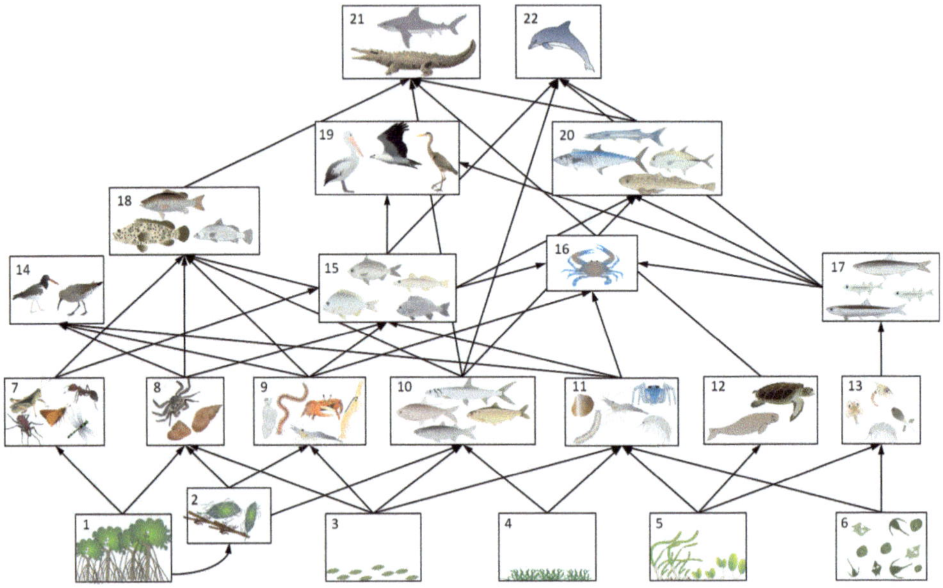

Fig. 9 Deluge Inlet food web. *1* mangroves; *2* mangrove detritus; *3* microphytobenthos; *4* green filamentous algae; *5* seagrass; *6* phytoplankton; *7* insects; *8* sesarmid crabs and mangrove snails; *9* upper intertidal benthic fauna; *10* phytodetritivorous fish; *11* lower intertidal and subtidal benthic fauna; *12* dugongs and sea turtles; *13* zooplankton; *14* wader birds; *15* macrobenthic carnivores (fish); *16* portunid crabs; *17* plantivorous fish; *18* large carnivorous fish; *19* – piscivorous birds; *20* piscivorous fish; *21* sharks and crocodiles; *22* dolphins (Figures and symbols courtesy of the Integration and Application Network, University of Maryland Center for Environmental Science (ian.umces.edu/imagelibrary/))

mangroves form the basis of food webs for bottom feeding organisms, phytoplankton perform the same role in food webs leading to planktivorous fish and invertebrates (Underwood and Kromkamp 1999; Bouillon et al. 2000) (Fig. 8).

Another important source of organic carbon for Deluge Inlet is terrestrial material seasonally transported from the Boyd catchment, and also from other rivers and creeks that run into Hinchinbrook Channel. The Herbert River (Fig. 1) is the largest river flowing into Hinchinbrook Channel, and a major source of sediment and organic matter in Hinchinbrook Channel, responsible for ~27 % of the total organic carbon pool (Alongi et al. 1998). Due to the hydrology of Hinchinbrook's waterways this material can be trapped in the Channel for long periods of time and so can penetrate into Deluge Inlet. Due of the hydrology of the Hinchinbrook Channel, this material is not efficiently flushed out of the area. In fact, because tides of similar amplitude enter the Channel at its north and south ends, currents in the area are weak and water residence time in the Channel can be as long as 50 days (Wolanski et al. 1990, 1999). Also, the surrounding mountains and dense mangrove forests provide shelter from the wind, further facilitating the retention of organic material in the system.

These sources of organic carbon support a food web of approximately five trophic levels in Deluge Inlet, with a number of different pathways of energy flow from primary producers to top level predators (Fig. 9) (Abrantes and Sheaves 2009). In the forested upper intertidal, detritus of mangrove origin and microphytobenthos support a range of invertebrate deposit feeders such as mud whelks, e.g. *Telescopium, Terebralia*, grapsid crabs, e.g. *Metopograpsus frontalis, Perisesarma messa*, and fiddler crabs, *Uca* spp. Lower in the intertidal and subtidal, MPBs, filamentous algae and seagrass and its epiphytes support a rich community of invertebrates including small bivalves (e.g. *Tellina* spp., *Gary* spp.), copepods, polychaetes, amphipods, caridean shrimps and juvenile penaeid prawns, as well as phytodetritivorous fish such as the mullets, *Mugil cephalus, Valamugil* spp., hairback herring, *Nematalosa come*, and gizzard shad, *Anodontostoma chacunda*. Seagrass also provides nutrition for green sea turtles, *C. mydas*, and dugongs, *D. dugon*. In the water column, estuarine phytoplankton forms the basis of the pelagic food chain, which includes zooplankton, planktivorous invertebrates such as swarming shrimp, *Acetes sibogae*, and fish such as herrings, *Herklotsichthys* spp., sardines, *Sardinella* spp., and glassfish, *Ambassis* spp.

Higher up in the food web, macrobenthic carnivores like bream, *Acanthopagrus* spp., ponyfish, *Leiognathus* spp., whiting, *Sillago* spp., mangrove jacks, *L. argentimaculatus*, and silverbiddies, *Gerres* spp. feed on benthic invertebrates from the different habitats, and are in turn prey for piscivores like the barramundi, *Lates calcarifer*, trevallies, *Caranx* spp., queenfish, *Scomberoides* spp., flathead, *Platycephalus fuscus*, and barracudas, *Sphyraena* spp. Finally, top predators such as

Irrawaddy dolphin, *Orcaella brevirostris*, and humpback dolphins, *S. chinensis*, crocodiles, *Crocodylos porosus*, and sharks e.g. bull sharks, *Carcharhinus leucas*, link the different energy pathways into a complex food web. Many bird species are also part of this food web, preying on intertidal invertebrates from different habitats (e.g. oystercatchers, plovers) and on fish (e.g. fish eagles, cormorants, pelicans). Therefore, while at lower trophic levels the different food chains are somewhat separated, with different habitats supporting somewhat distinct consumer communities, there is an amalgamation of these food chains at higher trophic levels (Abrantes and Sheaves 2009).

Nursery Grounds

Perhaps the most prominent ecological role of Deluge Inlet is the provision of nursery grounds for marine species. Although many species, such as gobies spawn in the estuary and complete their life-cycles there, many others have adult populations elsewhere and use Deluge Inlet as juvenile nurseries. For many of the smaller species nursery occupation only lasts for a few months (Robertson and Duke 1990), but some larger species, like mangrove jack, *L. argentimaculatus*, and estuary grouper, *Epinephelus coioides* and *E. malabaricus*, may remain in the estuary as juveniles for up to 5 years (Sheaves 1995a). The pattern of nursery ground use is very variable. Species like mangrove jack and estuary grouper have adults that remain offshore, species like barramundi, *L. calcarifer*, and bream, *A. pacificus*, spawn in the mouths of estuaries, hence their eggs and larvae are transported between estuaries by coastal currents (Garratt 1993; Sheaves et al. 1999), but return to use estuarine and freshwater parts of Deluge Inlet as adult habitats. Juveniles of others such as trevally, *Caranx ignobilis* and *C. sexfasciatus*, and queenfish, *Scomberoides commersonnianus* and *S. tala*, use estuaries as nurseries but also range across other coastal waters.

Marine fish use Deluge Inlet as a nursery for a variety of reasons. Perhaps the most important is the rich supply of food that occurs there. The juvenile phase of a fish's life requires the greatest amount of nutrients (Yanez-Arancibia et al. 1994), something that estuaries like Deluge Inlet can provide. Their position at the interface between land and sea means nutrient input is much higher than in habitats further offshore, while mangrove forests, seagrass beds and micro algae represent a large biomass of primary producers that convert these nutrients into organic carbon to support estuarine food webs. In addition, structurally complex habitats, like mangroves, provide refuges for juveniles, while mangrove roots, fallen timber and complex bank structures produce eddies and backwater areas where current flows are reduced, allowing small fish to reduce energy expended on swimming. Consequently, the availability of a variety of habitats within Deluge Inlet is an important component of its value as a nursery (Sheaves 2005, 2009), with juvenile fish able to access habitats that provide rich feeding, refuge from predation and reduced energy expenditure as they are required (Sheaves 2009).

While the provision of nursery values is important to juvenile fish, the presence of the nursery function has broader ecological consequences. Fish and crustacean assemblages in Deluge Inlet include a mix of estuary and marine spawned species, and it is the marine-spawned species that have the greatest influence on overall system functioning and dynamics because they undergo the most substantial seasonal shifts in abundance through recruitment and emigration. Consequently marine spawned species make the greatest contribution to changes in species mix and food web functioning over time. In addition, the movement of juveniles offshore is an important factor in the export of estuarine nutrients to offshore habitats, because juveniles migrating offshore carry energy and nutrients accumulated in their bodies during estuarine residence to offshore waters (Deegan 1993).

Connectivity

The use of a diversity of habitats for different life-history needs can only occur if there is free connection among different habitat units. This unimpeded connectivity is one of the key features supporting Deluge Inlet's near pristine status because it is one of the few estuaries in the region with no human constructed barriers. Almost all mainland estuaries have dams or weirs for water storage (Sheaves et al. 2007b), or bund walls that prevent high tides entering high intertidal areas, providing increased areas for grazing and plantation agriculture (Hyland 2002; Sheaves et al. 2007a). Connectivity occurs at a large range of scales (Sheaves 2009), and this high level of connectivity is a major contributor to nursery ground value (Nagelkerken et al. in review) and to Deluge Inlet's complex biodiversity. As well as allowing fish, such as the endangered Jungle Perch, *Kuhlia rupestris*, to move between its freshwater habitats and estuary spawning areas, connectivity allows juveniles to move to key habitats at specific times; for instance moving into mangroves at high tide for feeding and refuge (Sheaves 2005; Olds et al. 2012).

Anthropogenic Influences, Vulnerability and Pressure for Change: Towards 2050

The key resources of Deluge are its natural beauty, its biodiversity values and its fisheries resources. The clear waters, sandy substrates and mangrove forests of Deluge Inlet combine with a mountain backdrop (Fig. 4) to set it apart from most Australian mangrove estuaries. This natural beauty extends to a variety of lush marine habitats, such as mangroves, seagrass, sand flats, that support diverse flora and fauna, and makes the Inlet a popular ecotourism destination and a regular overnight anchorage for house boats touring Hinchinbrook Channel. The waterways of Deluge Inlet support a recreational line fishery for barramundi, *L. calcarifer*, mangrove jack, *L. argentimaculatus*, javelin fish, *Pomadasys kaakan*, trevally, *C. ignobilis*, and bream *A. pacificus*. Although it is closed to net fishing, Deluge Inlet supports an active commercial fishery for the mud crab, *Scylla serrata*, which is also targeted by recreational fishers. Given its location in a National Park, Marine National Park and the Wild River status of Boyd's Creek and surrounding wetlands, the future of Deluge Inlet and its ecosystems depends on continuation of that protection and the quality of governance under the associated legislation.

Vulnerability

At the moment isolation from the mainland, the lack of a resident indigenous community, and National Park and Wild River status, insulate Deluge Inlet from most human impacts. Currently impacts are light, with direct influences stemming mainly from boating, ecotourism and fishing (Table 1). Although the Inlet has been used by indigenous fishers for over 2,000 years (Campbell 1982), their visits are rare today.

Threats from Human Activity

Deluge Inlet is only accessible by boat, and at present boat traffic is relatively light, with an average of 1–2 boat visits on week days and 5–20 on weekends. This include small (<6 m) recreational and commercial fishing boats, and houseboats and yachts that often overnight in the mouth of the Inlet. Boat traffic has the potential to cause bank erosion and undermine bankside vegetation, although these effects are probably minor compared to more substantial erosion from regular seasonal floods. Other potential boat-related impacts stem from anchor damage, although these are probably only of minor consequence to the sandy substrate of the estuary.

More substantial impacts are likely to result from fishing pressure and aquaculture (Table 1). Bait collection by digging or pumping invertebrates only occurs at a low level, probably because most people are reluctant to leave their boats due to the Inlet's substantial crocodile population. Some bait collection is conducted using cast nets, but this tends to be much less common in Deluge Inlet than many nearby estuaries, possibly because a predominance of sandy substrates means shrimps (a major target of bait netters) are less common than in muddy estuaries (Sheaves and Johnston 2009).

Recreational anglers remove moderate numbers of fish from Deluge Inlet but because no commercial net fishery is allowed in Hinchinbrook Channel (Fisheries Regulation 2012) the current pressure on fish stocks is relatively low. A more substantial impact comes from commercial crabbing, for the mud crab *S. serrata*, which is allowed in Hinchinbrook Channel. At least one crab fisher works Deluge Inlet most days, contributing a large proportion of the boat traffic on week days. As well as potential impacts on banks from the boat traffic, and the removal of large numbers of crabs by the fishery, there is a substantial risk of "ghost fishing" due to lost crab traps. Lost crab traps can keep "fishing" for extended periods as trapped crabs and fish act as "bait" to attract more crabs and fish into the trap (Matsuoka et al. 2005). This is a particular concern for juvenile grouper that use Deluge Inlet as nurseries (Sheaves 1995a) and are particular susceptible to trapping (Sheaves 1995b).

Although no aquaculture is conducted in or around Deluge Inlet there are prawn and barramundi aquaculture ventures some 12 km away on land, and a sea-cage venture in Conn Creek, a coastal estuary arm of Hinchinbrook Channel, around 10 km from Deluge Inlet (Fig. 1). Even though these ventures are remote from Deluge Inlet they present a substantial threat to fish populations. In 2011 Severe Tropical Cyclone Yasi passed across the area severely damaging sea-cages in Conn Creek leading to the escape of an estimated 280 t of barramundi. Barramundi are major predators, so this escape represents a substantial increase in predation pressure in estuaries of the region, with the potential for severe ecosystem effects, such as trophic cascades where changes in predation affect numbers at lower trophic levels. There is also the potential for more pervasive effects because the escape of aquaculture stocks can introduce disease and/or parasites into the environment (Davenport et al. 2003) and compromise the genetic integrity of wild populations (Doupé and Lymbery 1999).

The central position of Deluge Inlet along Hinchinbrook Channel, and the wide, protected anchorage it provides makes it a regular site for house boats and yachts to overnight. Although these visits are common there is rarely more than one or two boats anchored there. Deluge Inlet is also visited by an ecotourism operator once or twice a week. Together, the fishing and visitor traffic have the potential to cause environmental degradation from the waste and rubbish they leave behind. However, most visitors appear to behave responsibly because it is rare to see rubbish floating in the waters of Deluge Inlet or trapped in bankside vegetation.

Increasing, and increasingly more affluent, populations in north-eastern Australia mean that, despite the protection afforded under environmental legislation, it is likely that Deluge Inlet will face elevated anthropogenic pressures

Table 1 Vulnerability assessment for Deluge Inlet

	Geomorphology/landform/ hydrology	Plants	Animals	Processes
Sea-level rise	Inundation of low-lying land, possibly mitigated by increases in wetland plant growth and sediment build up from increasing CO_2	Mangroves and seagrass can migrate, but lowland *Melaleuca* wetlands are at risk because the land they can occupy is restricted by the island's mountainous terrain	Most animals can adapt and will migrate with their habitats	Processes unaffected except where habitat availability changes
Risk	R_s: low	R_s: low	R_s: low	R_s: low
	R_l: uncertain	R_l: substantial	R_l: uncertain	R_l: uncertain
Climate variability	Extreme changes in rainfall could alter erosion pattern and sediment supply Reduced inputs of freshwater could change salinity profiles and lead to reduced water clarity	Reduction in total rainfall or extended dry periods could limit plant growth and alter habitat distribution Increased turbidity could limit seagrass distribution	Most animals are "estuarine" species pre-adapted to extreme conditions so should be relatively unaffected except in extreme cases	Large changes in freshwater input could alter bio-geochemical processes and change sediment and nutrient transport patterns
Risk	R_s: low	R_s: low	R_s: low	R_s: low
	R_l: uncertain	R_l: uncertain	R_l: uncertain	R_l: uncertain
Extreme events	Large cyclones can increase erosion temporarily	Large cyclones can severely damage mangroves and seagrass beds	Short-term direct impacts on animal populations, more extensive indirect impacts via damage to habitats and food sources	Short-term disruption of many processes
Risk	R_s: moderate	R_s: high	R_s: moderate	R_s: moderate
	R_l: increasing	R_l: increasing	R_l: increasing	R_l: increasing
Invasive species and range shifts	No effect	Isolation from the mainland makes introduction of invasive plants unlikely unless carried in by birds	Isolation from the mainland makes accidental introduction of invasive fish unlikely. Reports of introduced cane toads *Bufo marinus* on Hinchinbrook Island but no reports from Deluge Inlet	Little impact likely to processes given low likelihood of invasive species impacts
Risk	R_s: nil	R_s: low	R_s: low	R_s: low
	R_l: nil	R_l: unknown	R_l: increasing	R_l: unknown
Boat traffic	Increased boat traffic will increase bank erosion	Increased erosion of banks can undermine plants and inhibit colonisation of new banks by mangroves. Pollutants from fuel spillage and wastes can inhibit plant growth	Loss of habitat due to erosion can impact animal communities, as can increases in pollutants. Plastic and other rubbish can kill marine reptiles, birds and mammals	Impacts to habitats and ecosystems can flow through to impact processes
Risk	R_s: moderate	R_s: moderate	R_s: moderate	R_s: low
	R_l: increasing	R_l: increasing	R_l: increasing	R_l: increasing
Fishing (recreational)	Direct habitat damage from bait collection	Minor damage from anchoring and tying to mangroves	Direct removal of fisheries and bait species Substantial likelihood of extended damage from "ghost fishing" due to lost crab traps	Extensive removal of species can impact food webs and other processes
Risk	R_s: low	R_s: low	R_s: moderate	R_s: low
	R_l: increasing	R_l: increasing	R_l: increasing	R_l: increasing
Fishing (commercial)	Low except for boat-wash issues Little likelihood of increase because of wild river status and fisheries limitations	Little direct impact	Direct removal of fisheries species Substantial likelihood of extended damage from "ghost fishing" due to lost crab traps	Extensive removal of species can impact food webs and other processes

(continued)

Table 1 (continued)

	Geomorphology/landform/hydrology	Plants	Animals	Processes
Risk	R_s: moderate	R_s: low	R_s: moderate	R_s: low
	R_l: moderate	R_l: low	R_l: increasing	R_l: increasing
Aquaculture	Little impact	Little impact	Issues with escape of fish from sea-cages in landward estuaries of Hinchinbrook Channel. Potential for effects on wild genetics and increased pressure on resources if mass releases occur	Escaped fish can impact food webs and other processes
Risk	R_s: low	R_s: low	R_s: moderate	R_s: moderate
	R_l: low	R_l: low	R_l: increasing	R_l: increasing
Infrastructure development	Little likelihood because of National Park and Wild Rivers status	Little likelihood because of National Park and Wild Rivers status	Little likelihood because of National Park and Wild Rivers status	Little likelihood because of National Park and Wild Rivers status
Risk	R_s: low	R_s: low	R_s: low	R_s: low
	R_l: low	R_l: low	R_l: low	R_l: low

Text relates to exposure and capacity to adapt
R_s short-term risk, R_l long-term risk

over the next 30 years. The numbers of boats using Hinchinbrook Channel has been increasing over recent years, so although numbers able to camp within the National Park are strictly limited the day to day traffic is burgeoning. Consequently, the current impacts from fishing pressure and visits for ecotourism are likely to intensify, leading to the progressive degradation of estuarine condition that has led to widespread degradation of estuaries around the world (Lotze et al. 2006). One aspect of population growth that may lead to positive outcomes for near-pristine sites like Deluge Inlet is the increasing focus on environmental issues that it brings. For instance, although extensive development for farming and grazing (Ayukai 1998) has left the floodplain of the Herbert River with many environmental issues with the potential to secondarily impact even relatively isolated locations like Deluge Inlet, recent attempts to improve its coastal wetlands have been successful (Veitch et al. 2007). Despite local successes much more extensive measures are required to minimise potential impacts from development (Davis et al. 2013). Given increasing levels of environmental awareness in the community, impacts like increased sewage and urban pollution seem unlikely, as does any resumption of poorly controlled mining such as that caused by tin dredging in the Herbert River's tributaries during the 1940s (Wegner 1984). More worrying are plans for extensive port expansions along the Queensland coast. At present a sugar export facility operating at the southern end of Hinchinbrook Channel has minimal environmental impact, but this could change if the port was to be expanded and used for other exports such as minerals that would be likely to greatly increase the chances of chemical contamination of the area's waterways.

Threats from Nature

The major natural threat to Deluge Inlet comes from tropical cyclones. The most recent of these was Cyclone Yasi in 2011, which caused extensive damage to Deluge Inlet's mangroves. Perhaps the most substantial impact was on *Rhizophora* spp. As well as the damage from trees being blown over and broken, many of the standing trees eventually died because the ends of branches were damaged, meaning the apical leaf buds, where most of the growth of *Rhizophora* occurs, were lost. This prevented regrowth and led to the eventual death of many trees. Although severe, cyclone damage is part of the natural cycle of change that occurs in estuaries of the region, and is responsible for the particular form of forest that prevails.

Other threats come from climate change and sea level rise as a result of global warming (Table 1). Sea level rise is likely to cause progressive inundation of low-lying land; although this is likely to be mitigated to some extent by increases in mangrove growth and sediment build up due to increasing CO_2 levels (Gilman et al. 2008). Effects of sea level rise are likely to be minimal for Deluge Inlet's estuarine ecosystems. With no human imposed barriers, mangroves and seagrasses are free to migrate as sea levels change; with most animals unaffected because they are habitat associated and so will move with their habitats. In contrast, lowland *Melaleuca* wetlands adjacent to many intertidal areas are at risk because the land they can occupy is restricted by the island's mountainous terrain.

Most climate change effects are difficult to predict and depend on the direction and extent of changes in weather patterns. If changes in rainfall patterns are extreme they could alter erosion and sediment supply patterns leading to changes in geomorphology, hydrology and biogeochemistry. Reduced inputs of freshwater could change salinity profiles, leading to reduced water clarity, and alter biogeochemical processes and sediment and nutrient transport patterns. Reduction in total rainfall or extended dry periods could limit mangrove growth, while increased turbidity could limit seagrass distribution. In contrast, most marine animals are "estuarine" species, pre-adapted to extreme conditions (Sheaves 2012), so should be relatively unaffected by direct effects of climate change, with most impacts flowing from alterations to their habitats and sources of nutrition. Large increases in the frequency of large cyclones could lead to progressive shifts in mangrove forest structure if the interval between events was too short to allow recovery. Changes in temperature are likely to cause range extensions as more favourable temperatures allow animals to move to areas they were previously excluded from. This already appears to be happening, with regular reports from professional crabbers fishing in Deluge Inlet of catches of *Scylla olivacea*, a commercial crab previously confined to more northerly waters.

Conclusion

At present Deluge Inlet appears to remain close to its pristine condition, although anecdotal reports suggest that its fish resources were more abundant in the past. Key pressures and threats come from increases in boat visits and fishing due to increasing population and boat ownership, and the possibility of removal of the ban on commercial net fishing in Hinchinbrook Channel. There are opportunities for improvements if threats from fishing, boating and aquaculture can be reduced by tighter management in line with Deluge Inlet's National Park and Wild River status. In particular, at present recreational fishing effort and the number of boat visits are unmanaged. While this does not seem a substantial problem at current levels of activity, it may become so if boating and fishing activities increase. Additionally, the area has a history of poorly managed development. Risks associated with aquaculture ventures, introduction of disease and/or loss of genetic diversity in wild populations, were not considered sufficient to prevent development approvals in a high conservation value area in the cyclone belt. Clearly, good governance of the balance between increased environmental awareness and increasing anthropogenic pressures is the key to the maintenance and even improvement of Deluge Inlet's high ecological vales. With sensitive and effective management, there is no reason that the Inlet's current condition should not be maintained for the foreseeable future.

References

Abrantes K, Sheaves M (2009) Food web structure in a near-pristine mangrove area of the Australian Wet Tropics. Estuar Coast Shelf Sci 82:597–607

Alongi DM, Boto KG, Tirendi F (1989) Effect of exported mangrove litter on bacterial productivity and dissolved organic carbon fluxes in adjacent tropical nearshore sediments. Mar Ecol Prog Ser 56:133–144

Alongi DM, Ayukai T, Brunskill GJ, Clough BF, Wolanski E (1998) Sources, sinks, and export of organic carbon through a tropical, semi-enclosed delta (Hinchinbrook Channel, Australia). Mangrove Salt Marsh 2:237–242

Ayukai T, Miller D, Wolanski E, Spagnol S (1998) Fluxes of nutrients and dissolved and particulate organic carbon in two mangrove creeks in northeastern Australia. Mangrove Salt Marsh 2:223–230

Bird ECF, Hopley D (1969) Geomorphological features on a humid tropical sector of the Australian coast. Aust Geogr Stud 7:89–108

BOM (2012) Australian Bureau of Meterology. http://www.bom.gov.au/qld/

Boto KG, Alongi DM, Nott ALJ (1989) Dissolved organic carbon-bacteria interactions at sediment-water interface in a tropical mangrove system. Mar Ecol Prog Ser 51:243–251

Bouillon S (2011) Storage beneath mangroves. Nat Geosci 4:282–283

Bouillon S, Chandra Mohan P, Sreenivas N, Dehairs F (2000) Sources of suspended organic matter and selective feeding by zooplankton in an estuarine mangrove ecosystem as traced by stable isotopes. Mar Ecol Prog Ser 208:79–92

Bouillon S, Connolly R, Lee SY (2008) Organic matter exchange and cycling in mangrove ecosystems: recent insights from stable isotope studies. J Sea Res 59:44–58

Bridgewater PB, Cresswell ID (1999) Biogeography of mangrove and saltmarsh vegetation: implications for conservation and management in Australia. Mangrove Salt Marsh 3:117–125

Bunt JS (1979) A survey method for estimating potential levels of mangrove forest primary production. Mar Biol 52:123–128

Bunt JS, Bunt ED (1999) Complexity and variety of zonal pattern in the mangroves of the Hinchinbrook area, northeastern Australia. Mangrove Salt Marsh 3:165–176

Bunt JS, Williams WT, Duke NC (1982) Mangrove distributions in north-east Australia. J Biogeogr 9(2):111–120

Burrows DW (2003) The role of insect leaf herbivory on the mangroves *Avicennia marina* and *Rhizophora stylosa*. PhD thesis, James Cook University, Townsville

Campbell JB (1982) New radiocarbon results for north Queensland prehistory. Aust Archaeol 14:62–66

Chmura G, Anisfeld S, Chaoon D, Lynch J (2003) Global carbon sequestration in tidal, saline wetland soils. Glob Biogeochem Cycl 17:1111. doi:10.1029/2002GB001917

Clough B (1998) Mangrove forest productivity and biomass accumulation in Hinchinbrook Channel, Australia. Mangrove Salt Marsh 2:191–198

Davenport J, Black K, Burnell G, Cross T, Culloty S, Ekaratne S, Furness B, Mulcahy M, Thetmeyer H (2003) Aquaculture: the ecological issues. Blackwell Science Ltd., Malde, 89 pp

Davis AM, Lewis SE, O'Brien DS, Bainbridge ZT, Brodie JE (2013) Water resource development and high value coastal wetlands on the lower Burdekin floodplain, Australia. In: Wolanski E (ed) Estuaries of Australia in 2050 and Beyond. Springer, Dordrecht

Deegan L (1993) Nutrient and energy transport between estuaries and coastal marine ecosystems by fish migrations. Can J Fish Aquat Sci 50:74–79

Donato D, Kauffman J, Murdiyarso D, Kurnianto S, Stidham M, Kanninen M (2011) Mangroves among the most carbon-rich forests in the tropics. Nat Geosci 4:293–297

Doupé RG, Lymbery AJ (1999) Escape of cultured barramundi (*Lates calcarifer* Bloch) into impoundments of the Ord river system, western Australia. J R Soc West Aust 82:131–136

Duke NC, Bunt JS (1979) The genus *Rhizophora* (Rhizophoraceae) in north-eastern Australia. Aust J Bot 27:657–678

Duke NC, Bunt JS, Williams WT (1981) Mangrove litter fall in north-eastern Australia. I. Annual totals by component in selected species. Aust J Bot 29:547–553

Fisheries Regulation (2012) Fisheries Regulation 2008. Reprinted as in force on 21 September 2012. State of Queensland, Queensland Government, Brisbane, 655 p

Fratini S, Cannicci S, Vannini M (2000) Competition and interaction between Neosarmatium smithi (Crustacea: Grapsidae) and Terebralia palustris (Mollusca: Gastropoda) in a Kenyan mangrove. Mar Biol 137:309–316

Fratini S, Vannini M, Cannicci S, Schubart CD (2005) Tree-climbing mangrove crabs: a case of convergent evolution. Evol Ecol Res 7:219–233

Furukawa K, Wolanski E (1996) Sedimentation in mangrove forests. Mangrove Salt Marsh 1:3–10

Garratt PA (1993) Spawning of the riverbream, *Acanthopagrus berda*, in Kosi estuary. S Afr J Zool 28:26–31

Gilman EL, Ellison J, Duke NC, Field C (2008) Threats to mangroves from climate change and adaptation options: a review. Aquat Bot 89:237–250

Hamann M, Limpus C, Read M (2007) Chapter 15: Vulnerability of marine reptiles of the great barrier reef to climate change. In: Johnson J, Marshall P (eds) Climate change and the great barrier reef: a vulnerability assessment. GBRMPA, Townsville, pp 465–495

Hinchinbrook Wild Rivers Declaration (2007) http://www.ehp.qld.gov.au/wildrivers/pdf/hinchinbrook_declaration_07.pdf. National Water Commission, Australian Government, Canberra

Hyland SJ (2002) An investigation of the impacts of ponded pastures on barramundi and other finfish populations in tropical coastal wetlands, final report. QDPI, Brisbane

Jennerjahn T, Ittekkot V (2002) Relevance of mangroves for the production and deposition of organic matter along tropical continental margins. Naturwissenschaften 89:23–30

Johnston R, Sheaves M (2007) Small fish and crustaceans demonstrate a preference for particular small-scale habitats when mangrove forests are not accessible. J Exp Mar Biol Ecol 353(2):164–179. doi:10.1016/j.jembe.2007.05.039

Johnston R, Sheaves M (2008) Cross-channel distribution of small fish in tropical and subtropical coastal wetlands is trophic-, taxonomic-, and wetland depth-dependent. Mar Ecol Prog Ser 357:255–270. doi:10.3354/meps07297

Kanowski J, Hopkins MS, Marsh H, Winter JW (2001) Ecological correlates of folivore abundance in north Queensland rainforests. Wildl Res 28:1–8

Langley J, McKee K, Cahoon D, Cherry J, Megonigal J (2009) Elevated CO2 stimulates marsh elevation gain, counterbalancing sea-level rise. Proc Natl Acad Sci U S A 106:6182–6186

Lawler I, Parra G, Noad M (2007) Chapter 15: Vulnerability of marine mammals in the great barrier reef to climate change. In: Johnson J, Marshall P (eds) Climate change and the great barrier reef: a vulnerability assessment. GBRMPA, Townsville, pp 498–513

Lee Long W, Mckenzie L, Roelofs A, Makey L, Coles R, Roder C (1998) Baseline survey of Hinchinbrook region seagrasses. Research publication N°. 51. Great Barrier Reef Marine Park Authority, Townsville

Lopez GR, Levinton JS (1987) Ecology of deposit-feeding animals in marine sediments. Q Rev Biol 62:235–260

Lotze H, Lenihan H, Bourque B, Bradbury R, Cooke R, Kay M, Kidwell S, Kirby M, Peterson C, Jackson J (2006) Depletion, degradation, and recovery potential of Estuaries and coastal seas. Science 312:1806–1809. doi:10.1126/science.1128035

MacIntyre H, Geider R, Miller D (1996) Microphytobenthos: the ecological role of the "secret garden" of unvegetated, shallow-water marine habitats. I. Distribution, abundance and primary production. Estuaries 19:186–201

Marine Parks (2009) Marine Parks (Declaration) Regulation 2006. Reprinted as in force on 18 December 2009. State of Queensland, Queensland Government, Brisbane, 114 p

Marsh H, Lawler I (2000) Dugong distribution and abundance in the northern Great Barrier Reef Marine Park – November 2000. Research publication no. 77. Great Barrier Reef Marine Park Authority, Townsville

Matsui N (1998) Estimated stocks of organic carbon in mangrove roots and sediments in Hinchinbrook Channel, Australia. Mangrove Salt Marsh 2:199–204

Matsuoka T, Nakashima T, Nagasawa N (2005) A review of ghost fishing: scientific approaches to evaluation and solutions. Fish Sci 71:691–702

Micheli F (1993) Feeding ecology of mangrove crabs in north eastern Australia: mangrove litter consumption by *Sesarma messa* and *Sesarma smithii*. J Exp Mar Biol Ecol 171:165–186

Miller D, Geider R, MacIntyre H (1996) Microphytobenthos: the ecological role of the "secret garden" of unvegetated, shallow-water marine habitats. II. Role in sediment stability and shallow-water food webs. Estuaries 19:202–212

Mueller H, Ayukai T (1998) Concentration and molecular weight distribution of dissolved organic carbon in a mangrove creek in the Hinchinbrook area, Australia. Mangrove Salt Marsh 2:231–235

Nagelkerken I, Sheaves M, Baker R, Connolly RM (in review) The seascape nursery: a novel spatial approach to identify and manage nurseries for coastal marine fauna. Ecological Applications

Oakes JM, Connolly RM, Revill AT (2010) Isotope enrichment in mangrove forests separates microphytobenthos and detritus as carbon sources for animals. Limnol Oceanogr 55:393–402

Olds AD, Connolly RM, Pitt KA, Maxwell PS (2012) Primacy of seascape connectivity effects in structuring coral reef fish assemblages. Mar Ecol Prog Ser 462:191–203

Parra G (2007) Observations of an Indo-Pacific humpback dolphin carrying a sponge: object play or tool use? Mammalia 71:147–149

Peterson CH (1991) Intertidal zonation of marine invertebrates in sand and mud. Am Sci 79:236–249

Preen T (2000) Dugongs, boats, dolphins and turtles in the Townsville-Cardwell region and recommendations for a boat traffic management plan for the Hinchinbrook Dugong protection area. Research publication no. 67. Great Barrier Reef Marine Park Authority, Townsville

Quammen ML (1984) Predation by shorebirds, fish, and crabs on invertebrates in intertidal mudflats: an experimental test. Ecology 65:529–537

Robertson AI, Duke NC (1990) Recruitment, growth and residence time of fishes in a tropical Australian mangrove system. Estuar Coast Shelf Sci 31(5):723–743

Robertson A, Kirshner S, Smyth P, Charles S, Bates B (2006) Subseasonal-to-interdecadal variability of the Australian monsoon over north Queensland. Q J R Meteorol Soc 123:519–542

Schories D, Barletta-Bergan A, Barletta M, Krumme U, Mehlig U, Rademaker V (2003) The keystone role of leaf-removing crabs in mangrove forests of north Brazil. Wetl Ecol Manag 11:243–255

Sheaves M (1992) Patterns of distribution and abundance of fishes in different habitats of a mangrove-lined tropical estuary, as determined by fish trapping. Aust J Mar Freshw Res 43:1461–1479

Sheaves M (1995a) Large lutjanid and serranid fishes in tropical estuaries: Are they adults or juveniles? Mar Ecol Prog Ser 129:31–40

Sheaves M (1995b) Effect of design modifications and soak time variations on Antillean-Z fish trap performance in a tropical estuary. Bull Mar Sci 56:475–489

Sheaves M (2005) Nature and consequences of biological connectivity in mangrove systems. Mar Ecol Prog Ser 302:293–305

Sheaves M (2009) Consequences of ecological connectivity: the coastal ecosystem mosaic. Mar Ecol Prog Ser 391:107–115

Sheaves M (2012) Ecosystem equivalence and the ability to generalise: insights from global consistencies in mangrove fish assemblages. Mar Ecol Prog Ser 461:137–149

Sheaves M, Johnston R (2009) Ecological drivers of spatial variability among fish fauna of 21 tropical Australian estuaries. Mar Ecol Prog Ser 385:245–260

Sheaves M, Molony B (2000) Short-circuit in the mangrove food chain. Mar Ecol Prog Ser 199:97–109. doi:10.3354/meps199097

Sheaves M, Molony BW, Tobin AJ (1999) Spawning migrations and local movements of a tropical sparid fish. Mar Biol 133:123–128

Sheaves M, Johnston R, Abrantes K (2007a) Fish fauna of dry tropical and subtropical estuarine floodplain wetlands. Mar Freshw Res 58:931–943

Sheaves M, Johnston R, Molony B, Shepard G (2007b) The effect of impoundments on the structure and function of fish fauna in a highly regulated dry tropics estuary. Estuar Coast 30:507–517

Shoo LP, Williams Y (2004) Altitudinal distribution and abundance of microhylid frogs (*Cophixalus* and *Austrochaperina*) of north-eastern Australia: baseline data for detecting biological response to future climate change. Aust J Zool 52:667–676

Shoo LP, Williams SE, Hero JM (2005) Potential decoupling of trends in distribution area and population size of species with climate change. Glob Chang Biol 11:1469–1476

Stephenson PJ (1990) Layering in felsic granites in the main East pluton, Hinchinbrook Island, north Queensland, Australia. Geol J 25:235–336

Trenerry M, Werren G (1993) Possum assemblages in rainforest of the Carbine Tableland, NEQ, with special reference to *Hemibelideus lemuroids*. Mem Qld Mus 34:188

Underwood G, Kromkamp J (1999) Primary production by phytoplankton and microphytobenthos in estuaries. Adv Ecol Res 29:93–153

Veitch V, Burrows DW, Hudson D, Butler BM (2007) Removal of aquatic weeds from Lagoon Creek, Herbert catchment, north Queensland: trialling novel removal methods and demonstration of environmental benefits. ACTFR report no. 07/15, Australian Centre for Tropical Freshwater Research, Townsville

Wegner J (1984) Hinchinbrook Shire during world war two. In: Dalton BJ (ed) Lectures on north Queensland history, vol 4. James Cook University of North Queensland, Townsville, pp 213–229

Wetland Info (2012) Queensland Department of Environment and Resource Management and Queensland Wetlands Program. http://wetlandinfo.derm.qld.gov.au/wetlands.html. Accessed 20 Feb 2012

Wolanski E (2007) Estuarine ecohydrology. Elsevier, Amsterdam

Wolanski E, Mazda Y, King B, Gay S (1990) Dynamics, flushing and trapping in Hinchinbrook Channel, a giant mangrove swamp, Australia. Estuar Coast Shelf Sci 31:555–579

Wolanski E, Spagnol S, Ayukai T (1999) Field and model studies of the fate of particulate carbon in mangrove-fringed Hinchinbrook channel, Australia. Mangrove Salt Marsh 2:205–221

Yanez-Arancibia A, Lara Dominguez AL, Pauly D (1994) Coastal lagoons as fish habitats. Elsevier Oceanogr Ser 60:363–376

Recent, Rapid Evolution of the Lower Mary River Estuary and Flood Plains

David Williams

Abstract

The Lower Mary River is located 90 km to the east of Darwin, the capital city of the Northern Territory of Australia. The wetlands of the Lower Mary are a valuable ecological and economic resource supporting tourism and primary industry. The wetlands have had minimal disruption by human interaction and form a large component of the Northern Territory's coastal wetland systems. The top end of the Northern Territory is in the wet-dry tropics where the annual rainfall of 1,700 mm falls mainly between December and April. The region also experiences regular cyclonic activity during these periods. The wetlands and floodplains formed up to 5,000–8,000 years ago when sea levels stabilised. The coastal plains prograded rapidly and extensive mangrove forests and freshwater habitats formed. It has been evident that during the last 50 years the coast has receded, the estuarine channels have expanded and become deeper and wider and tributary channels have grown across the flood plains invading previous freshwater environments. The cut and no recover model of coastal retreat is proposed as the dominant process of coastal erosion allowing the tidal creeks to expand onto the floodplains and change the environment from dominantly freshwater to saltwater. It is forecast that over the next 40 years the Lower Mary River estuary may continue to grow and become similar to neighbouring estuaries and those of the north-facing coast of the Northern Territory.

Keywords

Erosion • Sedimentation • Evolution • Salinity • Cyclones • Wetlands • Sediment fluxes

D. Williams (✉)
Australian Institute of Marine Science (AIMS), 23 Ellengowan Drive,
Brinkin, NT 0811, Australia
e-mail: dk.williams@aims.gov.au

E. Wolanski (ed.), *Estuaries of Australia in 2050 and Beyond*, Estuaries of the World,
DOI 10.1007/978-94-007-7019-5_16,

Box 1

David Williams studied the Mary River estuary. The wetlands of the estuary are a valuable ecological and economic resource supporting tourism and primary industry, and have had minimal human impact. During the last 50 years the coast has receded, the estuarine channels have expanded and become deeper and wider, and tributary channels have grown across the flood plains invading previous freshwater environments. Thus the simplest management option for the Mary Estuary is managed retreat and adaptation.

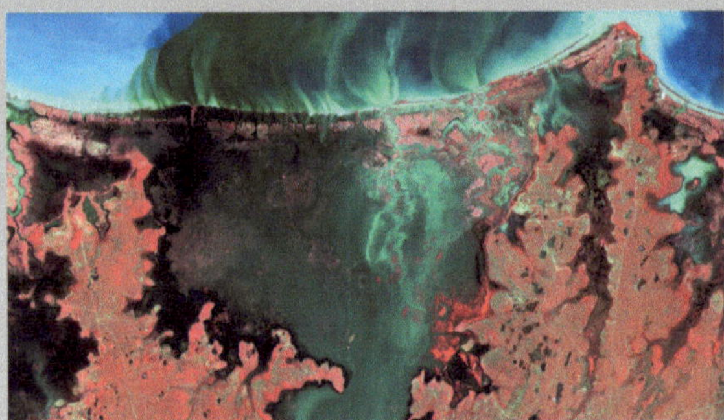

It is forecast that over the next 40 years the Lower Mary River estuary may continue to grow and become similar to neighbouring estuaries and those of the north-facing coast of the Northern Territory.

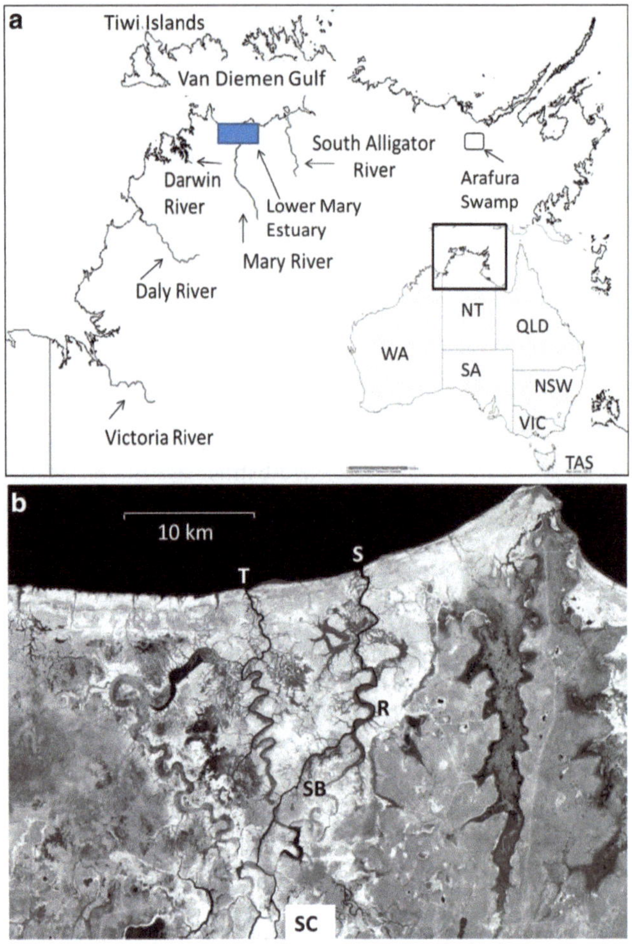

Fig. 1 (**a**) A location map of key rivers in the Northern Territory. (**b**) A LANDSAT view of the Lower Mary River flood plains. *T* is Tommycut Creek, *S* is Sampan Creek, *R* is Roonees Lagoon, *SB* is the S bends, *SC* is Shady Camp

Introduction

The estuary of the Lower Mary River (Figs. 1 and 2) is located 90 km to the east of Darwin in the Northern Territory. The coastline of the Lower Mary estuary forms part of the southern boundary of Chambers Bay in Van Diemen Gulf (Fig. 1). Van Diemen Gulf is a large shallow embayment with a tidal range of 4 m. The Mary River has a catchment area of 6,000 km^2 and flows northward into the Gulf.

The Lower Mary River is both a valuable ecologic and economic resource. It is close to Darwin, the capital city of the Northern Territory of Australia and the river provides a valuable recreational area, mainly for fishing but also for nature observers especially those who wish to view the iconic saltwater crocodile. The crocodile populations are among the densest in the Northern Territory if not the world (Webb et al. 1987). The Mary River floodplain is an important part of the network of over 10,000 km^2 of coastal freshwater wetlands in the Northern Territory. These areas form some of the last remaining significant areas of wetland in Australia unaffected by water extraction or other intensive agricultural development (Whitehead et al. 1990, p. 85; Whitehead and Saalfeld 2000, p. 496). The river has extensive wetlands that support a diverse range of vegetation and are a main habitat for migratory birds including magpie geese. The system also supports large yields of fish including barramundi, threadfin salmon and snapper (de Lestang and Griffin 2000). Parts of the wetlands are now used to condition cattle in the late dry season for the meat industry. The economic input of the Lower Mary River wetlands to the Northern Territory economy is estimated to be in excess of $AUD 43 million (McInnes 2003).

Like most of the coastal systems in the Northern Territory the area has not had major impacts from human populations. There are no large industrial or urban developments. The exception is that in the earlier part of the twentieth century Asiatic water buffalo, *bubalus bubalis*, were introduced and the numbers were not well controlled. The main impact of the water buffalo was that they create swim channels and some of these eroded and joined with other tributary channels. There has also been anecdotal evidence of professional fishing interests removing shoals that formed natural breaks between

Fig. 2 Photographs of (**a**) freshwater floodplain trees killed by saline intrusion, (**b**) breaches of the coastal cheniers, (**c**) mangrove stumps in the eroded shoreline, (**d**) initiation of tidal creeks through a breach in the shoreline chenier, (**e**) a bank slump in the main channel

pools in the river in an attempt to have better access but these stories have never been substantiated. These impacts may have helped accelerate erosion and the change from a predominant wetland system with only a minor network of estuarine channels but it cannot be ignored that other systems that have had a similar geomorphological beginning have evolved into large estuarine channel systems. The Lower Mary River may well provide a window into the processes of evolution of the other estuarine environments in the Northern Territory.

Climate and Geomorphology

The climate is described as wet-dry tropical. Average daily maximum temperatures are approximately 32 °C and the average daily minimum temperature is approximately 24 °C. The average annual rainfall is approximately 1,700 mm. The rainfall occurs mainly in the months from December to April. The rainfall is not evenly distributed across the season but often comes as the result of tropical low depressions quite often associated with cyclonic activity. On average the region experiences 1–2 cyclones per year.

Approximately 8,000 years ago the coastal valleys in the region were flooded as sea level continued to rise. At approximately 6,000 years ago when sea level stabilised at or close to its present level mangrove forests occupied much of what is now estuarine plain in a period termed the big swamp phase (Woodroffe et al. 1986). This stage was superseded as sea levels fell slightly (1–2 m) and the estuarine systems formed defined channels as vertical sedimentation continued to infill the estuary and mangroves were replaced by sedge/grassland. The coast prograded rapidly during this time and lines of chenier ridges formed along the coast forming effective barriers that precluded the spring tides from most of the floodplains. Extensive freshwater wetlands formed behind these coastal barriers.

The Mary River has a catchment area of 6,000 km^2 before it flows into the headwaters of the wetland system. From there the riverine channels become discontinuous forming numerous billabongs and extensive pool habitats. Billabong is an Australian term describing a section of still water adjacent to a river, cut off by a change in the watercourse. It is also sometimes used to describe large, usually freshwater, pools. This 60 km section of wetland is typical of what the entire system looked like before the tidal creek expansion. From Shady Camp (shown on Fig. 1b) it is a further 35 km to the coast. It is this section that has undergone rapid changes in recent time. It is worth noting that the wetland system extends over 100 km from the coast and that the bed levels of the pools in the system are below the level of the lowest ocean tides. It is the barriers between billabongs and pools in the system that prevents the tides from reaching as far inland as the neighbouring estuaries.

Hydrology and Tides

The Mary River is a mostly perennial stream only ceasing to flow in exceptionally dry years. An analysis of flow data from the Northern Territory Government Water Resources database shows that late in the dry season (June –November) stream flows are maintained by groundwater discharge and flow rates are low, typically less than 1 m^3/s. Some additional flows from groundwater discharge are added by tributary creeks but these flows are more than accounted for by evapotranspiration. The level in the wetland pools recedes throughout the dry season. During the wet season, which is typically December to March flows in the Mary River are often above 40 m^3/s. February and March had statistically the highest flow periods with flows more than 100 m^3/s for more than 50 % of the time. Bank full floods are in the order of 2,000 m^3/s and have an average recurrence interval of 4 years, i.e. bank full flow has a 25 % probability of being equalled or exceeded in any year. Tidal flows in Sampan Creek which is the main estuarine arm of the Lower Mary system has peak spring tide flows of around 600 m^3/s at the mouth and 200 m^3/s midway along the estuary. These flows are equalled or exceeded by flood water inflows during the wet season 10 and 40 % of the time respectively. Therefore during the wet season the estuary of the Lower Mary is significantly influenced by fresh water inputs due to floods. In the dry season there is no influence from fresh water and the salinity in the upper estuary near Shady Camp can exceed the salinity of the ocean in Van Diemen Gulf due to the effects of evapotranspiration. Depending on the time of the year the Lower Mary estuary may exhibit the properties of an inverse estuary or develop a salinity maximum zone similar to other estuaries in the region (Wolanski 1986; Williams 2006).

Data from the Darwin tide gauge, which has the longest period of record in the Northern Territory (1959 to date), show annual oscillations in mean sea level and this is shown in Fig. 3. The variations in mean sea level are mainly due to the strength of the North – North West wind during the monsoon period from December to April. The coastal shelf in this part of northern Australia is wide, approximately 100 km, and shallow, with depths often less than 30 m and winds push the water up against the coast. The sea bed gradient is very small and an increase in sea level of 0.1–0.2 m results in 1–2 km of coastal area inundated with water for longer periods of time making them susceptible to erosion. Figure 3 also reveals that the mean sea level is rising at a mean rate of approximately 5 mm/year. The linear regression through the annual mean sea levels has an r^2 value of 0.406 which indicates that a linear rise in sea level is an influencing process. The strength of the monsoon period which brings strong North-north westerly winds and tropical low depressions which can develop into cyclones is another dominant process that affect sea level in the region.

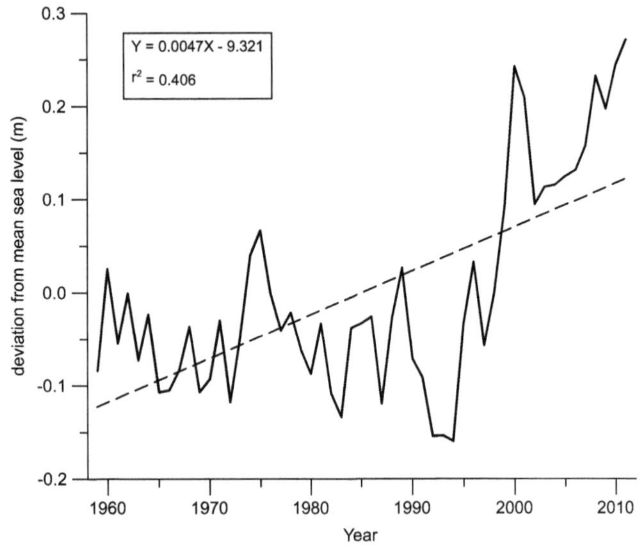

Fig. 3 Time-series plot of the annual mean sea level (*MSL*) in Darwin

Waves in Van Diemen Gulf are mainly wind driven and when the wind is from a northerly direction they impact directly on the coast. North westerly winds are somewhat protected from the Tiwi Islands but there is still a considerable fetch of 70 km. Simple wind fetch models show that for storm winds wave heights can reach 1.3–1.5 m and at high tide these will break over coastal cheniers. During stronger cyclonic winds wave heights could be in the order of greater than 3 m. No wave data exist but it is reasonable to assume that wave climate offers erosional forces during the wet season when storms are frequent and prolonged.

Geomorphology

Woodroffe et al. (1986) have described the geomorphological development of the coastal floodplains of the Alligator Rivers Region. It is logical to assume that other estuarine systems in the Northern Territory have developed in a similar way. For example the Victoria, Daly, Adelaide, South and East Alligator Rivers are large macro tidal estuarine systems where the channels reach over 100 km inland with only a minor reduction in tide levels. Only a few coastal wetland systems without defined estuarine channels remain in the Northern Territory. The Lower Mary River along with the Arafura Swamp in Arnhem Land is one of these systems. The Lower Mary estuary does not have such a well-developed main channel but, as shown below, the recent rapid expansion of the channel network over the past 70 years or so indicates that the Lower Mary estuary may now be developing in the same way as other estuarine systems. The neighbouring South Alligator River to the east of the Lower Mary River has also been showing increased tidal creek extension over the past 40 years (Cobb 1998).

Woodroffe and Mulrenna (1993) described the changes that the Lower Mary River has undergone over at least the past 70 years. A map of the expansion of the system using historic aerial photos from the Second World War coastal reconnaissance photos in the 1940s to aerial survey photos of the late 1980s was made and is shown in Fig. 4a. A later schematic map using Landsat TM satellite imagery from 2005 is also used to show that the system continues to expand and that coastal erosion especially to the west of Tommycut Creek, this is shown in Fig. 4b. The result of the expansion of the tidal creek network is that up to 17,000 ha of freshwater vegetation including large stands of Melaleuca paperbark forest have been destroyed, shown in Fig. 2.

A major disturbance to the integrity of the Mary River floodplain may have occurred when the Asiatic water buffalo were introduced on the floodplain as a source of meat and hides to promote a local industry. This was done in the early part of the twentieth Century and several abattoirs were set up near the wetlands. This industry ceased during the 1980s and the buffalo were left to run wild until they were mostly finally removed as part of a bovine tuberculosis eradication campaign. The main impact that buffalo had was the formation of swim channels. There have been mixed reports on the numbers of buffalo that were present on the floodplains at any one time. Whatever the numbers, whether high or low, the swim channels are large enough that they are represented on 1:100,000 scale topographic maps. The swim channels are formed when herds of buffalo wade and roll on the wet floodplains and scour out pools so they can immerse themselves in mud and water. Overtime the scour holes

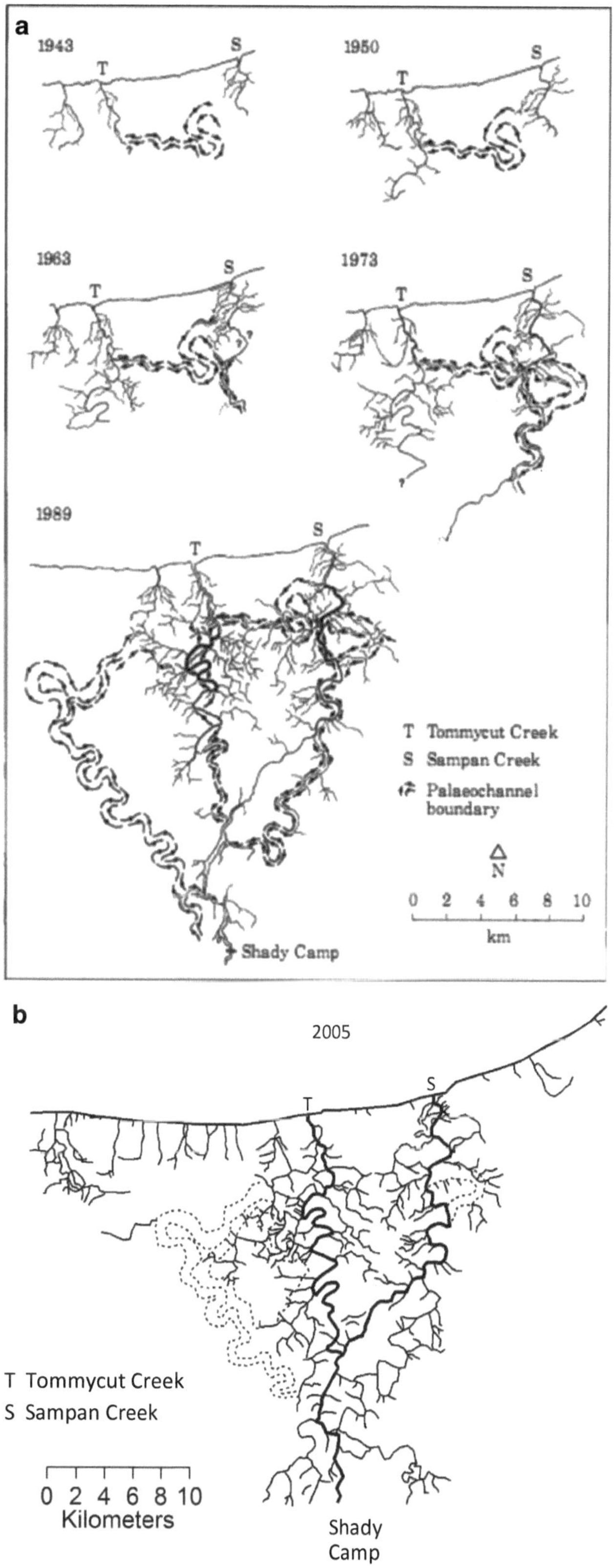

Fig. 4 (**a**) The pattern of tidal creeks from 1943 to 1989 in the Mary River flood plain (Reproduced from Woodroffe and Mulrennan, with permission). (**b**) The pattern of tidal creeks in 2005; note the numerous tidal creeks resulting from breaches in the shoreline chenier

become elongated and a similar in shape to narrow billabongs. On the topographic maps the swim channels are shown to radiate out from areas of slightly higher ground. Over time the newly expanding tidal tributaries have joined with the swim channels and this has provided a mechanism for the tidal channels to move across the floodplain at a faster rate. While this mechanism is one way it is not the only way tidal creeks have expanded.

Chappell and Grindrod (1983) discussed two mechanisms of coastal progradation which are thought to have oscillated during the Holocene period and been responsible for coastal development and the formation of cheniers which serve to protect the floodplain from the highest tides and pond freshwater. During the rapid progradation mode extensive mangroves forests exist on the coastal fringe. These mangrove forests help filter high frequency waves from impacting the coast and trap sediment. They also offer a suitable habitat for various molluscs and shell beds form along the coastal fringe. During cyclones many of the mangroves are denuded and the shell beds are washed up onto the coast. These shell beds form ridgelines known as cheniers and over time calcium dissolves from the shells and cements the surface and therefore forms a hard barrier. The level of the chenier is above the high tide level and so acts to inhibit inundation of the floodplain by tides. An ample sediment supply allows the coast to recover and prograde and the mangrove colonies re-establish. This is shown in Fig. 5a.

In the cut and recover mode there are initially less mangrove forest stands along the coast than before. Shell beds still exist along the coastline and are available to be deposited landward as storm surge deposits during cyclones. The coast recedes as sediment is eroded from the coast line and many mangroves are denuded but enough remain so as the colony can recover in the ensuing periods of relative calm. Eventually the coastline recovers, mangroves forests re-establish and the coast progrades slightly and shell beds re-establish. This process is shown in Fig. 5b.

However it is evident, and shown before in Fig. 2, that the coast is now eroding and receding and many breaches are forming in the shoreline and tidal creeks penetrating onto the floodplain. This is the cut and no recovery mode, which is shown in Fig. 5c. In this mode the initial phase is as per the cut and recover mode with a stand of mangroves along the coast and a coarse sediment supply of shell available. During cyclone activity the shell material is deposited and the mangroves are again denuded. However in this mode there is not a long enough calm period or available sediment supply for the coastline to recover. Ensuing cyclones and storms further erode the coast and the last of the mangroves are denuded. The coarse sediment supply is removed from the system and mud deposits form along the coast, forming a muddy environment that is not conducive for mollusc colonisation. The coast is now in a recession mode (Fig. 5c), it continues to erode and the chenier ridges formed previously also start to erode.

The cut and no recovery model helps to explain the widening and deepening of the mouth of the tidal creeks (Figs. 6 and 7). As successive storms impact the coast and flooding occurs, the cheniers have eroded and are not being replaced with additional sediment. The result is numerous breaches in chenier barriers (Fig. 2) have occurred and the high tides are more frequently inundating the coastal plain (Fig. 6). The frequent salt water inundation scalds the floodplain surface and kills the freshwater vegetation (Fig. 2), which then does not regrow, thus exacerbating the erosion and expansion inland of the creeks (Fig. 4).

As the coastline is now eroding, many remnant large mangrove trees now exist only as stumps (Fig. 2). The mangroves that remain are small and do not assist to protect the coast nor do they provide an environment where shelly substrates can form. Shell material that previously was broken down into calcareous sand is now also being eroded and is being replaced by mud (Fig. 5c).

During the coastal storms and cyclones mangroves are depleted from along the shore line. The sediment budgets change, the coast is breached and new tributaries start to form. These are halted at previous chenier lines but eventually these are breached (Fig. 2) and the tributaries grow. Both the constant action of the tides and the seasonal floods combine to force the creeks to grow.

The main channels have grown and deepened and have reached the full tide range of the ocean tides. Mangrove colonisation on the channel banks is rapid and thick forest stands develop. The channel banks are formed in fine cohesive mud and due to the large semi diurnal tides the banks remain saturated with water as there is insufficient time for the soil to drain out the saline groundwater. Cores have been taken from the channel banks and tested in the soil engineering laboratory at Charles Darwin University. The sediment cores failed all compressive tests and so exhibit no ability to bear heavy loads such as mangrove forests. The result is that after mangrove forest density and size reaches a limit where the exposed banks at low tide cannot bear the weight the bank slumps. The bank slump has the form of a slip circle failure and the result is a channel bank with near vertical sides. Spring tides regularly overtop the channel banks and flood the adjacent plain and when the water returns to the channel during the ebb tide a small water fall is formed (Fig. 8). The impact and turbulence created by the waterfall then undercuts the channel in a similar manner to gully formation. The head of the bank is then undercut and the bank retreats toward the floodplain. This process is continually repeated and the tributary channel that has been formed migrates across the floodplain initially perpendicular to the main source channel. Subsequent flood and ebb flow into the new tributary channel

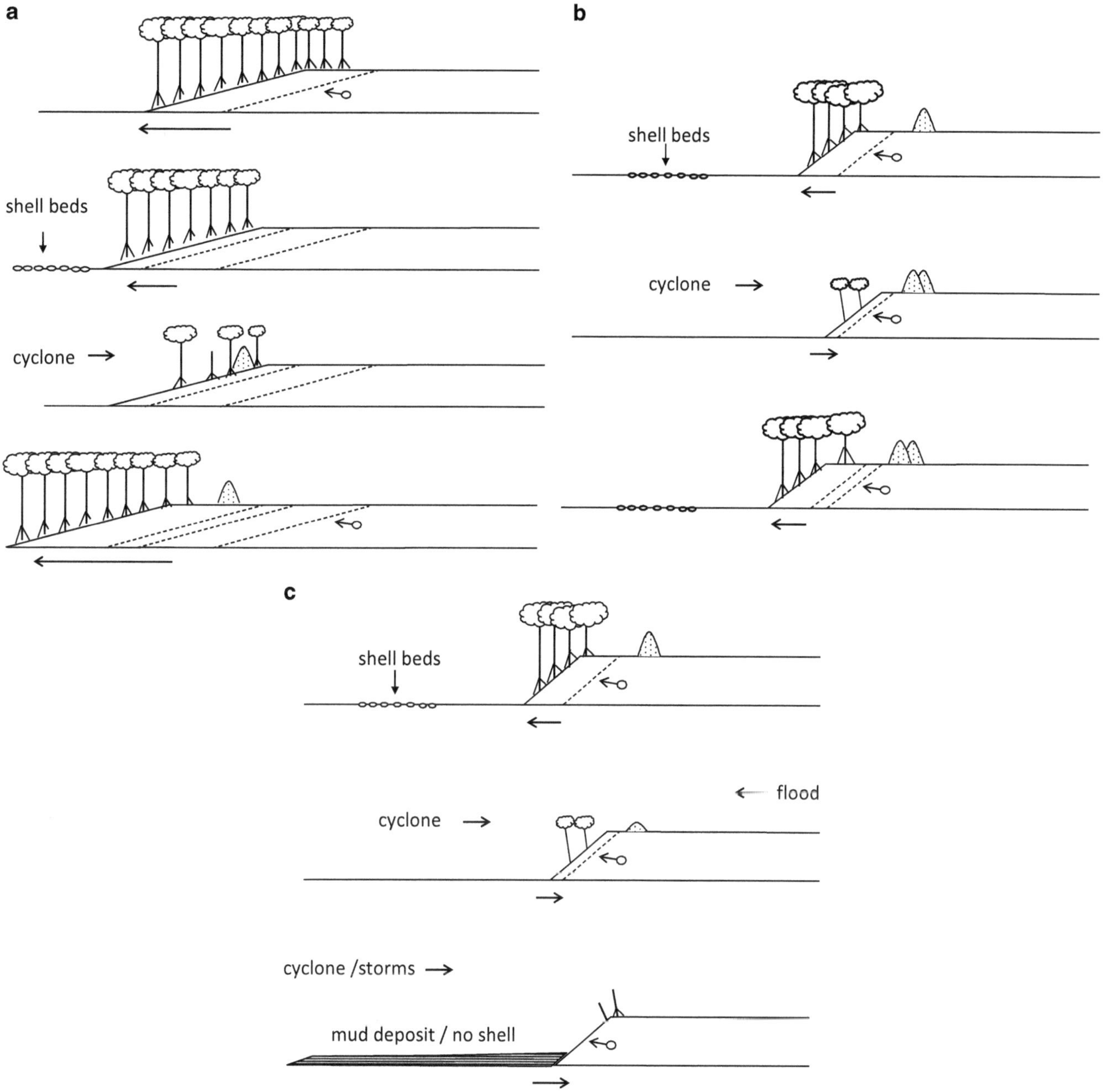

Fig. 5 (**a**) Rapid progradation mode. (**b**) The cut and recovery mode. (**c**) The cut and non recovery mode

shapes the form of the channel. After a period of time the bed level of the tributary channel approaches the bed level of the main channel (Fig. 6). In this way the tidal tributary also becomes a main tributary and new tidal creeks can be initiated from these channels creating a network of channels that spread across the floodplain surface (Fig. 4). Lottig and Fox (2007) described salt marsh systems in the United States where bank collapses were due to the death of vegetation causing bank instabilities. In the Lower Mary the banks at low tide cannot bear the weight of a developed mangrove forest and the banks slump with live vegetation. Over a short time period the mangroves that have slumped die but the strength of the tidal currents erode the bank slumps and disperse the sediment and the vegetation plays no further role in bank stability. The bank slumping not only initiates new tributary growth but also promotes widening of the channel and lateral migration. Analysis of satellite imagery of the Lower Mary estuary shows there are far more tributary channels formed on the outside of channel bends than on the inside of bends or on straight sections.

Observations in the field showed that headward growth of the channels was more rapid during the wet season when

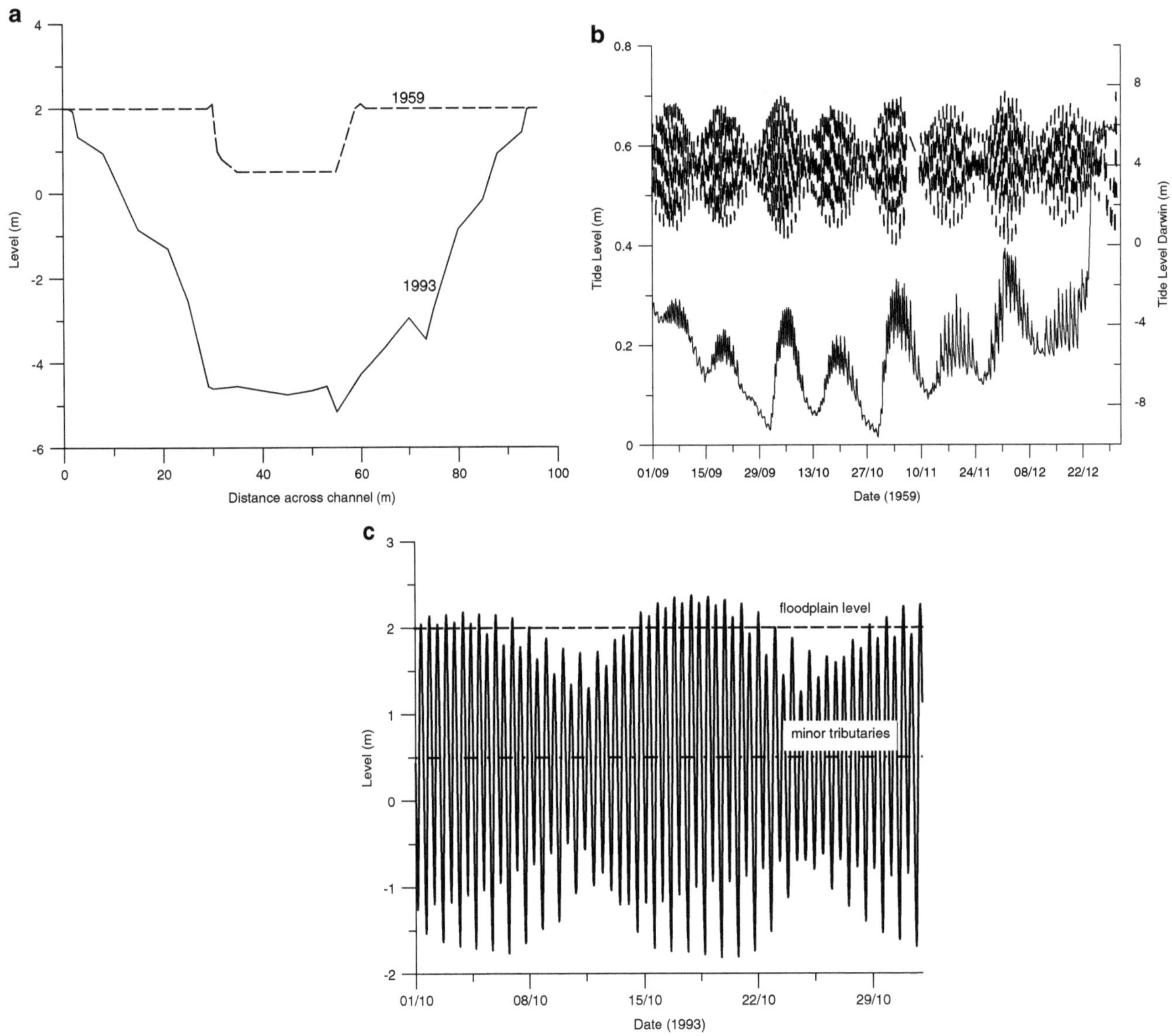

Fig. 6 (**a**) The change in cross-section at Roonees Lagoon from 1959 to 1993. (**b**) Time-series plot of the tides at Roonees Lagoon and Darwin in 1959. (**c**) Time-series plot of the tides at Roonees Lagoon in 1993

freshwater flowed in the channel as well as the periodic semi diurnal tide. Several soil samples were collected and subjected to dispersion tests (Carter and Gregorich 2006). All soil samples exhibited higher dispersion in fresh water than salt water. During the dry season the periodic inundation wets the soil and then has time until the next spring time to evaporate. This leaves behind salt in the soil. Over the course of the dry season the accumulated salt content increases and breaks down the structure of the soil. When the wet season arrives and the first fresh water flows occur, whether from rainfall direct at the site or overbank flow from floods, the salinised soil is dispersed causing an ever faster tidal creek expansion than gully head expansion and erosion alone. This mechanism of channel growth warrants further examination.

Tidal River Hydraulics

The main channel of the Lower Mary system has grown rapidly over time. Figure 6a shows the cross sectional form at Roonees lagoon, located 15 km from the coast. In 1959 the channel width was 25 m and the channel depth was 2.5 m. A survey taken at the same site in 1993 shows a channel that was then 100 m wide and up to 10 m deep.

The tidal range in 1959 had a maximum range of 0.3 m (Fig. 6b). These maximum ranges correlate with the spring neap tide cycle of the offshore tides. This indicates that there were barriers in the channel between the sea and the site. The tides could only enter the site during the larger spring tides. In 1993 the tidal range measured from an installed tide gauge

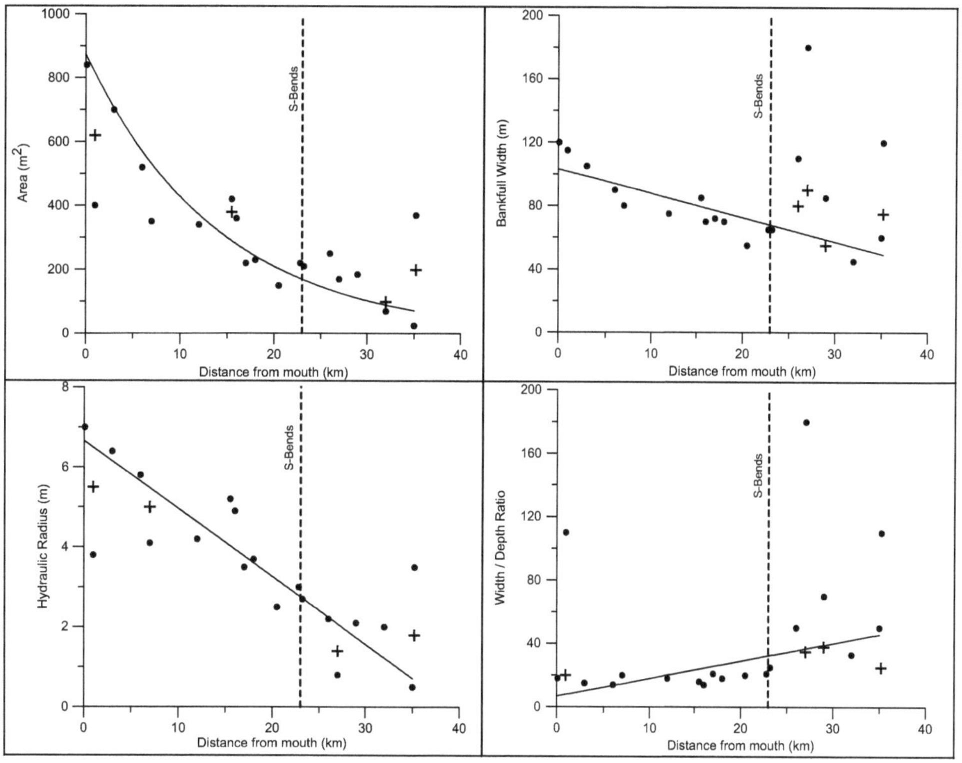

Fig. 7 The change in the geometric properties of Sampan Creek from 1993 (.) to 1996 (+). Only the significant changes in 1996 are shown; they show the channelization of the creek with the removal of the natural barriers and infilling of billabongs

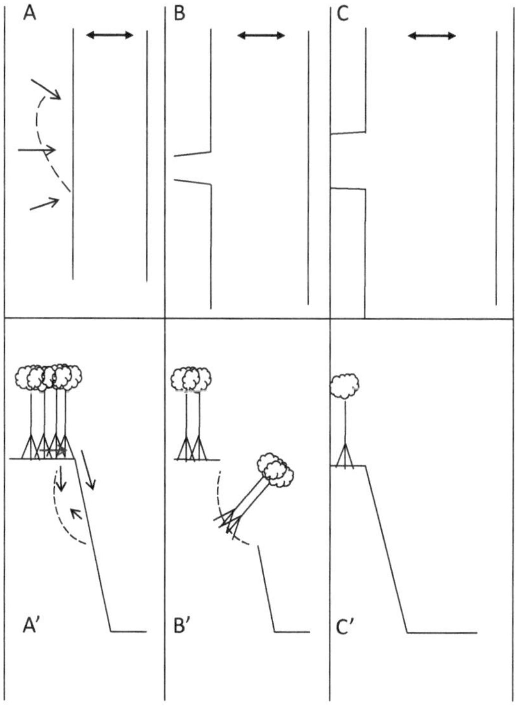

Fig. 8 Sketch of tidal creek initiation due to bank slumping. *A*. A *slip circle* bank collapse begins. *B*. The *circle* area fails and falls into the channel taking mangroves and water draining back after high tides continue to erode the area and make it larger. A channel forms at the head of the *slip circle* perpendicular to the main channel. *C*. The area continues to erode and the tributary channel deepens and the main channel widens due to the bank slumpage. *A′*, *B′*, *C′* is the order of occurrence for channel tributary and channel widening processes

was up to 4 m which is the same as the offshore tide. This clearly shows that any barrier between Roonees Lagoon and the sea had fully eroded by 1993 (Fig. 6c).

Large river floods also move through the system. The impact of floods has not been investigated in previous studies of the Lower Mary. Yet the impact is significant because with the increase in channel area, the drainage of river floods is now more efficient and the floodplain is inundated for less time (Fig. 9). This is likely to impact the ecology of the flood plains but no floodplain ecology studies have been carried out.

What Will the Estuary Look Like in 2050?

It has been hypothesised that a sea level rise by itself, ignoring the natural geomorphological changes presently occurring in the Mary Estuary, will invoke a new big swamp phase. But this hypothesis now appears unreasonable because of the evidence shown above that the channels are deepening and increasing in length, thus sediment is exported seaward and the system is not infilling. The export is enhanced by the breaching of the cheniers and this creates numerous side channels through which flood waters and their riverine sediment are exported offshore (Fig. 10).

More likely is that the coast will recede further until a concave shoreline is formed to ultimately resemble that of the Alligator Rivers (Fig. 1). At the moment the Mary River shoreline is quite linear. An analysis of map and satellite

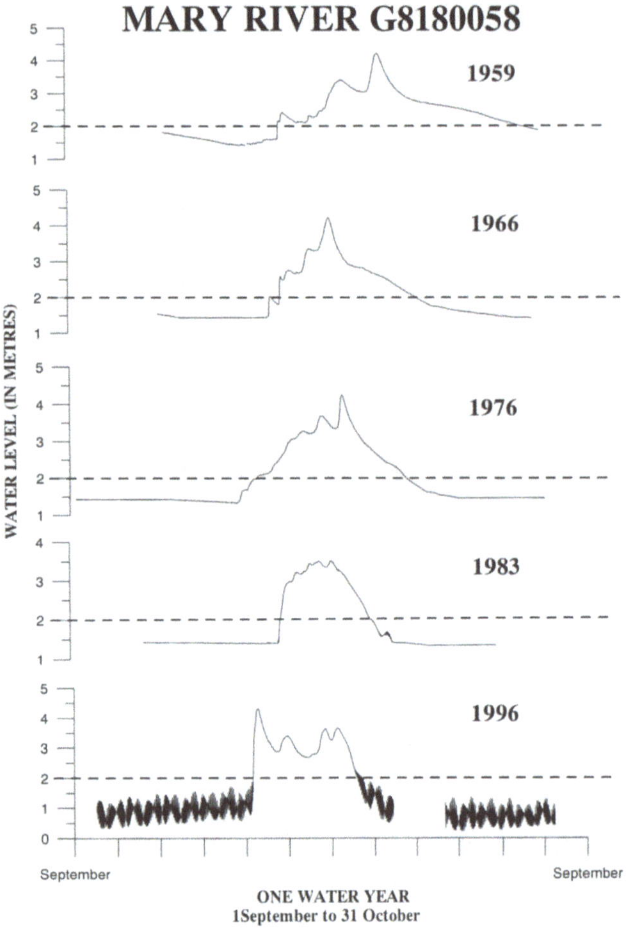

Fig. 9 Time-series plot of the water level at Shady Camp from 1959 to 1996. Note the increasing steepness of the recession of the hydrographs, implying that the widening and deepening of the channels enable more efficient drainage

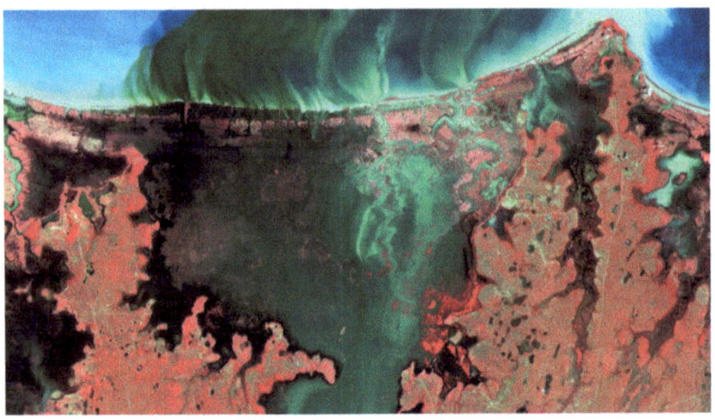

Fig. 10 LANDSAT image on 7th January 1997 during a river flood. Note the numerous breaches in the shoreline chenier

data show that all major rivers on the north facing shoreline of the Northern Territory discharge into the sea via concave shorelines.

Waves and currents will further erode the coast and the tidal tributaries will continue to grow. The estuary will probably connect through the upstream wetland making a continuous channel that may propagate up to 100 km inland. The geomorphology and tidal dynamics described above suggest no reason why this should not happen.

The only thing that may prevent this scenario to develop, or slow it down, is that climate change may lead to more intense rainfall (Hennessey et al. 2004) making erosion higher and coarse sediment loads increase which nourish barriers preventing tidal ingress. However this mechanism may not develop because increasing floods will further breach the cheniers (Fig. 10).

Engineering attempts, if any, to prevent this scenario are likely to fail. For instance, attempts have been made to halt salt water intrusion through the construction of barriers. One barrier at Shady Camp was installed over 20 years ago and is now being outflanked by tidal creeks (see the 2005 map in Fig. 4b). Other barriers have all failed being washed away during the wet season. The problem with barriers is that there is no high ground on either side of tidal creeks for the barriers to be effectively mated. The only restriction in the wetland where two headlands are close by is at Shady Camp and the width of the floodplain at this location is 5 km. Building a structure across this section of floodplain has not been attempted and would be very expensive. The floodplain soils are also soft muds that are 10–20 m thick and an effective barrier would have to be built so the structure would not be undermined.

Any such barrages would also have to allow the passage of floodwater and allow the passage of fish otherwise the important ecological role of Barramundi and other fish migration would be interrupted with dire effect. It may be possible to develop techniques of utilising native vegetation to provide effective barriers to prevent the growth of new tributary channels across the floodplain and these should be trialled. However it is unlikely that these techniques could be implemented in the main channel successfully due to the high cost of construction and maintenance.

It is reasonable to conclude that in the next 40 years the estuary will continue to change and start to function more like the neighbouring estuaries of the Adelaide and Alligator Rivers. These estuaries have wide channels that convey the tide over 100 km inland. Examination of maps and satellite images show extensive paleo channel networks on coastal floodplains of the Northern Territory that are narrower and more sinuous than the contemporary channels. As the modern channels have incised deeper into the floodplains they have remained more stable in location and extended further inland. These estuaries are mangrove lined and freshwater swamps and billabongs exist on the floodplains adjacent to the estuary. The swamps and billabongs are replenished each wet season through local rainfall and runoff and not necessarily from flooding from the main channel. The majority of estuarine channels on the north facing Northern Territory shoreline also exhibit these properties. The Arafura Swamp in Arnhem Land is similar to the Lower Mary River and previous studies by the author have shown that the estuarine channel of the Arafura Swamp is also extending.

While many of the values of the present Lower Mary River wetlands will be lost an environment that still has valuable recreational and biodiversity values will form. The Lower Mary is most likely following a similar evolutionary path to its neighbours. Thus the simplest management option for the Mary Estuary is managed retreat and adaptation.

References

Carter MR, Gregorich EG (2006) Soil sampling and methods of analysis, 2nd edn. Canadian Society of Soil Science. CRC Press, Boca Raton

Chappell J, Grindrod J (1983) Chenier plain formation in northern Australia. In: Thom BG (ed) Recent advances in Australian coastal studies. Academic, New York, pp 197–231

Cobb SM (1998) Channel extension and the geomorphology of tidal creeks in Kakadu National Park, Northern Territory. Thesis, Bachelor of Science Honours, University of Western Australia. Environmental Research Institute of the Supervising Scientist IR280

de Lestang P, Griffin RK (2000) Impacts of saline intrusion on Barramundi in the Mary River Region. Northern Territory Department of Primary Industry and Fisheries. Fishery Report No. 50. Northern Territory Government. ISBN:0724530681

Hennessey K, Page C, McInnes K, Walsh K, Pittock B, Bathols J, Suppiah R (2004) Climate change in the Northern Territory. CSIRO Consultancy report for the Northern Territory Government Department of Infrastructure, Planning and Environment. Climate Impact Group, CSIRO Atmospheric Research School of Earth Sciences, Melbourne University, Melbourne Australia

Lottig NR, Fox JM (2007) A potential mechanism for disturbance-mediated channel migration in a south eastern United States salt marsh. Geomorphology 86(2007):525–528

McInnes R (2003) Cost benefit analysis of Mary River salinity mitigation. Australian Greenhouse Office and Northern Territory Government, Canberra. ISBN ISBN 1920840257

Webb GJW, Manolis SC, Whitehead PJ (1987) Wildlife management: crocodiles and alligators. Surrey Beatty and Sons, Sydney

Whitehead PJ, Saalfeld K (2000) Nesting phenology of the Magpie Goose Anseranas semipalmatain monsoonal northern Australia: responses to antecedent rainfall'. J Zool (Lond) 251:494–508

Whitehead PJ, Wilson BA, Bowman DMJS (1990) Conservation of coastal wetlands of the northern territory of Australia: the Mary River floodplain. Biol Conserv 52:85–111

Williams D, Wolanski E, Spagnol S (2006) Hydrodynamics of Darwin Harbour. In: Wolanbski E (ed) The environment in Asia Pacific Harbours. Springer, Dordrecht

Wolanski E (1986) An evaporation-driven salinity maximum zone in Australian tropical estuaries. Estuarine Coastal Shelf Sci 22:415–424

Woodroffe CD, Mulrenna ME (1993) Geomorphology of the lower Mary River Plains, Northern Territory. Australian National University Northern Australian Research Unit and Conservation Commission of the Northern Territory, Darwin

Woodroffe CD, Chappell JMA, Thom BG, Wallensky E (1986) Geomorphological dynamics and evolution of the South Alligator Tidal River and Plains, Northern Territory. Australian National University North Australia Research Unit Mangrove Monograph No. 3 Darwin 1986

Index

E. Wolanski (ed.), *Estuaries of Australia in 2050 and Beyond*, Estuaries of the World,
DOI 10.1007/978-94-007-7019-5, © Springer Science+Business Media Dordrecht 2014

MIX
Papier aus verantwortungsvollen Quellen
Paper from responsible sources
FSC® C105338

If you have any concerns about our products,
you can contact us on
ProductSafety@springernature.com

In case Publisher is established outside the EU,
the EU authorized representative is:
Springer Nature Customer Service Center GmbH
Europaplatz 3, 69115 Heidelberg, Germany

Printed by Libri Plureos GmbH
in Hamburg, Germany